COMMUNICATION ACOUSTICS

COMMUNICATION ACOUSTICS

AN INTRODUCTION TO SPEECH, AUDIO AND PSYCHOACOUSTICS

Ville Pulkki and Matti Karjalainen
Aalto University, Finland

This edition first published 2015

Registered office
John Wiley & Sons Ltd, The Atrium, Southern Gate, Chichester, West Sussex, PO19 8SQ, United Kingdom

For details of our global editorial offices, for customer services and for information about how to apply for permission to reuse the copyright material in this book please see our website at www.wiley.com.

Library of Congress Cataloging-in-Publication Data
Pulkki, Ville.
Communication acoustics : an introduction to speech, audio, and psychoacoustics / Ville Pulkki and Matti Karjalainen.
pages cm
Includes index.
ISBN 978-1-118-86654-2 (hardback)
1. Sound–Recording and reproducing. 2. Hearing. 3. Psychoacoustics. 4. Sound. I. Karjalainen, Matti, 1946-2010.
II. Title.
TK7881.4.P85 2015
620.2–dc23

2014030757

A catalogue record for this book is available from the British Library.

Set in 10/12pt Times LT Std by SPi Publisher Services, Pondicherry, India

1 2015

To Vappu, Sampo, Raisa, *and HUT people*

To Sari, Veikko and Sannakaisa, *and Aalto Acoustics people*

Contents

About the Authors

Ville Pulkki has been working in the field of audio from 1995. In his PhD thesis (2001) he developed a method to position virtual sources for 3D loudspeaker set-ups after researching the method using psychoacoustic listening tests and binaural computational models of human hearing. Later he worked on the reproduction of recorded spatial sound scenarios, on the measurement of head-related acoustics and on the measurement of room acoustics with laser-induced pressure pulses. Currently he holds a tenure-track assistant professor position in Aalto University and runs a research group with 18 researchers. He is a fellow of the Audio Engineering Society (AES) and has received the AES Publication Award. He has also received the Samuel L. Warner memorial medal from the Society of Motion Picture and Television Engineers (SMPTE). He has a background in music, having received teaching from the Sibelius Academy in singing and audio engineering alongside instruction in various musical instruments. He has also composed and arranged music for many different ensembles. He enjoys being with his family, renovating his summerhouse and dancing to hip hop.

Matti Karjalainen (1946–2010) began his career as an associate professor of acoustics at the Helsinki University of Technology (TKK) in the 1980s. He maintained a long and prolific career as a researcher and visionary leader in acoustic and audio signal processing, both as a pioneer of Finnish language speech synthesis and developer of the first portable microprocessor-based text-to-speech synthesizer in the world. For ten years he was Finland's only university professor of acoustics, leading the Laboratory of Acoustics and Audio Signal Processing at the TKK (now Aalto University) until 2006. Some of his groundbreaking work included applying his expert knowledge of psychoacoustics to computational auditory models, as well as sophisticated physical modelling of stringed instruments utilizing the fractional delay filter design for tuning, a now standard technique in this field. Later in life, augmented reality audio and spatial audio signal processing remained among Matti's greatest research interests. For his achievements in audio signal processing Matti received the Audio Engineering Society fellowship in 1999, the AES silver medal in 2006 and the IEEE fellowship in 2009. On his 60th birthday he founded the Matti Karjalainen Fund, supporting young students into studying acoustics. Matti's share of the revenues from this book are routed to the fund. In May 2010, Matti passed away at home, survived by his wife, daughter and son Sampo, a well-known software designer living in the US.

Preface

The book *Kommunikaatioakustiikka* by Matti Karjalainen (1946–2010) has always been around during my research career in audio and psychoacoustics, starting from my PhD studies (1996–2001), through the periods when I was a postdoc (2001–2005), a senior researcher (2005–2012), and now during my tenure track professorship (2012–). I first used the book as a reference, as it summarized many relevant topics and provided good pointers on where to find more information. I have also been teaching the corresponding course at Aalto University (the university formerly called Helsinki University of Technology), first during Matti's sabbatical years, then sick leaves, and regularly after his passing away. It was my and many other people's opinion that the book was great, but it did not have a counterpart written in English. Matti himself knew this, and he worked on a translation, a of which he completed about 30% in 2002, including the preface that follows.

For a long time I thought that I should finish Matti's work, as it would benefit people in the fields of audio and speech. However, I also understood that it would be quite a hard job. The final motivation came from my university, which stated that all MSc-level teaching should have course material in English as well, starting from autumn 2015. So, in autumn 2013, I decided to complete the book. To ensure that I would really do it, I proposed the book to Wiley, since I understood that I needed a deadline. I also thought that international distribution would benefit the propagation of the book. The book project meant a period of 10 months where I worked so much that I felt that my hands were stuck permanently to my laptop.

The book grew by about 30% from the original Finnish book, as I added quite a bit more material on audio techniques and updated many parts of the book. Consequently, the subject matter of the book might be too large for a single-semester course. Teachers are encouraged to leave some chapters out, as the whole book might be too much information to be digested in one go. I shall be updating the companion web page of this book with sound examples and other material to help teachers of such courses.

The book covers many fields within acoustics, and without great help from many professionals in the field, the book would be less detailed and less complete. First of all, I received very kind help in translating and updating the text from my PhD students Marko Takanen (Chapter 13), Teemu Koski (Chapter 19), and Olli Rummukainen (Section 11.7). Juha Vilkamo and Marko Takanen also provided text and figures from their PhD theses. The following professionals have read and commented on, or otherwise helped with, the project: Paavo Alku, Brian C. J. Moore, Mikko Kurimo, Ville Sivonen, Nelli Salminen, Ilkka Huhtakallio, Cleopatra Pike, Catarina Hiipakka, Alessandro Altoè, Mikko-Ville Laitinen, Søren Bech, Archontis Politis,

Olli Santala, Sascha Disch, Tapio Lokki, Lauri Savioja, Hannu Pulakka, Richard Furse, Unto K. Laine, Vesa Välimäki, Javier Gómez Bolaños, Cumhur Erkut, Damian Murphy, Simon Christiansen, Jesper Ramsgaard, Bastian Epp, Athanasios Mouchtaris, Nikos Stefanakis, Antti Kelloniemi, Kalle Koivuniemi, Ercan Altinsoy, Lauri Juvela, Symeon Delikaris-Manias, Tapani Pihlajamäki, Antti Jylhä, Tuomo Raitio, Martti Vainio, Gaëtan Lorho, Mari Tervaniemi, Antti Kuusinen, Jouni Pohjalainen, Christian Uhle, Torben Poulsen, Davide Rocchesso, Nick Zacharov, and Thibaud Necciari. Luis R. J. Costa worked on removing the worst Finglishisms in the book, making them into more readable English expressions.

I, of course, hope that the book is successful, and new editions come out in time. With that in mind following the tradition started by Brian C. J. Moore in his *Introduction to Psychology of Hearing*, I would hereby like to open a similar contest. A prize of a box of Finnish chocolate confections will be awarded to the reader who spots the most errors in this edition, and writes to me to point them out. Game on!

I hope you will enjoy reading the book, and that you will find it beneficial in your research work and studies.

Ville Pulkki
Otaniemi, Espoo, Finland
May 2014

Preface to the Unfinished Manuscript of the Book

The origins of this book date back to 1980 at the Helsinki University of Technology (HUT). I started lecturing the course 'Communication Acoustics' (in Finnish), based on a collection of material from previous lecturers, in the tradition of engineering psychoacoustics influenced by some US and German textbooks.

The first encounter with principles on how we perceive sound was a very inspiring one, not only because of the need to traverse through a vast amount of experimental results on the functioning of hearing and gradually structure the underlying principles, but also because of having to start to generate ideas on how this could be simulated by computational means. This knowledge has gradually developed in various research projects, and the present course on 'Communication Acoustics' is a much more mature version than the first one.

Finishing a Finnish textbook for the course after 18 years was the proper time to look at the possibility of rewriting and extending it into an English version. This task was motivated by finding that there was no modern textbook covering the wide field of communication by sound and voice, especially from the point of view of engineering psychoacoustics. To fill a bit of this gap, this book was written. Sabbatical leave during 1999–2000 from my HUT duties allowed me to do most of this writing.

The importance of engineering psychoacoustics has been growing rapidly since the late 1980s. There are several reasons for this. The rapid development of digital audio is one of the most important driving forces. Audio coding and related questions of sound quality showed engineers and scientists how essential knowing the principles of human auditory perception is.

Not only hearing-related knowledge and know-how but also knowledge of a wide variety of other disciplines involving sound and voice are needed to work succesfully in advanced research and development. Three cornerstones, in my view, are essential for such success: (physical) acoustics, auditory perception, and (digital) signal processing. The first tells us how physical systems behave, the second how we behave when capturing sound and its content, and the last is the primary tool to implement modern sound technology.

This textbook is written mainly for engineering-oriented readers as a general tutorial to a wide range of topics important in R&D on modern sound technology. On the other hand, it is written with the objective of providing this knowledge to a much wider audience as well. The reader may focus on the subtopic of interest, in most cases without having to digest all the prior material in full detail. Also, formal theories from mathematics are not extensive and may be skipped, at least in the first reading. One important function of the book is to provide pointers to relevant literature on each subtopic. In this sense, it may also have a modest role as a handbook of references.

As I have already said, the scope of the book is wide, and it tries to be comprehensive but not deep in every subject. The first chapters present an overview of acoustics, signal processing, speech, and audio. These are more like introductions to the most important concepts. Readers who are familiar with them may skip them. To some degree, understanding these topics and concepts is necessary in the rest of the book. The main part of the book is related to the human auditory system, its function and properties, as well as to modelling for research and applications.

I have been lucky in having help and support from many people. My students have provided feedback on my lectures and the Finnish version of this book.

This is the first English version of the book, and I hope, in addition to proving useful to as many readers as possible, that readers are willing to provide feedback to help me improve this book in the future.

Matti Karjalainen
Fiskars, Finland
January 2002

Introduction

Efficient use of sensory functions and communication has been one of the most important factors in the evolution and survival of animals in nature. Especially for the highest forms of evolution, vision and hearing are the two main modalities to support this view in a complementary way. Visual information, based on the laws of optics, reflects the environment in a geometrically appropriate and reliable manner, while auditory sensing and perception, based on the laws of acoustics, are less dependent on physical constraints such as obstacles between an observer and objects to be perceived. Vision often dominates audition, especially if an object is clearly visible or moving, while hearing may capture important information even when there are no perceivable visual events. Sensory integration, i.e., fusion of information from different modalities into a coherent percept, is characteristic of living species. Only when senses provide conflicting cues must they compete for contribution to the final percept.

Two main ways of utilizing the auditory sense are to sense *orientation* in the environment and for *communication* between subjects. The former activity can be found among early phases of animal evolution. As an example, the sound events shown in Figure I.1 bring information to the subject about the surroundings. The sounds caused by the shoes of the horse imply the type of terrain, the wind sounds bring information about the weather, and the sounds caused by animals, even from visually obscure locations, report their presence, action, and location. Any of the sounds may reach conscious attention and may startle the subject, with resulting reactions.

Species with highly advanced and specialized hearing abilities have evolved, for example the echolocating bat. This animal sends chirps - frequency sweeps - and receives their reflections from surrounding objects. Auditory analysis of the echoes enables the bat to construct an image-like representation of its environment for navigation, even in fast flying. Many animals have a very sensitive, accurate, or specialized auditory system. The hearing system is important as an early warning indicator of dangerous situations or as an aid for hunting.

Sound is an excellent means for communication. Uttering of sounds is an easy way to warn others or to express the internal state of the subject, such as emotions, action plans, etc. Gestures and facial expressions are useful only when there are no limitations for visual

Communication Acoustics: An Introduction to Speech, Audio, and Psychoacoustics, First Edition.
Ville Pulkki and Matti Karjalainen.

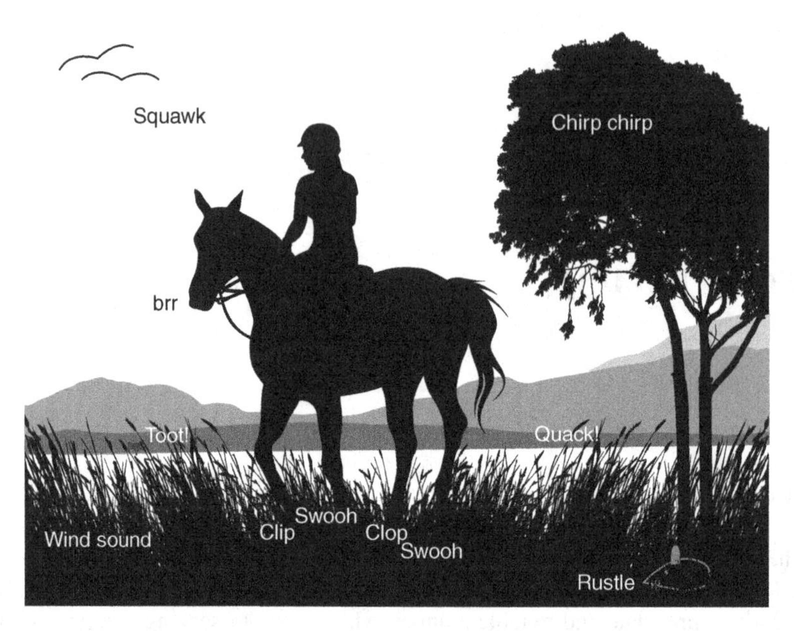

Figure I.1 Environmental orientation; a situation where information from external objects is carried by sound. The listener can both localize the sound sources and also decipher the cause of the sound.

communication. Sound can, in favourable conditions, carry relatively far and propagate around a visually opaque object. One of the major disadvantages of sounds and voice is that they do not leave a permanent physical trace like a footprint in sand. Thus, animals are not able to use sounds as message records to transfer sound-based information in time.

Orientation and communication by sound are activities inherent to human beings as well. Orientation is often instinctive, without conscious attention. We receive continuously a multitude of sound information, but most of these data remain outside of our consciousness. Sounds that are unexpected or otherwise in the focus of attention can be analysed and memorized in more detail over long periods of time. If a sound is annoying, disturbing, or just so loud that it can be harmful to the hearing of a subject, it is called *noise*.

The human being has evolved into a being with more advanced communication abilities than other living beings. Voice production evolved towards *speech* and *spoken language*. Prerequisites for this were the development of organs for speech production and the auditory ability to analyse complex voice signals that carry linguistic and conceptual information and knowledge. Only later did man discover systematic ways to store linguistic information in written form. Even today there are spoken languages without a corresponding written language.

Speech is a fast and flexible way of expressing conceptually structured information, emotions, and intentions, as illustrated in Figure I.2. A spoken message consists of linguistic and non-linguistic information. Linguistic information is built of basic units (phonemes) and their combinations (words, phrases, sentences). Non-linguistic features, such as speaker identity and pitch - expressing emotions, are an integral part of speech and may even change the interpretation of linguistic content. Speech contains a lot of redundancy that is, multiple ways of coding the same information in order to function properly in adverse acoustic environments, and it

Figure I.2 Speech communication in different situations between subjects or from the presenters to the audience. The acoustic waves carry the information either directly from the presenter to the listener or through an audio system.

Figure I.3 Musical communication with electronic sound reinforcement. The audience responds acoustically to the band by clapping hands and with their voices.

is not dependent on the visibility of the speaker. A fundamental requirement for successful communication is a common code – a common language or dialect and a common conceptual model of the world.

Humans have developed another important type of communication by sound: *music*. It is not primarily for conveying linguistic and conceptual information but rather for evoking aesthetic and emotional experiences, as in Figure I.3. Music may, however, also carry strong symbolic meanings between subjects that share common musical associations to experiences and events in their cultural or social life.

Human beings were not satisfied with the limitations of acoustic communication where a long distance was a problem and no physical trace of sound was left to convey a message in time. The first sound-recording devices were based on mechanical principles. Only through discoveries in electricity and electronics did the techniques of recording and long-distance communication of sound and voice become everyday utilities. The first devices to extend the communication range were the *telephone* and the *radio*, as shown in Figure I.4. Acoustic waves were converted to corresponding electrical signals by a microphone. Weak signals from a microphone were strengthened using electronic amplifiers. By compensating for losses

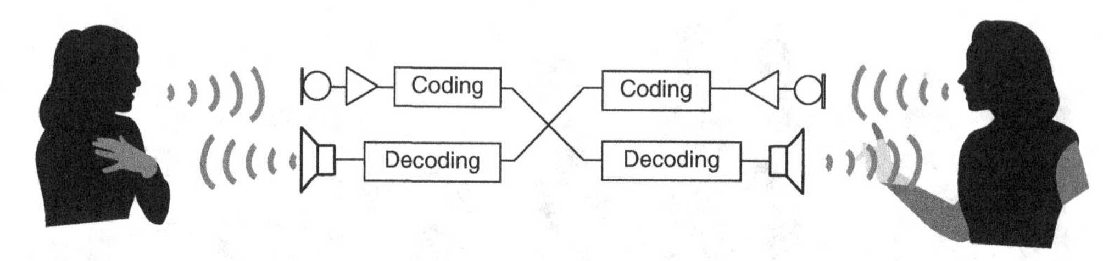

Figure I.4 Speech communication through a technical transmission channel.

in telephone lines by amplification, it became possible to transmit speech over any distance. The radio was invented for wireless broadcasting over a long range from a transmitter.

Mechanical cutting using sound waveforms enabled the first recording and playback by the *phonograph* and the *record player*. Electronic amplification improved the quality of sound and made it louder. A step forward was the *tape recorder*, in which the only remaining mechanical function was moving the magnetic tape. Finally, digital signal processing and computers have enabled storage of sound as bits on digital media, even without any moving parts. Digital documents are, in principle, perfect, in the sense that they can be copied and stored infinitely without any loss of information. Digital signal processing further enabled *digital audio* and *speech processing* to store and transmit signals economically using *audio and speech coding*, where the number of bits needed is reduced by an order of magnitude. In spite of rapid digitalization, the interface to humans still remains non-digital. Analogue components are needed: microphones and amplifiers to capture sound; amplifiers together with loudspeakers or headphones for sound reproduction, to make signals audible and loud enough.

Two very recent major steps in communications are the *Internet*, which is a data network to provide all forms of digital information, and the cellular wireless networks for *mobile communications*. Both of them enable, especially when they are integrated, access to new formats of multimedia, including sound and voice in their most advanced forms. Wireless networks allow such communication in most parts of the world, anytime, for a majority of people.

Early in the history of sound reproduction, one of the goals was to create a realistic spatial impression. Two-channel *stereo* was adopted in the 1960s to provide a better sound image and a more natural sound colour perception with two ears than was possible with monophonic reproduction that is, with a single channel. Different multi-channel systems with or without elevated loudspeakers have been proposed, and nowadays a wide variety of systems is available for spatial sound reproduction. Advanced techniques for headphone listening are also available.

Generally, *digital audio* means all methods of sound recording, processing, synthesis, or reproduction where digital signal processing and digital processors are utilized. *Perceptually-based audio techniques* take into account the human resolution of different attributes in sound perception, such as in frequency, time, or space. These techniques are utilized when there are some limitations in the audio communication channel, such as the transmission bandwidth and latency, characteristics of microphones and loudspeakers, and when the spatial composition of the reproduction system is not ideal. The target is to deliver audio with optimal quality; the resources in the communication channel are allocated as well as possible so as to provide the best possible experience.

Digital technology also enables *man–machine communication by voice*, as illustrated in Figure I.5. Messages can be conveyed from or to various devices and computers in everyday life using our most natural means of communication, the spoken language. *Speech synthesis* means

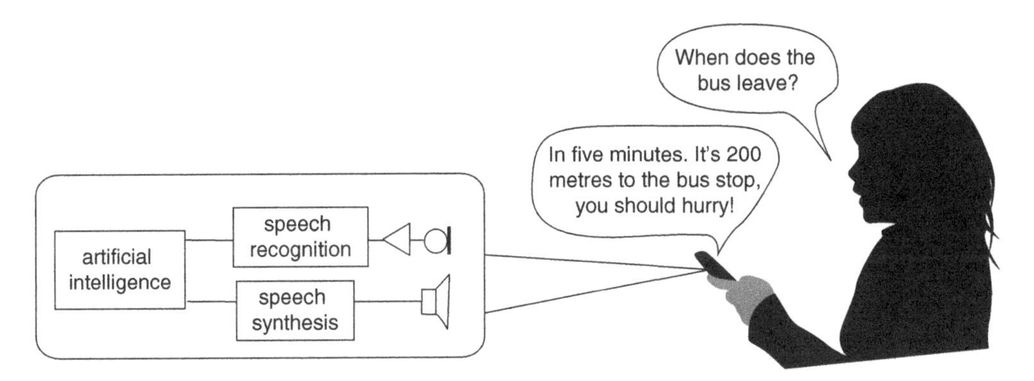

Figure I.5 Man–machine communication by voice.

machine generation of spoken messages. Automatic *speech recognition* enables machines to 'understand' simple or complex speech messages from human speakers. This is a demanding technology to compose in a way that humans feel natural and comfortable with it. In most advanced forms of voice communication, a *natural language processing* capability and a conceptual world view must be given to the computer, or it must be given the ability to acquire and learn them. The first successful voice-based man–machine interfaces emerged in the 2010s, when the voice-based remote control of TV, different automated telephone services, and voice control systems for personal devices appeared on the market.

1

How to Study and Develop Communication Acoustics

The rest of this book describes facts, theories, and models of communication by sound and voice – both human and machine. The objective of basic research in this field is to better understand how we communicate, while the engineering goal is to develop and use technology to make this communication more versatile and powerful. In both cases, the challenge is to understand a variety of topics and to solve problems by approaching them from several points of view.

1.1 Domains of Knowledge

Different phenomena in nature, living organisms, society, and technology obey different laws and exhibit different properties and thus require different scientific concepts to study them. Even when phenomena appear similar, a detailed interpretation of specific laws to be applied may be quite different. That is why we can have (and actually must use) different *domains of knowledge*.

In some problems it is enough to have a look at just one aspect of reality (one domain), but with complicated multidisciplinary problems, many domains must be mastered. In complex cases, such as the topic of this book, we are forced to know at least the basics of several domains of knowledge in order to master the field properly. Some of these domains are more basic, while some are important from a methodological point of view in research and others from an application or a practical point of view. For a modern approach to sound and voice communications, we deal in this book with the following domains of knowledge:

- *Acoustics* and the fundamentals of physical sound deal with the interaction between physical objects that form the basis for sound and vibration phenomena. This domain answers questions on how sound is generated by sound sources, how it propagates or is absorbed, how it behaves in a closed space such as a room, auditorium, or a concert hall, and so on.
- *Signal processing* comprises the theories and techniques on how signals carrying information are generated, transmitted, analysed, or transformed by humans, in nature, or using technical

Communication Acoustics: An Introduction to Speech, Audio, and Psychoacoustics, First Edition.
Ville Pulkki and Matti Karjalainen.

devices. In this domain, the basic physical details are of less importance, and abstraction from physical interaction to (typically) one-directional input–output causal relations is emphasized.
- *Speech and audio* are concerned with the fundamental characteristics of spoken and audio signals, and the specific techniques for communication with them. *Spoken language* and *speech or language technology* are of great importance, although in this book we will discuss them only briefly. *Music*, *music acoustics*, and *music technology* also belong to this category.
- *Physiology* and the *psychophysics (psychoacoustics) of hearing* study how our auditory system works both from a physiological and a functional point of view. This domain includes several important subtopics, one of which is *spatial hearing*, which deals with how we localize sound sources and perceive sound environments.
- *Mathematics* and *computer science* are the sciences that provide the general methodologies required to get the formal understanding necessary to model and realize complex communication processes.
- *Engineering applications* is the field that utilizes a wide variety of techniques for sound and voice communications. This includes traditional acoustics design, control of noise, speech technology (speech analysis, synthesis, coding, recognition), audio technology (recording, production, reproduction), multimedia sound (integration of sound with other modalities) virtual acoustics, and technical audiology for helping with hearing problems, etc.

Research in these and related topics is often scattered and separated. One of the main objectives of this book is to provide the reader with a broad overview, without digging too deeply into details of any subtopic, since there are more specialized books and publications available for that purpose. Before starting this journey of facts, theories, and models, a characterization of related research methodologies and how they evolve is given.

1.2 Methodology of Research and Development

Scientific and engineering knowledge of communication processes has developed over the last hundreds, even thousands, of years. First there were beliefs that sounded rational, but a closer investigation found them to be false or only partially true. Some early thinkers and experimenters, such as Pythagoras, turned out to be very successful and influential (Hunt, 1992). Pythagoras studied the behavior of strings and understood the inverse relation between pitch and string length in ancient Greece during the sixth century B.C. Related to the harmony (consonance) of sounds, he assumed that the whole universe followed simple number rules of harmony. In the 1700s experimental studies in, and mathematical theories of, the physical sciences improved our understanding on how, for example, sound propagates in the air. Sound sources were better understood, and Hermann von Helmholtz had a modern view on the functioning of the human ear (von Helmholtz, 1954) more than a century ago. Since then, electronic and digital communications have forced scientists and engineers to learn more deeply the secrets of signals and the information conveyed by them.

There are three basic ways a scientist or engineer may acquire knowledge about a system or process, such as communication by sound and voice:

- *Experimentation* with, say, concrete physical or psychophysical, systems. This has been the standard approach used by scientists and engineers, augmented by practical or theoretical

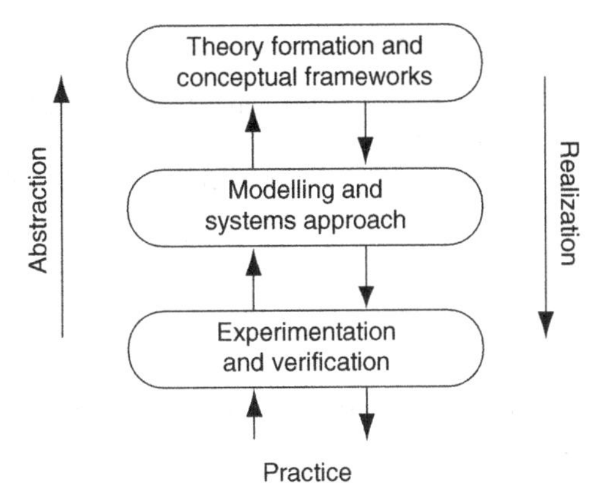

Figure 1.1 Abstraction and realization processes in research through experimentation, modelling, and theory formation.

thinking. Concrete experimentation involves a large number of details, which makes seeing general laws behind the multitude of phenomena difficult, but on the other hand it helps to verify that the thinking is not too far from reality.

- *Modelling* of an object or phenomenon by means of constructing another kind of system that simulates the original phenomenon or object but is easier to experiment with. The first modelling techniques were based on constructing concrete, physical objects as models. The modern approach, using computer-based simulation, has expanded the capability of modelling techniques dramatically.
- *Theory formation* is a conceptual approach to formulating general principles inherent in a specific field of research. A theory has higher generality and potentially the widest applicability compared to experimentation and modelling techniques. On the other hand, very general theories are hard to apply to complex and specific cases in practice; they rather help and direct the thinking process.

Many forms of scientific and engineering methodologies can be formulated as different combinations of the three approaches. Figure 1.1 characterizes these as levels of the abstraction process as well as realizing abstractions back towards practice. *Abstraction* means getting rid of details that are not essential for modelling or understanding a specific object or principle, and concretization (realization) is the opposite process of adding details in order to approach reality or practical construction of a system.

Recently, in this age of computers being applied everywhere, the three approaches of experimentation, modelling, and theory formation have started to merge and integrate. Experiments are often carried out using computational models before the final validation against practice. The difference between theory and computational models is also becoming less distinctive, since advanced computer models, including artificial intelligence and logic programming, may enable high-level theoretical inference.

Progress in research and practical applications is based on continuous iteration between the levels of abstraction, aiming at improved models and theories that better meet the challenges of reality.

1.3 Systems Approach to Modelling

A *systems approach* to studying a specific object means abstraction and modelling which is based on the finding that very many (if not all) objects in reality exhibit some general properties that can be described using *systemic concepts*. The target may be of a physical origin, a society of living beings, or a human-made technical artefact, and yet some general concepts may apply to all of them.

Figure 1.2 shows a very simple diagram and Figure 1.3 a more complex one that may be understood as systemic descriptions of something, not specific to any domain of knowledge until so related. Based on such descriptions, one may compile a list of concepts representing systemic aspects. The reader may wish to propose another set of systemic concepts, or disagree with the authors about the usefulness of such concepts. The list presented here is not intended as a basis of a 'general systems theory', rather it orients the reader to the way the authors have tried to formulate the organization and content of this book.

- *Element*: an entity that is considered not to have an internal systemic representation. This means that (a) we have an abstract element, for example in mathematics, (b) we don't want to pay attention to an element's internal complexity, or (c) we are not able to represent the element in more detail due to its internal complexity. A typical element representing an entity without an internal description but with an input–output relationship is the *black-box* diagram of Figure 1.2.
- *Relation*: something that represents a connection or link, like an interaction, similarity, or distance, between elements, objects, or systems. Types of relations will be mentioned for different systemic concepts below. A *property* can be understood as a relation of an object to

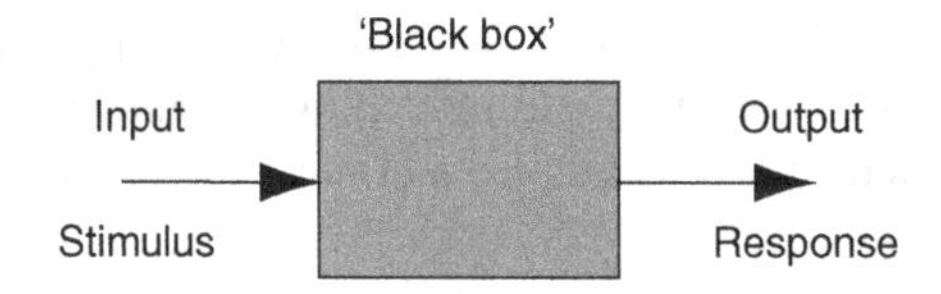

Figure 1.2 A black-box element with an input–output relationship.

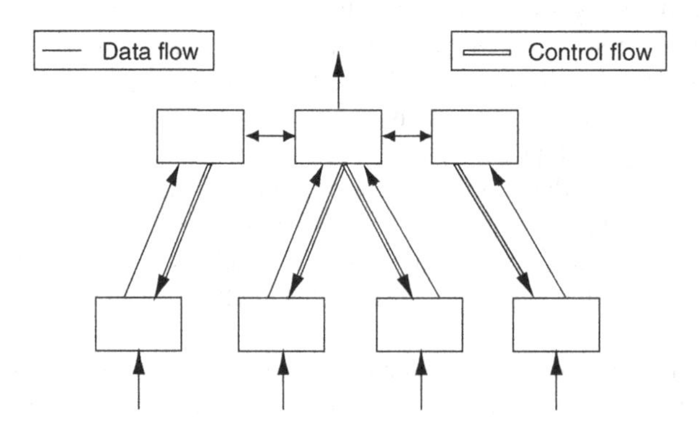

Figure 1.3 A block diagram characterizing a system that consists of elements, data, and control flow links between them, and a hierarchical-type (two-level) organization.

a general reference, such as a classification system. Causal relations are discussed with the concept of function(ality) below.

- *Structure*: represents a set of objects and relations between them that constitute a relatively permanent aspect of the system under study. The structure may define a spatial, temporally ordered, or a more abstract composition of entities and relations. Structures are often hierarchical (see below) through *parts* and *part-of* relations, i.e., an entity consists of parts that together form the (whole) entity.
- *Function(ality)*: an aspect that is complementary to structure, i.e., it represents relations between changing properties of the entity under study. *Causal relations*, meaning relations between cause and effect, stimulus and response, or input and output, are particularly important since the functionality of physically realizable systems is based on them, requiring that the (causal) effect cannot appear before the cause in time.
- *Event*: an activity in a system that occurs within a specific time span. A *stream* is a sequence of events.
- *State*: a concept that describes how a system 'memorizes' the history of its behaviour. Its reaction (output) to external causes (inputs) depends also on its internal state (unless it is a memoryless system).
- *Object*: a generic concept that has been used in many ways, for example as a target of observation or manipulation. From computer science the concept has gained a more systematic usage – object-oriented methodology – where an object is understood as an entity having an internal state and functionality through type-dependent functions.
- *Type* and *class*: concepts that are based on the similarity or dissimilarity of entities. Entities that share a common property or properties are categorized in the same type or class from the viewpoint of this property. In object-oriented methodology, a class is a definition for the structure and behaviour of its *instances*, i.e., objects for computation based on the class definition.
- *System*: a very general and generic concept to describe any object that has enough systemic properties, although in specific system theories it may be given a precise meaning. It typically refers to an abstract representation where the specific domain of knowledge, such as acoustics, is deliberately left aside and systemic concepts are used instead.
- *Control*: goal-driven behaviour. Living species have evolved and learned functionalities with target behaviour towards more successful life in their environment. Advanced machines are programmed or they adapt to their working conditions in order to meet some *optimal* criteria. *Feedback* is an important principle of control where the response is compared with the goal in order to improve control.
- *Process*: refers to (often controlled or goal-driven) the functionality of a system. Processes and processing are typically described as input–output relationships or data- and control-flow diagrams.
- *Organization*: represents a higher-level structure and functionality, the result of a goal-driven process. *Self-organization* refers to the ability of a system to improve its organization towards more advanced functionality. Real self-organization is more demanding and complex than simple *learning* and adaptation.
- *Hierarchy*: the *parts* and *part-of* relations in structures that constitute systems common in nature. A hierarchy in an organization means a level-like control structure where 'lower-level' units realize detailed activities and 'higher-level' units perform more strategic control, the flow of control taking place 'vertically' between levels. Figure 1.3 depicts a strictly hierarchical two-level system where control flows from top to bottom and data from the bottom

up. A *heterarchical* system is an organization principle where control is distributed and functional objects act in several organizations, representing different levels in different sub-organizations. A strict hierarchy may be very efficient in optimal conditions, but a heterarchy is more versatile, especially if its elements are able to adapt to new roles when needed.
- *Data* and *information*: essential entities used by advanced systems and organizations to enable goal-driven functionalities through *communication*. Data are raw material for information, extracted from signals by observation – or by technical means in modern society. Information is a generic term for meaningful data that an intelligent system can utilize. *Knowledge* is conceptually organized information, and *language* is the primary means of communicating it – between humans. Intelligent machines are starting to achieve lower levels of knowledge and (natural) language.

Intelligent systems – animals, humans, and advanced machines – are able to create an information-based model, an internal representation of their environment, in order to successfully carry out their tasks. Due to the complexity of reality, such an internal representation can never be complete in full detail. *Overcomplexity* of target objects and environments leads to the need for optimal representations and information processing strategies. We encounter overcomplexity in many forms in the theories and models of this book. Sometimes using a simple model is sufficient although we know the structure and function of the object in detail. Often, at the frontiers of science, the best we know and can do is to use an undercomplex model that is far from perfect but still useful. Overcomplexity may be seen, for example, as statistical behaviour of a system, leading to a probabilistic model. A more powerful model may be able to represent the system deterministically. Many systems are inherently complex, e.g., chaotic, in a way that strongly limits the possibility of modelling and behaviour prediction.

For creatures of biological origin, advanced information processing is the result of continuous evolution through mutation and selection. The use of language and conceptual thinking has helped humans to survive in nature. An integral part of this evolution is information technology and engineering, including books and newspapers, telephone, radio and TV, computers, and mobile communications. Computers and human-engineered technical systems are already superior to humans in strictly logic-based computation tasks. *Pattern recognition* tasks are where humans still strongly outperform machines. It remains to be seen how technology will evolve, and what forms of human and machine information systems will exist in the future.

1.4 About the Rest of this Book

Communication by sound and voice is a many-sided and multidisciplinary topic. Thus, it is not possible to write a single book that covers all related aspects in depth. As discussed above, many subfields of science and technology must be combined to compose an up-to-date view of the whole field. This textbook emerged from a need to cover most of this field rather than concentrating on a single subtopic. More specialized publications can be found when more detailed information is needed.

1.5 Focus of the Book

The title of this book, *Communication Acoustics: An Introduction to Speech, Audio, and Psychoacoustics*, reflects the broad and interdisciplinary approach taken by the authors. It is, in some aspects, quite close to traditional presentations that are entitled something like

Communication Acoustics, which actually is the name of the Finnish book by Matti Karjalainen from which this textbook emerged.

The most central topic of this book is related to auditory perception. Every communication channel consists of a source, a channel, and a receiver. In our case the ultimate receiver is the human auditory system. This is found to be the most challenging link by far in the communication chain. Transmission channels and many sound sources can be formulated by physical and technical models more easily. Human speech or music as sound sources are also quite involved, since a full understanding of these phenomena requires including cognitive aspects too. The receiver, especially the human auditory system, is complex and intricate, even starting from the peripheral parts. Due to this complexity, the auditory aspects in communication by sound and voice have often been omitted or greatly simplified, especially in engineering-oriented textbooks. This simplification or omission is also due to the fact that formal and instrumental knowledge of auditory functions is only emerging.

Another motivation for placing special emphasis on the auditory aspects is that perception of sound is almost always the ultimate reason to study sound, since physically sound carries only little energy and thus has a negligible effect on nature or technical systems. The effects of sound and voice on man, through the hearing organ, are why we pay so much attention to the phenomenon called sound. The importance and emphasis of auditory perception does not mean, however, that other topics and disciplines are less essential in the communication chain. It just means that perception requires special effort to meet its complexity and importance.

The motivation for including basic concepts of physics and signal processing in this book is that, although there exist numerous specialized books on these topics, such knowledge is a prerequisite for understanding the communication chain from the source to a human receiver. The reader is presented with some mathematics and semiformal theory on these topics. If the reader finds them difficult to follow, acquiring at least an intuitive view of these concepts is recommended, at least for the first reading.

The hope is that after reading this book the reader can work on development of techniques based on knowledge of the perceptual mechanisms. This goal is approached by introducing the physical, signal-processing, and psychophysical background to the sound technologies, where a human listener is involved. These technologies involve all audio and speech techniques, certain noise-related techniques, hearing aids, and audiology.

The scope of the book is relatively broad, much broader than can be covered fully in a single volume. Consequently, this means that in the depth dimension of knowledge, this book is shallow, really just scratching the surface of existing literature. References in this book are intended to help the reader to find more specialized sources of information and topics that are omitted from this text.

1.6 Intended Audience

Every book is written for some audience, at least in the minds of the authors themselves. In the case of this book, the targeted audience is expected to be relatively broad and possibly quite inhomogeneous. The approach here is probably more from an engineering sciences point of view than any other. Semiformal presentation of facts, theories, and models, as well as knowledge that may be used for practical applications, comes from this starting point. The authors are convinced that such an approach will become more common in other fields and technical disciplines as well. The modelling paradigm discussed above, especially computational modelling, will become increasingly necessary in order to master such phenomena as auditory perception and modelling of complex communication situations. This is believed to be true both in basic research and in the development of engineering applications.

References

Hunt, F.V. (1992) *Origins of Acoustics*. Acoustical Society of America.
von Helmholtz, H. (1954) *On the Sensation of Tone*. Dover Publications.

2

Physics of Sound

The word *sound* in English has two confusingly different meanings. It may refer to the physical phenomenon or to the subjective percept. The old philosophical dilemma ponders, 'If a tree in the forest falls down and there is no observer, does it make a sound?' Based on experimental evidence and by making a clear distinction between these two meanings, we may say that a falling tree causes a physical sound event that can be recorded and analysed every time, but it does not make any sound in the sense of an auditory perception if there is no subject to hear the event.

In this chapter, we look at the physical side of the concept of sound. Without a physical basis there can be no sound event; the emergence of sound requires a physical substrate, and its perception a physiological one. Although complicated in practice, the physics of sound has a basis that is well formulated mathematically and is thus a widely studied topic in science and engineering. We do not go deeply into physical acoustics here, but rather we present an overview of the most basic concepts necessary or helpful to understand communication by sound and voice. Finding more specialized textbooks and publications is easy, and this chapter includes references to many such sources for more information. This overview serves to refresh the memory of those who have already studied physical acoustics and as a starting point for those who have not.

2.1 Vibration and Wave Behaviour of Sound

Sound, from a physical point of view, is a wave physically propagating in a medium, usually air, and in most cases is caused by a vibrating mechanical object. Exceptions can also be found, such as the electrical discharge of lightning. Some sources of sound waves, interesting within the scope of this book, are the speech organs of a speaker, the vibrating plates and the air column of a guitar body, and the diaphragm of a loudspeaker.

Sound radiated from machines is often considered undesirable sound, *noise*, that can degrade the performance of human hearing if it is exposed to too loud a sound and for a long time. Even not-so-loud sound may be annoying and disturbing, for example while sleeping or when concentrating on a specific task. On the other hand, noise-like sounds also carry information

Communication Acoustics: An Introduction to Speech, Audio, and Psychoacoustics, First Edition.
Ville Pulkki and Matti Karjalainen.

that helps us to orientate and manage our complex environments. It is good to realize that the energy level of sound is almost always so low that its physical effects can be neglected, but its effects on humans and animals are the reason we are interested in sound.

2.1.1 From Vibration to Waves

The event that causes a sound or vibration is called *excitation* or the *source*. A sound wave or vibration can propagate in a physical *medium*, it may be boosted by *resonance* effects and attenuate due to *losses* that transform it into other energy forms (mostly to heat). Sound waves and vibrations can be explained as an alternation between two forms of energy, potential energy and kinetic energy, starting from a simple case and approaching a general case of sound propagation.

2.1.2 A Simple Vibrating System

There are three basic variables that are used to describe the state of a physical particle:

- *Position* or *displacement* from a reference position. Let us denote this by y.
- *Velocity*, the time derivative of position or the ratio of the difference in position and interval in time t in which the change in position occurs, $v = \mathrm{d}y/\mathrm{d}t = \dot{y}$.
- *Acceleration*, the time derivative of velocity or the second derivative of position, $a = \mathrm{d}v/\mathrm{d}t = \mathrm{d}^2y/\mathrm{d}t^2 = \ddot{y}$.

Note that for a particle of non-zero mass and size, there are three rotational variables that may be important in some other contexts.

A mass (Figure 2.1a) and a spring (Figure 2.1b) can be combined to make the simplest possible vibrating system, as shown in Figure 2.1d. The force acting on the mass due to acceleration is

$$F = ma = m\ddot{y}, \tag{2.1}$$

where F is force [kg m/s^2] = [N, Newton], m is *mass* [kg] and a is acceleration, showing the linear dependence of force and acceleration. For an ideal spring we can write

$$F = -\mathrm{K}y, \tag{2.2}$$

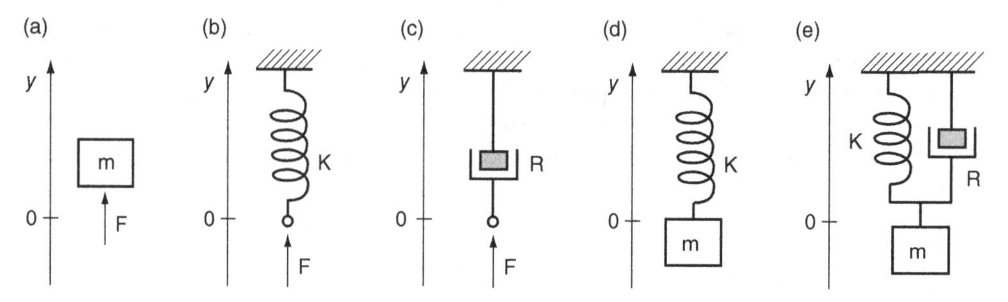

Figure 2.1 (a) A mass, (b) a spring, (c) a dashpot, (d) a mass–spring system and (e) a damped mass–spring system.

where K is the *spring constant* [N/m] and y is the displacement from the equilibrium position when the force is zero. Figure 2.1c shows a *dashpot* that represents energy losses in a vibrating system, typically converting energy to heat, which, in an ideal case, can be expressed as

$$F = -Rv = -R\dot{y}. \tag{2.3}$$

Here, R is a coefficient that relates force F and velocity v linearly. This linear relation does not hold, for example, in the case of mechanical friction.

Figure 2.1e represents a simple vibrating system with a damping element. Assuming that there are no external forces acting on the system, we can write

$$m\ddot{y} + R\dot{y} + \mathrm{K}y = 0 \tag{2.4}$$

since the sum of the forces must be zero. If there are no losses ($R = 0$) and the mass is initially displaced by $y = A$, the system starts to vibrate according to the expression

$$y(t) = A\cos\omega_0 t = A\cos 2\pi f_{\mathrm{res}} t, \tag{2.5}$$

where the maximum displacement A is called the *amplitude*, f_{res} is the *frequency* (*characteristic frequency, resonance frequency, eigenfrequency*), and $\omega_0 = 2\pi f_{\mathrm{res}}$ is the corresponding *angular frequency* of vibration. Curve a) in Figure 2.2 characterizes the oscillation of Equation (2.5). If the initial state is zero displacement but non-zero velocity, the equation has function $\sin\omega_0 t$ instead of $\cos\omega_0 t$. In either case, such a vibration is called a sinusoidal oscillation.

The eigenfrequency of a mass–spring system can be computed from its parameters:

$$f_{\mathrm{res}} = \frac{1}{2\pi}\sqrt{\frac{\mathrm{K}}{m}}. \tag{2.6}$$

For any periodic oscillation the relation between frequency f [Hz, hertz] and period T [s, seconds], for the duration of one oscillation, is

$$f = 1/T. \tag{2.7}$$

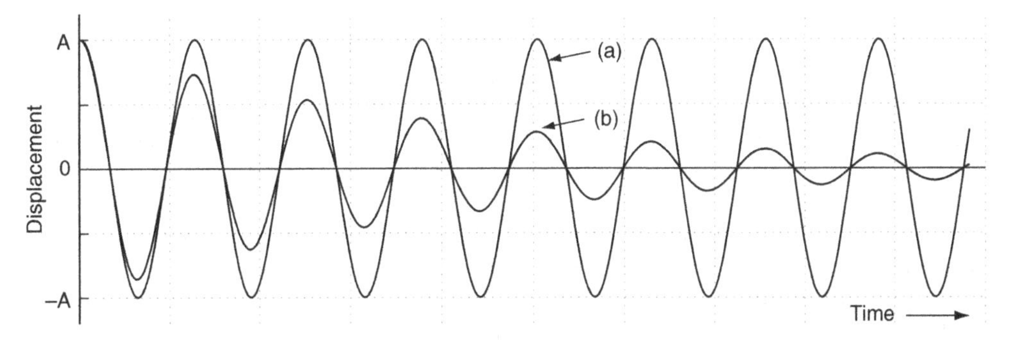

Figure 2.2 The displacement of a simple mass–spring system as a function of time for (a) the lossless case and (b) the lossy (damped) case.

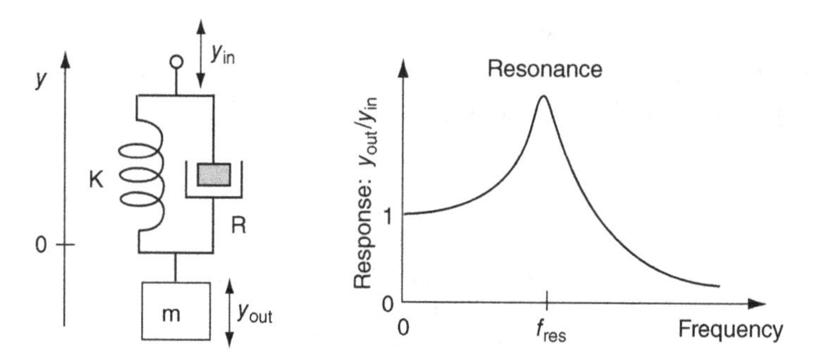

Figure 2.3 The resonance phenomenon in a mass–spring system. If the upper end of the string is moved sinusoidally (excitation), the movement of the mass (response) is increased around the resonance frequency (eigenfrequency).

In practice, losses damp the oscillation of real mass–spring systems, as a result of which the amplitude of the oscillation decreases exponentially in time as

$$y(t) = Ae^{-\alpha t}\cos(\omega_{\mathrm{d}}t + \phi) = A(t)\cos(\psi(t)), \tag{2.8}$$

where α is a damping coefficient and ω_{d} is the angular frequency of the damped oscillation. $A(t)$ is the *amplitude envelope* and $\psi(t)$ the *instantaneous phase*. The damped oscillation is characterized by curve b) in Figure 2.2. Losses often mean energy conversion to heat, but the energy may also be transferred to another kind of oscillation, such as electrical vibration (see Section 4.1).

Mechanical vibration is typically the source of acoustic waves, and the radiation of sound also drains some of the energy from the mechanical vibration system.

2.1.3 Resonance

The simple mass–spring combination oscillates most easily at its eigenfrequency given in Equation (2.6) or in its vicinity. If the mass–spring damping system of Figure 2.3 is excited by a constant-amplitude sinusoidal movement at different frequencies at the top end of the spring, the response – the amplitude of the mass oscillation – follows the resonator response curve also shown in the figure. When the vibration frequency is near f_{res}, we say that the system is in *resonance* or that it resonates. Such a system is called a *resonator*.

Resonance is a phenomenon found frequently in physical systems. It may be desired, undesirable, or even harmful, depending on the case. If the radiation of sound of a musical instrument is too weak, then building a body or sound board with stronger resonances may help to increase the loudness of the sound source. At the same time, the resonance colours the sound which, if properly designed, can make it more appealing. On the other hand, resonances may amplify noise or even cause a machine to malfunction due to strong resonant behaviour. In such a case, damping helps to reduce undesired or dangerous vibration or harmful noise.

A simple but important type of resonator in acoustics is the *Helmholtz resonator*, which is shown in Figure 2.4. The air inside a closed volume V, due to its compressibility, acts as a mechanical spring, and the moving air in the tube or opening above behaves like a mass. Together they make a mass–spring system that works as an acoustic resonator. For instance, by properly blowing into the opening of an empty bottle, this effect can be easily demonstrated.

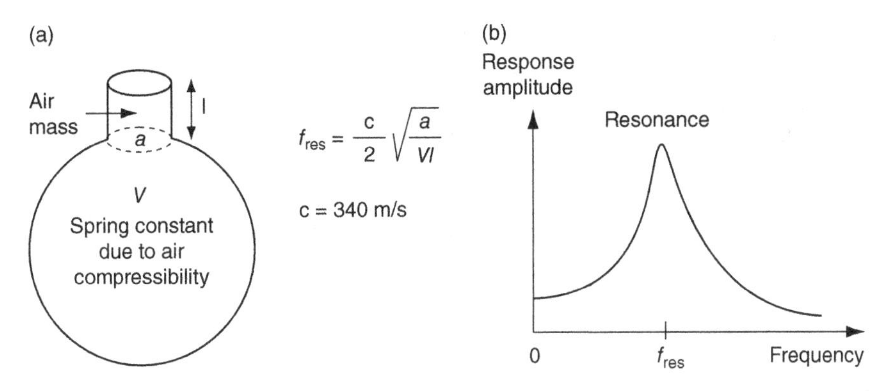

Figure 2.4 Helmholtz resonator: (a) principle of structure and (b) response (e.g., pressure variation in the bottle) as a function of frequency due to (external pressure) excitation.

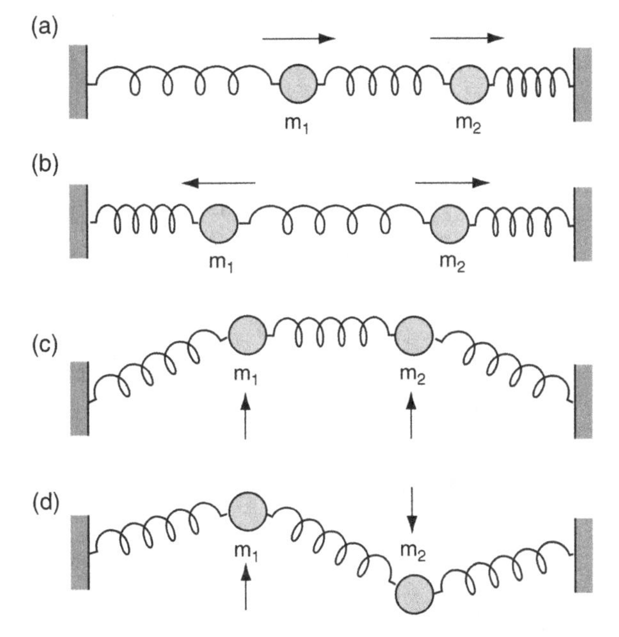

Figure 2.5 (a and b) Longitudinal and (c and d) transversal (vertical) vibration modes in a spring-coupled, two-mass system.

The principle of Helmholtz resonance takes different shapes. The body of the violin or the guitar is a good example where the air hole(s) and the body result in the lowest resonance frequency of the instrument. A bass-reflex type of loudspeaker enclosure adds a desired boost to the low-frequency response of the loudspeaker (see Section 4.1.1). The interior of a car cabin, when a window is slightly opened while driving fast, may strongly amplify turbulent noise at low or very low (infrasound) frequencies.

2.1.4 Complex Mass–Spring Systems

When several masses are coupled through springs, the resulting system shows more complicated oscillatory behaviour. Figures 2.5a and 2.5b illustrate a case where two masses, coupled together by springs and attached to a fixed (non-movable) support, vibrate in the horizontal direction. Such movement is called *longitudinal* vibration. The case of the same two spring-coupled masses now moving in the vertical direction is depicted in Figures 2.5c and 2.5d. The movement occurs perpendicular to the coupling spring direction and is called *transversal* vibration.

In both cases each mass shows one degree of freedom to vibrate. If not limited to move in some direction, each mass can move in three dimensions, thus exhibiting three degrees of freedom (independent components of vibration).

If we expand the system to include three masses, we will have longitudinal vibrations as shown in Figure 2.6a–c, or the vertical vibrations of Figure 2.6d–f.

2.1.5 Modal Behaviour

Since masses coupled through springs cannot vibrate independently, the resonance behaviour of such systems is more complex than in the case of the simple mass–spring system of Figure 2.3.

Figures 2.5 and 2.6 characterize vibration patterns that are called *normal modes* or *eigen-modes*, or simply *modes*. At and near a mode frequency (= resonance frequency), the vibration

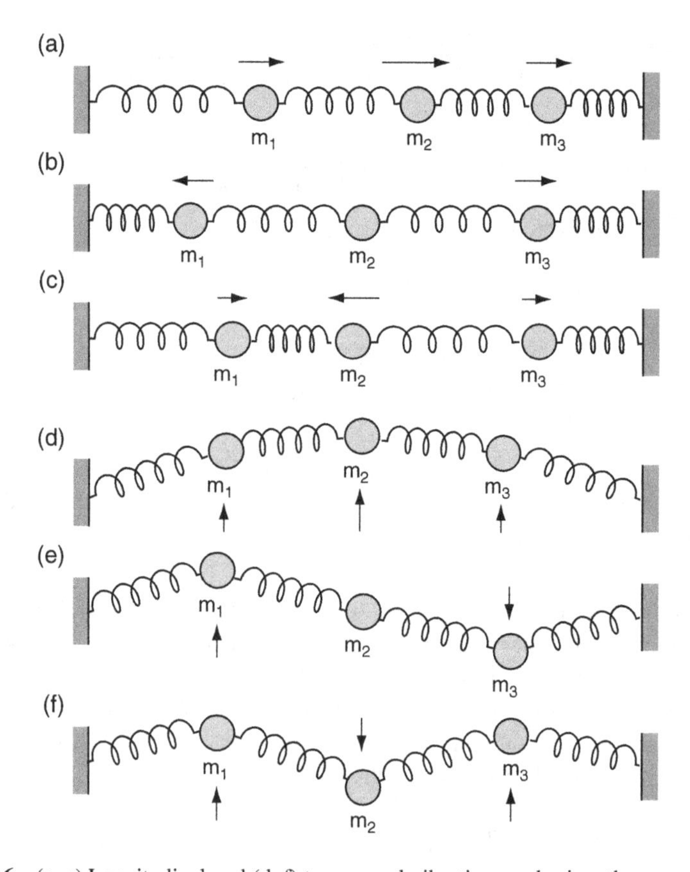

Figure 2.6 (a–c) Longitudinal and (d–f) transversal vibration modes in a three-mass system.

response grows in amplitude when the system is excited, say, by moving one of the terminal supports.

Such modal behaviour is characteristic of simple and complex resonators. As will be explained below, for example the string, the air column, the membrane of a musical instrument, or the human vocal tract show a multitude of mode frequencies. Three-dimensional systems, such as rooms and concert halls, show the same kind of modal behaviour.

Modal frequency vibrations, such as those characterized in Figures 2.5 and 2.6, are only a small subset of possible vibrations in these systems. The special role of modes is that all other vibrations can be expressed as linear combinations of the modes, or, in other words, as movements where the modal oscillations at each point are each summed together with proper amplitude scaling.

The infinitely growing complexity of mass–spring systems is understood when analysing the behaviour of a vibrating string, such as the guitar string, supported at both ends. Figure 2.7 illustrates the eight lowest modes of such a string. We can consider the string as a continuum of infinitely small masses and springs. This means that, in theory, there are infinitely many modal frequencies (see also Sections 2.3 and 2.4.4).

2.1.6 Waves

In a spatially distributed homogeneous mass–spring medium, a vibratory movement (excitation, source) causes a wave to propagate from the source. Figure 2.8a characterizes a transversal one-dimensional wave starting to propagate from a moving source. The medium may be, for example, a rope that is moved rapidly up and down by hand. Transversal means that the movement of the particles in the medium is perpendicular to the direction of the wave propagation.

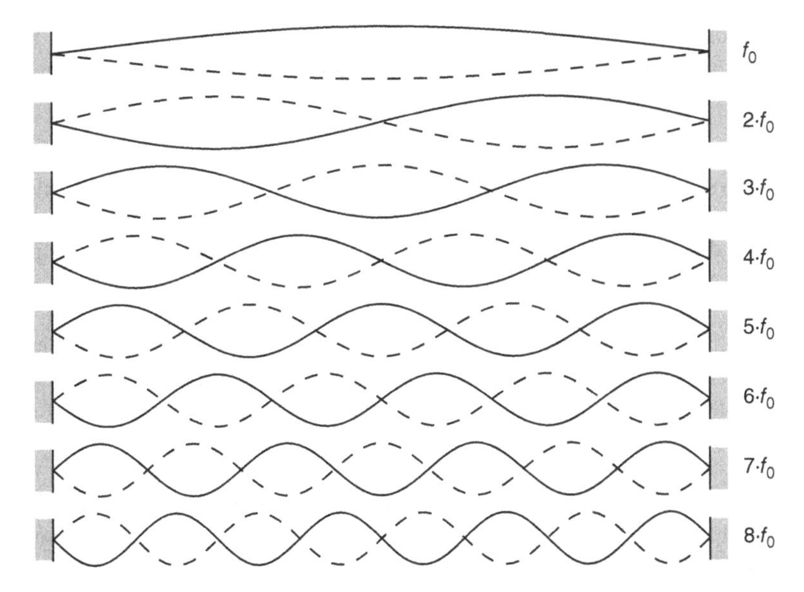

Figure 2.7 The eight lowest resonance modes of a vibrating string. In each mode, the string vibrates between the ends shown by solid and dashed lines. Nodes are points where the vibration amplitude is zero.

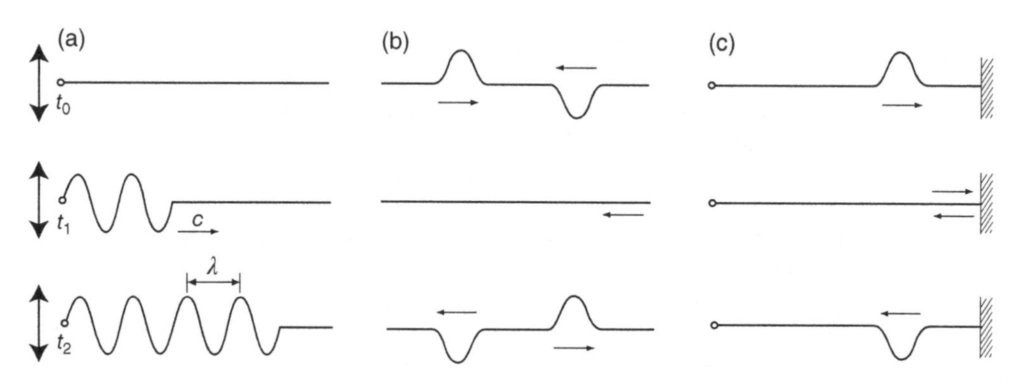

Figure 2.8 (a) Wave propagation in a continuously distributed mass–spring chain, such as a rope, (b) the passing of two wave components travelling in opposite directions, and (c) the wave reflection at a fixed termination.

In Figure 2.8c, the mass–spring medium (such as a rope) is terminated rigidly. The discontinuity of the medium causes a reflection of the wave at the termination. As will be shown below, any discontinuity of the medium can result in full or partial reflection of a wavefront.

In mathematical terms, a wave in a homogeneous and lossless medium follows a simple rule called the *wave equation*:

$$\ddot{y} = c^2 y''. \tag{2.9}$$

Here y is displacement, c is the velocity of propagation of the wavefront, $\ddot{y}$ is the second time derivative of displacement equalling $\partial^2 y/\partial t^2$, and y'' is the second position derivative of displacement equalling $\partial^2 y/\partial x^2$.

One-dimensional wave propagation, as characterized in Figure 2.8, obeys the general solution of the wave equation for a wave propagating in a homogeneous medium,

$$y(t,x) = g_1(ct - x) + g_2(ct + x), \tag{2.10}$$

where y is a physical wave variable (such as displacement), c is the propagation velocity of the wave, t is time, and x is a position coordinate. This equation states that any wave, here $y(t,x)$, can be expressed as a sum of two waveforms, g_1 travelling in the negative x-axis direction and g_2 in the positive x-axis direction. This travelling-wave solution was published by d'Alembert in 1747.

Figure 2.8b illustrates an interesting case where two waves of the same form but opposite polarity pass each other in opposite directions. It may first appear illogical that the waves do not cancel each other totally. Remember, however, that the energy of the waves cannot disappear; it is just in kinetic form in the middle situation and reappears as potential energy after passing.

For a periodic waveform, the distance between equal phase points in the wave (for example, peaks) is called the *wavelength*. The wavelength λ is related to the frequency f and the wave propagation velocity c by

$$\lambda = c/f \tag{2.11}$$

The frequency f in hertz [Hz] is the number of oscillations per second. A more general case, such as sound propagating in air, is a three-dimensional phenomenon. Concepts related to wave behaviour are discussed further in Section 2.3.

When a sound wave encounters a discontinuity, such as a new medium, the simple wave equation, Equation (2.10), for undisturbed propagation of a waveform is not valid anymore. Often, the case is similar to the one in Figure 2.8c where the termination of a medium causes reflection of a wave. In this specific case, the displacement is bound to be zero at the termination, so that the reflecting wave component must be equal to the arriving wave but opposite in sign.

The vibration modes of the string with a distributed mass–spring in Figure 2.7 can also be explained through wave propagation and reflection. It can be shown that two sinusoidal wave components having the same mode frequency but travelling in opposite directions on the string result in a *standing wave* where no net energy transfer takes place. Each subwave reflects back from the terminations and the vibratory energy remains 'in place'. As Figure 2.7 shows, there are points of maximum vibration as well as points of zero (or minimal) vibration called *nodes*.

2.2 Acoustic Measures and Quantities

Acoustics is the branch of physics studying mechanical vibrations. Acoustics uses a set of concepts, variables, quantities, and measures to characterize waves, fields, and signals. Some of the physical measures and quantities most important in the field of communication acoustics are mentioned here. Hearing-related concepts are discussed in later chapters.

2.2.1 Sound and Voice as Signals

When the value of a physical variable is registered at a specified spatial point as a function of time, a signal is obtained. Signals and signal processing discussed in more detail in Chapter 3.

We first need to define some concepts that are needed to characterize sounds as physical phenomena:

- A *pure tone* is a sinusoidally varying sound signal, such as the cosine-form vibration of Equation (2.5) and Figure 2.2a. Thus, it consists only of a single frequency component. Although generating an ideal pure tone is not possible in practice, it can be approximated and is a useful abstraction.
- A *combination tone* consists of a set of pure tones called *partials*, each having its own frequency, amplitude, and phase.
- In *periodic sound signals*, the waveform repeats itself, and the signal consists of partials that have a harmonic relationship. The lowest frequency is called the *fundamental frequency*, often denoted f_0. Partials are called *harmonics*, and their frequencies are integer multiples of the fundamental, $f_n = nf_0$. Most musical instruments in Western music generate harmonic or almost harmonic signals.
- *Non-periodic sounds* may consist of discrete frequencies that are not in harmonic relationships or of a continuous distribution of partial frequencies. In the latter case the signal sounds noise-like.

The 'strength' or 'intensity' of a signal can be characterized by many measures. For a periodic signal, such as a sine wave, the maximum value of the waveform, called the *amplitude*, is often used. If the signal is not symmetric about the abscissa, the negative peak value may have a larger absolute value than the positive and is sometimes used as the amplitude value. Also, if the average value of a signal is not zero – this is often referred to as the DC value (direct current value, due to the analogy from electrical engineering) – the amplitude may be expressed as the maximum deviation from the average (DC) level.

Another common measure is the *root mean square* or *RMS value*, also called the *effective value*. It is defined, for example, for pressure $p(t)$ as

$$p_{\mathrm{rms}} = \frac{1}{t_2 - t_1}\sqrt{\int_{t_1}^{t_2} p(t)^2 \, \mathrm{d}t}, \tag{2.12}$$

where the time range of integration can be over one period for a periodic signal or a long enough – ideally infinite – time span for non-periodic signals. For a pure tone (sinusoidal signal), the *peak value* $\hat{p} = \sqrt{2}\, p_{\mathrm{rms}}$.

2.2.2 Sound Pressure

The most important physical measure in acoustics is *sound pressure*. Pressure, in general, is the force applied per unit area in a direction perpendicular to the surface of an object. Its unit is the Pascal [Pa] = [N/m^2]. Sound pressure, on the other hand, is the deviation of pressure from the static pressure in a medium, most often air, due to a sound wave at a specific point in space. Sound pressure values in air are typically much smaller than the static pressure. Sounds that human hearing can deal with are within the range of $20 \cdot 10^{-6}$ to 50 Pa.

Sound pressure is also an important measure due to the fact that it can be measured easily. A good condenser microphone (see Section 4.1.2) can transform sound pressure into an electrical signal (voltage) with high accuracy.

2.2.3 Sound Pressure Level

Because sound pressure varies over a large range in Pascal units, using a logarithmic unit, the *decibel* [dB], is more convenient. The ratio of two amplitudes, A_1 and A_2, in decibels is computed from

$$L = 20 \log_{10}(A_2/A_1), \tag{2.13}$$

and it is used widely in acoustics, electrical engineering, and telecommunications. Some decibel values that are worth remembering are given in Table 2.1.

Table 2.1 Some values in decibels worth remembering.

Ratio	Decibels	Ratio	Decibels
1/1	0		
$\sqrt{2} \approx 1.41$	$\approx 3.01 \approx 3$	$\sqrt{1/2} \approx 0.71$	$\approx -3.01 \approx -3$
2/1	$\approx 6.02 \approx 6$	1/2	$\approx -6.02 \approx -6$
$\sqrt{10} \approx 3.16$	10	$\sqrt{1/10} \approx 0.316$	−10
10/1	20	1/10	−20
100/1	40	1/100	−40
1000/1	60	1/1000	−60

The concept of the decibel in acoustics is used in a special way. If the denominator A_1 in Equation (2.13) is a fixed reference, decibels are then absolute *level* units. The reference sound pressure $p_0 = 20 \cdot 10^{-6}$ Pa is used so that the *sound pressure level* (SPL) L_p [dB] is

$$L_p = 20 \log_{10}(p/p_0) \tag{2.14}$$

This value of p_0 is selected so that it roughly corresponds to the threshold of hearing, the weakest sound that is just audible, at 1 kHz. Human hearing is able to deal with sounds in the range 0–130 dB, from the threshold of hearing to the threshold of pain (for more details, see Chapters 7 and 9). SPL values in dB are more convenient to remember than sound pressure values in Pascals.

SPL values can be converted to sound pressure from the inverse of Equation (2.14):

$$p = p_0\, 10^{L_p/20}. \tag{2.15}$$

2.2.4 Sound Power

Sound power P, like physical power in general, is defined in watts [W] as the physical work done in one second. In acoustics, sound power is considered to be the property of a sound source radiating energy along with the sound wave it creates.

Only a fraction of the primary power of a sound source is transformed into acoustic power. The *efficiency* η of a sound source is

$$\eta = P_\mathrm{a}/P_\mathrm{m}, \tag{2.16}$$

where P_a is the radiated sound power and P_m is the primary power, such as mechanical or electrical power.

Sound power can also be expressed using a logarithmic measure. The *sound power level* L_W in decibels is defined as

$$L_W = 10 \log_{10}(P/P_0), \tag{2.17}$$

where P is sound power [W] and P_0 is the reference power of $1 \cdot 10^{-12}$ W. Note the coefficient 10 for power instead of 20 for pressure. This follows simply by expressing power P in terms of pressure p, $10 \log_{10}(P_1/P_2) = 10 \log_{10}(p_1^2/p_2^2) = 20 \log(p_1/p_2)$, in Equation (2.14).

2.2.5 Sound Intensity

Sound intensity I [W/m^2], as a physical measure, is defined as the sound power through a unit area, describing the flow of sound energy. Mathematically speaking, it is a vector – it has magnitude and direction, as will be further discussed in Section 14.4.7. *Sound intensity level* L_I is defined as

$$L_I = 10 \log_{10}(I/I_0), \tag{2.18}$$

where the reference $I_0 = 1 \cdot 10^{-12}$ [W/m^2].

2.2.6 Computation with Amplitude and Level Quantities

When two or more sounds contribute to a sound field, they can affect the total field in different ways. At any single moment in time, the pressure values add up. This linear superposition of waves is valid in air at normal sound levels. Non-linearities may appear elsewhere, for example in sound reproduction systems.

The resulting amplitude or RMS values and level quantities can be categorized as follows:

1. Sound sources are *coherent* if they or their partials have the same frequencies. Depending on their phase difference they can:
   - add constructively if they have the same phase;
   - add destructively if they are in opposite phase; or
   - in other cases the result depends on the amplitudes and phases of the components.
2. Sound sources are *incoherent* if their frequencies do not coincide, in which case the powers of the signals are summed.

When two coherent signals with the same amplitude A are added constructively, the resulting signal has the amplitude $2A$, which means an increase in level of $20\log_{10}(2A/A) \approx 6$ dB. If two incoherent signals with the same amplitude are added, the resulting level will increase by $10\log_{10}(2A^2/A^2) \approx 3$ dB.

In the general case of incoherent sounds with sound levels L_1 and L_2, the resulting level, based on the addition of powers, will be

$$L_{\text{tot}} = 10\log_{10}\left(\frac{P_1 + P_2}{P_0}\right) = 10\log_{10}\left(10^{L_1/10} + 10^{L_2/10}\right). \tag{2.19}$$

2.3 Wave Phenomena

Sound waves behave similarly in gases and liquids. Such a medium is called a *fluid*. In an ideal fluid, only longitudinal wave propagation is possible, there will be no transversal propagation.

Each medium where sound waves propagate has a characteristic *sound velocity*. For air, the most important medium in acoustics, the sound velocity c_{air} depends on temperature:

$$c_{\text{air}}(T) = 331.3 + 0.6T, \tag{2.20}$$

where T is temperature in °C and velocity is in m/s. This approximation of c_{air} is valid for typical room temperatures.

When a vibrating body generates a sound field in the surrounding fluid, the geometry of the field depends on the form of the sound source and its size compared to the wavelength. In practical cases, the sound field is so complex that it cannot be solved analytically and can only be approximated numerically. With some simplifying assumptions, the characteristics of a radiated sound field can, in many cases, be understood quite easily. The two simplest idealizations are spherical and planar wave fields.

2.3.1 Spherical Waves

A pulsating sphere as a sound source emits a *spherical wave field* that propagates from the source at the velocity of sound c, as illustrated in Figure 2.9. The sound pressure in the wave is inversely proportional to the distance r from the mid-point (symmetry point) of the source,

$$p(r) \propto q/r. \tag{2.21}$$

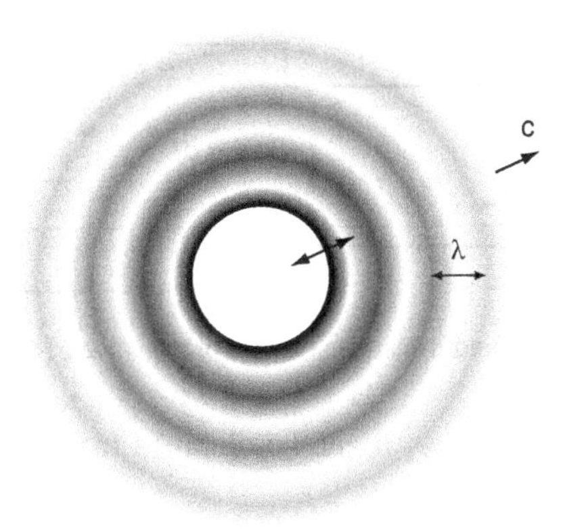

Figure 2.9 Wave propagation from a spherical, sinusoidally vibrating source. The greyscale outside the spherical source denotes the pressure of the sound field; a darker colour means a higher value. The disturbances propagate at the velocity of sound c, and the distance between pressure maxima is the wave length λ.

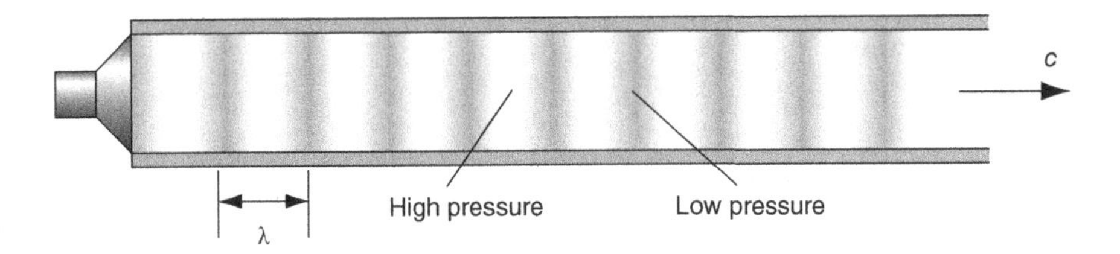

Figure 2.10 Plane wave propagation in a homogeneous tube.

In this formula, q is the volume velocity, a concept defined later. Note that the size of the sphere does not have an effect on the wave field, although a larger sphere is a more efficient radiator. As a mathematical idealization, a *point source* is a useful abstraction.

Any form of sound source that is small in dimension compared to the wavelength and vibrates quite homogeneously can be approximated as a spherical or point source. Thus, at low frequencies such sources, like typical loudspeakers or human speakers, are practically spherical wave sources.

2.3.2 Plane Waves and the Wave Field in a Tube

Another important special case of wave fields is a *plane wave*. A large and homogeneously vibrating planar surface emits a plane wave. In a lossless medium, the planar wavefront preserves its waveform (see Figure 2.8) and propagates without attenuation.

In a homogeneous tube, as shown in Figure 2.10, at frequencies where the cross-sectional dimensions are smaller than the wavelength, only a plane wave can propagate.

Two physical variables are used to fully describe a wave in a tube: sound pressure p and *volume velocity* q in m^3/s. The volume velocity characterizes the flow of the medium through a cross-sectional area in unit time. The relation between pressure and volume velocity is

$$p = Z_a q, \tag{2.22}$$

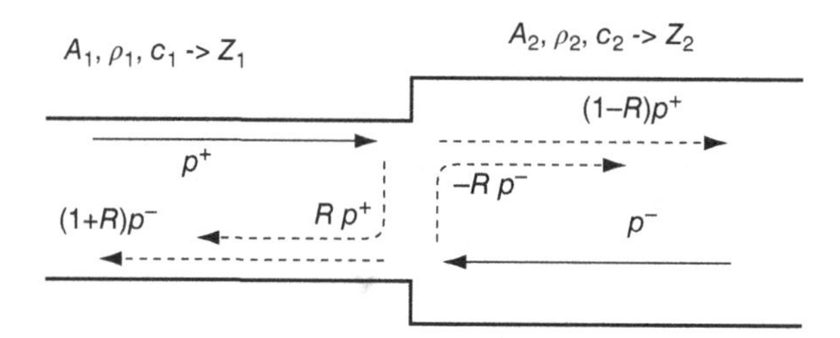

Figure 2.11 Reflection and transmission of a plane wave at a discontinuity in a tube.

where Z_a is the *acoustic impedance*. It can be obtained from the physical properties of the medium as

$$Z_a = \rho c / A, \tag{2.23}$$

where ρ is the density of the medium [kg/m^3], c is sound velocity, and A is the cross-sectional area [m^2].

Another related concept is the *characteristic impedance* Z_0, which is an inherent property of the medium in which the wave is travelling. It is defined as the ratio of pressure and particle velocity in a plane wave,

$$Z_0 = \rho c. \tag{2.24}$$

If the cross-sectional area or any acoustic parameter of the medium in a tube changes at any position so that the acoustic impedance Z_a changes, the simple propagation of a wavefront is disturbed. At such a discontinuity, an arriving wave is split so that part of it reflects back and part of it propagates through the discontinuity. In Figure 2.8c, the wave in a rope reflects back completely due to the fixed termination. In the tube of Figure 2.11, when a plane wave p^+ meets a change in acoustic impedance from Z_1 to Z_2, the wave component p_r^+ that reflects back is

$$p_r^+ = R p^+, \tag{2.25}$$

and the component p_f^+ that propagates through the junction is

$$p_f^+ = T p^+ = (1 - R) p^+. \tag{2.26}$$

Here, R is the *reflection coefficient*

$$R = \frac{Z_2 - Z_1}{Z_2 + Z_1} \tag{2.27}$$

and $T = 1 - R$ is the *transmission coefficient*. Figure 2.11 also shows a wave p^- propagating to the left and being reflected by the factor $-R$ and transmitted by the factor $T = 1 + R$. From Equations (2.25)–(2.27), it can be concluded that if $Z_1 = Z_2$, which means that there is a perfect impedance match, the reflection coefficient $R = 0$ and no reflection occurs. If impedances Z_1 and Z_2 are very different (impedance mismatch), a strong reflection occurs.

For a tube terminated in a rigid wall, the impedance of the termination $Z_2 >> Z_1$, the impedance of the tube, which implies from Equation (2.27) that $R \to 1$. When the termination is an open end, the impedance Z_2 is very small (at low frequencies) and $R \to -1$.

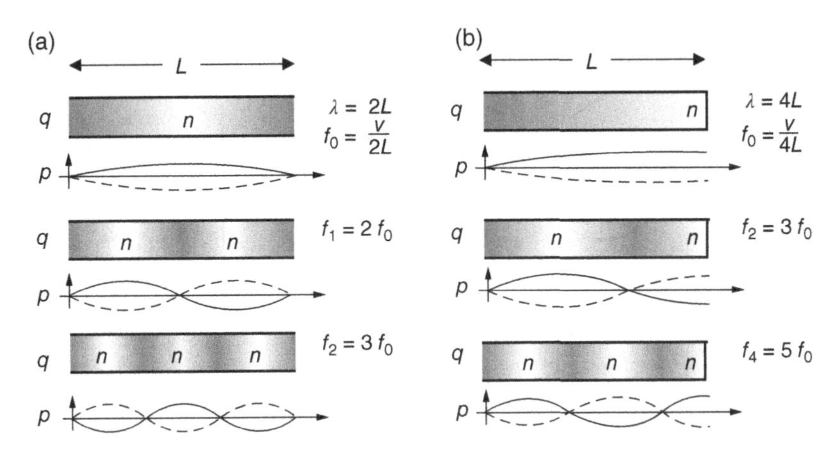

Figure 2.12 The behaviour of the lowest modes in a tube with (a) open ends and (b) one end closed. The pressure p of each mode is shown as a curve below the tube, and the volume velocity q is shown as a colouring of the tube, with darker colour shades denoting higher velocity. n means a nodal point where the volume velocity is minimal. Adapted from Rossing *et al.* (2001).

Table 2.2 The velocity of longitudinal waves in some media.

Medium	Velocity [m/s]
Air (20 °C)	343
Helium	970
Water	1410
Steel	5100
Glass	12000–16000

In a tube of finite length, the terminations reflect waves back so that standing waves appear, resulting in modes and resonance frequencies. Figure 2.12 depicts the lowest modes in a tube that is left open at both ends and in another tube that is closed at one end. The first case is approximated, for example, in the flute, generating all harmonics, and the second case in the clarinet, where (low order) even harmonics are weak due to lack of resonance.

2.3.3 Wave Propagation in Solid Materials

The behaviour of solid matter deviates from that of fluids since transversal (shearing) forces are possible in solids, which results in the possibility of transversal waves forming in addition to longitudinal waves. The velocity of longitudinal waves in several media is listed in Table 2.2.

Transversal waves appear, for example, in a string under tension (Figures 2.7 and 2.13) or in a bar (Figure 2.14). In a non-stiff string (a wire where negligible force is needed to bend it), the velocity c_t of the transversal wave depends on the string tension T in N and mass density μ in kg/m as

$$c_t = \sqrt{(T/\mu)}. \tag{2.28}$$

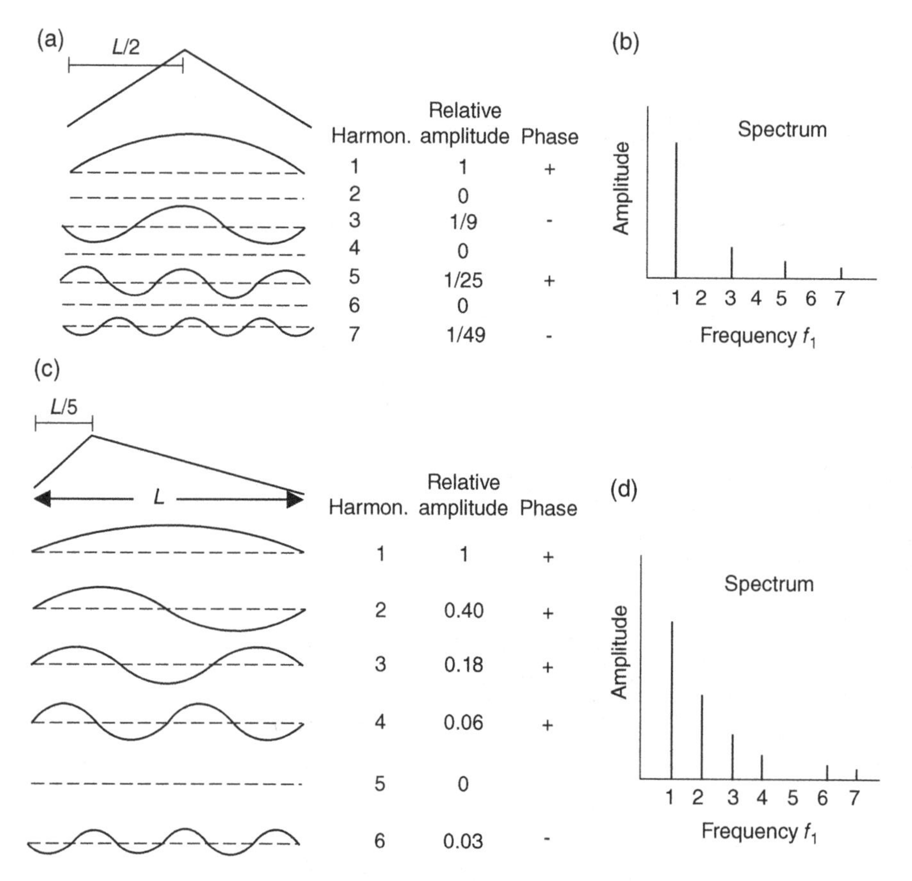

Figure 2.13 (a) The initial displacement of a string when plucked in the middle and (b) the harmonic content of the vibration. When the string is plucked asymmetrically (one fifth of L from the end), the initial displacement is shown in (c) and the harmonic components in (d). Source: Rossing *et al.* 2001.

Figure 2.13 depicts the behaviour of a string when plucked (a) in the middle and (c) one fifth of the length of the string from the end. The spectra consist of harmonic components, but with every Nth harmonic missing for $N = L/L_{\rm pp}$, where L is the length of the string and $L_{\rm pp}$ is the plucking point distance from the end.

An example of transversal waves that resemble string behaviour but which are not purely one-dimensional and harmonic in spectral content is wave propagation in a bar. Due to bending stiffness, the transversal waves are *dispersive*, implying that different frequencies propagate at different velocities (higher frequencies propagate faster than lower ones). This results in modal frequencies that do not have a harmonic relationship. Figure 2.14a illustrates the lowest modes in a free bar and Figure 2.14b the case where the bar is clamped at one end. The resulting sounds are strongly inharmonic. Some degree of inharmonicity can also be found in string instruments, especially in the piano at low frequencies, due to the stiffness of the strings.

Membranes and plates are also often used as sound sources. For example, the membrane of a drum under tension is a two-dimensional equivalent of the vibrating string. Some of the lowest modal patterns of a circular membrane are shown in Figure 2.15. The patterns show

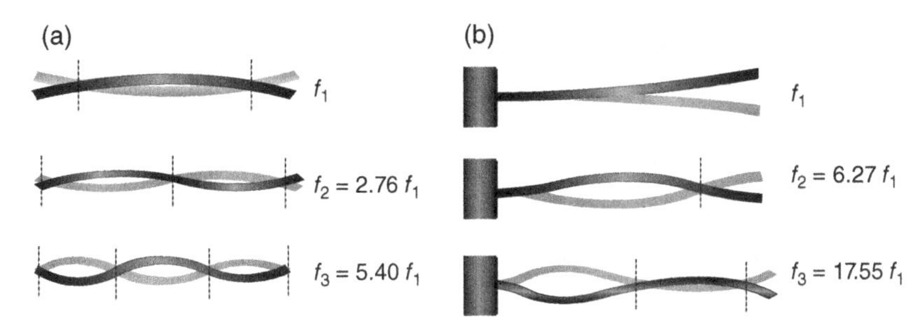

Figure 2.14 Modal patterns and frequencies of a bar: (a) freely vibrating (unsupported) and (b) rigidly clamped at one end. The nodes of vibration are shown with vertical dashed lines. Adapted from Rossing *et al.* (2001).

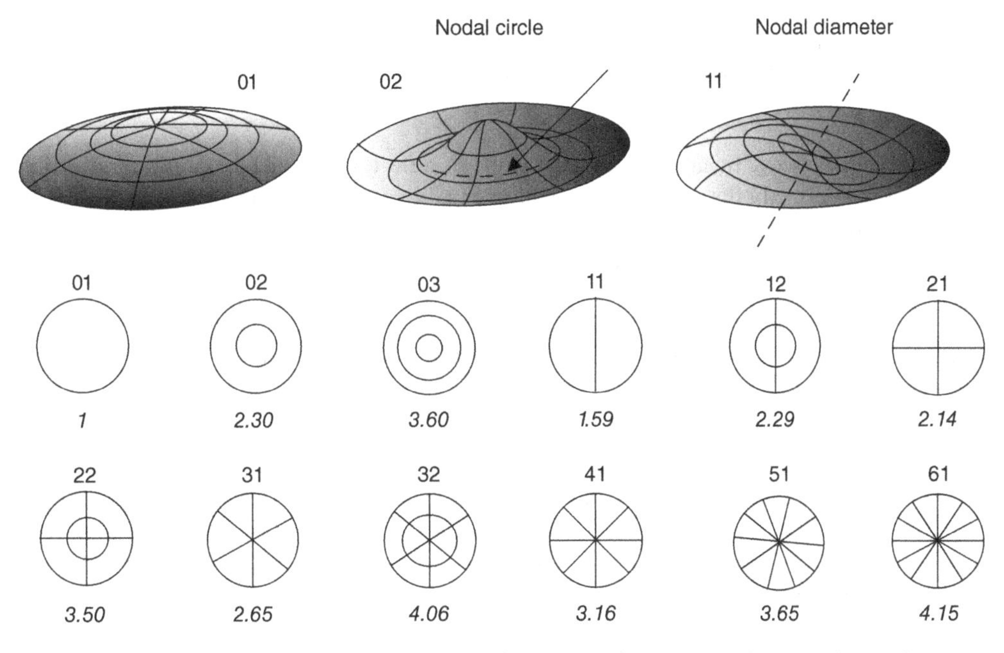

Figure 2.15 Modal vibration of a circular membrane supported at the edge: membrane shapes at moments of maximal displacement (top row) and nodal patterns and corresponding relative resonance frequencies (two lowest rows). Adapted from Rossing *et al.* (2001).

different symmetries, both circular and diametral. Once again, the modal frequencies are not in harmonic relationships, and it is characteristic that the density of mode frequencies increases towards high frequencies.

2.3.4 *Reflection, Absorption, and Refraction*

When a wavefront encounters the surface of another medium, such as a hard wall, a fraction of the wave energy reflects back and the rest propagates into the other medium or transforms into thermal energy. Figure 2.16 illustrates these phenomena when a plane wave hits a wall with

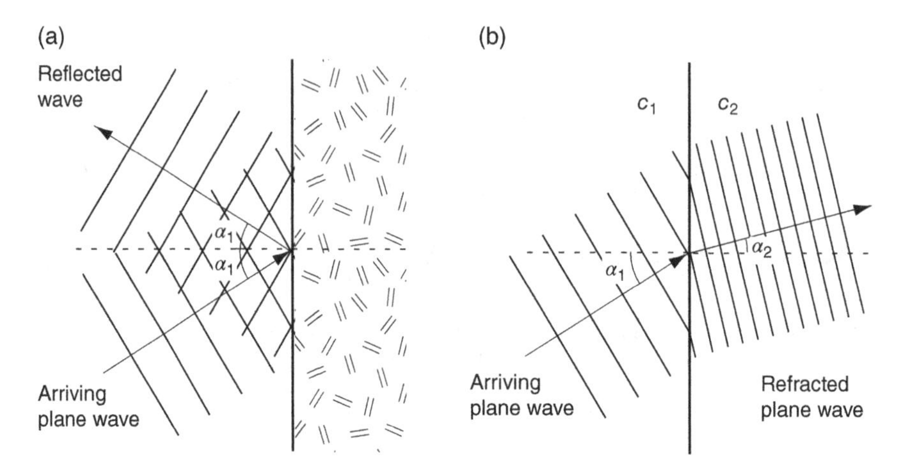

Figure 2.16 (a) Reflection of a plane wave at a hard surface, and (b) refraction (bending) of the wavefront when entering another medium with a different sound velocity.

an incidence angle α_1. In an ideal case, the angle of reflection is equal to α_1. If the reflecting surface is not perfectly flat, spreading of the wave in directions around this mirror image angle will occur (diffuse reflection).

The fraction of sound energy that is not reflected is absorbed from the sound field of the first medium. The ratio of absorbed energy to the incident energy is called the *absorption coefficient*. When the reflection of the wave is described by the coefficient R, as defined in Equation (2.25), the absorption coefficient a is obtained from the equation

$$a = 1 - |R|^2. \tag{2.29}$$

Refraction is the phenomenon of the wavefront bending when entering a medium with a different sound velocity. Figure 2.16b characterizes this when the sound velocity is lower in the second medium. According to Snell's law, $c_1 \sin\alpha_2 = c_2 \sin\alpha_1$, where c_1 and c_2 are sound velocities as defined in Figure 2.16b.

2.3.5 Scattering and Diffraction

When a wave interacts with a geometrical discontinuity like a boundary, *scattering* occurs. Reflection can be interpreted as the simple case of *backscattering* from a boundary layer. More general scattering in acoustics is called *diffraction*. It is a complex phenomenon that is difficult to analyse and model. A solid object or its edge acts as a kind of secondary source. Huygens' principle interprets a wavefront as a set of directed secondary sources that maintains a regular wave propagation. A geometrical discontinuity disturbs this propagation and makes diffraction the secondary observable source.

Figure 2.17 shows a typical case where a noise barrier is used to mask noise sources, such as vehicles on a highway, to decrease disturbance in the environment. At high frequencies, where the wavelength is small compared to the barrier, shadowing due to the barrier is efficient. At low frequencies, however, the sound waves propagate around the barrier so that the desired masking remains relatively small.

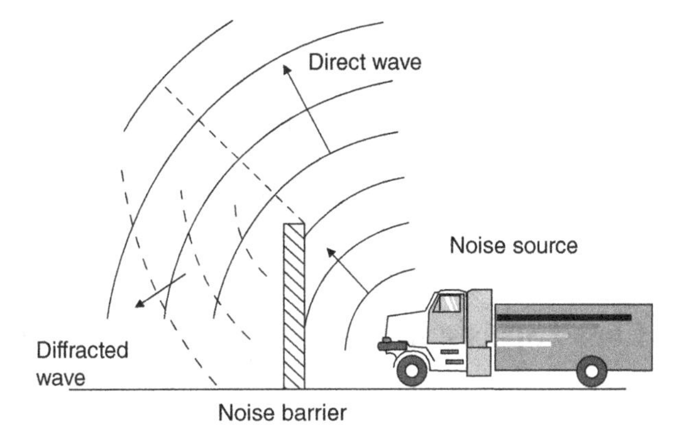

Figure 2.17 Diffraction of sound at the edge of a sound barrier.

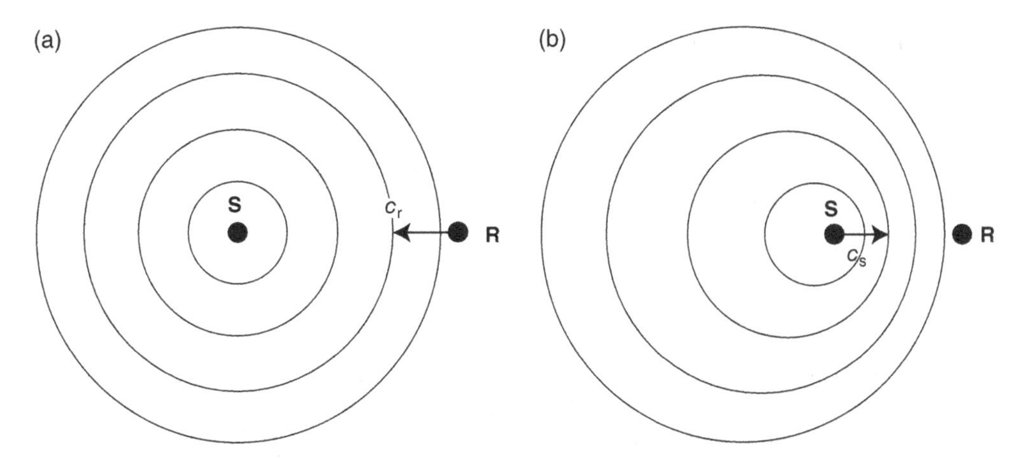

Figure 2.18 Doppler effect (a) when the receiver moves and the source is immobile, (b) when the receiver is immobile and the source moves.

2.3.6 Doppler Effect

Imagine a free-field condition with a single receiver and a single sound source emitting one sinusoid with frequency f. If the distance between the source and the receiver diminishes with time, meaning that the source moves towards the receiver or vice versa, as shown in Figure 2.18, each successive wave is emitted from a position closer to the receiver than the previous wave. Therefore, the time between the arrival of successive waves at the observer decreases, causing an increase in the frequency compared to the static situation.

Conversely, if the wave source moves away from the observer, each wave is emitted from a position farther from the observer than the previous wave, thus increasing the arrival time between successive waves and reducing the frequency. The distance between successive wavefronts is increased, so the frequency of the sinusoid is lower. This can be observed when a vehicle producing a harmonic tone (such as an ambulance) passes by at high speed, the perceived pitch of the tone lowers rapidly when the vehicle goes past the observer. This phenomenon is called the *Doppler effect*.

The frequency of the sinusoid observed by the receiver f can be computed as

$$f = \left(\frac{c + c_r}{c + c_s}\right) f_0, \tag{2.30}$$

where the velocity of the receiver is c_r, which is positive if the receiver moves towards the source and negative in the other direction. The velocity of the source is c_s, which is positive if the source moves away from the receiver and negative in the other direction.

2.4 Sound in Closed Spaces: Acoustics of Rooms and Halls

A major part of communication by sound and voice takes place in spaces constrained by walls or surfaces, such as living rooms, auditoria, offices, concert halls, etc. The surfaces reflect impinging sound waves, and in a fraction of a second there are thousands of reflections which result in reverberation that gradually decays. Reflections and reverberation make sound and voice louder in the more distant parts of the room. They may also render the sound more 'colourful' and pleasant if the room has proper acoustics but, on the other hand, may make it difficult and annoying to communicate if the acoustic properties are not well matched to the form of communication. Background noise is another common reason for problems in communication.

In this section, we study the basic physical properties of closed space acoustics – rooms, auditoria, and concert halls. Subsequent sections will discuss some related aspects, and finally in Chapter 17 we study the subjective and objective quality factors, especially those related to performing spaces.

2.4.1 Sound Field in a Room

In a room with simple geometry, sound propagation from a source to a receiver (listener) can be characterized as shown in Figure 2.19. The first wavefront arrives along the direct path (if the source is visible), and soon after this the first reflections from walls, ceiling, and floor arrive.

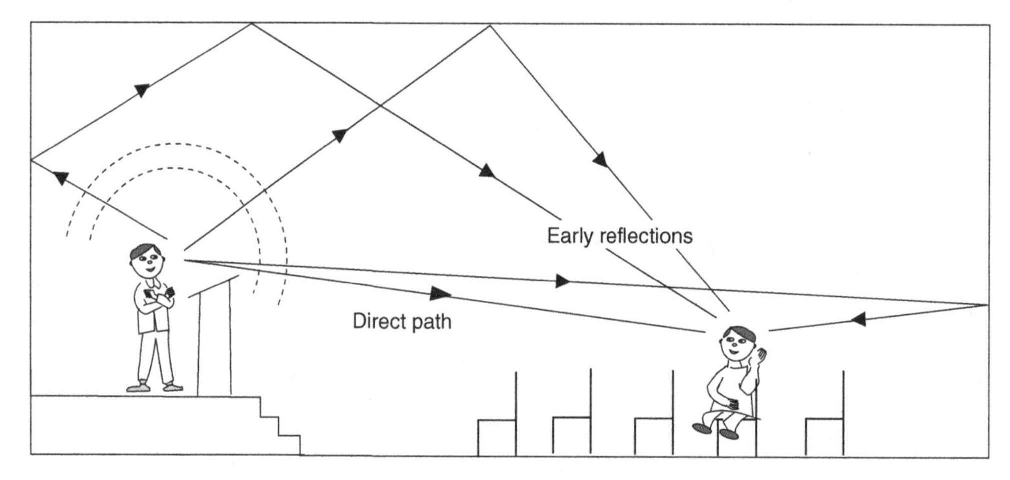

Figure 2.19 Paths of direct sound and the few first reflections in a rectangular room from a speaker to a listener, assuming specular reflections from walls.

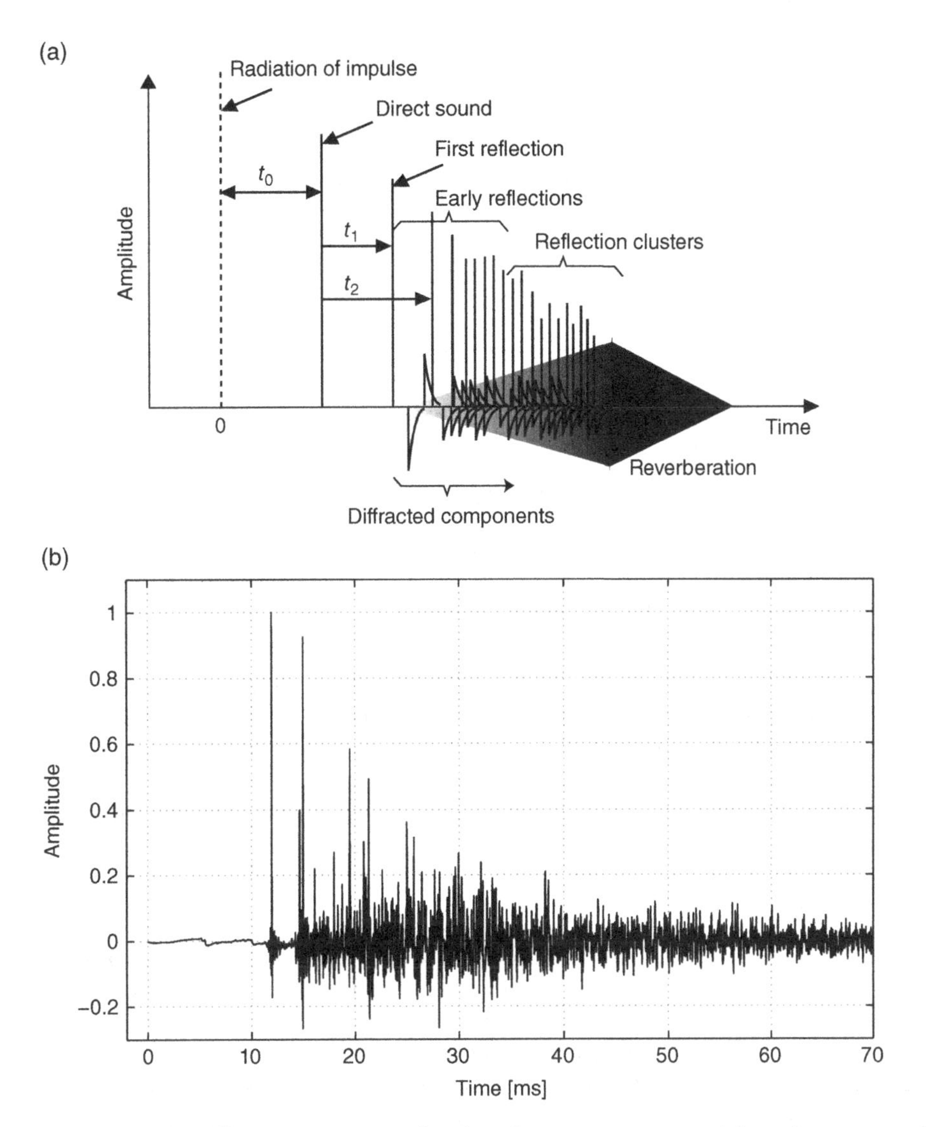

Figure 2.20 (a) The different components of an impulse response measured from the source to the receiver in the case of Figure 2.19, consisting of direct sound, early reflections, diffracted components, reflection clusters, and reverberation. (b) The impulse response measured with a laser-induced pressure pulse in a listening room.

If the source sends an impulse-like sound, the response at the point of the receiver is as depicted in the *reflectogram* of Figure 2.20. Direct sound is followed by early reflections, diffracted components, and then by reverberation, where individual reflections may not be separately visible. Early reflections, for speech up to about 60 ms and for music up to about 100 ms, increase the loudness of sound. Early reflections also strongly contribute to spatial perception, such as in the estimation of room size and source distance. Late reverberation, if too loud, decreases intelligibility of speech or fast passages of music.

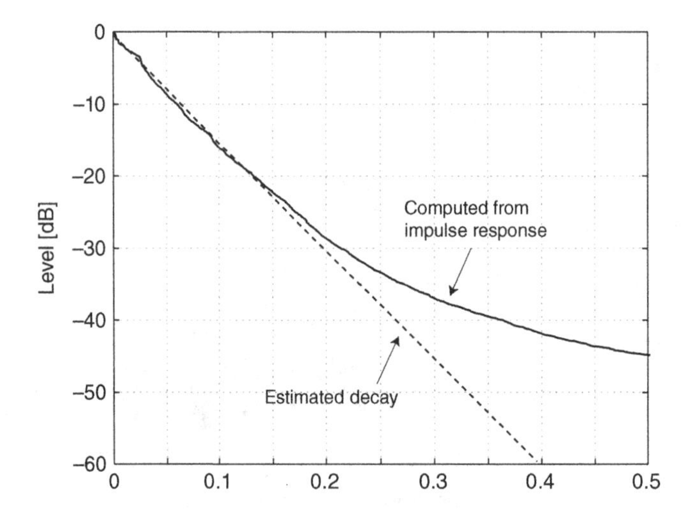

Figure 2.21 Decay of reverberation computed from an impulse response. Due to noise in the measurement, the computed curve does not follow exponential decay. The extrapolated reverberation time T_{60}, the time taken for the response to decay 60 dB, is approximately 0.4 s.

For comparison with the principle in Figure 2.20a, Figure 2.20b plots a measured impulse response of a relatively damped listening room. Note that the reflections have the same polarity as the direct sound, but the diffracted components may have either positive or negative polarity, depending on the geometry (Svensson *et al.*, 1999). Also, the reverberation has both positive and negative polarities. This is also shown in the measured impulse response. Measured room impulse responses can be downloaded, for example, from OpenAIR (2014).

2.4.2 Reverberation

Figure 2.21 plots the decay of the sound pressure level in a room, assuming that a steady-state sound field of white noise is first injected into the room and then interrupted at moment t_0. The curve is plotted on a logarithmic dB scale against a linear time scale. The decay of the sound energy is exponential, which is characteristic of many physical resonator systems (see Equation (2.8)), and corresponds to linear decay on a logarithmic decibel scale.

The single most important parameter describing the acoustics of a room is the *reverberation time*. It is defined as the time period T_{60} during which the sound pressure level decays 60 dB, and it can be estimated from Sabine's formula (Sabine, 1922):

$$T_{60} = 0.161\frac{V}{S}, \tag{2.31}$$

where V is the volume of the room [m^3] and S is the total *absorption area* of the room surfaces. There are other formulations, such as Eyring's formula, that can be more accurate in specific conditions (Kuttruff, 2000). Notice that in many textbooks symbols S and A are used in roles opposite to those here. Here, area is A for consistency with other formulas. The absorption area of room surfaces can be computed from

$$S = \sum a_i A_i \tag{2.32}$$

Table 2.3 Absorption coefficients of materials in different frequency bands.

Frequency	125	250	500	1000	2000	4000
Glass window	0.35	0.25	0.18	0.12	0.07	0.04
Painted concrete	0.10	0.05	0.06	0.07	0.09	0.08
Wooden floor	0.15	0.11	0.10	0.07	0.06	0.07

by summing the product of the absorption coefficient a (Equation (2.29)) and the surface area A in m^2 over each surface i. There exist tables giving the absorption coefficient for different materials as a function of frequency, for example, in octave bands, as shown in Table 2.3. Note that each person inside a room adds to the absorption area S by approximately 0.5 m^2.

Recommendable T_{60} values are about 2 seconds for a large concert hall, about 1.4 seconds for a chamber music hall, 0.5–1.0 seconds for a speech auditorium, and about 0.35 seconds for a listening test room. The acoustic parameters for concert halls and auditoria, and their measurements, are discussed further in Section 17.9.2.

2.4.3 *Sound Pressure Level in a Room*

The amplifying effect of reflections and room reverberation can be understood easily by considering the summation of direct and reverberant sound. The direct sound pressure is inversely proportional to the distance between source and receiver, as stated in Equation (2.21). The level of the reverberant field is approximately constant in the whole room, and it can be considered to be approximately incoherent with the direct sound. Based on the summing of two power levels, the total sound pressure level L_p in dB will be

$$L_p = L_W + 10\log_{10}\left(\frac{Q}{4\pi r^2} + \frac{4}{S}\right), \tag{2.33}$$

where L_W is the sound power level (Equation (2.17)) of the source in dB, Q is the directivity of the source, r is the distance of the source from the receiver in metres, and S is the absorption area in m^2 of room surfaces (see Equation (2.32)). The directivity Q of an omnidirectional source (a source that radiates equally in all directions) is 1.0 and for other sources it can have larger or smaller values, depending on the direction.

The first term inside the logarithm of Equation (2.33) corresponds to the intensity caused by the direct sound, it being proportional to directivity and inversely proportional to the area of a sphere of radius r. The second term corresponds to the reverberant field, which is inversely proportional to the absorption area of the room. Note that this expression says that the reverberant field is equal in the whole room. This formula is derived from a statistical formulation of reverberation and it may not be a valid approximation for complex room geometries.

Figure 2.22 characterizes the SPL dependency on the distance r separately for direct sound, reverberant sound, and the total sound level. The direct sound pressure level decreases by 6 dB for every doubling of the distance. At a certain distance, called the *reverberation distance* or the *radius of reverberation*, the direct and reverberant fields have the same level, and this is

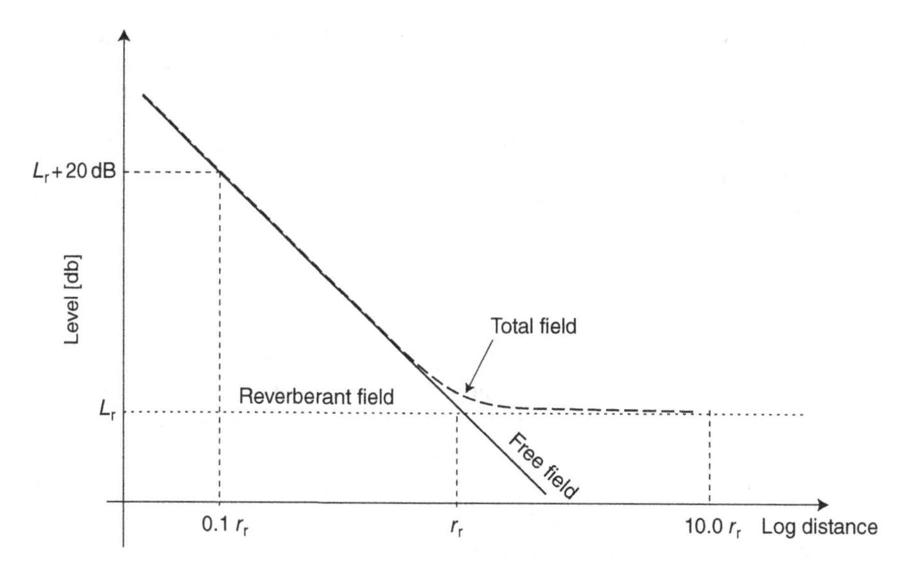

Figure 2.22 The theoretical behaviour of the sound pressure level in a room for a steady-state sound as a function of the source–receiver distance due to (a) a direct sound in a free field (solid line), (b) a reverberant field (dotted line), and (c) the total sound field (dashed line). The distance r_r is the reverberation distance, and L_r is the sound pressure level caused by the reverberant field.

obtained from Equation (2.33) by equating the two terms inside the logarithm and solving for r_r:

$$r_r = \frac{1}{4}\sqrt{\frac{QS}{\pi}}. \tag{2.34}$$

Beyond the reverberation distance the total field remains approximately constant, so that the perceived loudness does not essentially decrease, but the sound is perceived to be more reverberant since the direct sound level decreases.

2.4.4 Modal Behaviour of Sound in a Room

While Figures 2.19 and 2.21 characterize the temporal evolution of a sound field in a room, another picture is obtained by looking at sound transfer properties in the frequency domain. A hard-walled room can be understood as a three-dimensional resonator where a large number of modal frequencies can be found in the audible frequency range. For a rectangular room with hard walls, the frequencies of normal modes can be computed analytically from

$$f(n_x, n_y, n_z) = \frac{c}{2}\sqrt{\left(\frac{n_x}{L_x}\right)^2 + \left(\frac{n_y}{L_y}\right)^2 + \left(\frac{n_z}{L_z}\right)^2}, \tag{2.35}$$

where c is the sound velocity; L_x, L_y, and L_z are the dimensions of the room in the three directions; and integer variables n_x, n_y, and n_z are given all the combinations of the values 0, 1, 2, . . . The lowest modes correspond to plane-wave standing wave resonances between opposite walls, such as with $n_x = 1$, $n_y = 0$, $n_z = 0$, in Equation (2.35). Cross-modes, such as for $n_x = 1$, $n_y = 1$, and $n_z = 1$, are not as easy to conceive or visualize.

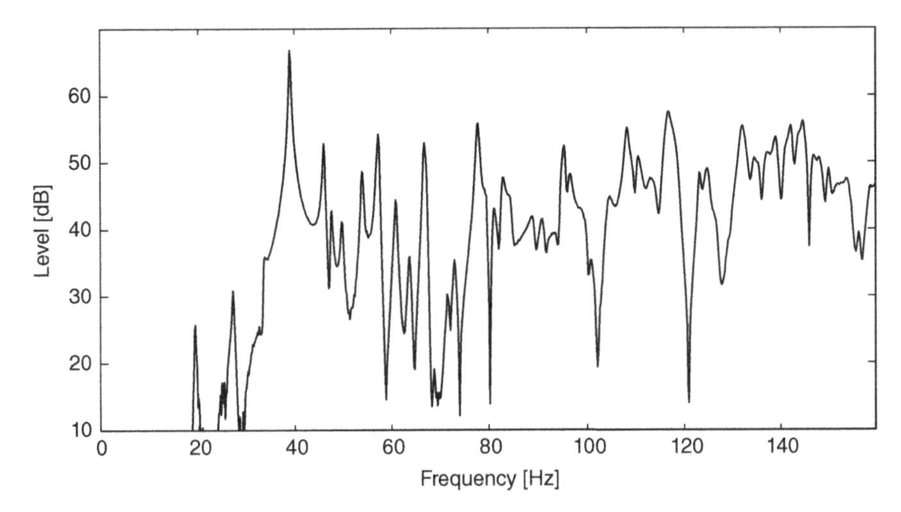

Figure 2.23 Magnitude response measured in a hard-walled room from a source to a receiver. (Below a strong mode at 39 Hz the level of modes is lower due to the roll-off of the loudspeaker response.) Towards higher frequencies the modes start to overlap more and the field is characterized as diffuse.

The density of modes increases rapidly as a function of frequency, and is proportional to the square of the frequency. Note that the density of modal frequencies d_f is approximately constant for one-dimensional acoustic systems, such as strings (Figure 2.13) and tubes (Figure 2.12), directly proportional to the frequency ($d_f \propto f$) for two-dimensional acoustic systems, such as a membrane (Figure 2.15), and proportional to the square of the frequency ($d_f \propto f^2$) for three-dimensional acoustic systems, such as rooms.

Figure 2.23 plots the magnitude response measured in a hard-walled room. The magnitude response (see Section 3.2.5) describes how each frequency component is emphasized or attenuated (in dB) when transferred from the source to the receiver. The lowest modes are quite separate peaks, but at higher frequencies the modes start to overlap and fuse. A transition frequency between these regions is called the *critical frequency* f_c (Schroeder, 1987), defined as

$$f_c = K\sqrt{T_{60}/V}, \tag{2.36}$$

where $K = 2000 \ldots 4000$, T_{60} is the reverberation time, and V is the volume of the room. Above the critical frequency the acoustic field is said to be *diffuse*. A diffuse field is composed of a large number of wavefronts travelling in all directions with equal probability, contributing to the high density of reflections and the high density of modes. Thus, a diffuse field can be modelled by statistical means, as the group behaviour of modes in the frequency domain and reflections in the time domain.

2.4.5 Computational Modelling of Closed Space Acoustics

The acoustic behaviour of real rooms, auditoria, and concert halls is very complex. This can be understood when estimating the possible number of degrees of freedom for vibration in the sound field. At the highest audible frequencies the wavelength is about 1 cm. Each such cube has three degrees of freedom to vibrate. Thus, a hall with dimensions $20 \times 40 \times 10$ metres has about $3 \cdot 2000 \cdot 4000 \cdot 1000 = 24{,}000{,}000{,}000$ degrees of freedom to vibrate! Only for very

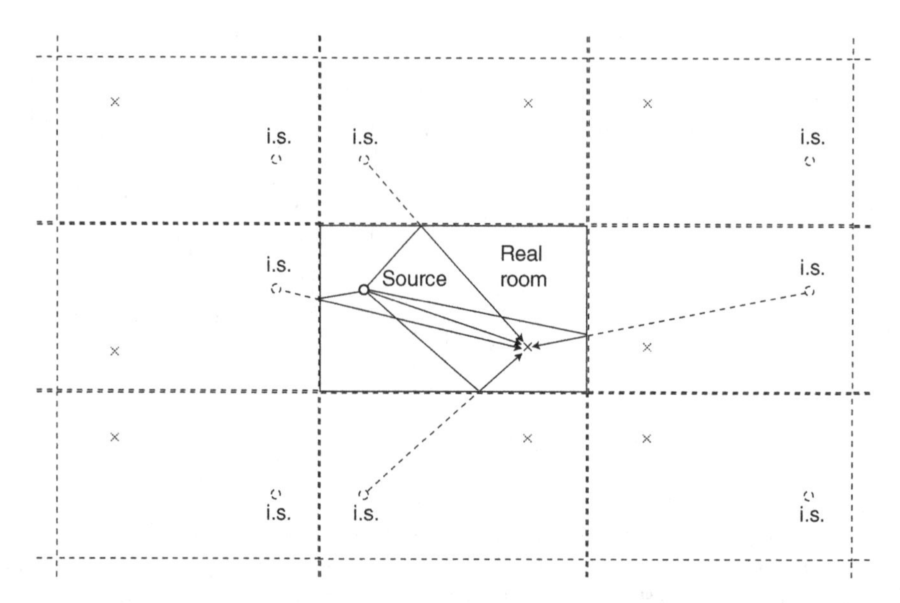

Figure 2.24 The principle of image source modelling. The mirrored rooms surrounding the real room and the related image sources (i.s.) are computed to find the paths of sound waves arriving at the receiver point in the real room. Only four first-order reflections are illustrated.

simple geometries, such as a shoebox with hard walls, can the field behaviour be computed by analytical means.

To meet the needs of computing and designing acoustics for complex spaces, a number of approximate numerical techniques have been developed. A conceptually simple technique is based on the *image source method*. Figure 2.24 illustrates a rectangular room with a source and a receiver. According to wave reflection laws, as illustrated in Figures 2.16 and 2.19, the reflections can be modelled as coming from image sources in image rooms. Each reflection in the real room corresponds to penetrating a wall between image rooms. Each path has an attenuation due to the $1/r$ distance law and damping for each reflection, and the wave will be delayed by the propagation time of the wave from the (image) source to the receiver. The frequency-domain room response in Figure 2.23 has been computed using the image source method as applied to a shoe-box-shaped room. If the room is not rectangular, the method can still be used, although the visibility of each image source has to be validated in each case (Kristiansen *et al.*, 1993).

Another commonly used numerical computational technique is the *ray-tracing* method. A large number of 'sound rays', analogous to light rays, are sent from the source in different directions. Each ray is traced through reflections from surfaces until the energy of the ray is below a certain threshold. The receiver can be modelled as a volume, such as a sphere or a cube, and each traced ray that hits it adds to the response record. The ray-tracing technique is statistical, but if the number of rays is sufficiently large and the receiver is small enough, the result may yield a good approximation of the room response.

In a room with a complex shape, both image-source and ray-tracing methods are problematic at low frequencies. Features like corners, object edges, and curved surfaces exhibit phenomena, such as diffraction, that are not easily simulated with these techniques. Different element-based techniques are more accurate in such cases. The *finite-element method* (FEM) (Thompson, 2006; Zienkiewicz and Taylor, 1977) assumes the space is a set of finite-sized elements having mass, spring, and damping properties. Given the geometry of a room and the sound source(s), the sound field at any point can be solved.

A variation of the element method is the *boundary-element method* (BEM) (Gumerov and Duraiswami, 2005), where only the boundaries of the space are taken into account in the first phase, and later the sound field at any point in the space can be solved. One more element-oriented technique is the *finite-difference time-domain method* (FDTD) (Botteldooren, 1995). The problem with all element-based methods is that the number of elements grows very rapidly when the wavelength of the highest frequency of interest decreases.

Summary

This chapter presented an overview of fundamental concepts in physical acoustics that are considered important in understanding communication by sound and voice, including the wave behaviour of sound in a free field, at material boundaries, and in closed spaces. References are given for further study on these issues, if needed.

While in physical acoustics such properties of sound as spatial distribution, two-way interaction between subsystems, energy behaviour, and specific physical variables are important, in the next chapter we make abstractions that simplify these issues. In a signal processing approach, we are more interested in looking at sound signals as abstract variables and their relationships as transfer functions and one-way interactions.

Further Reading

To better understand the general fundamentals of acoustics and wave equations, the reader is referred to such books as Beranek and Mellow (2012) and Morse and Ingard (1968).

Physical acoustics contains many subfields and requires concepts that are not discussed above. General *linear acoustics* deals with sound and vibration in fluids (gases and liquids) and solid materials with different geometries and structures, assuming that the condition of linearity (Section 3.2.1) is valid. If an acoustic system is non-linear, it is typically much more difficult to study. The subfield of *non-linear acoustics* deals with waves having high pressure levels, such as *shock waves* (Pierce, 1989; Raspet, 1997) and *cavitation* (Lauterborn, 1997).

Among other popular topics in acoustics research are *underwater acoustics*, also called *hydroacoustics* (Crocker, 1997, part IV); *ultrasound*, which studies sound above normally audible frequencies (Crocker, 1997, part V); *infrasound*, which concerns itself with sounds below normally audible frequencies (Gabrielson, 1997); *noise control* (Crocker, 1997, part VIII); *room*, *building*, and *architectural acoustics*, including *concert hall acoustics* (Ando, 2012; Barron, 2009; Beranek, 2004); *acoustics of musical instruments* (Fletcher and Rossing, 1998) and the *singing voice* (Sundberg, 1977); and *acoustic measurement techniques* (Beranek, 1988; Crocker, 1997, part XVII).

References

Ando, Y. (2012) *Architectural Acoustics: Blending Sound Sources, Sound Fields, and Listeners*. Springer.
Barron, M. (2009) *Auditorium Acoustics and Architectural Design*. Taylor & Francis.
Beranek, L. (1988) *Acoustical Measurements*. Acoustical Society of America.
Beranek, L. (2004) *Concert Halls and Opera Houses: Music, Acoustics, and Architecture*. Springer.

Beranek, L. and Mellow, T. (2012) *Acoustics: Sound Fields and Transducers*. Elsevier Science.

Botteldooren, D. (1995) Finite-difference time-domain simulation of low-frequency room acoustic problems. *J. Acoust. Soc. Am.* **98**(6), 3302–3308.

Crocker, M.J. (ed.) (1997) *Encyclopedia of Acoustics, vol. 1-4*. John Wiley & Sons.

Fletcher, N.H. and Rossing, T.D. (1998) *The Physics of Musical Instruments*. Springer.

Gabrielson, T.B. (1997) *Infrasound*. John Wiley & Sons, pp. 367–372.

Gumerov, N.A. and Duraiswami, R. (2005) *Fast Multipole Methods for the Helmholtz Equation In Three Dimensions*. Elsevier Science.

Kristiansen, U., Krokstad, A., and Follestad, T. (1993) Extending the image method to higher-order reflections. *Acta Acustica*, **38**(2–4), 195–206.

Kuttruff, H. (2000) *Room Acoustics*. Taylor & Francis.

Lauterborn, W. (1997) *Cavitation*. John Wiley & Sons, pp. 263–270.

Morse, P.M. and Ingard, K.U. (1968) *Theoretical Acoustics*. McGraw-Hill.

OpenAIR (2014) A library of room and environmental impulse responses in a number of spatial audio formats. `http://www.openairlib.net/`.

Pierce, A.D. (1989) *Acoustics, An Introduction to Its Physical Principles and Applications*. Acoustical Society of America.

Raspet, R. (1997) *Shock Waves, Blast Waves, and Sonic Booms*. John Wiley & Sons, pp. 329–340.

Rossing, T.D., Moore, F.R., and Wheeler, P.A. (2001) *The Science of Sound* 3rd edn. Addison-Wesley.

Sabine, W.C. (1922) *Collected Papers on Acoustics*. Harvard University Press.

Schroeder, M.R. (1987) Statistical parameters of the frequency response curves of large rooms. *J. Audio Eng. Soc.*, **35**(5), 299–306.

Sundberg, J. (1977) *The Acoustics of the Singing Voice*. Scientific American.

Svensson, U.P., Fred, R.I., and Vanderkooy, J. (1999) An analytic secondary source model of edge diffraction impulse responses. *J. Acoust. Soc. Am.*, **106**, 2331–2344.

Thompson, L.L. (2006) A review of finite-element methods for time-harmonic acoustics. *J. Acoust. Soc. Am.*, **119**(3), 1315–1330.

Zienkiewicz, O.C. and Taylor, R.L. (1977) *The Finite Element Method* volume 3. McGraw-Hill.

3

Signal Processing and Signals

Sound and voice cause vibration or wave propagation in a medium. If we register the value of a vibration of a wave field variable in a spatial position as a function of time, the result is a sound *signal*. This can be performed using a microphone or a vibration sensor, resulting in an electrical signal that can be processed and stored. Sound signals in electrical form can also be reconverted to sound by using loudspeakers.

Signal processing is the branch of engineering that provides efficient methods and techniques to analyse, synthesize, and transform signals. This chapter presents briefly signal processing fundamentals with regard to sound and voice signals.

3.1 Signals

Signals that use electrical or electronic circuits and work with signal values on a continuous scale are called *analogue signals* and methods that process them are called *analogue signal processing*. Such signals are, in most cases, considered to be continuously observable in time and thus are called *continuous-time signals*. If such a continuous-time signal is sampled properly at specific moments in time, a *discrete-time signal* is obtained. When these samples are further converted to discrete numbers, the result is called a *digital signal*. Methods and techniques to cope with such number sequences are called *digital signal processing*, or *DSP* for short.

3.1.1 Sounds as Signals

A signal, such as a wave or vibration variable as stated above, is a function of time that can be represented or approximated in different ways. A sound signal can be any of the following:

- *Mathematical function*, for example a sinusoidal signal, a *pure tone*

$$y(t) = A\sin(2\pi ft) = A\sin(\omega t), \tag{3.1}$$

Communication Acoustics: An Introduction to Speech, Audio, and Psychoacoustics, First Edition.
Ville Pulkki and Matti Karjalainen.

where A is the amplitude or maximum deviation from zero, f is the frequency or the number of vibration cycles in a second, ω is the angular frequency, and t is time. Another example is a *noise* signal

$$n(t) = \text{rand}(t), \tag{3.2}$$

where rand(·) is a function that yields a randomized value for each time moment.

- *Discrete-time numeric sequence*, for example

$$x(n) = \begin{bmatrix} 0.1 & 2.2 & 3.5 & 4.0 & 3.1 & -0.9 & 2.1 & 0.5 & -1.1 & -2.1 & -0.8 & 0.2 \end{bmatrix}, \tag{3.3}$$

where n is a discrete-time index. In matrix notation, the sequence in Equation (3.3) is typically given as a column vector instead of a row vector.

- *Graphical presentation*, for example Figure 3.1, where the mathematical signals of Equations (3.1) and (3.2), the numerical sample sequence of Equation (3.3), a short interval

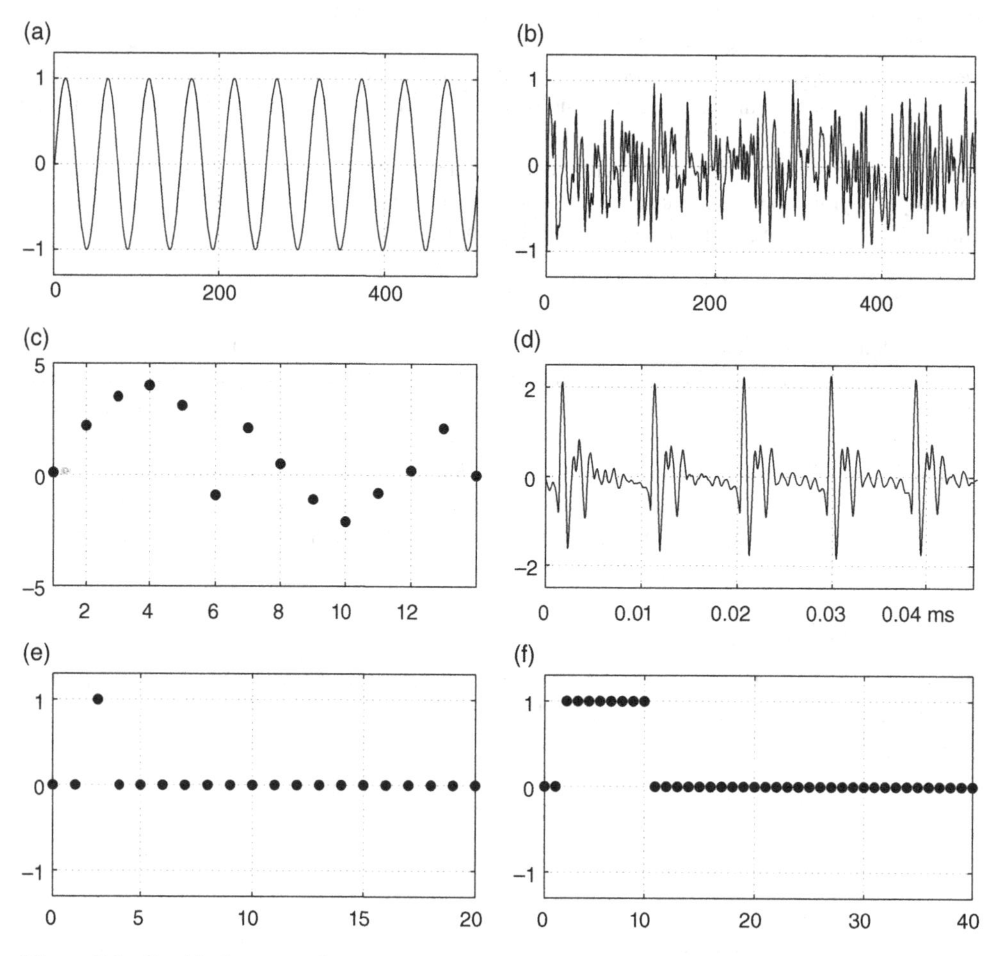

Figure 3.1 Graphical presentations of signals: (a) a sinusoidal or a pure tone signal, (b) random noise, (c) a discrete-time sample sequence, (d) a portion of recorded speech signal, (e) the unit impulse at time moment $n = 3$, and (f) a discrete-time pulse of finite duration.

of a recorded vowel speech signal, a unit impulse $\delta(n - n_0)$, $n_0 = 2$, and a pulse of finite duration are presented. In the figures, the horizontal axis is either time or the sample index and the vertical axis is the value of the signal variable.

The first two mathematically expressed signals are continuous in time while the sample sequence, the unit impulse, and the pulse are discrete-time signals. All signals in Figure 3.1 are, in fact, discrete-time sample sequences in the computer memory, although some of them are plotted as continuous curves.

3.1.2 Typical Signals

- *Pure tone*: as in Equation (3.1)
- *Amplitude-modulated tone*:

$$p(t) = A\,[1 + m\,\sin(\omega_m t)]\,\sin(\omega_0 t), \tag{3.4}$$

  where $m \in [0...1]$ is the modulation index, ω_m is the frequency of modulation, and ω_0 is the angular frequency of the tone.
- *Frequency-modulated tone*:

$$p(t) = A\,\sin[\omega_0 t + k\,\sin(\omega_m t)t], \tag{3.5}$$

  where k is the width of modulation.
- *Tone burst*: a tone which has been set to zero outside the time span $[t, t + \Delta t]$.
- *Sine wave sweep*: where a sine wave is generated with instantaneous frequency that glides linearly or logarithmically from an initial value to a final value, for example over the entire audible range.
- *Chirp signal*: a signal where a fast frequency sweep is carried out.
- *Unit impulse* or the Dirac delta function $\delta(t)$ or $\delta(n)$: a signal that has the value zero everywhere else except at the temporal position zero, where it has the value one, i.e., $\delta(n) = 1$ when $n = 0$. It holds that

$$\int_{-\infty}^{\infty} \delta(t)\mathrm{d}t = 1. \tag{3.6}$$

- *Pulses*: examples are *Gaussian waveforms* or wavelets, and pulse trains made of them.
- *White noise*: a signal where the average spectrum is flat.
- *Pink noise*: a signal that has a spectrum with a 3-dB/octave decay towards high frequencies.
- *Uniform masking noise*: a signal that has a similar spectrum to pink noise but flattens at frequencies below 500 Hz.
- *Modulated noise* (amplitude and frequency modulation) and *noise bursts*.
- *Harmonic tone complexes*:

$$p(t) = \sum_n A_n \sin(n 2\pi f_0 t + \phi_n), \tag{3.7}$$

  where A_n is the amplitude of each harmonic, f_0 is the *fundamental frequency* of the tone complex, and ϕ_n is the starting phase of each harmonic. All partials of the complex are thus integer multiples of the fundamental frequency, or have a common denominator.

- *Complex combination sounds*:

$$p(t) = \sum_i A_i \sin(2\pi f_i t + \phi_i), \tag{3.8}$$

where f_i are the arbitrarily chosen frequencies, and ϕ_i the starting phases of the partials.
- *Sawtooth wave*: a signal that, in the time domain, has linear rises and subsequent steep drops, thus having a shape reminiscent of the teeth of a saw. It can be mathematically written as

$$p(t) = t \mod T, \tag{3.9}$$

where $T = 1/f$, mod is the modulo operator, and f is the frequency of repetition of the sawtooth wave.
- *Triangle wave*: a signal that, in the time domain, has alternating linear rises and falls. It can be expressed mathematically as

$$p(t) = T/2 - |(t \mod T) - T/2|, \tag{3.10}$$

where $T = 1/f$ and f is the frequency of repetition of the triangle wave.
- *Square wave*: a signal that, in the time domain, has alternating steep rises and falls that are evenly spaced. Mathematically it can be expressed as

$$p(t) = \text{sign}[(t \mod T) - T/2], \tag{3.11}$$

where $T = 1/f$, f is the frequency of the square wave, and the sign function returns -1 for a negative argument and $+1$ for a positive argument. Sawtooth, triangle, and square waves have a harmonic spectrum, and they can be expressed using Equation (3.7).

3.2 Fundamental Concepts of Signal Processing

Signal processing includes a set of methods that are important for understanding communication by sound and voice. Among these are, for example, linear time-invariant (LTI) systems and processes and the Fourier transform and related signal analysis and synthesis, including spectrum analysis. A special topic in digital signal processing is digital filtering, an efficient method to implement LTI systems. The short presentation of these methods below is intended to refresh the memory of those who have already studied these topics and as a brief overview for those who have not. The mathematics here are kept simple, and the formulas may be skipped entirely by concentrating on the text and the graphical examples if the reader is unfamiliar with such mathematics.

In addition to the basic signal processing, this section includes an overview of some adaptive and learning computational methods (evolutionary computation), such as hidden Markov models.

3.2.1 Linear and Time-Invariant Systems

In signal processing, we are typically interested in the input–output relationship of the system under study (see the black-box formulation in Section 1 of the Introduction). If the output signal $y(t)$ of a system as a function of time depends only on the input signal $x(t)$, their relationship can be expressed generally as $y(t) = h\{x(t)\}$. A system is *linear* and *time invariant* (LTI) if the following is true:

$$h\{\mathrm{a}x_1(t) + \mathrm{b}x_2(t)\} = \mathrm{a}\,h\{x_1(t)\} + \mathrm{b}\,h\{x_2(t)\}, \tag{3.12}$$

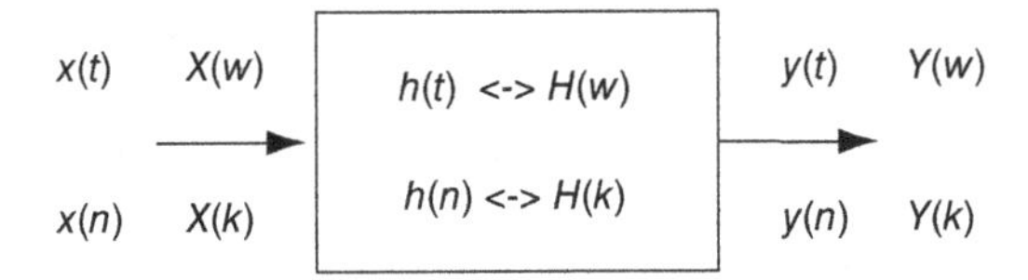

Figure 3.2 The linear and time-invariant (LTI) system as a black box with the related mathematical concepts in the time domain (lower case symbols) and in the frequency domain (upper case symbols).

where a and b are two constants and $x_1(t)$ and $x_2(t)$ are two input signals. In words, the response of an LTI system to the sum of two input signals is equal to the sum of the responses to the individual inputs separately, and the corresponding input and output signal values can be scaled linearly by a constant gain factor. Furthermore, if we apply an input to the system now or T seconds from now, the output will be identical except for a time delay of T seconds. The analysis and implementation of LTI systems is typically easier and more efficient than for systems that do not have or approximate this property. An LTI system also guarantees that it does not create any new frequency components that do not exist in the input signal, i.e., it does not generate non-linear distortion. Thus, a pure tone (a sinusoidal signal) remains a pure tone when propagating through an LTI system.

An LTI system can be represented in the *time domain* with its *impulse response* $h(t)$ or $h(n)$ and in the *frequency domain* with its *transfer function* $H(\omega)$ or $H(k)$. While the *time domain* is a more intuitive way of representing signals, the *frequency domain* has specific useful properties that are discussed throughout the book.

The system may be represented graphically as a black box, as shown in Figure 3.2. Variable t in these formulations refers to time as a continuous-valued variable and n is used as a discrete-time index variable. In the transfer functions, ω is the angular frequency used in continuous-time signals and k is a similar discrete frequency index variable corresponding to the discrete-time representations.

Examples of systems where linearity is of importance are audio recording and reproduction equipment. Amplifiers can be designed to have low non-linear distortion, but, for example, loudspeakers at relatively high power levels and at low frequencies can be highly non-linear. Linearity is a goal in signal processing, although in some tasks, such as perceptually motivated processing, non-linear processing methods are necessary.

3.2.2 Convolution

The relation between the input and output signals of an LTI system can be expressed mathematically using the convolution operation denoted by $*$ and defined as

$$y(t) = x(t) * h(t) = \int_{-\infty}^{+\infty} x(\tau)\, h(t-\tau)\, \mathrm{d}\tau \qquad \text{or} \tag{3.13a}$$

$$y(n) = x(n) * h(n) = \sum_{i=-\infty}^{+\infty} x(i)\, h(n-i). \tag{3.13b}$$

The first expression is for continuous-time signals and systems, and it is called the *convolution integral*, while the second, the *convolution sum*, is for discrete-time signals and systems. For practical signals, the limits of t and n in the integral and the sum are finite.

Although formally simple, convolution is a surprisingly complicated operation to comprehend. It may also be a computationally demanding operation if the response sequences are long. Using the LTI concept, if an arbitrary input signal is considered to be a sequence of impulses with different amplitudes, the output is the combined response of the responses to all of these input impulses.

The impulse responses $h(t)$ or $h(n)$ for a real-time system are always *causal*; that is, $h(\cdot) = 0$ when t or $n < 0$. This means that a physically realizable system does not have information on the future values of its input signal.

If a system does not meet the LTI condition of Equation (3.12), it is *time-variant* if its response properties change as a function of time, or it is *non-linear* if its transfer properties depend on the signal passing through it. In both cases the system can generate new frequency components that do not exist in the input signal. The analysis, modelling, or synthesis of such systems is typically substantially more difficult than for an LTI system. A slightly non-linear system can be approximated using a linearized model if the error is tolerable.

A good example of a highly non-linear and time-variant system is human hearing. As a result, modelling of the auditory system is a complex task, and only some of its peripheral parts may be modelled in the LTI sense.

3.2.3 Signal Transforms

A useful mathematical approach in signal processing is based on transforming signals into another form that makes processing or interpreting them easier. A particularly useful set of transforms yields a mapping between the time- and frequency-domain representations, the Fourier transform described below being the most important one.

Mathematical tools that are needed for frequency-domain representations are *complex numbers* and *complex-valued functions*. A complex number c is composed of a real part x and an imaginary part y written as

$$c = \mathrm{Re}\,\{c\} + \mathrm{j}\,\mathrm{Im}\,\{c\} = x + \mathrm{j}\,y, \tag{3.14}$$

where $\mathrm{Re}\,\{\cdot\}$ means the real part of, $\mathrm{Im}\,\{\cdot\}$ the imaginary part of, and j, often also denoted by i, is the imaginary unit $\mathrm{j} = \sqrt{-1}$. A fundamental equation for operating with complex numbers is the *Euler relation* for complex exponentials

$$e^{\mathrm{j}\phi} = \cos\phi + \mathrm{j}\sin\phi \tag{3.15}$$

that ties the phase angle ϕ to the real and imaginary components:

$$c = x + \mathrm{j}y = |c|e^{\mathrm{j}\phi} \tag{3.16a}$$

$$|c| = \sqrt{x^2 + y^2} \tag{3.16b}$$

$$\angle\, c = \arg\{c\} = \phi = \arctan(x/y), \tag{3.16c}$$

where $|\cdot|$ means the absolute value of or the magnitude of and $\angle$ and $\arg\{\cdot\}$ mean the phase or argument of.

3.2.4 Fourier Analysis and Synthesis

The analysis of an LTI system is mathematically simplified if it is represented in the frequency domain, and the signals are described as functions of frequency instead of time. This can be done using the *Fourier transform*

$$X(\omega) = \mathcal{F}\{x(t)\} = \int_{-\infty}^{+\infty} x(t)\, e^{-\mathrm{j}\omega t}\, \mathrm{d}t \tag{3.17a}$$

$$X(k) = \mathcal{F}_{\mathrm{d}}\{x(n)\} = \sum_{n=0}^{N-1} x(n)\, e^{-\mathrm{j}k(2\pi/N)n}. \tag{3.17b}$$

The transform in Equation (3.17a) is valid for continuous-time signals and systems while Equation (3.17b) is for discrete-time cases. The reader is referred to a standard textbook, such as Proakis (2007), for a more thorough discussion on the mathematical details and applications of the Fourier transform.

The continuous-time and discrete-time transform operators are denoted here by $\mathcal{F}\{\cdot\}$ and $\mathcal{F}_{\mathrm{d}}\{\cdot\}$. The latter form, Equation (3.17b), is called the *discrete Fourier transform* (DFT). It is defined for a finite length sequence ($n = 0 \ldots N - 1$), and it can be computed very efficiently using the *fast Fourier transform* (FFT) (Proakis, 2007). The FFT is applicable for periodic signals, one period being the index range of the summation similar to Equation (3.17b).

An interpretation of what the Fourier transform does is that it is a correlation (a kind of similarity comparison, defined by Equation (3.26a)) of the signal $x(\cdot)$ to be transformed with a pair of sinusoids, a sine and a cosine, together expressed as a complex exponential $e^{-\mathrm{j}\omega t}$ or $e^{-\mathrm{j}k(2\pi/N)n}$. When applying this rule to each frequency (ω or n), the result represents the frequency content of $x(\cdot)$. See Figure 3.3 for an example.

The inverse transforms for Equations (3.17a) and (3.17b) are

$$x(t) = \mathcal{F}^{-1}\{X(\omega)\} = \frac{1}{2\pi} \int_{-\infty}^{+\infty} X(\omega)\, e^{\mathrm{j}\omega t}\, \mathrm{d}\omega \tag{3.18a}$$

$$x(n) = \mathcal{F}_{\mathrm{d}}^{-1}\{X(k)\} = \frac{1}{N} \sum_{k=0}^{N-1} X(k)\, e^{\mathrm{j}k(2\pi/N)n}, \tag{3.18b}$$

which map the signal from the frequency domain back to the time domain. The transforms in Equations (3.17a) and (3.17b) may be interpreted as *Fourier analysis* and the transforms in Equations (3.18a) and (3.18b) as *Fourier synthesis*. Figure 3.3 illustrates how a sawtooth waveform is constructed as a linear combination of its sinusoidal components. Any signal that meets certain continuity requirements can be represented with arbitrary precision as a sum of sinusoidal components of different frequencies. These components are often called *partials*.

An important advantage of using the Fourier transform is that it converts the computationally expensive convolution in the time domain into a much simpler multiplication in the frequency domain.

$$\mathcal{F}\{x(t) * y(t)\} = X(\omega) \cdot Y(\omega) \tag{3.19a}$$

$$\mathcal{F}_{\mathrm{d}}\{x(n) * y(n)\} = X(k) \cdot Y(k), \tag{3.19b}$$

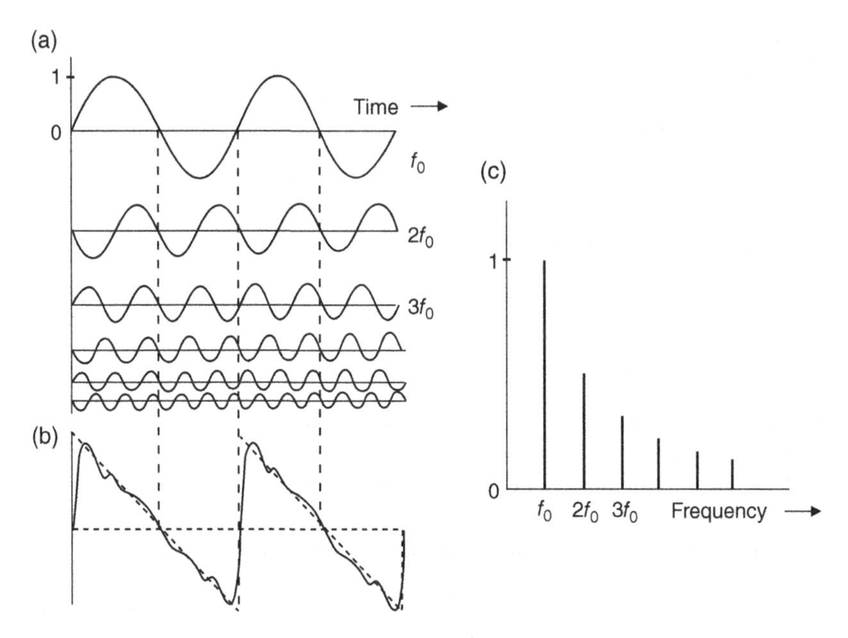

Figure 3.3 The decomposition of a sawtooth waveform into its sinusoidal frequency components. Putting together these components is called Fourier synthesis, and breaking up a waveform into its components is called Fourier analysis.

where the lower case symbols x and y denote time signals and the upper case symbols X and Y their frequency-domain transforms. Note that the definition of the Fourier transform involves the assumption of infinite periodicity of the signals $x(n)$ and $y(n)$. However, by appending the signals with sequences of zeros, the equivalent of a linear time-domain convolution is nevertheless obtained. The process of appending zeros to the start or end of the signal is often called *zero padding*. Having accounted this for, the FFT and the inverse fast Fourier transform (IFFT) can be utilized in the efficient computation of convolution in the following way:

$$x(t) * y(t) = \mathcal{F}^{-1}\{X(\omega) \cdot Y(\omega)\} = \mathcal{F}^{-1}\{\mathcal{F}\{x(t)\} \cdot \mathcal{F}\{y(t)\}\} \tag{3.20a}$$

$$x(n) * y(n) = \mathcal{F}_{\mathrm{d}}^{-1}\{X(k) \cdot Y(k)\} = \mathcal{F}_{\mathrm{d}}^{-1}\{\mathcal{F}_{\mathrm{d}}\{x(n)\} \cdot \mathcal{F}_{\mathrm{d}}\{y(n)\}\}. \tag{3.20b}$$

3.2.5 Spectrum Analysis

Representing signals in the frequency domain, as decompositions into their frequency components, is important not only as a powerful signal processing technique but also because it resembles the way the human ear analyses signals. Audio signals that we are able to hear as sounds are thus often analysed by means of frequency transforms in order to find their audible features and cues. The result of the Fourier transform (Equations (3.17a) and (3.17b)) is complex-valued. Such a complex transform can be equivalently expressed by a pair of separately plotted *spectra*, the *magnitude spectrum* on the logarithmic decibel scale

$$|X(\omega)|_{\mathrm{dB}} = 20 \log_{10} |X(\omega)| \tag{3.21a}$$

$$|X(k)|_{\mathrm{dB}} = 20 \log_{10} |X(k)|, \tag{3.21b}$$

and the *phase spectrum*

$$\varphi(\omega) = \angle X(\omega) = \arg\{X(\omega)\} \tag{3.22a}$$

$$\varphi(k) = \angle X(k) = \arg\{X(k)\}. \tag{3.22b}$$

In principle, $\varphi(k)$ can be unequivocally solved from $\mathrm{Im}\{X(k)\}$ and $\mathrm{Re}\{X(k)\}$ using Equation (3.16c). Unfortunately, the expression for the phase spectrum cannot be expressed explicitly. In many computer languages, such as in Matlab, $\varphi(\mathrm{k})$ = angle(X(k)) = atan2(Im {X(k)} , Re {X(k)}).

For sound signals, the concept of *spectrum analysis* typically refers to a representation of the magnitude spectrum only, because, as is well known, the auditory system is relatively insensitive to the phase of a signal. The motivation to use the logarithmic dB scale for the magnitude spectrum comes from the fact that we perceive the level of a signal more logarithmically than linearly. The dB scale is also appropriate for the graphical representation of spectra. However, as will be shown in later chapters, the detailed behaviour of the auditory system deviates from the simple Fourier spectrum analysis in many ways.

The phase spectrum $\varphi(\cdot)$ from Equations (3.22a) and (3.22b) is cyclically limited between $-\pi$ and π, exhibiting discontinuous jumps between these boundaries. If a continuous phase spectrum is desired, a *phase unwrapping* operation is necessary.

Often a more useful representation than the phase function itself involves the *group delay* τ_g and the *phase delay* τ_p

$$\tau_\mathrm{p}(\omega) = -\varphi(\omega)/\omega, \tag{3.23a}$$

$$\tau_\mathrm{g}(\omega) = -\mathrm{d}\varphi(\omega)/\mathrm{d}\omega. \tag{3.23b}$$

The phase delay represents the delay of a frequency component when propagating through a system. Group delay, as the frequency derivative of the phase, describes the delay of the modulation, such as the amplitude envelope, of a frequency component. From the point of view of the auditory system, the group delay is the most relevant of these phase representations.

In practice, spectrum analysis must be localized in time, since the spectral properties of sound signals typically vary over time. *Windowing* is used to accomplish this by multiplying the signal with a *window function* which is then Fourier analysed:

$$X(\omega) = \int_{t_\mathrm{b}}^{t_\mathrm{e}} w(t)\, x(t)\, e^{-\mathrm{j}\omega t}\, \mathrm{d}t \tag{3.24a}$$

$$X(k) = \sum_{n=n_\mathrm{b}}^{n_e} w(n)\, x(n)\, e^{-\mathrm{j}k(2\pi/N)n}, \tag{3.24b}$$

where $w(t)$ is a window function (weight function) that is non-zero only in the time span denoted by the limits of integration and summation in Equations (3.24a) and (3.24b) and zero elsewhere.

Some frequently applied window functions are the Hamming, Hann (a.k.a. Hanning window), Blackman, and Kaiser windows (Mitra and Kaiser, 1993). Note that cropping a span from a signal and zeroing elsewhere corresponds to using a rectangular window. Figure 3.4 illustrates an example where a sinusoidal signal is analysed with different windows, including rectangular and Hamming windows, with and without synchrony to the periodicity of the sine wave.

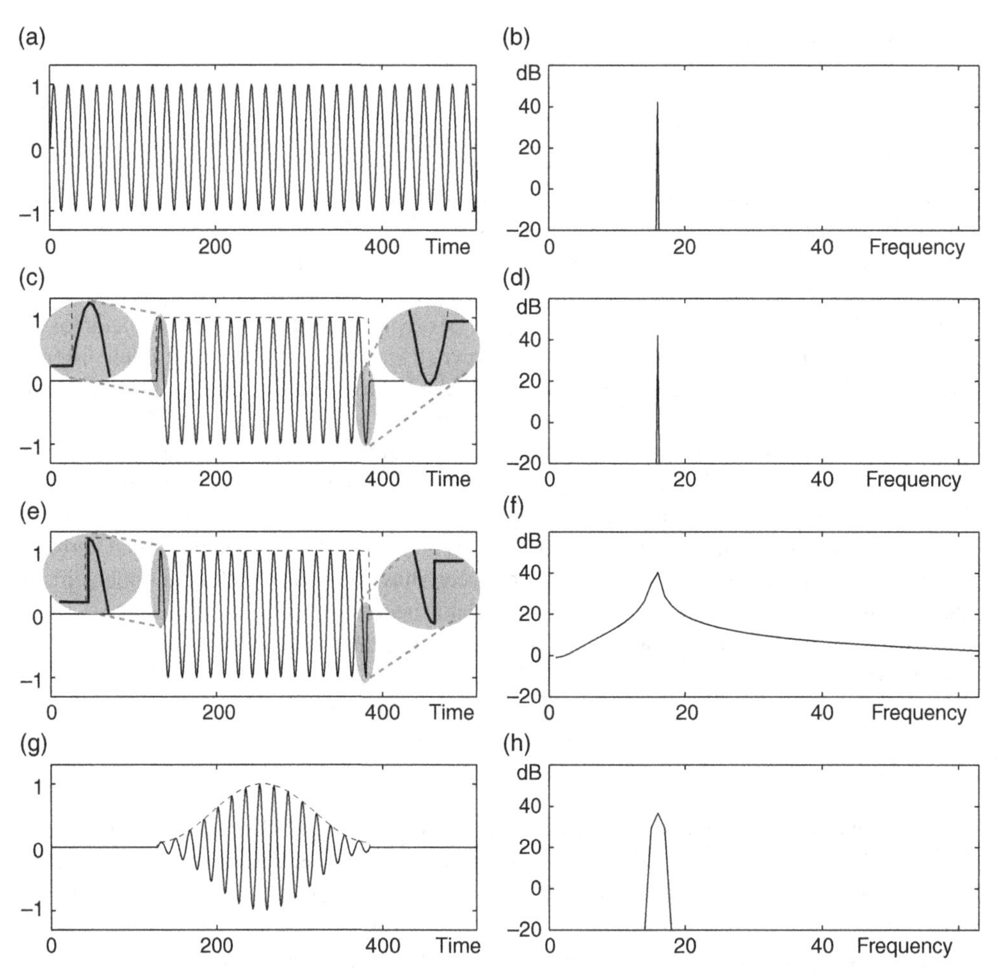

Figure 3.4 Spectrum analysis using the Fourier transform and windowing. (a) A sine wave and (b) its magnitude spectrum with a very long window. (c) A rectangular window that is synchronized with the periodicity of the sine wave. The ending of the sinusoid in the window and the starting of it join together circularly continuously in amplitude and slope. (d) The corresponding spectrum with no artefacts. (e) A rectangular window that is not in periodicity synchrony with the signal, showing as a discontinuity in amplitude between the starting and ending positions. (f) The resulting spectrum that shows the spreading of the spectrum. The Hamming window (dashed envelope line) in (g) always removes most of the spectral spreading far from the peak, but the spectrum peak itself will be broadened, as shown in (h).

The selection of a window function is always a compromise between spectral and temporal resolution. The longer the window, the better the spectral resolution and the worse the temporal resolution, and vice versa. Theoretically, the resolutions in time (Δt) and frequency (Δf) are bound by the equation

$$\Delta t \cdot \Delta f \geq 0.5. \tag{3.25}$$

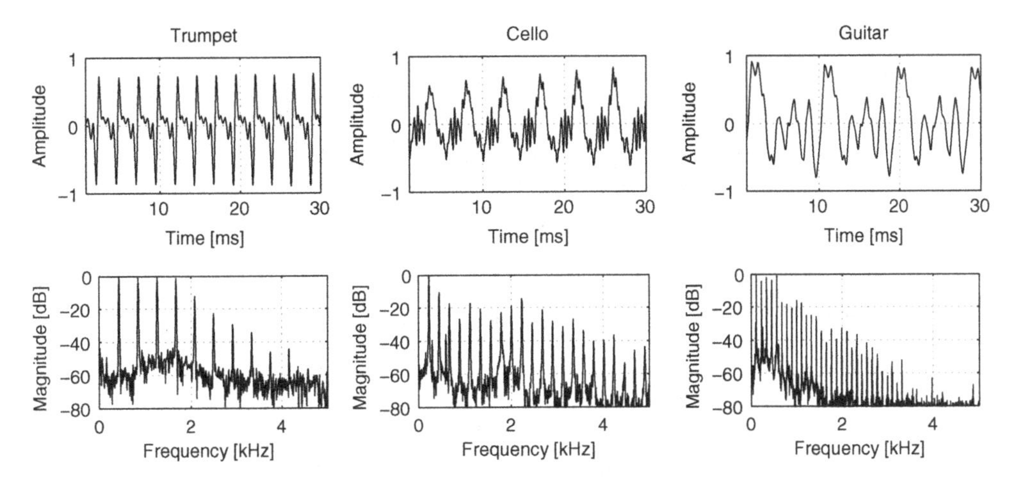

Figure 3.5 The waveforms and spectra of three instrument sounds.

For audio signals, the length of the window is often motivated by the time-vs.-frequency resolution of the auditory system. If only a single window length is used, the value is typically selected to be between 10 and 30 milliseconds, which, using Equation (3.25), limits the frequency selectivity to between 50 and 17 Hz. Noting the non-linear scale of auditory perception, the selection of the window length is especially relevant for analysis in the lowest frequencies.

Examples of real audio signals of musical instruments and their magnitude spectra are shown in Figure 3.5. See Figure 5.7 on page 87 for spectra of vowels and consonants.

3.2.6 Time–Frequency Representations

A time–frequency representation is obtained, for example, by applying a window, such as a 20-ms Hamming window, to a portion of a signal, taking the Fourier transform, and then moving to the next portion of the signal and repeating the procedure. This spectral sampling interval, called the *hop size*, is typically about 10 ms for analysing speech and other audio signals. Such a frame-based analysis of spectra is called *short-time Fourier analysis*, and its graphical representation is called a *spectrogram*.

Since a spectrogram is a three-dimensional mapping – the magnitude level as a function of time and frequency – it cannot be illustrated by a single curve. One typical graphical representation is as an intensity map where the grey shade or colour at each point stands for the magnitude (see Figure 3.6). Another representation is the 'waterfall' or mesh plot with a set of curves, as shown in Figure 6.5 on page 105.

Short-time Fourier analysis is a special case of a *time–frequency representation*. Other choices are *wavelet analysis* and *Wigner distributions*. In wavelet analysis (Cohen, 1995; Vetterli and Kovacevic, 1995), the frequency and time resolution are not uniform but vary so that at high frequencies the time resolution, is better with a coarser frequency resolution, and vice versa at low frequencies. A method related to wavelet processing is the constant-Q transform (Holighaus *et al.*, 2013; Schörkhuber *et al.*, 2013), where the signal is divided into time-frequency tiles of equal area, but the bandwidth increases with frequency to maintain a constant Q (Equation (3.30)). The temporal length of the tiles decreases with increasing frequency, which makes the processing a bit more complicated. However, perfect reconstruction processing has been achieved with constant-Q methods.

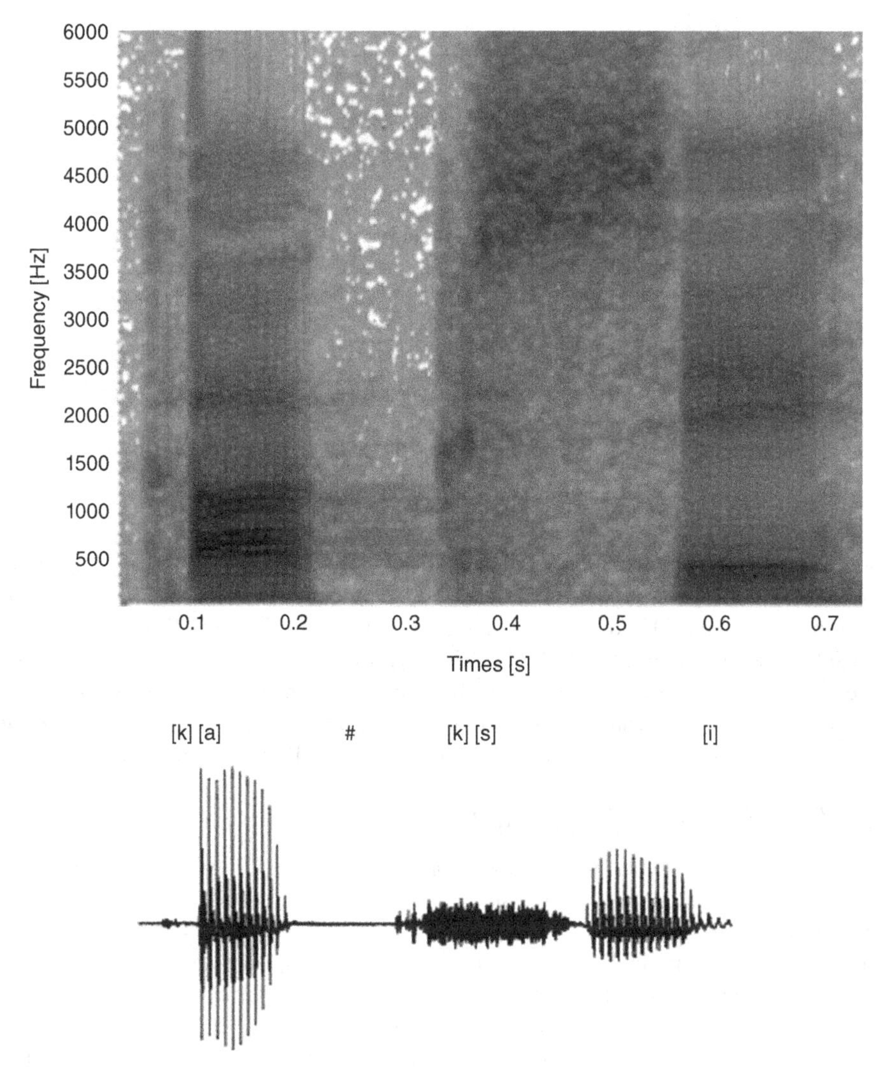

Figure 3.6 The spectrogram of a spoken word (top) aligned in time with the original waveform (bottom). Higher magnitude levels are shown with a darker shade of gray in the spectrogram. The voiced parts the formant resonances are darker horizontal stripes, while in the noise-like fricative part only relatively high frequencies have noticeable energy. Matti Karjalainen utters the word 'kaksi', meaning 'two' in Finnish, which he often used to test audio and speech techniques. The uttered phones are shown, where '#' denotes silence.

3.2.7 Filter Banks

A *filter bank* is a set of band-pass filters that is fed the same input, where the centre frequencies of the filters vary over desired ranges of frequencies. It thus separates a broadband input signal into multiple time-domain narrowband signals, which are called sub-bands. This is similar to the functioning of hearing, as will be discussed later in Chapter 7.

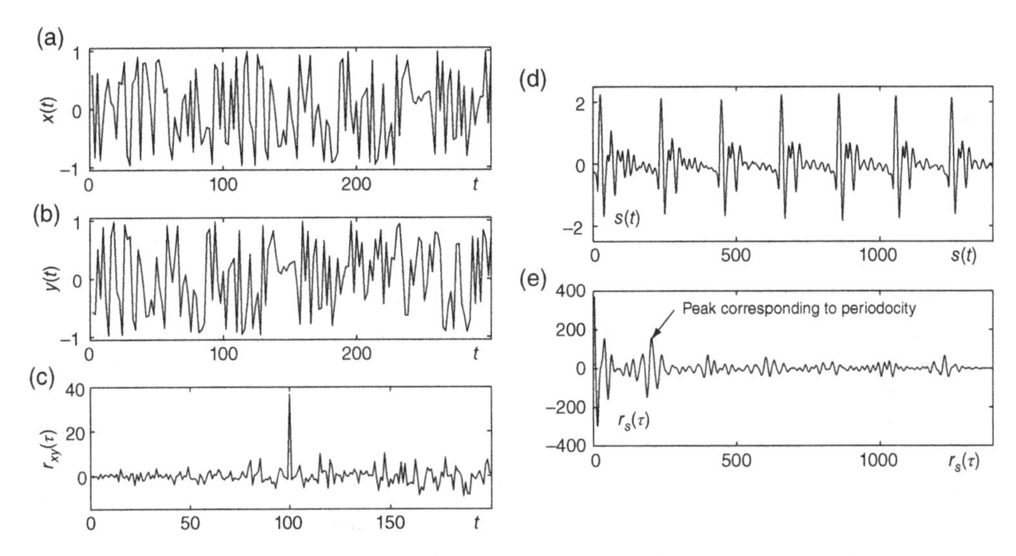

Figure 3.7 Examples of cross- and autocorrelations: (a) A random signal $x(t)$, (b) the same delayed signal $y(t) = x(t - 100)$, and (c) the cross-correlation $r_{xy}(\tau)$ showing a peak that indicates the time delay. (d) A voiced speech signal $s(t)$, and (e) its autocorrelation $r_S(\tau)$ indicating the periodicity (the inverse of the fundamental frequency).

Filter banks have a number of uses. For example, when used as an equalizer, the signals in sub-bands are amplified according to the desired equalization curve, and they are then combined to form the equalized signal. Since sometimes *perfect reconstruction* is desired, the filters must be designed so that the original signal can be recovered when the sub-band signals are combined. Filter banks have various applications in audio effects, audio coding, and spatial audio reproduction, as will be discussed later in Chapter 15.

3.2.8 Auto- and Cross-Correlation

The similarity of two waveforms can be analysed by computing their *cross-correlation*

$$r_{xy}(\tau) = \int_{-\infty}^{+\infty} x(t)\, y(t+\tau)\, \mathrm{d}\tau \tag{3.26a}$$

$$r_{xy}(k) = \sum_{i=0}^{N-1} x(i)\, y(i+k). \tag{3.26b}$$

Figures 3.7a–c depict a case where the cross-correlation of a signal and its time-delayed counterpart indicate the amount of delay. A special case of correlation is *autocorrelation*, where a signal is compared to itself in order to find the periodicity, or repeatability, of the waveform. In this case $x(\cdot) = y(\cdot)$ in Equations (3.26a) and (3.26b). In the example of Figures 3.7d–e, the autocorrelation of a voiced speech signal has a peak corresponding to its periodicity. The autocorrelation function is periodic, showing maxima also for integer multiples of the fundamental period.

3.2.9 Cepstrum

The *cepstrum* (Oppenheim and Schafer, 1975) is a transform that shows some resemblance to autocorrelation. It is computed as the inverse Fourier transform of the logarithmic magnitude spectrum,

$$c_x(t) = \mathcal{F}^{-1}\{\log |\mathcal{F}\{x(t)\}|\}. \tag{3.27}$$

This resemblance can be understood from the fact that the logarithmic magnitude spectrum is a real-valued function and that the inverse Fourier transform is an operation very similar to the Fourier transform, as is evident from Equations (3.17a) and (3.18a). The computation of the cepstrum thus treats the magnitude spectrum similarly to a time-domain signal. The result can be thought to be 'the spectrum of a magnitude spectrum curve'. The logarithm of the magnitude spectrum provides a particular representation for differentiating specific processes affecting a speech spectrum: the harmonic response of the glottis is represented by a fast-changing, periodic log-spectrum, while the formants produced by the vocal tract are represented by a longer and smoother envelope and convey the information of which phoneme is uttered. The inverse-Fourier-transform operator projects these two envelopes to different points in the cepstrum $c_x(t)$, providing a representation that is practical for processes such as speech recognition.

3.3 Digital Signal Processing (DSP)

Digital signal processing (DSP) means discrete-time numerical processing of signals (Mitra and Kaiser, 1993; Oppenheim *et al.*, 1983; Strawn, 1985). If a signal to be processed is originally in analogue form, i.e., continuous in time and amplitude, it must first be converted to a number sequence by *analogue-to-digital conversion* (A/D-conversion). Conversely, a digital signal (a number sequence) can be converted back to continuous-time form by *digital-to-analogue conversion* (D/A-conversion). Signal processing itself is carried out by a *digital signal processor*, which can be a special digital circuit, a programmable digital signal processor, or a general purpose processor or computer. A generic structure for such a system is shown in Figure 3.8.

3.3.1 Sampling and Signal Conversion

When converting between analogue and digital signals, the *sampling theorem* (or the *Nyquist theorem*) requires that the sampling rate must be at least twice as high as the highest signal component to be converted. Otherwise *aliasing* will occur, whereby signal components of frequency higher than the *Nyquist frequency* (half of the sampling rate) will be mirrored to below the Nyquist frequency, thus distorting the signal. To avoid aliasing, an A/D-converter normally includes a low-pass filter that yields enough attenuation above the Nyquist frequency.

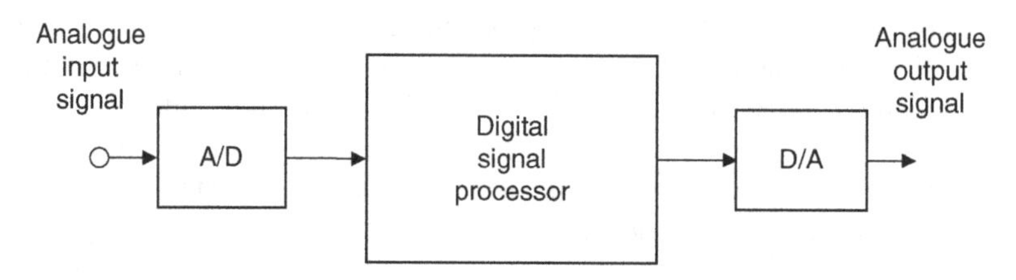

Figure 3.8 A block diagram for the digital signal processing of analogue signals, including an A/D-converter, a digital signal processor, and a D/A-converter.

D/A-conversion also typically includes a low-pass filter (reconstruction filter) to make the output continuous in time and free from frequencies above the Nyquist frequency. This means that digital signal processing deals with band-limited signals.

Sampling rates common in audio technology are 44.1 kHz (compact disc), 48 kHz (professional audio), 32 kHz (less demanding audio), and for very demanding audio 96 kHz or even 192 kHz. Speech technology uses the sampling rate 8–16 kHz. The telephone bandwidth of 300–3400 Hz requires a sampling rate of about 8 kHz.

Numerical samples from A/D-conversion may be coded in various ways. The most straightforward representation is to use *PCM coding* (pulse-code modulation). Each sample is *quantized* into a binary number where the number of bits implies the precision of the result. Figure 3.9 illustrates the principle of such a conversion using four bits, which corresponds to 16 levels. Sample values of an analogue signal are mapped onto binary numbers so that there are 2^n discrete levels when the number of bits is n.

Quantization with finite precision generates an error called *quantization noise*. The *signal-to-noise ratio* (SNR, see Section 4.2.6) describes the level of the signal compared to the level of noise, and for quantization noise it improves by 6 dB for each added bit, so that 16 bits, often used in audio, yield a maximum SNR of about 96 dB. Because the dynamic range of the auditory system is about 130 dB, even more than 22 bits may be needed.

Digital signal processing has many advantages compared to analogue techniques. It is predictable, and a single DSP processor may be programmed to compute any DSP program that does not exceed its processing capacity. For example, real-time spectrum analysis can be implemented using the FFT.

3.3.2 Z Transform

The *z transform* is a fundamental mathematical tool for describing LTI systems in digital signal processing. The z transform of a digital signal (sample sequence) $x(n)$ is

$$X(z) = \mathcal{Z}\{x(n)\} = \sum_{n=-\infty}^{\infty} x(n)\, z^{-n} \tag{3.28}$$

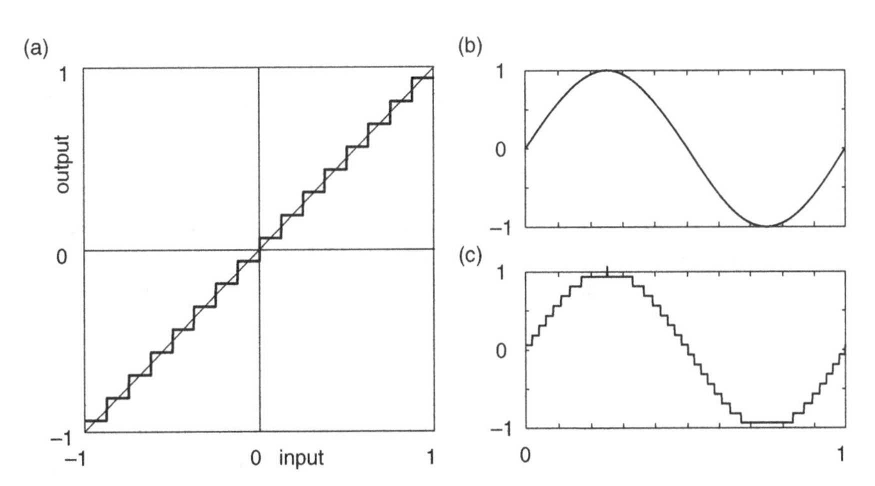

Figure 3.9 Quantization and PCM representation of a sinusoidal signal with 4 bits (16 levels): (a) characteristics of the quantization curve, (b) sinusoidal analogue input, and (c) quantized signal waveform.

The complex variable z used in the transform is related to the *unit delay* between two adjacent samples as

$$\mathcal{Z}\{x(n-1)\} = z^{-1}X(z). \tag{3.29}$$

3.3.3 Filters as LTI Systems

A filter modifies the magnitude or phase spectrum of an input signal. Often, filters are applied to attenuate some frequencies and leave others untouched. In most cases, the processing also changes the phase spectrum. In some cases, the filters do not change the magnitude spectrum, but only the phase spectrum.

There are several types of filters:

- A *low-pass filter* is one that leaves low-frequency signals unmodified and attenuates signals with frequencies higher than the *cutoff frequency*. In the design of such filters, the response is often set to be 0 dB for frequencies in the *passband* (frequencies below the cutoff frequency), −3 dB at the cutoff frequency, and to lower values in the *stopband* (frequencies higher than the cutoff). The response in the stopband depends on the filter type and design.
- A *high-pass filter* is the opposite of a low-pass filter, meaning that the passband is located at frequencies higher than the cutoff and the stopband correspondingly at lower frequencies.
- A *band-pass filter* leaves a band of frequencies unmodified and attenuates all other frequencies. It can be implemented as a combination of a low-pass and a high-pass filter in a cascade. The *bandwidth* Δf is defined as the difference between the upper and lower cutoff frequencies $\Delta f = f_u - f_l$. The Q value of a band-pass filter is then defined as

$$Q = f_c/\Delta f. \tag{3.30}$$

  A high Q value thus implies a narrow band-pass filter, and vice versa.
- A *band-reject filter* is the opposite of a band-pass filter: it removes a certain band of frequencies and leaves the rest unmodified. It can thus also be implemented as a combination of low-pass and high-pass filters, but this time in parallel.
- An *all-pass filter* leaves the amplitude of all frequencies unchanged, but changes their phase relationships.
- *Arbitrary-response filters* are designed to have an arbitrary response both in magnitude and in phase.

Most commonly, filters are implemented as digital filters using DSP structures, or as analogue filters using electronic circuits. Many acoustic phenomena can be interpreted as filters. For example, the effect of atmospheric absorption of sound is effectively a low-pass filter, and the acoustic effect of a room can also be considered as a filter with an arbitrary response.

3.3.4 Digital Filtering

Digital filtering (Haykin, 1989; Jackson, 1989; Parks and Burrus, 1987) is a fundamental technique in digital signal processing. The input–output relationship of any band-limited LTI system (Figure 3.2) can be represented and implemented using a digital filter. Since algorithms for computing digital filters can be well optimized, this approach is useful when simulating or solving LTI engineering problems.

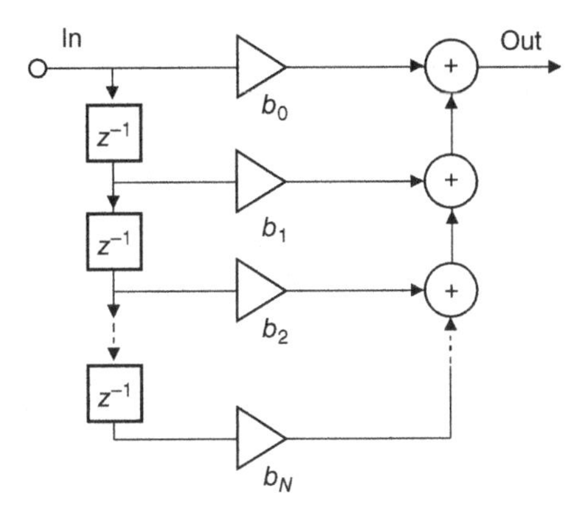

Figure 3.10 FIR filtering as a signal-flow graph. Elements z^{-1} represent unit delays, and multipliers b_n correspond to the coefficient values (tap coefficients) of the impulse response.

The two main types of digital filters are the finite impulse response *(FIR) filter* and the infinite impulse response *(IIR) filter.*

An FIR filter is computed using the principle in Figure 3.10 as a linear combination of delayed versions of the input signal; that is, by a convolution of the input and the impulse response of the filter. Each block z^{-1} is a unit delay, and the tap coefficients b_n before summation are directly the coefficient values of the impulse response $h(n)$. The transfer function of an FIR filter is simply

$$H_{\text{FIR}}(z) = \sum_{n=0}^{N-1} b_n z^{-n} = b_0 + b_1 z^{-1} + \cdots + b_{N-1} z^{-(N-1)} \tag{3.31}$$

FIR filter design and signal processing with them is relatively straightforward, but FIR filters may be computationally expensive.

The transfer function of an IIR filter is

$$H_{\text{IIR}}(z) = \frac{\sum_{n=0}^{N-1} b_n z^{-n}}{1 + \sum_{p=1}^{P-1} a_p z^{-p}} = \frac{b_0 + b_1 z^{-1} + \cdots + b_{N-1} z^{-(N-1)}}{1 + a_1 z^{-1} + \cdots + a_{P-1} z^{-(P-1)}} \tag{3.32}$$

One typical signal-flow graph formulation of an IIR filter, the direct form II, is depicted in Figure 3.11. It is different from an FIR filter in that there are feedback paths through multipliers a_n. An IIR filter may be unstable if it is not designed properly, which means that it can produce arbitrarily large output values for a finite input signal if a frequency is exponentially amplified in the feedback structure. The stability criterion for an IIR filter is that the poles of the transfer function, that is, the roots of the denominator polynomial of Equation (3.32), must be inside the unit circle on the complex plane $|z| < 1$. The zeros (roots of the numerator) don't have such a limitation.

3.3.5 Linear Prediction

A signal, or the system that has generated it, can be modelled in the LTI sense using *linear prediction* (LP) (Markel and Gray, 1976). In the literature the term *linear predictive*

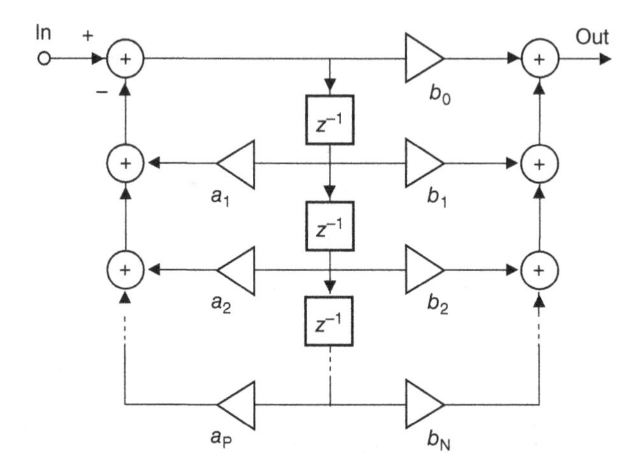

Figure 3.11 IIR filtering as a signal-flow graph. Coefficients b_n correspond to non-recursive FIR-like substructures (Figure 3.10), and a_n are the coefficients for recursive feedback.

coding and especially its abbreviation LPC is used as a general concept, but a more systematic approach is to use the term *linear prediction* for modelling and LPC only for coding purposes where LP analysis is quantized or otherwise coded. LP modelling also has other names, such as *autoregressive (AR) modelling* in estimation theory.

In LP modelling, we can imagine that the signal to be analysed is generated by an IIR system (Figure 3.11), where in the numerator of Equation (3.32) the coefficient $b_0 = 1$ and all the other coefficients $b_n = 0$ for $n \geq 1$. This is called an *all-pole* type IIR system. *LP analysis* yields optimal values in the least mean square sense for the denominator polynomial coefficients a_p, so that the resulting all-pole IIR filter system is the best one for predicting a new signal sample as a linear combination from the previous samples.

The most frequently used form of LP analysis is the *autocorrelation method*, where the coefficients a_p are solved from a linear matrix equation (normal equations)

$$\begin{bmatrix} r_0 & r_1 & r_2 & \cdots & r_{P-1} \\ r_1 & r_0 & r_1 & \cdots & r_{P-2} \\ r_2 & r_1 & r_0 & \cdots & r_{P-3} \\ \vdots & \vdots & \vdots & \ddots & \vdots \\ r_{P-1} & r_{P-2} & r_{P-3} & \cdots & r_0 \end{bmatrix} \begin{bmatrix} a_1 \\ a_2 \\ a_3 \\ \vdots \\ a_P \end{bmatrix} = \begin{bmatrix} r_1 \\ r_2 \\ r_3 \\ \vdots \\ r_P \end{bmatrix}, \tag{3.33}$$

where P is the order of LP analysis, that is, the order of the all-pole filter, and r_k are autocorrelation coefficients $r_x(k)$ from

$$r_x(k) = \sum_{i=0}^{N-1-k} x(i)\,x(i+k) \tag{3.34}$$

for the signal frame under study consisting of N samples.

The IIR filter $1/A(z)$ is called the *synthesis filter*, where $A(z) = 1 - \sum a_p z^{-p}$ and a_p are the coefficients from Equation (3.33). The FIR filter $A(z)$ itself is called the *inverse filter*. If it acts on the original signal, a *residual signal* that is spectrally flattened (whitened) is obtained.

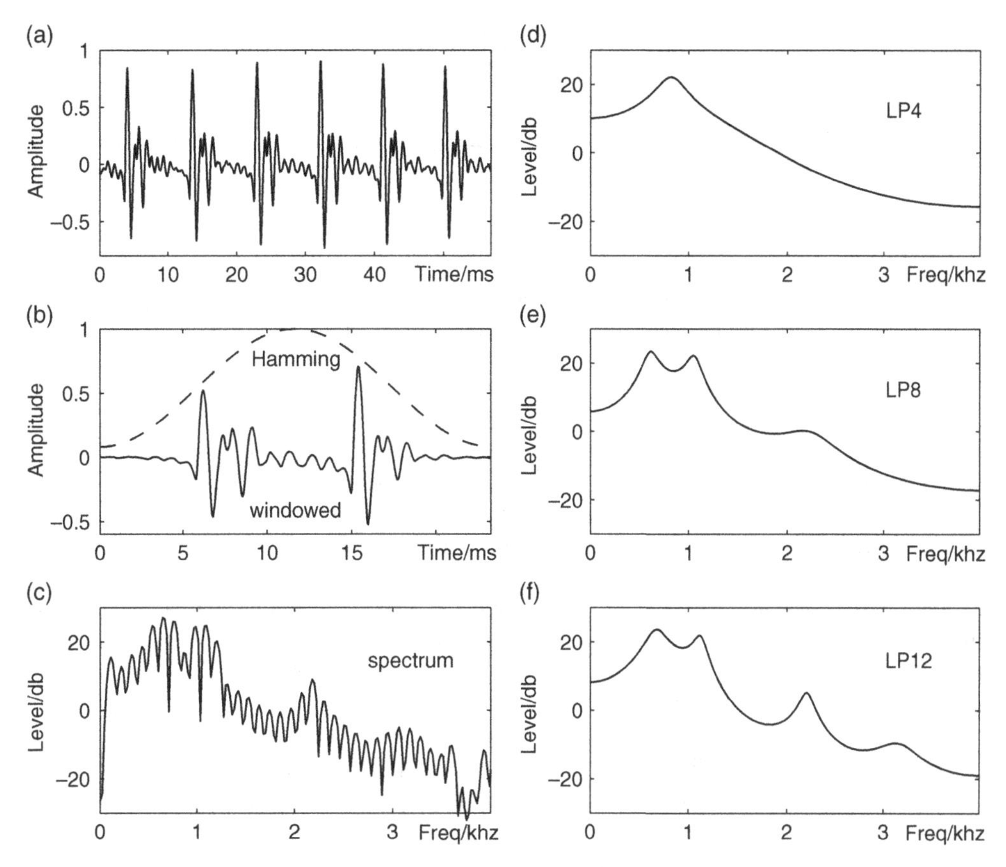

Figure 3.12 (a) A voiced speech signal (vowel) as a function of time. (b) The Hamming window (dashed line) and a Hamming-windowed frame of the signal (solid line), (c) its Fourier spectrum, and linear prediction spectra for LP orders of (d) 4 , (e) 8, and (f) 12. A sampling rate of 8 kHz has been used.

Naturally, if this residual is used as the excitation signal for filter $1/A(z)$, the original signal is synthesized, hence the name synthesis filter. In Section 5.3 we discuss the modelling of speech using source–filter models, whereby linear prediction is a natural choice and an effective technique.

LP analysis makes it possible to easily compute *spectral envelopes* or the *LP spectrum* that is, to remove the spectral fine structure of a speech or audio spectrum. Especially for speech signals, LP analysis is an effective way to separate the source (excitation) and filter (vocal tract transfer properties) in a source–filter model (see Figure 5.11 on page 91). Figures 3.12d–f illustrate LP spectra computed from the speech signal in Figure 3.12a, first windowed to obtain the signal in Figure 3.12b. As a reference, a Fourier spectrum of the speech frame is shown in Figure 3.12c. When the LP order is increased, the LP spectrum resolution improves. The LP order 8 (Figure 3.12e) already yields a fairly good approximation, and orders 10–12 are considered high enough for speech with a sample rate of 8 kHz. More generally, an order equal to the sample rate in kHz plus two is recommended for speech, since it is enough to represent speech formants and the general shape of the spectrum.

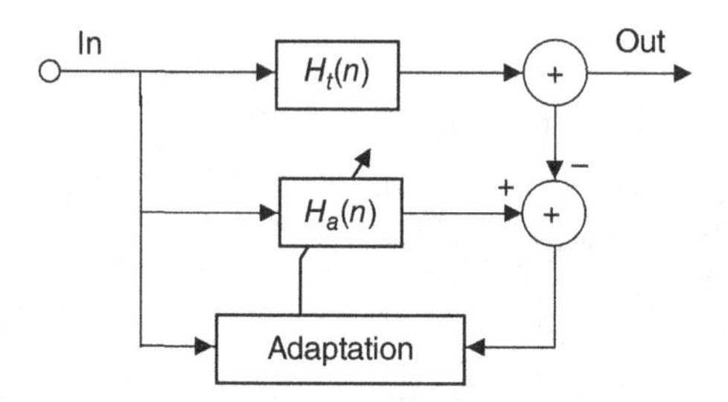

Figure 3.13 An example of adaptive filtering, where filter $H_a(n)$ adapts to the input–output relationship of target system $H_t(n)$, thus modelling the target system.

3.3.6 Adaptive Filtering

Digital filters with constant coefficients are an efficient and flexible approach for modelling LTI systems and signals. However, real world signals and systems can change their parameters as a function of time. There is a need for signal processing systems and algorithms that are able to adapt to changing conditions, or that are able to learn or can be trained to behave in a desired manner. Often there exists a rule that can be applied to change the parameters of a DSP system to adapt to the environment, or there are examples of desired behaviour that can be taught to the system. In such cases it would be preferable if the DSP system could automatically adapt to the external conditions. Sometimes it is enough to do the adaptation or learning only once or every now and then in a controlled manner.

In *adaptive filtering* (Haykin, 1989, 2005; Widrow and Stearns, 1985), the goal is that a signal processing function using digital filters can adapt to properties of the input signal in a meaningful way. Technically, adaptation means that the transfer function of the filter is controlled by changing the filter coefficients. Thus, the filter is no longer time-invariant. This time-variance problem can be formulated in different ways.

- A digital filter can be adapted so that the output signal is as close as possible to a target signal, so that their difference, for example in the least-squares sense, is minimized. After successfully adapting a filter to transform the input of a target system to the output of the target system, the filter can be used as a model of the target system (see Figure 3.13).
- When the input and output of the adapted system are interchanged, the adaptive filter attempts to make an inverse filter of the system to be modelled. The inverse filter can be used in series with the system so that modelled to equalize the total response so that it is close to the ideal one.
- An adaptive filter can be formulated to predict future values of the input signal by minimizing the prediction error. This is close to the idea of linear prediction, as described above.
- Adaptive filtering can also be used to cancel noise in a signal, thus enhancing signal quality. An important subproblem is *echo cancellation* (Sondhi *et al.*, 1995), where an echo in a system, acoustic or electronic, is attenuated.

3.4 Hidden Markov Models

Statistical modelling is used extensively in speech processing in particular, where a speech signal is interpreted as a sequence of units, each unit consisting of one or more states and having transition probabilities between states. Typically, a separable unit of speech is a speech

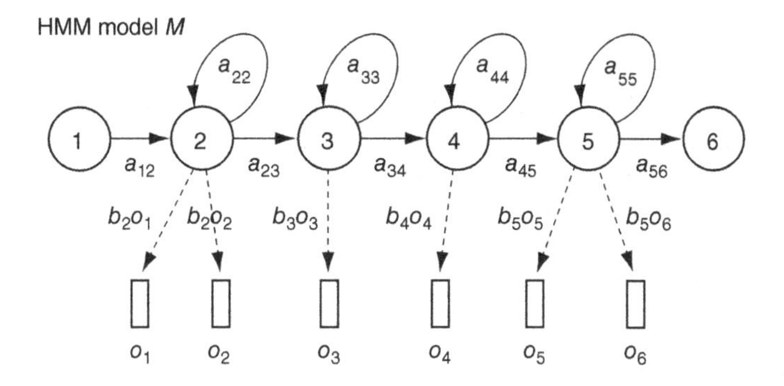

Figure 3.14 An example of a hidden Markov model having six states.

sound, for instance a phone, and a state in a phone represents a statistically stable segment in it, say the beginning, middle, or the end part. In speech recognition (Section 16.3) the problem is how to find such a sequence of units, given a speech signal or feature vectors describing it. A *hidden Markov model* (HMM) (Lee, 1989) is a popular learning method to do this.

An example of an HMM is illustrated in Figure 3.14 as a state transition diagram consisting of six states, $X = 1, 2, \ldots 6$, and transition probabilities a_{ij}. Some of the transitions return to the same state but others proceed to the next. If the model is used as an event generator, each state transition emits an observation vector $O = o_i$. In our case, traversing through states $X = 1 \ldots 6$, the model emits an observation vector sequence $o_1 \ldots o_6$. Here, the initial and final states do not emit observation vectors and states $X = 2$ and $X = 5$ emit two observations due to one transition returning to the same state. Such a model may be applied to represent any size of speech or sound event: a phone (Section 5.2), a syllable, a word, a longer message, or a sound or a passage in music.

If an HMM is used for recognition rather than generation of observations, the problem is that only observations, typically the feature vectors analysed from a speech signal, are known. The joint probability that the known observation $O = o_1, o_2, \ldots o_N$ is emitted by model M and state sequence $X = 1 \ldots 6$ is the product $P(O, X|M) = a_{12}b_2(o_1)\, a_{22}b_2(o_2) \ldots$. A simple way to use HMMs in pattern recognition is to select the recognition result for the model M that gives the highest probability $P(O, X|M)$ for the feature vector sequence $O = o_i$. The name 'hidden' Markov model comes from the fact that the state sequence X_i is hidden, not known.

3.5 Concepts of Intelligent and Learning Systems

There are several useful formulations for learning principles that may be combined to develop complex signal and information processing. *Artificial intelligence* is a commonly used term for the development of information and knowledge processing systems that try to mimic what is considered human intelligence. Some specific topics in intelligent and learning systems are the following:

- *Artificial neural networks* (ANNs) (Haykin, 1994; Kohonen, 1990; Lippmann, 1987; Luo and Unbehauen, 1997) are signal and information processing techniques that learn or are trained to acquire a targeted behaviour, typically an input–output relationship of interest. Some principles of ANNs are found by trying to simulate the behaviour of biological neural nets, but the similarity is not very strong, and true neural networks, for example in the brain, may work in quite different ways. Learning ability and simple self-organization are, however, important properties that ANNs can provide.

- *Pattern recognition* is the general idea of reducing complexity of signals in order to find the essential information carried by them (Duda *et al.*, 2012). Pattern recognition is typically a classification (categorization) process whereby irrelevant and redundant information is discarded and a signal is represented by features or classifications. Neural networks and hidden Markov models are among the most popular methods of pattern recognition, especially in speech recognition.
- *Fuzzy systems* are used to form logically operable representations for recognition and control but without the true/false dichotomy of classical logic (Wang, 1994). Using words and concepts that have a flexible yet quantitative interpretation allows for making fuzzy inferences and control.
- *Knowledge-based systems* utilize different paradigms of artificial intelligence, the idea being to process information in a way that resembles human conceptual processing (Hayes-Roth *et al.*, 1983). *Rule-based systems* are often combinations of logic processing and object methodology (Gupta *et al.*, 1986). *Expert systems* are knowledge engineering systems that try to simulate human expert capabilities.
- *Genetic algorithms* and *evolutionary systems* tend to mimic the principles of biological evolution (Goldberg and Holland, 1988). Genetic algorithms simulate evolution through mutation and selection of the best candidates for further evolution towards an optimal solution of a given problem.

Summary

The purpose of this chapter has been to gather concepts that are fundamental in signal processing. This topic has become a major cornerstone both for understanding human communications and for developing engineering solutions to improve communications. Signal processing and its applications will be present throughout the later chapters of this book in one form or another.

Further Reading

Since this book is not particularly about signal processing, this chapter serves only as a brief introduction to the subject, and also familiarizes the reader with the notation used in the field. There are numerous more detailed introductions and textbooks on signal processing, for example (Mitra and Kaiser, 1993; Strawn, 1985; Oppenheim and Schafer, 1975; Proakis, 2007; Steiglitz, 1996; Templaars, 1996).

There are many important topics and applications on signal processing that the interested reader can study elsewhere. The reader is encouraged to read more on signal processing as applied to *audio techniques* in general in Zölzer (2008). There is also a rich literature on adaptive and learning systems, as well as in information processing by artificial intelligence, as has been referred to above.

References

Cohen, L. (1995) *Time–Frequency Analysis*. Prentice Hall.

Duda, R.O., Hart, P.E., and Stork, D.G. (2000) *Pattern Classification*. John Wiley & Sons.

Goldberg, D.E. and Holland, J.H. (1988) Genetic algorithms and machine learning. *Machine Learning*, **3**(2), 95–99.

Gupta, A., Forgy, C., Newell, A., and Wedig, R. (1986) Parallel algorithms and architectures for rule-based systems. *ACM SIGARCH Computer Architecture News*, **14**(2), 28–37.

Hayes-Roth, F., Waterman, D.A., and Lenat, D.B. (1983) Building expert systems. *Teknowledge Series in Knowledge Engineering*.

Haykin, S. (1989) *Modern Filters*. Macmillan.

Haykin, S. (1994) *Neural Networks, A Comprehensive Foundation*. Macmillan College Publishing.

Haykin, S. (2005) *Adaptive Filter Theory*. Pearson Education.

Holighaus, N., Dorfler, M., Velasco, G.A., and Grill, T. (2013) A framework for invertible, real-time constant-Q transforms. *IEEE Trans. Audio, Speech, and Language Proc.*, **21**(4), 775–785.

Jackson, L.B. (1989) *Digital Filters and Signal Processing*. Kluwer Academic.

Kohonen, T. (1990) The self-organizing map. *Proc. of IEEE*, **78**(9), 1464–1480.

Lee, K.F. (1989) *Automatic Speech Recognition, the Development of the SPHINX System*. Kluwer Academic.

Lippmann, R.P. (1987) An introduction to computing with neural nets. *IEEE ASSP Mag.*, **4**, 4–22.

Luo, F.L. and Unbehauen, R. (1997) *Applied Neural Networks for Signal Processing*. Cambridge University Press.

Markel, J.D. and Gray, A.H. (1976) *Liner Prediction of Speech Signals*. Springer.

Mitra, S. and Kaiser, J. (eds) (1993) *Handbook of Digital Signal Processing*. John Wiley & Sons.

Oppenheim, A.V. and Schafer, R.W. (1975) *Digital Signal Processing*. Prentice-Hall.

Oppenheim, A.V., Willsky, A., and Young, I. (1983) *Signals and Systems*. Prentice-Hall.

Parks, T.W. and Burrus, C.S. (1987) *Digital Filter Design*. Wiley.

Proakis, J.G. (2007) *Digital Signal Processing: Principles, Algorithms, and Applications*, 4th edn. Pearson Education.

Schörkhuber, C., Klapuri, A., and Sontacchi, A. (2013) Audio pitch shifting using the constant-Q transform. *J. Audio Eng. Soc.*, **61**(7/8), 562–572.

Sondhi, M.M., Morgan, D.R., and Hall, J.L. (1995) Stereophonic acoustic echo cancellation–an overview of the fundamental problem. *IEEE Signal Proc. Letters*, **2**(8), 148–151.

Steiglitz, K. (1996) *A Digital Signal Processing Primer*. Addison-Wesley.

Strawn, J. (ed.) (1985) *Digital Audio Signal Processing: An Anthology*. William Kaufmann.

Templaars, S. (1996) *Signal Processing, Speech and Music*. Swets & Zeitlinger.

Vetterli, M. and Kovacevic, J. (1995) *Wavelets and Subband Coding*. Prentice-Hall.

Wang, L.X. (1994) *Adaptive Fuzzy Systems and Control: Design and Stability Analysis*. Prentice-Hall.

Widrow, B. and Stearns, S.D. (1985) *Adaptive Signal Processing*. Prentice-Hall.

Zölzer, U. (2008) *Digital Audio Signal Processing*. John Wiley & Sons.

4

Electroacoustics and Responses of Audio Systems

Electroacoustics is a topic that connects acoustics and electrical engineering. Some physical phenomena can be used to convert an electrical signal – a voltage or a current – into a sound, or vice versa. Electroacoustic devices, particularly microphones, loudspeakers, and headphones, are essential components in speech communication, audio technology, and multimedia sound. Electroacoustic devices are also important in acoustic measurements and scientific research on various aspects of sound.

Sound reproduction and processing of audio signals has developed over more than a century to a stage in engineering where many of the components in a reproduction or signal processing chain can be made practically perfect from a perceptual point of view, implying that imperfections in the reproduction cannot be noticed.

Digital signal processing in particular has improved the price/quality ratio of audio recording and playback compared to analogue techniques, practically replacing them. Analogue electronic components, such as amplifiers, can also be made practically perfect. However, some electroacoustic components of the audio reproduction chain, especially loudspeakers and their connection to room acoustics, are more problematic. Headphone reproduction is also not straightforward if a very accurate sound field inside the ear canal is desired. These issues motivate the use of signal processing to enhance the responses of audio devices. This chapter will review the electroacoustics of loudspeakers and microphones, the measurement of system responses, basic properties of the responses, and the equalization of the systems.

4.1 Electroacoustics

4.1.1 Loudspeakers

Transforming electrical signals into acoustic sound is done using a loudspeaker or headphones. A *loudspeaker* (Borwick, 2001; Colloms, 1997) consists of one or more driver elements, normally built into an enclosure box. A loudspeaker must be driven by an amplifier that is able to supply the necessary electrical power, and a part of this energy is converted into sound energy.

Communication Acoustics: An Introduction to Speech, Audio, and Psychoacoustics, First Edition.
Ville Pulkki and Matti Karjalainen.

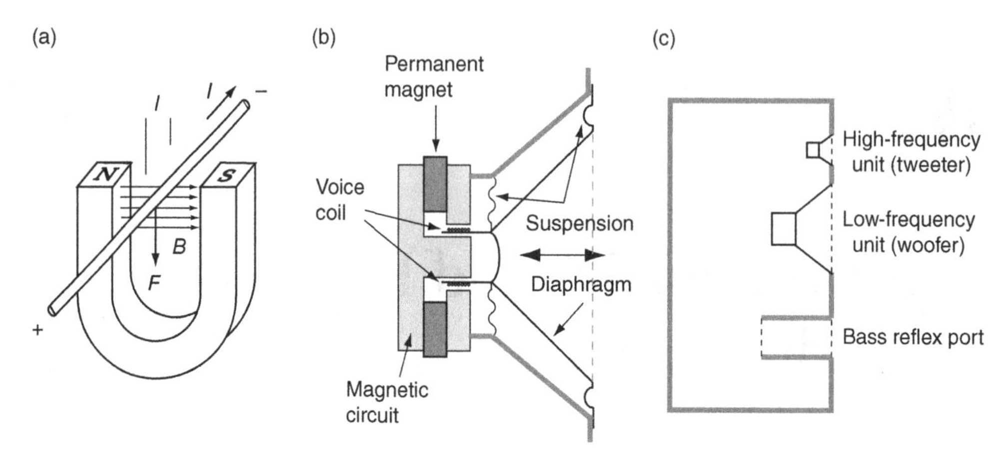

Figure 4.1 The working principle of the electrodynamic loudspeaker: (a) the current I through a wire in a magnetic field B is affected by the mechanical force $F = BI\ell$, (b) the structure of a typical loudspeaker driver element, and (c) the construction of a typical two-way loudspeaker with a bass-reflex type of enclosure.

Most loudspeaker driver units are based on the *(electro)dynamic* principle shown in Figure 4.1a. If a wire in a magnetic field of strength B carries a current I, a mechanical force $F = BI\ell$ acts to move the wire. Here, ℓ is the effective length of wire in the magnetic field. The basic structure of a dynamic loudspeaker driver is illustrated in Figure 4.1b. A cylinder-shaped voice coil–containing many turns of wire – can move in the air gap of a permanent magnet with a strong magnetic field. A voltage from a power amplifier, applied to the voice coil, results in a current through the coil, and this current results in the force specified by the equation above. The voice coil is attached to a cone-shaped diaphragm that is suspended so that it can easily be moved longitudinally by the force from the voice coil but not in other directions. The mechanical movement of the diaphragm makes the air move, thus resulting in an acoustic wave.

The sole element of Figure 4.1b is not enough for good sound reproduction. The first problem is that at low frequencies the sound pressure difference around the edge of the diaphragm will be cancelled ('short-circuited') since the distance is much less than that of a wavelength, and the element radiates very little sound. To prevent cancellation of the pressure difference, the element must be attached to an enclosure or a baffle. The two most common solutions are *closed box* and *bass-reflex* loudspeakers. The first, as the name implies, is an enclosure that is closed except for a hole for the driver element. The second, the bass-reflex construction, is characterized in Figure 4.1c. The enclosure contains a tube (or hole) that makes it behave like a Helmholtz resonator (Figure 2.4) in a specific, low frequency range. The advantage of this resonance is that it adds sound radiation and helps to boost this low frequency range.

The next problem to be solved in high-quality reproduction is that a single driver element is very difficult to design such that it can cover the entire audio frequency range, have a distortion-free power handling capacity, and have directivity of sound radiation in a desired manner. A common solution to this problem is to divide the frequency range of interest into sub-bands and to use a separate element for each subrange. In order to have an appropriate match to the wavelengths of radiated sound, a low-frequency element (woofer) has large dimensions while a high-frequency element (tweeter) has small dimensions. High-quality loudspeakers often have a third element, a mid-range driver, to cover middle frequencies, such as 1–4 kHz. Each element is designed to handle well its subrange and is fed through a crossover filter network that attenuates those frequencies in the full-range signal that are not intended for this specific driver element. When the radiation from the subrange elements is summed in the air, the full-range transfer function is obtained.

The efficiency of a loudspeaker to transform electrical energy in to acoustic sound is low, typically below 1%. If electrical power of about 50 W is fed to a speaker, the radiated acoustic remains below 1 W. In normal listening conditions this is, however, loud enough, as can be computed from Equation (2.33).

One more important acoustic property of a loudspeaker is its *directivity pattern*. In an anechoic room, we are only interested in the direct radiation to the listener, but in normal spaces with wall reflections and reverberation the listener receives all these delayed signal counterparts as well. In a relatively reverberant room, it is desirable to use a loudspeaker that has some directivity; that is, it radiates more sound along the main axis than in other directions, especially at middle to high frequencies. The colouration of sound depends both on the loudspeaker and the room properties in a complex way.

For carefully controlled experiments and acoustic studies, a special room called the *anechoic chamber* is often used. The walls of such a chamber are covered with highly absorptive material, practically eliminating all audible reflections and reverberation so that the direct radiation of a sound source can be measured or listened to.

Headphones

Another way to transform an electrical audio signal into acoustic sound is by using *headphones* (also called *earphones*) (Borwick, 2001). Headphones were designed initially to isolate radio-communication operators from external acoustic noise. However, they are used nowadays in portable personal audio equipment and high fidelity systems. Furthermore, they are also important for research purposes, especially when the sound in each ear must be controlled individually.

The transducer principle typically used in headphones is that of the electrodynamic loudspeaker, resembling the structure of Figure 4.1b but built on a smaller scale. An electrodynamic transducer for headphones has larger impedance than a loudspeaker, thus requiring less handling power. Another technique used in the highest-quality headphones is the *electrostatic principle*, which is the inverse of the condenser microphone working principle described in the next section. The electrostatic principle is sometimes also used to make high-quality loudspeakers, especially at mid-to-high frequencies.

The construction principle for headphones varies considerably. Small headphone elements are designed to be inserted in the ear canal or attached at its entrance. Larger headphones are either supported by the pinna (*supraaural*) or constructed as cup structures to cover the external ear entirely (*circumaural*).

Acoustically, the transfer function of a headphone to the eardrum depends on the acoustic coupling of the headphone to the ear, which can be explained by the pressure chamber principle (Borwick, 2001). Due to non-optimal sealing of the headphone cushions to the head or to the ear, the sound pressure partially leaks out from the space between the headphone and the ear (the pressure chamber). Such leakage influences the transfer function at low frequencies but the effect is minimal at middle and high frequencies.

The construction of large headphones depends on what compromise is made between sound isolation and the effect of leakage. To isolate the ear from external noise, the headphone may be *closed*. In this case, the transfer function of the headphone is sensitive to leakage. Alternatively, in *open* headphone design, the headphone is not sealed, and the sound from the headphone leaks out in a controlled manner. This makes the transfer function insensitive to leakage from the joint between the headphone cushion and the head. However, with open headphone design there is no isolation from external noise.

4.1.2 Microphones

The function of a *microphone* is to transform a sound signal (typically pressure) propagating in the air into a corresponding electrical signal (voltage) (Eargle, 2004). A good microphone should convert a wide range of frequencies, say from 20 Hz to 20 kHz, with a flat frequency response, low distortion and noise, and have a desired directional pattern.

Many conversion principles have been applied in microphones, but many of them, for example the carbon microphones used earlier in telephones, have become obsolete. The most important technology nowadays is the *condenser microphone* and its variants. Figure 4.2 illustrates the basic physical principle and a typical construction principle of a high-quality condenser microphone used for acoustic measurements and research purposes. The sound pressure variations in the air make one electrode of the condenser move while the other is kept fixed. If an electric charge is present between the electrodes, a change in distance between the electrodes will change the voltage between the electrodes. When this weak voltage signal is amplified inside the same microphone unit, the signal can be fed through a cable to audio equipment over reasonable distances. In the construction of Figure 4.2, the electric charge on the electrodes is provided from an external voltage source. A variant of the condenser microphone, the *electret microphone*, is based on using an electret material, which has electric charge trapped within, between the electrodes so that no external voltage is needed. A built-in amplifier is required here as well. Electret microphones have found widespread use especially in portable devices.

Another popular type of microphone is the *dynamic microphone*. It is based on the electrodynamic principle that was described in the context of the dynamic loudspeaker but now used to convert movement into an electrical signal. When a wire in a magnetic field moves, a voltage

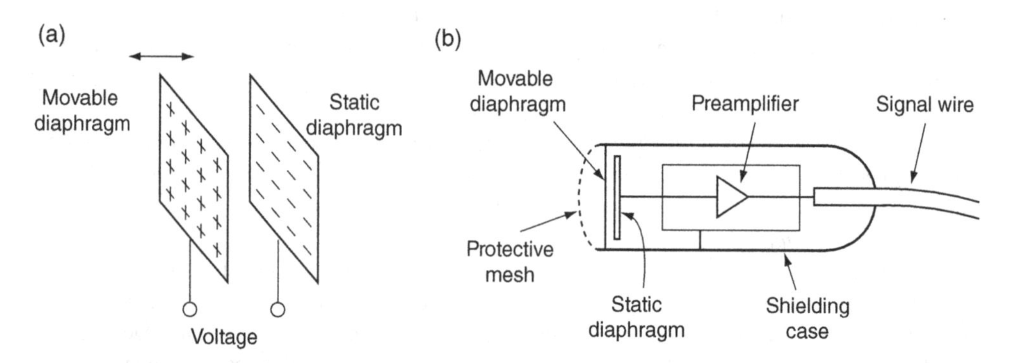

Figure 4.2 The condenser microphone: (a) the working principle and (b) a typical construction.

is induced between the terminals of the wire. When the diaphragm in Figure 4.1b moves, the voltage across the voice coil is proportional to the velocity of the diaphragm.

A *directional pattern* is an important property in microphones. The basic structure of Figure 4.2b is a *pressure microphone* that has an omnidirectional pattern, which means it is equally sensitive in all directions. If a microphone is sensitive to the particle velocity in air, the directional pattern is a dipole that reacts at the main axis from front and back but not from the sides. A *cardioid microphone*, which is a useful compromise, is maximally sensitive from the front and minimally sensitive from the rear. It is also possible to construct highly directional microphones that are sensitive only to sound coming from a narrow spatial angle. *Microphone arrays* together with signal processing can be used to make systems where the directional pattern of the resulting signal is controllable electronically. The directional patterns of microphones will be discussed later in this book in Sections 14.4.6, 18.5, and 19.5.2.

4.2 Audio System Responses

4.2.1 Measurement of System Response

As already mentioned, LTI systems can be characterized by an impulse response $h(t)$. The measurement of acoustic or audio systems can be easy or complicated, depending on the circumstances. The simplest case is one where a perfect impulse can be generated with high enough power to overcome the noise in the system. In general, for system measurements,

$$y(t) = x(t) * h(t) + n(t), \tag{4.1}$$

where $y(t)$ is the measured signal, $x(t)$ is the input signal, and $n(t)$ is the noise present in the system. If a Dirac impulse $\delta(t)$ is used as the input signal, $x(t) = \delta(t)$, and if the energy of $\delta(t)$ is much higher than the energy of the noise, the measured result is directly $y(t) \approx h(t)$. This approach can be applied, in some cases, to electric circuits or digital devices.

The case with acoustic and audio systems is often more complicated. The measurement of the impulse response of a room is an important case where perfect impulses are hard to generate. Unfortunately, a loudspeaker cannot produce impulses with high enough power. Balloon pops, hand claps, explosions, and electric sparks may be used to generate impulse-like sounds, all of which have some less than ideal properties, such as low repeatability and directional patterns deviating from the perfect, omnidirectional one. A high-power pulsed laser can also be used to generate impulses for acoustic measurements (see the example plot in Figure 2.20). Laser-induced pulses are repeatable and have an ideal omnidirectional pattern, although the cost of the device is, naturally, much higher, and using such devices is hazardous (Gómez Bolaños *et al.*, 2013).

The problem in producing impulses is the requirement of creating instantaneously very high signal amplitudes. For example, the maximum movement allowed by the diaphragm of the loudspeaker limits the peak amplitude to being relatively low. An engineering solution is then to use a known excitation signal $x_0(t)$ where the temporal length of sound is finite and still contains all the frequencies of interest. In this manner, the power of the excitation signal can be made much higher than the power of noise, and we can assume $n(t) \approx 0$. The impulse response can then be computed using *deconvolution*, which is most easily implemented in the frequency domain:

$$h(t) = \mathcal{F}^{-1}\frac{\mathcal{F}\{y(t)\}}{\mathcal{F}\{x_0(t)\}}. \tag{4.2}$$

The input signal $x_0(t)$ can be chosen to be a type of noise, such as the maximum-length sequence (Rife and Vanderkooy, 1989), or Golay code (Golay, 1961). Alternatively, linearly swept sinusoids (Berkhout *et al.*, 1980), or a known white-noise sequence (Borwick, 2001) can be used. Logarithmically swept sines can also be used, which allow for the simultaneous measurements of the distortion components (Farina, 2000). Here, the excitation signal $x_0(t) = a\sin[\omega(t)t]$, where the value of $\omega(t)$ is logarithmically changed from the lowest to the highest frequency to be measured. This means that a change in the frequency of the sweep is slow at first, and it accelerates exponentially. Typically, the temporal length of the sequence varies from between a few seconds to a minute. In principle, the longer the sweep, the better the signal-to-noise ratio obtained. However, the non-linear properties of the system limit the length of the sequences that can be used; the impulse responses may show, for example, time-smearing of the impulses if the sweeps used are too long. The swept-sine method also allows free control over the distribution of frequencies in the measurement, which is often desired since the frequency scale of human hearing is more logarithmic than linear.

4.2.2 Ideal Reproduction of Sound

Perfect reproduction of sound can be defined in terms of the transmission channel, by requiring that it is an LTI (linear and time-invariant) system where any input waveform is repeated perfectly at the output. For this ideal case, the transfer function $H(\mathrm{j}\omega) = 1$ (see Section 3.2.1). In practical audio we may relax this definition of perfect sound reproduction.

Since all physical systems are causal, the output is at least slightly delayed due to the finite speed of acoustic or electronic propagation of the signals. The permissible delay caused by a device may, in some sound reinforcement situations, be critical, and cannot exceed, say, 10 ms. However, when reproducing sound signals, for example when playing music from a recording, the delay can be seconds without major problems. Such a delay allows for the relaxation of the definition of perfect sound reproduction.

Another aspect that relaxes the constraints of the definition is that the signal level is allowed to change during storage or transmission, because this can be compensated for using gain control, and often a modifiable reproduction level is very desirable. When delay is permitted and the level at the output is allowed to differ from that at the input, the transfer function for ideal reproduction is $Y(\mathrm{j}\omega) = G\,e^{-\mathrm{j}\omega T}$, where $X(\mathrm{j}\omega)$ is the input, $Y(\mathrm{j}\omega)$ is the output signal, and G is a gain factor.

4.2.3 Impulse Response and Magnitude Response

The transfer properties of an audio signal channel are usually most often characterized by the *magnitude response*. This is commonly called the *frequency response*, but the concept *magnitude response* is used here since it is the proper, well-defined, and logical term.

Figure 4.3 shows several response plots of a small, two-way loudspeaker. Figure 4.3c depicts the magnitude response on a linear frequency scale. In audio, as well as in hearing-related

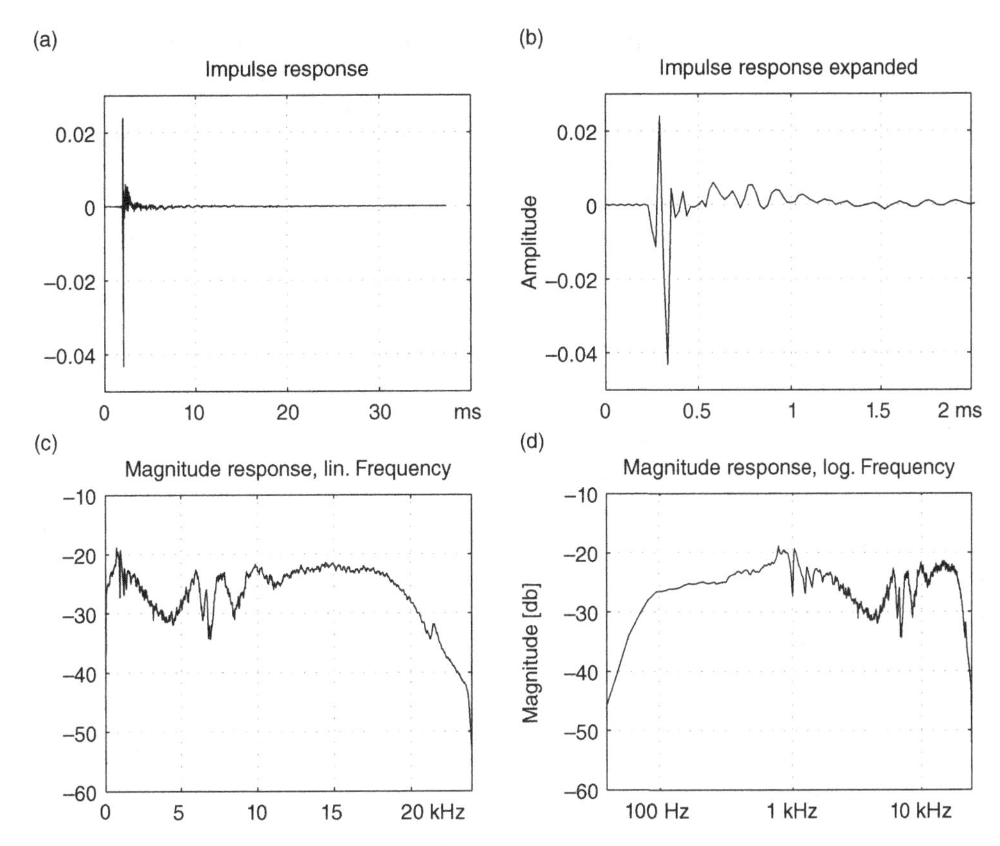

Figure 4.3 The measured responses of a small, two-way loudspeaker: (a) impulse response, (b) first two milliseconds of the impulse response, (c) magnitude response on a linear frequency scale, and (d) magnitude response on a logarithmic frequency scale.

topics in general, the linear frequency scale is found inappropriate since it does not portray the inherently wide frequency scale of auditory perception well enough. A logarithmic frequency scale, like the one in Figure 4.3d, is found to be well suited to audio and is used in most cases for magnitude response plots.

Deviations from a flat magnitude response within the audio range of 20 Hz–20 kHz change the perceived 'colour' or timbre of sound (Toole, 1986). However, close to the upper and lower frequency limits, the response flatness is less critical than in the mid-frequency range. The magnitude response of the loudspeaker of Figure 4.3 varies by ±5 dB in the range 80 Hz–18 kHz, which is easily noticed in direct comparison to the flat response, but without a reference most listeners cannot perceive the deviations when listening to music through the loudspeaker. In a practical listening environment, the interaction of the loudspeaker and the room acoustics makes the situation much more complex. Thus, in listening experiments, eliminating improper room acoustics is important by using loudspeakers in an anechoic chamber or by using headphone reproduction in most critical cases, or by using a specially designed listening room with carefully controlled acoustics. This will be discussed more in Section 14.3.2.

To provide an example of an audio system with a non-ideal response, Figure 4.3 (a and b) plots the measured impulse response of a loudspeaker under study. The close-up of the impulse response shows that the main part of the response is compact in time, within less than a millisecond, although some 'ringing' or post-oscillation can still be found after 10 ms of the main response. The magnitude response is plotted on both linear and logarithmic frequency scales, of which the logarithmic corresponds more to the human auditory scale. The magnitude response of the loudspeaker is relatively flat under 800 Hz, and at frequencies between 1 kHz and 10 kHz, a clear dip is seen in the response. Quality requirements for audio reproduction, especially from the auditory viewpoint, are discussed further in Chapter 17 about sound quality.

4.2.4 Phase Response

The transfer function of an LTI system, that is the Fourier transform of the impulse response, can be represented as a combination of the magnitude and phase response (see Equations (3.21a) and (3.22a)). While the magnitude spectrum is easy to interpret and relevent to audio systems, interpreting the phase spectrum is not as straightforward and also not as important, from a perceptual point of view, although in some cases the correct reproduction of phase is also important, as will be shown in Section 11.5.2.

A frequency-independent delay corresponds to a phase that is linearly dependent on frequency. Naturally, both the phase delay (Equation (3.23a)) and the group delay (Equation (3.23b)) are constants in such a case. The group delay is considered more relevant from the perceptual point of view than the phase delay. Since the auditory system is relatively insensitive to minor phase response degradations, it is often omitted in audio system measurements and specifications.

Figure 4.4 illustrates the group delay of the two-way bass-reflex loudspeaker that is analysed in Figure 4.3. The deviations from a constant average delay are interesting. Notice the growth of the group delay below about 200 Hz. This is due to the high-pass filtering characteristics of

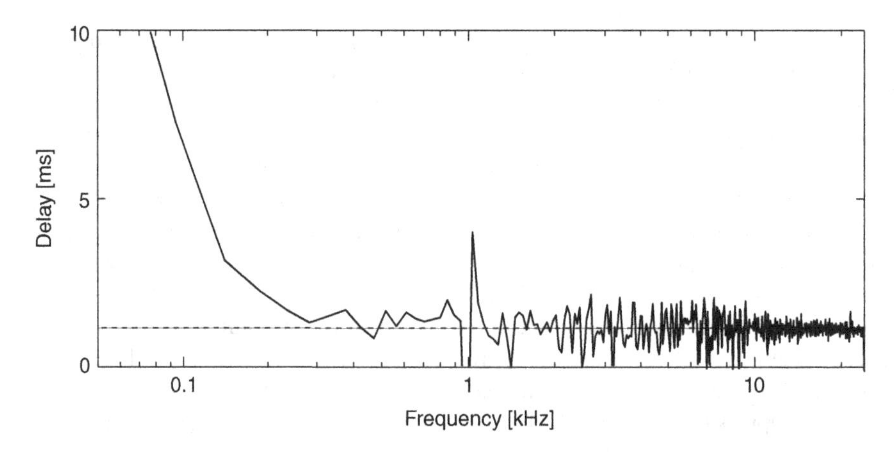

Figure 4.4 The group delay of a two-way bass-reflex loudspeaker on a logarithmic frequency scale.

the loudspeaker. For example, a bass-reflex loudspeaker behaves as a fourth-order high-pass filter at around and below the lower cutoff frequency.

4.2.5 Non-Linear Distortion

Any deviation from an ideal response in audio reproduction can be called *distortion*. The non-ideal magnitude and phase response is called *linear distortion*. In a linear (and time-invariant) system, the output contains only the same frequency components as the input, scaled by the magnitude response. In a non-linear system, however, new frequency components appear. Such behaviour is called *non-linear distortion*.

When a sinusoidal signal is fed to a non-linear system, the output may show harmonic distortion, where the harmonic components are integer multiples of the input frequency. If the harmonic spectrum of the output consists of component amplitudes $A(i)$, the *harmonic distortion* (HD_n) d computed up to the nth order of harmonics is the relation of the RMS value of the harmonic components to the RMS value of the signal

$$d = \frac{\sqrt{\sum_{i=2}^{n} A(i)^2}}{\sqrt{\sum_{i=1}^{n} A(i)^2}} \tag{4.3}$$

This is often expressed in a percentage measure $100 \times d$ %. Distortion may also be described as decibel levels of each harmonic or harmonics combined in relation to the linear response. The behaviour of distortion typically depends on the frequency and the level of the test signal. For a typical loudspeaker, the distortion increases at low frequencies due to the larger displacement of the bass-frequency element diaphragm.

Another example of the different non-linear distortion measures is the *total harmonic distortion* (THD), which is a measure that contains all the signal components that the fundamental frequency generates, including noise and possible non-harmonic components. Non-harmonic frequencies are those that are not integer multiples of the fundamental frequency. These also include *subharmonic* components, the frequencies of which are the fundamental frequency divided by an integer.

Natural sounds practically always contain harmonic or nearly harmonic components. Thus, harmonic distortion by itself is perceived as natural, a quality that makes sounds richer than pure tones. A reasonable amount of extra harmonics due to system non-linearities may improve perceived quality, for example, in the sound of an electric guitar. What makes system non-linearities sound undesirable is when non-harmonic components appear, increasing roughness (see Section 11.3) in the sound. Technically speaking, the phenomenon is called *intermodulation distortion* (IM). Any two sinusoidal signal components create sum and difference frequencies $f_{n,m} = nf_1 \pm mf_2$, where n and m are integers. Intermodulation distortion is measured by feeding two sinusoids of different frequencies and analysing the level of the intermodulation components.

There are many kinds of technical measures for distortion in audio and speech systems. Especially if the distortion mechanism is sufficiently complex, characterizing how annoying it is perceptually is very difficult. The current methods used to measure non-linear distortions are reviewed by Klippel (2006). Subjective listening tests are then the best way to evaluate the overall quality or its specific attributes. Since this may be tedious, objective measures based on models of auditory perception have been developed.

4.2.6 Signal-to-Noise Ratio

Undesired sounds that disturb the desired sound are called noise. Often, noise specifically means a random, unstructured signal. A straightforward technical measure of quality due to disturbing noise is the *signal-to-noise ratio* (SNR). It is defined in decibel units as

$$\mathrm{SNR} = 10 \log_{10} (P_\mathrm{S}/P_\mathrm{N}) \tag{4.4}$$

where P_S is the power of the desired signal and P_N is the power of noise. Which signal and noise powers are used depends on the appropriate temporal frame used, since for a long, time-varying signal a segmental is naturally better than a single value. As will be shown later, the SNR should be interpreted from an auditory point of view to make it meaningful and general.

4.3 Response Equalization

The unavoidable deviations from ideal audio reproduction, especially in loudspeakers and headphones, can be compensated for, at least partly, by signal processing. This compensation is called *equalization*, and a device or algorithm that performs equalization is called an *equalizer* (EQ). Equalizers are often used manually, by adjusting sound until it is perceived with the desirable timbre (sound colour) or some other quality, or equalization may be included in the reproduction system to automatically compensate for some degradation.

The most common form of equalization is the correction of the magnitude response of a reproduction device or system. Such a *magnitude response equalizer*, or simply EQ, if manually adjustable, may be used to control (boost or attenuate) the response (gain) of frequency bands, such as octave, third-octave, or other bandwidths, in the frequency range. This equalizer is sometimes called the *frequency equalizer*, which is somewhat misleading since it does not correct frequencies but response levels. The control knobs in the EQ are often arranged so that they show the correction in decibels as a function of frequency, giving rise to the name *graphic equalizer*.

Another type of equalizer is the *parametric equalizer*, which is used to adjust the level of a desired frequency band, its bandwidth, and centre frequency. An efficient way to compensate for magnitude response errors of LTI systems is the use of digital filters. If the original impulse response is measured to be $h(n)$, the EQ filter to flatten it is the *inverse filter* with the z transform $H^{-1}(z) = 1/H(z)$. When such a filter is connected in cascade with the original system, the resulting response will, in theory, be the ideal flat magnitude response and with constant delay.

In practice, there are always limitations to how appropriate rigorous equalization is. Figure 4.5 illustrates an example of a loudspeaker response and several equalized responses (Karjalainen *et al.*, 1997). Higher orders of EQ filters flatten the response increasingly, finally practically flattening the equalized response. However, one must remember that the equalized magnitude responses shown are valid only along the main axis of the loudspeaker in a free field, and that in other directions the response may be worse than the original unequalized one. In addition, equalization increases the movement of the loudspeaker diaphragm at amplified frequencies, which may also amplify some non-linear distortion components. Furthermore, in some cases, such frequency-specific amplification may also amplify unwanted noise originating from some earlier part of the signal chain.

Precise equalization of the response, however, is important for many experimental studies on auditory perception. Not only loudspeakers but also headphones should be carefully

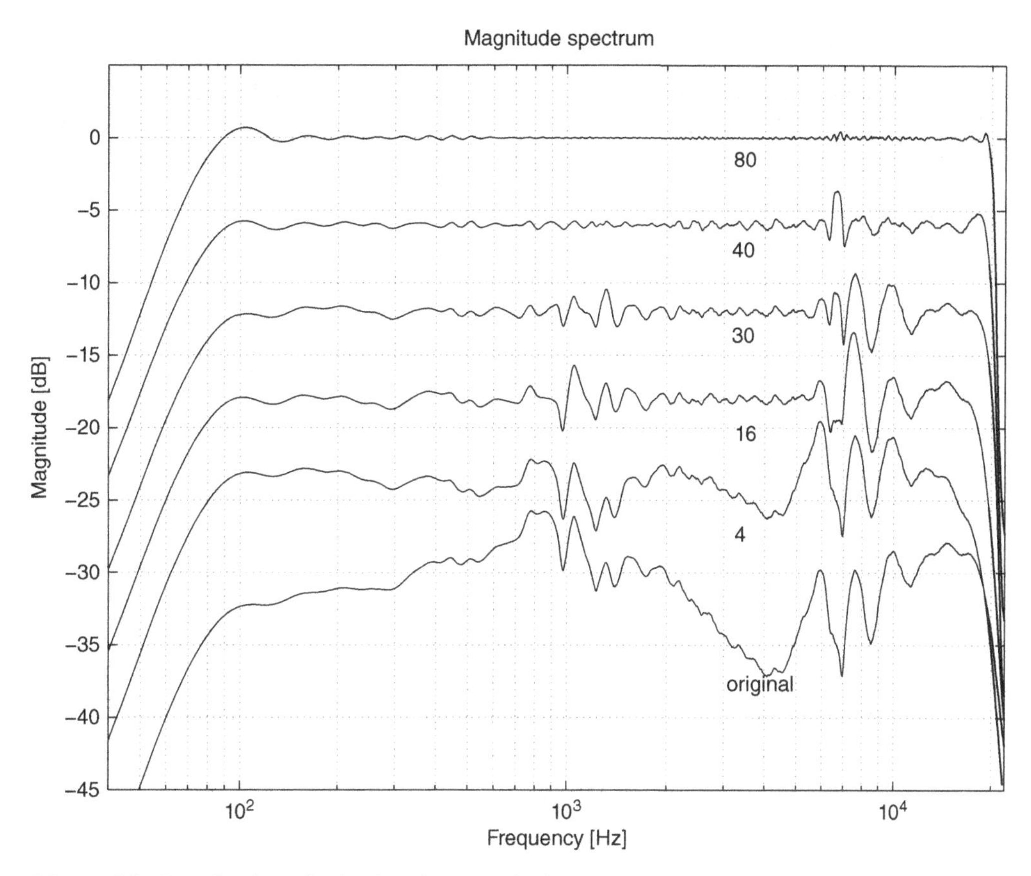

Figure 4.5 Equalization of a loudspeaker magnitude response using digital filtering. The original measured response is at the bottom and equalized responses for different filter orders are above it.

equalized to yield appropriate signals in the ear canals of a listener, for example, when studying spatial hearing phenomena. Compensation for non-linear distortion is more difficult than for linear response correction (Klippel, 2006), and the methods for non-linear EQ have not been developed very far.

Summary

Audio devices, especially electroacoustic components, often have a non-ideal response. The linear distortion created by such systems, in other words the deviations from the flat, unit magnitude response and from the frequency-independent group delay, can be measured simply by presenting a known sound signal to the system and by comparing the output to the input. A number of methods to measure the non-linearities of the systems via the distortion have been proposed, and a few of them were discussed here. Linear distortions may be equalized relatively simply using digital filters, although the equalization may amplify some non-linear effects of the system. Non-linear distortions are much harder to compensate for.

Further Reading

A more profound introduction to loudspeakers and microphones can be found in Borwick (2001), Colloms (1997), or Eargle (2004). The measurement and equalization of acoustic and audio systems are discussed in more detail in Borwick (2001), and Klippel (2006).

References

Berkhout, A.J., de Vries, D., and Boone, M.M. (1980) A new method to acquire impulse responses in concert halls. *J. Acoust. Soc. Am.*, **68**(1), 179–183.

Borwick, J. (2001) *Loudspeaker and Headphone Handbook*. Taylor & Francis US.

Colloms, M. (1997) *High Performance Loudspeakers*, 5th edn. John Wiley & Sons.

Eargle, J. (2004) *The Microphone Book: From Mono To Stereo To Surround – A Guide To Microphone Design and Application*. Taylor & Francis.

Farina, A. (2000) Simultaneous measurement of impulse response and distortion with a swept-sine technique *Audio Eng. Soc. Convention 108* AES.

Golay, M. (1961) Complementary series. *IRE Trans. Information Theory*, **7**(2), 82–87.

Gómez Bolaños, J., Pulkki, V., Karppinen, P., and Hæggström, E. (2013) An optoacoustic point source for acoustic scale model measurements. *J. Acoust. Soc. Am.*, **133**(4), 221–227.

Karjalainen, M., Piirilä, E., Järvinen, A., and Huopaniemi, J. (1997) Comparison of loudspeaker equalization methods based on DSP techniques *Audio Eng. Soc. Convention 102*. AES.

Klippel, W. (2006) Tutorial: Loudspeaker nonlinearities – causes, parameters, symptoms. *J. Audio Eng. Soc.*, **54**(10), 907–939.

Rife, D.D. and Vanderkooy, J. (1989) Transfer-function measurement with maximum-length sequences. *J. Audio Eng. Soc.*, **37**(6), 419–444.

Toole, F.E. (1986) Loudspeaker measurements and their relationship to listener preferences: Part 2. *J. Audio Eng. Soc.*, **34**(5), 323–348.

5

Human Voice

The acoustic communication mode specific to human beings is *speech*. It is the original type of linguistic communication, substantially important for everyday life. Before the emergence of written language, speech was also a way to keep information alive through oral repetition, from one subject to another and from one generation to the next.

The speech communication chain is characterized in Figure 5.1. This model pays attention to the *source* → *channel* → *receiver* structure of communication, but also to the generative process within the speaker and the analysis chain at the receiver. The speaker combines a linguistic message (such as words to be said) with non-linguistic information (rhythm, stress, and intonation of the words) at neural processing levels, constructs motoric control signals for the speech organs, and produces a spoken message conveyed by an acoustic waveform. The communication channel can be any medium, such as a pressure wave in the air, a wired or wireless telephone channel, voice over internet (VoIP), radio, or storage (= temporal transfer) by a recording device. Finally, the receiver is a subject who, in favourable conditions, is able to uncover a meaningful representation from the content of the message. Peripheral hearing is first used to extract a general auditory representation, and then more speech-specific processes are involved to decode the linguistic contents. Sometimes the non-linguistic content can be more important than the linguistic one.

Note that not only the channel but also the source and the receiver may be technical systems based on speech synthesis and recognition (see Sections 16.2 and 16.3), as long as some true speech communication takes place. In this chapter, the focus is first on speech production from both physical and signal processing points of view. The receiving function, implying the perception of speech, as well as the influence of the channel on it, will be discussed in later chapters.

5.1 Speech Production

Speech, as a means of linguistic communication (O'Shaughnessy, 1987), is a uniquely human process that is not found among other species of the world. It is so self-evident an ability that we often don't notice how complex and delicate it is until something goes wrong with it.

Communication Acoustics: An Introduction to Speech, Audio, and Psychoacoustics, First Edition.
Ville Pulkki and Matti Karjalainen.

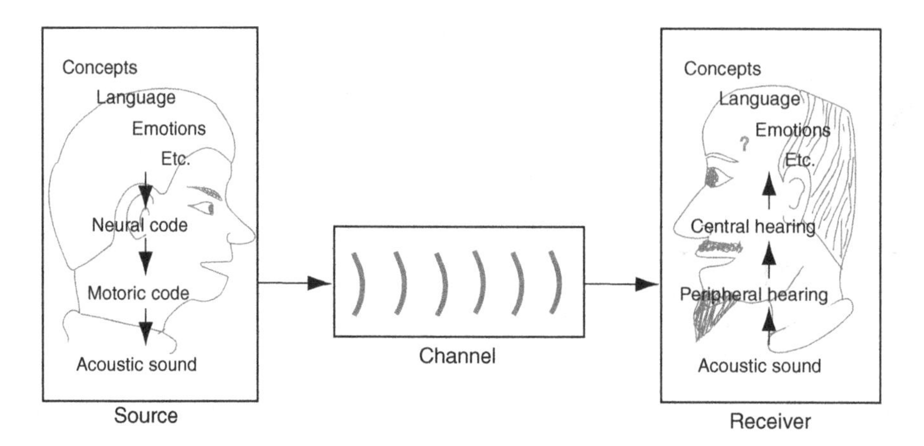

Figure 5.1 Speech communication chain: *source* → *channel* → *receiver*.

The production of a speech signal starts from a need to express the source subject's conceptual, emotional, or other internal representations, as shown in Figure 5.1. A linguistic structure (if it exists) is encoded, adding non-linguistic code (if any). This is a neural process, which results in motoric actions of the speech production organs. Finally, a speech signal is the acoustic output from the speech production organs.

Not much is known about the detailed brain processes at the highest levels of speech formation. What is known much better is the lowest level of a source subject's function, the acoustic theory of speech production (Fant, 1970).

5.1.1 Speech Production Mechanism

The acoustics of speech production have been subjected to extensive research. As early as at the end of the 18th century, it was shown experimentally by Christian Kratzenstein and Wolfgang von Kempelen that the generation of speech sounds can be explained using an acoustic–mechanic model (Flanagan, 1972; Schroeder, 1993).

Figure 5.2 illustrates a cross-sectional view of the human speech organs. The names of most of the important elements and positions used to characterize speech sounds by place of articulation are also given, and they are described below. The speech production mechanism will also be discussed in Section 5.3 from the point of view of modelling. In the next subsections, a brief review of the main functions of the speech organs is presented.

5.1.2 Vocal Folds and Phonation

The source of voiced speech is the airflow that is generated by the vocal folds located in the *larynx* (see Figure 5.2). The *vocal folds* are two horizontal tissues with an elastic membrane coating called mucosa (see Figure 5.3). The positioning and tension of the vocal folds are adjusted by the attached muscles. The orifice between the vocal folds is called the *glottis*. The area of the glottis is at its widest during breathing. During speech production, the glottal area varies temporally between zero and its maximum value.

When air is forced to flow from the lungs through the glottis, the vocal folds start to vibrate, closing and opening the glottis almost periodically. The formation of voiced sounds through

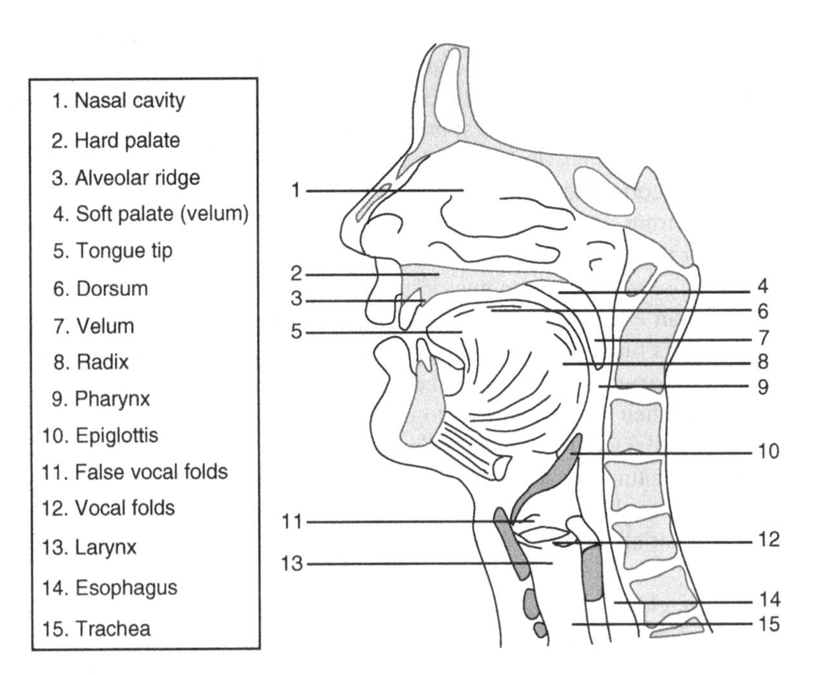

Figure 5.2 Cross-sectional view of the human speech production organs with anatomic names of the organs.

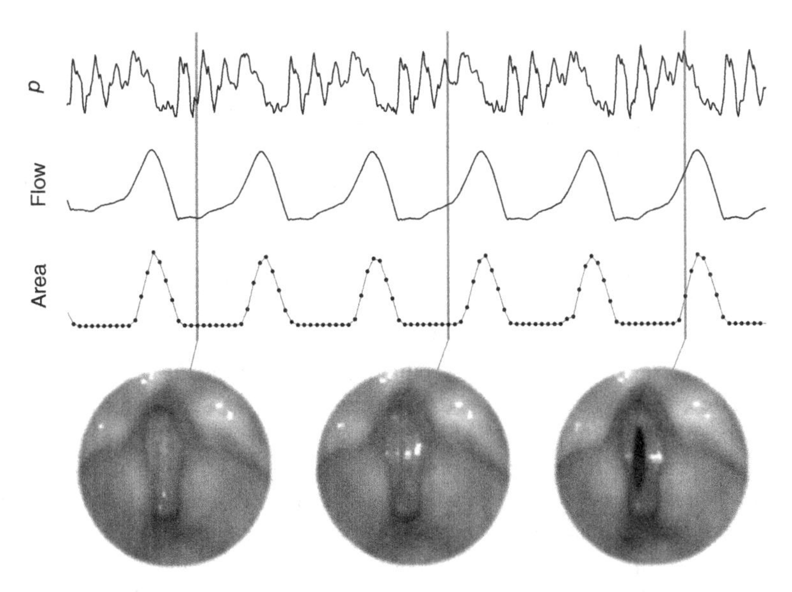

Figure 5.3 The vocal folds during phonation of a vowel imaged with a high-speed camera. The area of the orifice has been measured graphically and is shown as a time-dependent signal. The glottal flow estimated with inverse filtering from the pressure p signals is also shown. The vertical lines denote the temporal positions of image captures. Courtesy of Hannu Pulakka.

this vibration is called *phonation*. During speech production, humans are able to control the vibration mode of the vocal folds, hence varying the characteristics of the airflow pulses that are generated at the glottis as an acoustic excitation of voiced speech. Phonation types can be divided into three different categories: breathy, modal (or normal), and pressed (Alku and Vilkman, 1996). In *breathy phonation*, the area of the glottis is a smooth, almost sinusoidal function of time. The corresponding airflow pulse is also smooth in time, with a low-frequency emphasis in its spectrum. Breathy phonation typically has a non-zero leakage of air through the vocal folds over the entire glottal period. In *modal phonation*, the glottal area and the corresponding flow pulse are more asymmetric between the opening and closing of the glottis (see Figure 5.3 for an example of glottal flow during phonation). In *pressed phonation*, the closing phase shortens further in time, hence producing an excitation pulse that has more high frequencies in its spectrum. In addition to these three basic phonation types, humans are also capable of adjusting their glottal function to produce, for example, a *creaky voice* and *whispering*. In the former, the vibration mode of the vocal folds is irregular, involving two different modes that alternate in time, a phenomenon called *diplophonia*, and the duration of the closed phase is typically relatively long. *Whispering* is a speech production mode in which the vocal folds do not vibrate at all, and the sound excitation is produced by an exhaled airflow from the lungs.

The frequency of glottal oscillation is one of the most important acoustic parameters of voiced speech, the fundamental frequency f_0. In conversational speech, the average value of f_0 is about 120 Hz for males and about 200 Hz for females. Humans are, however, capable of producing much larger f_0 values: for a soprano singer, for example, f_0 can go up to 1500 Hz (Klingholz, 1990). The glottal waveform in voiced speech is discussed later in this chapter.

5.1.3 Vocal and Nasal Tract and Articulation

The wave from the glottis during phonation does not radiate directly out – it will propagate through the *pharynx* right above the larynx and then through the *oral cavity* above the tongue and possibly through the *nasal cavity*. These paths are called the *vocal tract* and the *nasal tract*. These acoustic tubes or cavities have an important impact on the final voice that radiates through the lips or the nostrils of a speaker. The process of controlling this acoustic process is called *articulation*.

The vocal tract from the glottis to the lips is about 17 cm in length for an adult male (Rabiner and Schafer, 1978), and about 14.5 cm for females (Klatt and Klatt, 1990). In the middle of the vocal tract, the tip of the soft palate called the *velum* opens or closes the sideway through the nasal tract to the nostrils. In normal production of speech this is open only for nasalized sounds. The effect of these two tracts is to determine the acoustic transfer function in such a way that the spectral properties of the radiating voice can convey speech information through a rich set of distinct sounds. From a signal processing point of view, the vocal and nasal tracts act as filters with controllable resonances that emphasize specific frequencies. These resonances are called *formants*, which are one of the most important cues in speech. The tracts may also create *antiformants*, dips in the spectrum; in other words, zeros in the transfer function.

The articulation position is primarily determined by the tongue, but also by the lips, the jaw, and by the opening or closing of the port to the nasal tract. The tongue is a complex and delicately controlled bundle of muscles that can move forwards and backwards as well as up and down, and its tip has a moving role of its own. Based on the positions of these speech production organs the resulting cross-sectional shape of the vocal tract (and the coupling or

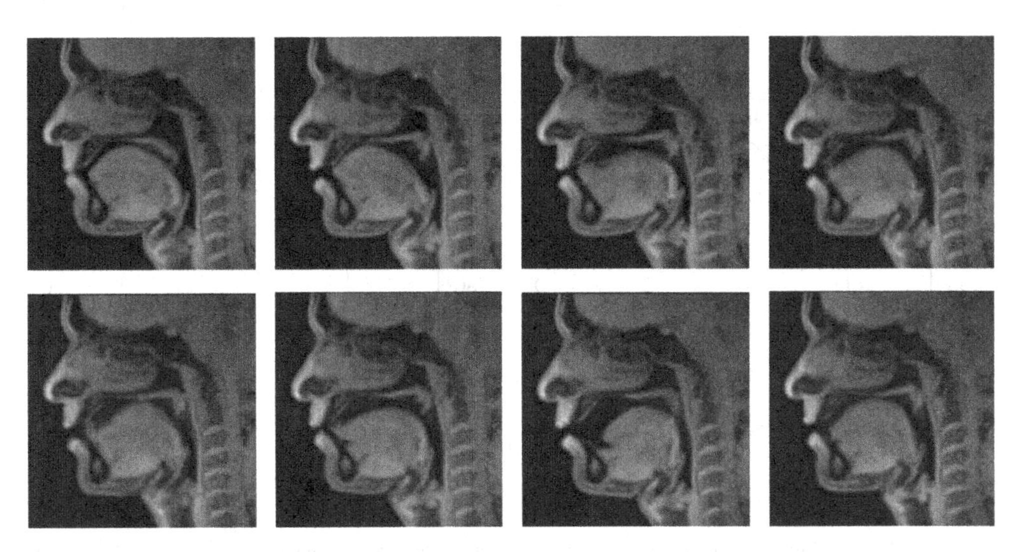

Figure 5.4 Snapshots of the movements of the speech organs during natural speaking. The original video (Niebergall *et al.*, 2011) has been measured using real-time magnetic resonance imaging (Uecker *et al.*, 2010). Licensed under the Creative Commons Attribution-Share Alike 3.0 Unported license. Copyright: Max-Planck Institute.

uncoupling of the nasal tract) determine the acoustic transfer function and the filtering effect on the glottal waveform to yield the final *voiced speech* sound. Figure 5.4 shows snapshots of the movements of the speech organs recorded using real-time magnetic resonance imaging. The movements of the tongue between the snapshots are large. In many cases the position of the tongue is close to some part of the roof of the mouth, and notice also the opening and closing of the nasal tract.

In addition to voiced sounds, speech signals contain *unvoiced speech* and mixed sounds with both voiced and unvoiced components. One type of unvoiced sound is created by frication, where the vocal tract has a *constriction*, a nearly closed point, where the airflow velocity is forced to become very high and turbulent, resulting in noise signal generation. Another case is the generation of explosive transient-like sounds where the vocal tract is suddenly opened with a sudden release of pressure. These sounds are called *plosives*, which may be either unvoiced or voiced; consider the difference between /p/ and /b/.

The processes, of human phonation and articulation can be studied by several means. One is the direct observation of the movements of the articulatory organs. With functional magnetic resonance imaging, it is possible to estimate how the cross-sectional shape of the vocal tract behaves during speech, as shown in Figure 5.4. The phonatory function, or the vibration of the vocal folds, is difficult to see due to their position, but, for example, with high-speed digital imaging it is possible to obtain a relatively accurate observation of how they vibrate, as shown in Figure 5.3. Other means of study are, for example, to use a *contact microphone* attached externally to the larynx or an *electroglottograph* to measure the external electrical conductivity changes in the skin close to the larynx caused by the opening and closing of the vocal folds (Colton and Conture, 1990). A further possibility is to analyse the speech waveform, as registered by a high-quality microphone, and to use inverse modelling techniques to derive an estimate of the the glottal waveform (Alku, 2011).

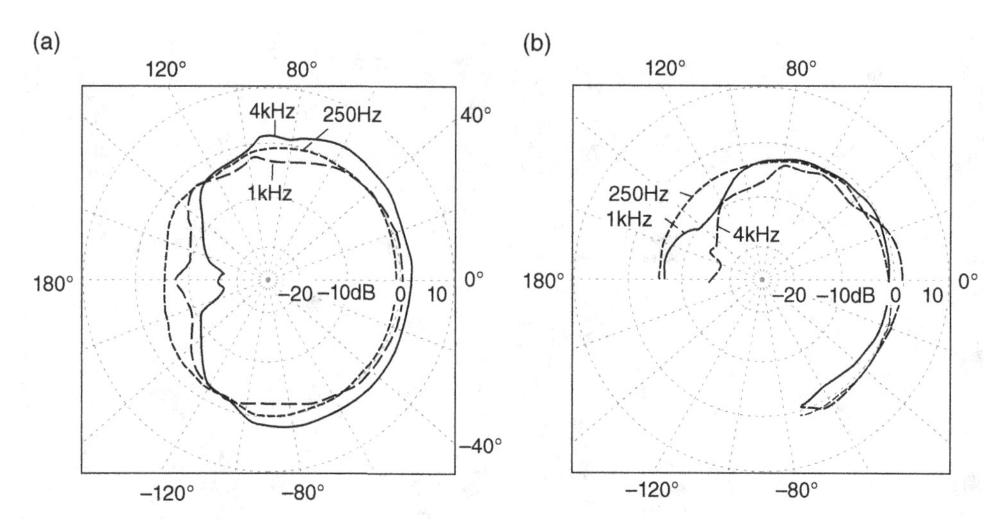

Figure 5.5 The directional patterns of the mouth of a dummy head measured at different frequencies (a) in the horizontal plane and (b) in the median plane.

5.1.4 *Lip Radiation Measurements*

The radiation of sound from the lips to the environment is also interesting in the context of speech communication (Flanagan, 1960). As the range of frequencies that humans can produce ranges from below 100 Hz to several kilohertz, it can be assumed that the directional pattern of a speaker changes with frequency prominently. At frequencies where the wavelength is much larger than the size of the head, the directional pattern of a human speaker obviously is a unitary sphere. The patterns have been measured with a dummy head with a mouth, and they are illustrated in Figure 5.5. As can be seen, the patterns are relatively constant for all frontal directions. In rear directions, the directions that exceed about 100° from the frontal direction, the sound is attenuated by a few dB for the 250-Hz case and somewhat more at higher frequencies.

5.2 Units and Notation of Speech used in Phonetics

Different languages vary in the form and number of basic units that are used to generate speech messages. Since there is temporal continuity of articulatory motion of the speech production mechanism, each unit is influenced by the preceding and following units. Therefore, speech is never composed of units in isolation but of a continuous flow of signal states. On the other hand, the structure of spoken language cannot be understood without some kind of discrete units that are concatenated into speech signals conveying linguistic information. This unit structure is reflected, more or less regularly, in written language. In languages such as English and French the written form (orthography) and phonetic transcription may often be quite different, while in languages at the other extreme, such as Finnish and Turkish, there is almost a one-to-one correspondence between the written and spoken forms.

Spoken languages exhibit an enormous variation in speech units and their combination. Languages can be grouped into *language families*, such as *Indo-European*, *Dravidian*, and *Uralic* languages (SIL, 2014). In different countries and areas, a language, such as English, may show remarkable variation, so that different *dialects* may vary radically within a single country.

Individual speakers have unique features and their own spoken language undergoes variation and evolution. This makes it difficult to describe speech formally as well as to develop systems and applications for spoken language technology.

Phonetics is the science that has developed ways to analyse and describe speech units and their features and structures composed of speech units (Ashby and Maidment, 2005). Various textual or symbolic notations exist to characterize how the units in speech are pronounced, both for practical guidance in dictionaries and for scientific purposes. An attempt to make a universal notation for any speech sound in any language has been taken in the *International Phonetic Alphabet* (IPA) (International Phonetic Association, 2014). The IPA notation is used in the examples within this book.

Since the symbols used in the IPA – including a large set of symbols for phonetic units, markers for so-called suprasegmental features, additional notational features called diacritics, etc. – are not well suited to computerized representation, different modifications have been developed with simplified notation. For example, IPA-ASCII applies only to ASCII characters. Other such symbol sets are SAMPA (Wells, 1997), where speech units specific to a single language are designed individually, and Speech Synthesis Markup Language (W3C, 2004), which is an XML-based (extensive markup language) markup language for assisting the generation of synthetic speech in Web and other applications.

The units of a spoken language can be defined at different structural and abstraction levels and based on different categorization criteria:

- *Phoneme*: an abstract unit of speech, a class that includes all such instances that don't cause meaning to change if used instead of another member of the same phoneme class. In this sense, the phoneme is a minimal linguistic unit. When a phoneme is referred to in text, the symbols are enclosed in a pair of slashes, as in /i/.
- *Phone* (speech sound): a concrete unit of speech sound, including details of producing the sound. Based on context, the properties of phones vary due to *coarticulation*, which is the continuity of articulatory movements and ease of pronounciation causing the change. Also based on context, a phoneme can be pronounced differently, and the alternative pronunciations for a phoneme are called *allophones*. Sequences of phones constitute syllables that are further composed into words, phrases, and so on. When the text discusses phones, they are referred to by enclosing them within square brackets, as in [i].
- *Diphone*: a unit providing another view to a sequence of phones. While a phone covers the time span from the previous phone boundary to the next, a diphone is understood as the time span from the middle of a phone to the middle of the next one, including the phone transition in the centre of the diphone. A phone boundary is typically the moment of the fastest transition between phones or the beginning of a transition. The duality of phones and diphones reflects the fact that some phones are quite static, carrying information in their central part, while in other cases the transition (diphone nucleus) carries the main phonetic information.
- *Triphone*: a temporal unit that covers two diphones. Longer units include more coarticulation within them, which is favourable in speech technology applications (recognition and synthesis), but on the other hand the combinatory explosion of the number of possible units makes them more problematic than short units. Triphones are utilized in automatic speech recognition, as will be discussed in Section 16.3.
- Phones can often be divided further into shorter *speech segments*, such as the transition phase, steady state phase, silence, burst of noise, and sudden onset.

Languages can typically be transcribed using a few tens of phoneme classes. The basic division of the classes is into the categories of *vowels* and *consonants*. A vowel is a speech sound where the vocal tract is relatively open without major constrictions. Vowels are produced relatively independently of their context. Consonants vary more in the manner of production. These two groups of phonemes are characterized below for the case of the English language.

5.2.1 Vowels

Vowels and consonants are combined in a language in a way that creates its rhythmic pattern, such as syllabic structure, where vowels often are kinds of anchor points to which consonants or consonant clusters are attached. Vowels can be classified using a few articulatory features:

- *Front–back position* of articulation as a measure of tongue position. For example, in the words *bet* and *but*, the only change in articulation is that the position of the tongue moves from the front in *bet* to the back in *but*.
- *Open–close dimension* (openness) of articulation specifies the up–down position of the tongue. The closer the tongue is to the palate, the more the vowel is 'closed'.
- *Rounded–unrounded* shape of the opening of the lips. If a vowel is pronounced first with lips [laterally] wide and then with rounded lips, the perceived vowel changes, as for the vowels in the words *but* and *caught*.

The number of vowel categories varies in different languages, ranging from two to a number higher than ten, depending somewhat on the classification system used (WALS, 2014). Table 5.1 shows 12 vowels in American English and Figure 5.6 shows a mapping of them with the features of articulation. The table does not include *diphthongs*, which are also known as gliding vowels. They refer to two adjacent vowel sounds occurring within the same syllable. Examples of vowel spectra are shown in Figure 5.7 for a male speaker. Notice the clear formant structure in the spectra and also the time-domain signals with a clear repetitive structure.

5.2.2 Consonants

Consonant sounds of a language contrast to vowels by being, in general, less intense, more context-dependent (more coarticulated), and produced with a more constricted vocal tract or by special sound-generation mechanisms. The manner of articulation is the configuration and interaction of the articulators when making a speech sound. The articulators include speech organs, such as the tongue, lips, and palate. The consonants in the English language can

Table 5.1 The most common vowels in American English.

IPA symbol	Examples	IPA symbol	Examples
i	beat	ɪ	bit (busy)
e	bait	ɛ	bet
æ	bat	ɑ	cot
ɔ	caught	o	coat
ʊ	book	u	boot
ʌ	but	ə	*a*bout

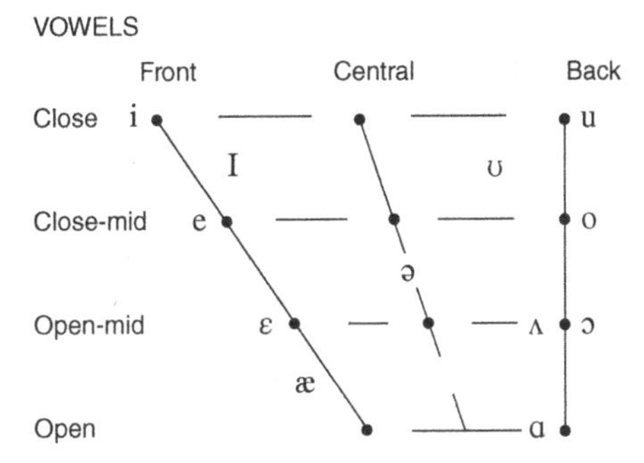

Figure 5.6 IPA chart of American English vowels (not including diphthongs). The front–central–back dimension denotes the position of the tongue, and the close–open dimension indicates how close the tongue is to the roof of the mouth. The symbols on the left of the vertical line have the mouth in an unrounded opening of the lips and those on the right in a rounded one. Adapted from http://www.langsci.ucl.ac.uk/ipa/ipachart.html, available under a Creative Commons Attribution-Sharealike 3.0 Unported License. Copyright ©2005 International Phonetic Association.

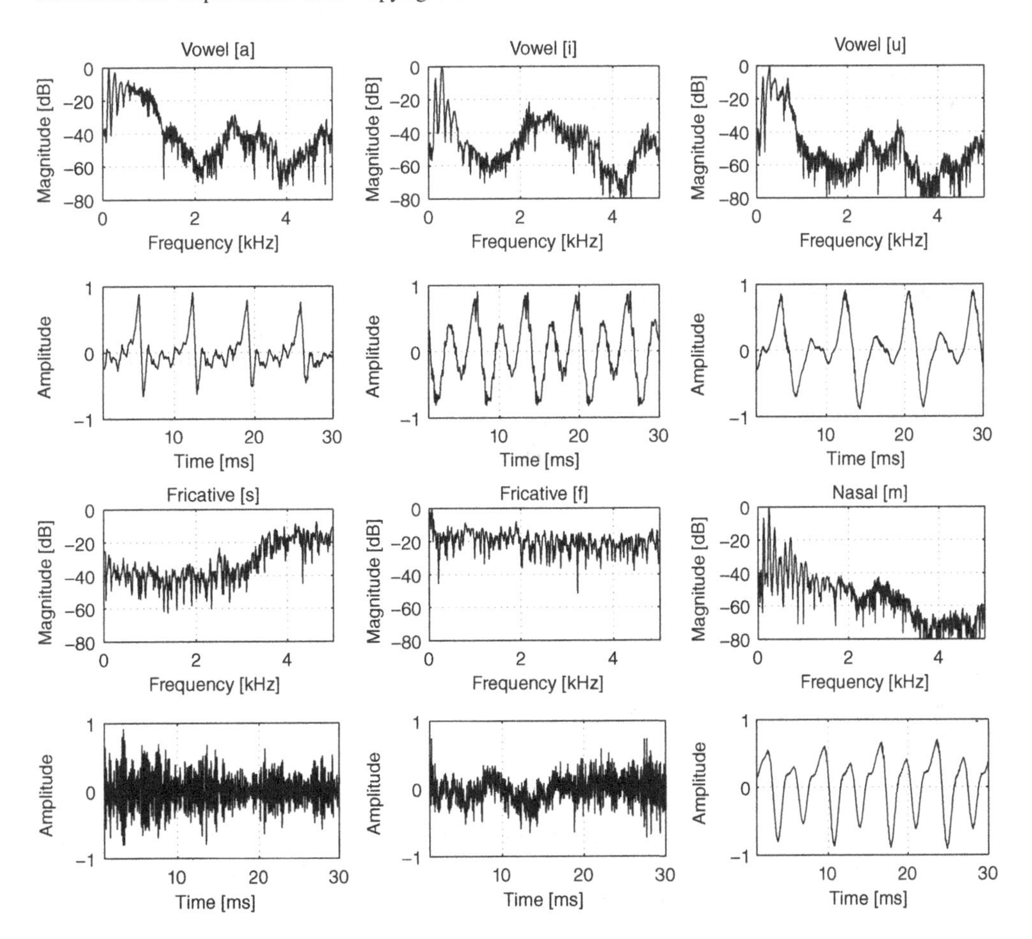

Figure 5.7 Spectra and waveforms of three vowels and three consonants.

be classified, based on the manner of articulation, into the following classes, which are not necessarily orthogonal to each other:

1. *Plosives* /p, b, t, d, k, g/: The vocal tract and nasal tract are completely closed causing a short silence /p, t, k/, or almost completely closed causing a faint voiced sound /b, d, g/. When the tract is opened, a short impulsive sound, or burst, is generated. The impulsive sound can be either noise-like or an exponentially decaying harmonic tone complex. The term *stop* consonant is often used interchangeably with *plosive consonant.*
2. *Nasals* /n, m, ŋ/: The vocal tract is closed, but the soft palate opens the nasal tract. The acoustic characteristics of both nasal and vocal tracts affect the sound. In most languages, including English, nasals are voiced.
3. *Trills* /r/: In trills the articulator vibrates rapidly (with a frequency of about 20–25 Hz) against the place of articulation. The only trill in English is /r/ (such as in *roar*), where the tongue vibrates against the alveolar ridge for about two to three vibrations. The known trills are voiced, and the vibrations cause an effect resembling amplitude modulation.
4. *Fricatives* /f, v, θ, ð, s, z, ʃ, ʒ, h/: The vocal tract is almost closed, because the articulator is brought close to the position of articulation. The narrow passage causes turbulant airflow, which is called frication. The frication causes a noisy sound, which is modified by the resonances of the vocal or nasal tract. The fricatives may be voiced or unvoiced.
5. *Approximants* /j, w, ɹ/: The approximants are similar to fricatives, but the articulators do not come close enough to generate frication.
6. *Laterals*: In lateral consonants the airstream proceeds along the sides of the tongue. English has only one lateral, /l/ (lateral approximant), where the tongue comes close to the alveolar ridge with voiced excitation.

In addition to the manner of articulation, the consonants can also be classified according to the position of articulation. As already mentioned, during the articulation the vocal tract is either totally or partially closed by the articulator, and the consonants articulated at the same position belong to the same class. The most common consonants in American English are shown in Table 5.2 with both their manner and position of articulation. The positions of articulation are shown in Figure 5.8. Examples of consonant spectra are shown in Figure 5.7 for a male speaker. Notice the clear patterns in the spectra and also the time-domain signals with noisy characteristics for unvoiced [s] and [f], and the repetitive structure with voiced [m].

5.2.3 Prosody and Suprasegmental Features

Speech signals exhibit features that are not strictly bound to the segmental structure. Such features are called *suprasegmental* or *prosodic* features. The concept *prosody* refers to the behaviour of prosodic features in general. The set of prosodic features comprises the following:

- *Intonation*. Many phonemes have voiced excitation, which always has the fundamental frequency f_0. The speakers actively vary f_0, or pitch, to convey the utterance information. These variations are called intonation. It can be used for a range of purposes, depending on language, such as indicating emotions and attitudes or signalling the difference between statements and questions. Intonation is also used to focus attention on important words of the spoken message and to help regulate conversational interaction. In tone languages, such as

Table 5.2 Table of common American English consonants.

IPA symbol	Example	Manner	Voiced	Position
j	you	approximant	yes	palatal
w	wow	approximant	yes	labial–velar
ɹ	red (American dialect)	approximant	yes	alveolar
l	lull	approximant	yes	lateral
r	roar	trill	yes	alveolar
m	my	nasal	yes	bilabial
n	none	nasal	yes	alveolar
ŋ	hang	nasal	yes	velar
f	fine	fricative	no	labiodental
v	valve	fricative	yes	labiodental
θ	thigh	fricative	no	dental
ð	though	fricative	yes	dental
s	say	fricative	no	alveolar
z	zoo	fricative	yes	alveolar
ʃ	show	fricative	no	postalveolar
ʒ	measure	fricative	yes	postalveolar
h	how	fricative	no	glottal
p	pot	plosive	no	labial
b	bib	plosive	yes	labial
t	tot	plosive	no	alveolar
d	did	plosive	yes	alveolar
k	kick	plosive	no	velar
g	gig	plosive	yes	velar
tʃ	church	affricate	no	alveopalatal
dʃ	judge	affricate	yes	alveopalatal

Chinese, pitch changes associated with syllables carry prominent linguistic information. This means that changing the pitch changes the phone, although other acoustic features remain constant.

- *Stress*. Some words, or syllables of words, are often given greater emphasis in spoken language. Stress can be implemented in different ways, depending on the speaker and on the language. The stressed syllable may have a higher or lower pitch than non-stressed syllables, which is called *pitch accent*. Other features for stress include *dynamic accent*, where the level of sound is raised during the stress, *qualitative accent* meaning differences in place or manner of articulation, and *quantitative accent*, where the length of a syllable is increased.
- *Rhythm* and *timing*. A differentiating factor between spoken languages is how words and phonemes are divided in time. Some languages allocate similar time for each syllable, while others divide the time based on other linguistic units. Quantitative stresses also affect the rhythm. On the other hand, some languages differentiate between short and long sounds, and these are called *quantity languages*. For example, the Finnish words /tuli/ and /tuuli/ (fire and wind) differ primarily in the duration of the stressed vowel.

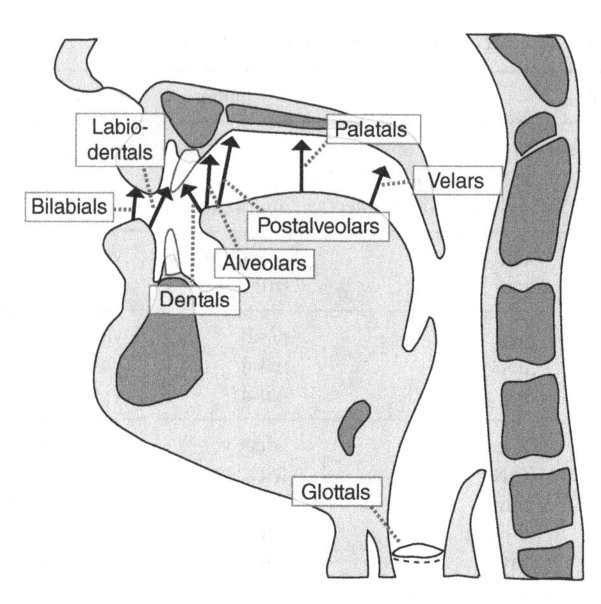

Figure 5.8 The positions of articulation of consonants and the corresponding movement of articulators.

5.3 Modelling of Speech Production

The acoustic production of speech signals can be approximated and modelled mathematically fairly easily (Fant, 1970). Based on such modelling, it is possible to derive signal processing algorithms that are found to be very useful in speech technology, like in speech synthesis and coding (Rabiner and Schafer, 1978). It also helps us to understand speech mechanisms.

Figure 5.9 illustrates an 'engineering-oriented' mechanical–acoustic model of speech production. The lungs are seen as a 'pressure compressor' that is the driving force for the glottal mass–spring oscillator or for frication and transient generation in vocal tract constrictions. The vocal and nasal tracts form an acoustic resonator or filter system, a kind of complex modulator or encoder of speech sounds. From the lips and the nostrils the signal may radiate and propagate to a listener or a microphone in a technical speech communication system.

The acoustically abstracted presentation in Figure 5.9 can be reduced further into the *circuit analogy* of Figure 5.10 that consists of three cascaded (= coupled in series) subsystems: the *excitation* (generator), a *transmission line* (a *two-port*), and a *radiation impedance* (acoustic radiation load). Acoustic variables that describe the functioning of the system may be the *sound pressure* p and the *volume velocity* u at each point of the system. In Figure 5.10, the variables are p_s (pressure from the lungs), u_g (volume velocity at the glottis), as well as u_m and u_n (volume velocities at the lips and the nostrils, respectively). If the behaviour within the vocal or nasal tract is of interest, these blocks must be described as distributed wave propagation subsystems.

In Figure 5.11, speech production is finally reduced into a *signal model*, the so-called *source–filter model*. In voiced signals (in phonation), the source is the glottal oscillation and in unvoiced sounds the source is frication noise created in a constriction or a transient sound generated by rapid release of pressure when a vocal tract closure is opened.

When the acoustic variables in Figure 5.9 are selected to be the sound pressure and the volume velocity, the natural choices for their electric analogies in this case are voltage and

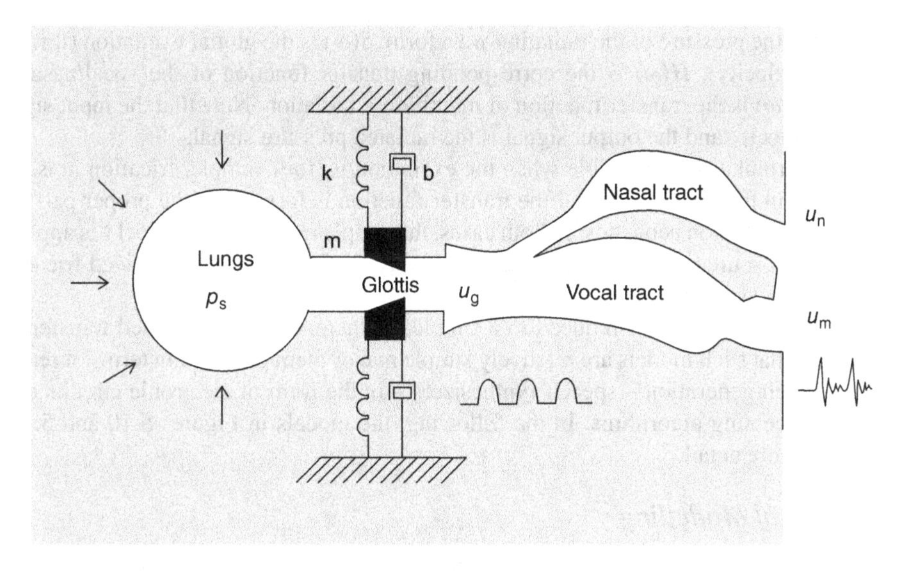

Figure 5.9 The mechanical–acoustic model of speech production.

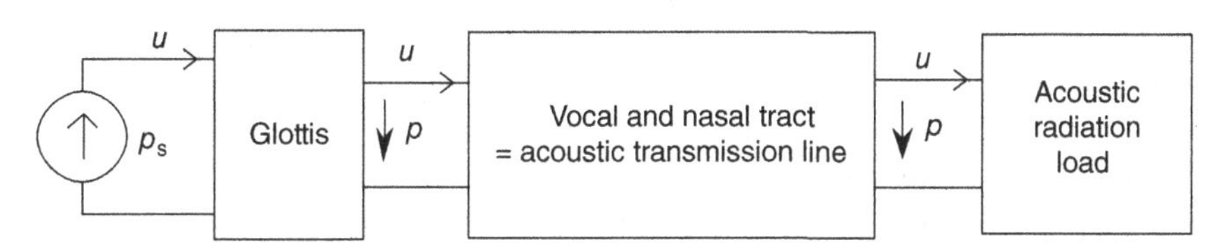

Figure 5.10 A circuit model of speech production: p denotes pressure (p_S is the source pressure from the lungs) and u is volume velocity.

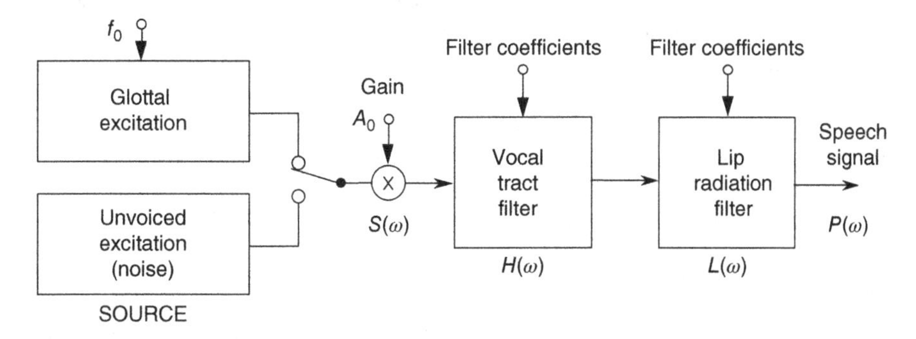

Figure 5.11 A signal model (source–filter model) of speech production.

current, respectively. In signal modelling (Figure 5.11), we do away with this dual variable modelling and consider only a single transfer function, for example, with the Fourier transform formulation:

$$P(\omega) = S(\omega) \cdot H(\omega) \cdot L(\omega), \tag{5.1}$$

where $P(\omega)$ is the pressure of the radiating waveform, $S(\omega)$ is the glottal excitation (for example, volume velocity), $H(\omega)$ is the corresponding transfer function of the vocal/nasal tract system, and $L(\omega)$ is the transfer function of mouth/nose radiation. Note that the input signal is the volume velocity and the output signal is the radiated pressure signal.

A similar formulation is possible when the excitation is, for example, frication noise from a constriction in the vocal tract, and the transfer function is formed by the proper part of the vocal tract and radiation acoustics. In both cases, the simple model of Figure 5.11 is applicable if the excitation source is selected according to the type of excitation. For voiced fricatives a mixed-source model is needed where voicing and noise are mixed properly.

Now the acoustic system is reduced to a simple, mathematically formulated transfer function. The fact that such models are relatively simple makes them practical in terms of realizing synthetic speech generation – speech synthesizers – in the form of electronic circuits or digital signal processing algorithms. In the following, the models in Figures 5.10 and 5.11 are discussed in more detail.

5.3.1 Glottal Modelling

Glottal oscillation, the almost periodic opening and closing of the vocal folds, is the main excitation in voiced sounds. Glottal functioning has been modelled mathematically in several ways, for example as a simple mass–spring system as in Figure 5.9 (Flanagan, 1972), as a spatially distributed mass–spring system (surface-wave transmission line), or just as experimental functions of time for volume velocity (Fant *et al.*, 1985). Figure 5.3 on page 81 plots the volume velocity at the glottis during static phonation of a vowel. The glottal waveform typically resembles a half-wave rectified sinusoidal signal with asymmetry, so that the opening phase is relatively slow and the closing phase is more abrupt, as seen in the figure.

Figure 5.3 also plots the pressure wave at the lips for the vowel that is generated. Notice that the main excitation is due to the closing event of the glottis, since this contains most of the excitatory energy at the formant frequencies. Recognize also that the glottal function is a non-LTI (not linear and time-invariant) system since its parameters – for example, the acoustic impedance in the glottal orifice – vary with glottal closing and opening. Thus, Equation (5.1) is not valid in a strict sense. Since the interaction of the vocal tract and the vocal folds is relatively weak, such a simplified and linearized model is useful as a first approximation.

There is no phonation in unvoiced sounds; that is, the vocal folds don't vibrate. Frication or pressure release bursts are the excitation signals in such speech sounds.

5.3.2 Vocal Tract Modelling

The vocal tract acts as an acoustic transmission line where the glottal excitation propagates to the lips. While the uvular link to the nasal cavity is open, the coupling of the nasal tract creates nasal or nasalized sounds. The shape of the tract is determined by the position of the tongue and its tip, the opening of the jaw, and the shape of the lip opening. From an acoustic point of view, the transmission properties of the vocal tract are due to the cross-sectional area function of the tract $A(x)$, where x is the position in the glottis–lips dimension. According to Equation (2.23), the acoustic impedance at each point is $Z_a(x) = \rho c / A(x)$; that is, it is inversely proportional to the area.

Figure 2.10 and Equation (2.22) on page 27 show us that a plane wave travels in the tube without modification as long as the area is constant, but at any discontinuity of $A(x)$ a part of the wave is reflected back and the rest propagates onwards. The simplest tract shape is one

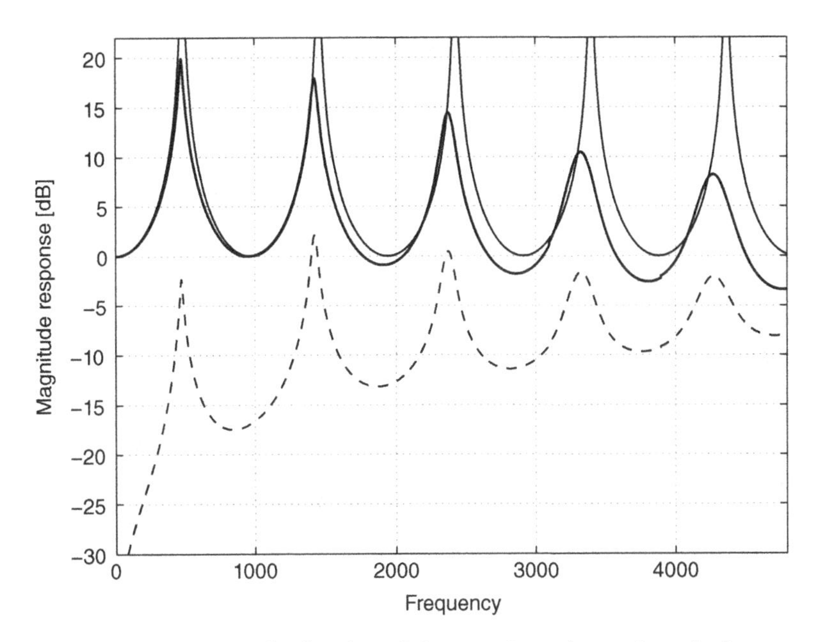

Figure 5.12 Solid lines: the transfer function of the neutral vocal tract (magnitude response of the volume velocity transfer from the glottis to the lips). An idealized, lossless case is plotted by the thin line, where the lines continue to positive infinity at resonance frequencies. The thick line represents a simulation with losses due to yielding walls, friction, and thermal loss. The dashed line shows the transfer function from the glottal velocity to pressure after lip radiation. Adapted from Rabiner and Schafer (1978).

where $A(x)$ is constant from the glottis to the lips. This is approximately true for the *neutral vowel*, something like /ə/. While the glottal termination has a high acoustic impedance and the lip opening a low impedance, the acoustic system for the neutral vowel corresponds to the quarter-wavelength resonator characterized in Figure 2.12b on page 29. The magnitude response $H(\omega)$ of an ideal lossless tract, that is, the ratio of the glottal volume velocity $U_{\mathrm{g}}(\omega)$ to the lip volume velocity $U_{\mathrm{l}}(\omega)$, is of the form

$$H(\omega) = \frac{U_{\mathrm{l}}(\omega)}{U_{\mathrm{g}}(\omega)} = \frac{1}{\cos(\omega l/\mathrm{c})}, \tag{5.2}$$

where l is the length of the tract and c is the speed of sound. $H(\omega)$ is characterized in Figure 5.12 whereby the formant resonances for a 17-cm long vocal tract (typical for an adult male) are located at frequencies

$$f_n = 500 \cdot (2n - 1), \qquad n = 1, 2, \ldots \tag{5.3}$$

that is, at $f_1 = 500\,\mathrm{Hz}, f_2 = 1500\,\mathrm{Hz}, f_3 = 2500\,\mathrm{Hz}$, and so on. The formant resonances of the ideal tract in Equation (5.2) have infinite amplitude (the thin line in Figure 5.12), while with frequency-dependent losses due to both vocal tract and lip radiation, the formants follow the bold line, as discussed by (Rabiner and Schafer, 1978). Finally, the transfer function from the glottal flow to the pressure signal after lip radiation is shown in the same figure with a dashed line.

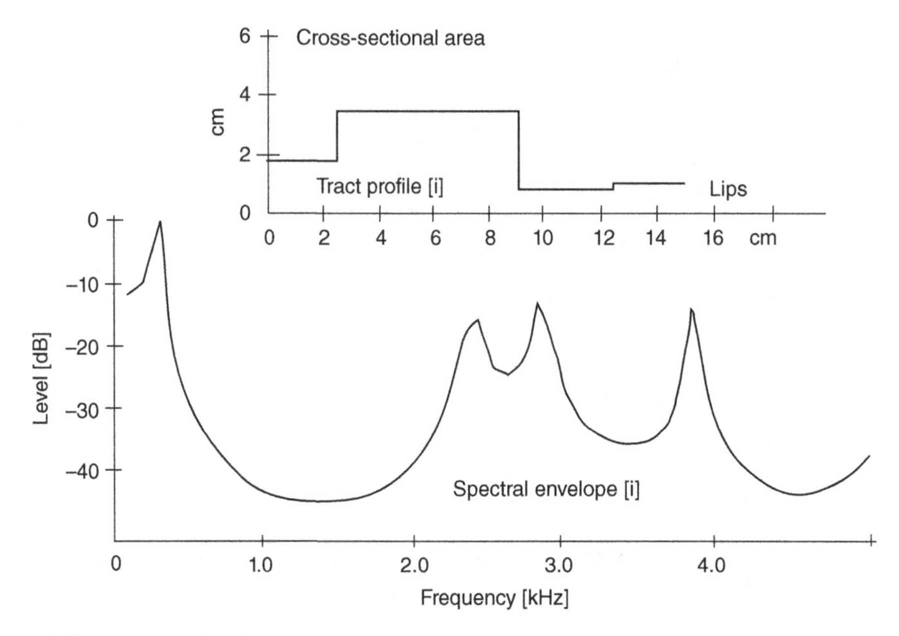

Figure 5.13 An approximation of the geometry of the vocal tract during the phonation of /i/ made with four tubular sections. The lower diagram shows the simulated magnitude spectrum of the response of the approximated tract.

For real speech signals the tract can take many shapes, and this results in a multitude of vocal tract transfer functions. For each shape the positions of the formant resonances take specific positions. The spectral shape, and especially the frequencies of the lowest formants, then carry the information for the identity of the speech sound to the listener. As an example, the shape of the vocal tract when pronouncing the vowel [i] is modelled. The tract is approximated with four concatenated cylinders of different lengths and diameters, as shown in Figure 5.13. In reality, the shape of the tract is, naturally, continuous. However, this approximation already gives quite a realistic result. The modelling takes into account the spectrum of the glottis excitation, the magnitude response of the model of the tract, and also the radiation from the lips. The magnitude spectrum of the resulting transfer function is shown in the figure, which matches nicely with the magnitude spectrum of [i] shown in Figure 5.7.

5.3.3 Articulatory Synthesis

Since the vocal tract can be considered an acoustic transmission line, it can be simulated computationally based on the idea of Figure 5.11 and acoustically on that of Figure 5.9. The vocal tract (and the nasal tract for nasalized sounds) can be approximated by a cascade of transmission line segments, as shown in Figure 5.13. This is called the *Kelly–Lochbaum model* (KL-model) (Kelly and Lochbaum, 1962), where the vocal tract cross-sectional profile $A(n)$ is mapped onto reflection coefficients $k(n)$:

$$k(n) = \frac{A(n+1) - A(n)}{A(n+1) + A(n)} \tag{5.4}$$

The principal advantage of articulatory synthesis is that the dynamics of the speech production systems are included in a natural way. There are, however, practical problems associated with getting data on vocal tract parameters from real speakers and with getting the complex, non-linear behaviour of the glottal oscillation and its coupling to the vocal tract acoustics. Articulatory modelling and synthesis remains an important research topic for understanding the underlying phenomena, but it may not be the most practical way to realize practical speech synthesizers.

5.3.4 Formant Synthesis

The source–filter model of Figure 5.11 can be used to approximate the generation of any speech sound. Thus, there is no necessity to model the details of the vocal tract, instead a black-box model realizing the spectral structure and related formant resonances of speech sounds is applicable. Each formant is created by a second-order filter section as a part of the overall synthesis filter structure.

The basic configuration of formant speech synthesis is the parallel filter structure shown in Figure 5.14. A finite number of second-order low-pass filters (typically 4–5) are in parallel in the structure, and it can be used to synthesize any phones of a language. A rich set of control parameters is necessary. Each formant has to be controlled by the frequency, the amplitude level, and the Q-value (sharpness of the resonance). Voiced and unvoiced excitation sources are needed, as in Figure 5.11.

Although the method is general in principle and can be used to produce any speech sound, it is not widely used in speech synthesis. The task of generating the control information for the synthesizer is demanding, and other solutions have replaced the technique in many applications. The synthesis methods which are in wide use will be discussed in Section 16.2.

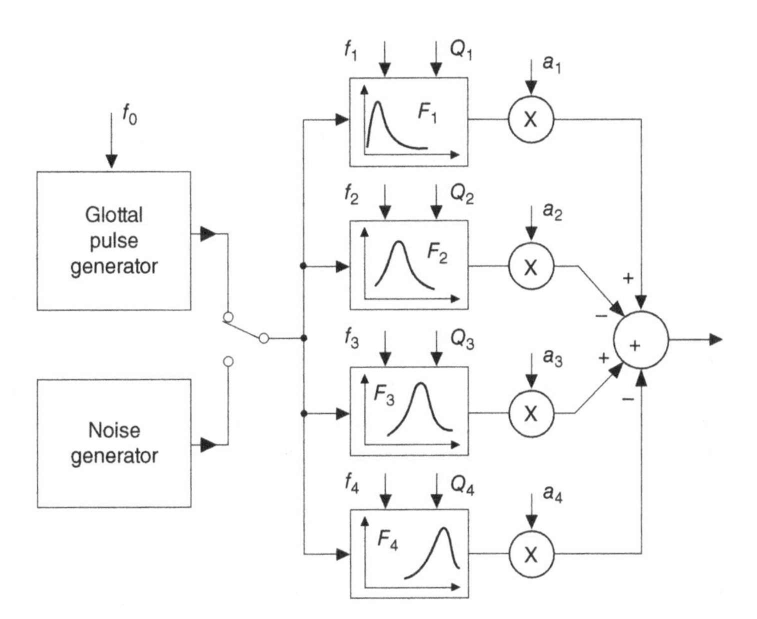

Figure 5.14 Formant synthesis with a parallel formant synthesizer. $F_1 - F_4$ are second-order resonators for individual formants, f_0 is the fundamental frequency of voiced excitation, and $f_1 - f_4$, $Q_1 - Q_4$, and $a_1 - a_4$ show the frequencies, Q-factors, and gains for each formant, respectively.

5.4 Singing Voice

There are a vast number of singing styles, which very creatively use the speech organs to generate different sounds. The styles vary from opera to reggae, from rap to lied, and from joik to chanson. There is no clear border between speech and singing, but some general remarks can be made about the differences (Rossing *et al.*, 2001).

- Singing has, in most cases, a controlled and easily definable pitch, which shifts often in rapid shifts from one frequency to another according to the melody and rhythm of the musical piece. In speech prosody, the changes in pitch can also be rapid, but the pitch typically changes smoothly all the time, while in singing the pitch is often relatively constant during each note. Quite often, each phone corresponds to a single note, although it can be divided to last for several notes.
- The spectrum of the vowels is quite often different from normal speech. The first harmonic or some other harmonics may be louder in singing than in speech. This effect is accomplished with such changes as lowering the larynx, opening the jaws more, advancing the tip of the tongue, and protruding the lips (Rossing *et al.*, 2001).
- Vibrato is used in singing much more often than in speech. When singing with vibrato, the pitch and/or the level is modulated at a rate close to 7 Hz (Sundberg, 1977). The depth and modulation frequency depend on the artist and on the musical style.
- The fundamental frequency may range more widely in singing than in speech. When the range of pitches in speech prosody is about one octave, the range may extend two octaves in singing (Rossing *et al.*, 2001).

Classical singing has perhaps been studied the most of all singing styles (Rossing *et al.*, 2001). The style has evolved from before the use of electrical reinforcement of sound, and the primary task of the singer is to produce a esthetic singing that can be heard when the accompanying symphony orchestra plays loudly. Singers are able to reinforce some of the formants of the vocal tracts into the *singer's formant*, which boosts by more than 20 dB the frequencies near 2–3 kHz (Sundberg, 1977). This makes the singing audible in such a scenario.

Another commonly known phenomenon is that trained soprano singers may produce such high pitches that the harmonics are sparse in frequency, so that the formant structure is not present in the signal. This makes high-pitched vocals unintelligible. Some female singers are also able to change the frequencies of the formants to match those of the harmonics of the note (Sundberg, 1977). This does not make the vocals more intelligible, but it raises the total level of sound, which might be desirable, for example, in opera singing.

An extreme example of the human ability to use speech organs artistically is *overtone singing* (a.k.a. throat singing), where the vocal folds produce a low-pitched sound with relatively high-level harmonics. The vocal tract is then used to produce very sharp resonances, which selectively amplify some of the harmonics (Bloothooft *et al.*, 1992). By changing the resonance frequencies, the singer creates melodies.

Summary

This chapter has been devoted to human voice as a communication signal; how it is produced by humans and technical systems. The modelling aspect is used here to characterize the general principles that are, in most cases, common to human voice production, the acoustic generation of sound, and to the electronic and computer means of generating voice in technical devices.

Further Reading

These topics are well studied in the existing research, and literature on them is rich, so in this chapter only the surface has been scratched, providing a tutorial-like presentation and a handbook-style overview. Further reading can be found easily in the following references. Recommended references for further reading on speech communication and technology are, for example, Deller *et al.* (2000); Fant (1970); Flanagan (1972); Markel and Gray (1976); O'Shaughnessy (1987); Rabiner and Schafer (1978); Santen *et al.* (1997); Stevens (1998) and Titze (1994). A more concise introduction to phonetics can be found in Ashby and Maidment (2005).

References

Alku, P. (2011) Glottal inverse filtering analysis of human voice production: a review of estimation and parameterization methods of the glottal excitation and their applications. *Sadhana*, **36**(5), 623–650.

Alku, P. and Vilkman, E. (1996) A comparison of glottal voice source quantification parameters in breathy, normal and pressed phonation of female and male speakers. *Folia phoniatrica et logopaedica*, **48**(5), 240–254.

Ashby, M. and Maidment, J. (2005) *Introducing Phonetic Science*. Cambridge University Press.

Bloothooft, G., Bringmann, E., Van Cappellen, M., Van Luipen, J.B., and Thomassen, K.P. (1992) Acoustics and perception of overtone singing. *J. Acoust. Soc. Am.*, **92**, 1827–1836.

Colton, R.H. and Conture, E.G. (1990) Problems and pitfalls of electroglottography. *J. Voice*, **4**(1), 10–24.

Deller, J.R., Proakis, J.G., and Hansen, J.H. (2000) *Discrete-Time Processing of Speech Signals*. IEEE.

Fant, G. (1970) *Acoustic Theory of Speech Production*. Mouton de Gruyter.

Fant, G., Liljencrants, J., and Lin, Q.C. (1985) A four-parameter model of glottal flow. *Speech Transmission Laboratory Quarterly Progress and Status Report, Royal Institute of Technology, Stockholm*, **4**, 1–13.

Flanagan, J.L. (1960) Analog measurements of sound radiation from the mouth. *J. Acoust. Soc. Am.*, **32**, 1613–1620.

Flanagan, J.L. (1972) *Speech Analysis, Synthesis and Perception*. Springer.

International Phonetic Association (2014) Homepage. `http://www.arts.gla.ac.uk/IPA/ipa.html`.

Kelly, J.L. and Lochbaum, C.C. (1962) Speech synthesis *Proc. Fourth Int. Congr. Acoust.*

Klatt, D.H. and Klatt, L.C. (1990) Analysis, synthesis, and perception of voice quality variations among female and male talkers. *J. Acoust. Soc. Am.*, **87**, 820–857.

Klingholz, F. (1990) Acoustic recognition of voice disorders: A comparative study of running speech versus sustained vowels. *J. Acoust. Soc. Am.*, **87**, 2218–2224.

Markel, J.D. and Gray, A.H. (1976) *Linear Prediction of Speech Signals*. Springer.

Niebergall, A., Uecker, M., Zhang, S., Voit, D., Merboldt, K.D., and Frahm, J. (2011) Real-time MRI–speaking `http://commons.wikimedia.org/wiki/File:Real-time_MRI_-_Speaking_(English).ogv`.

O'Shaughnessy, D. (1987) *Speech Communication–Human and Machine*. Addison-Wesley.

Rabiner, L.R. and Schafer, R.W. (1978) *Digital Processing of Speech Signals*. Prentice-Hall.

Rossing, T.D., Moore, F.R., and Wheeler, P.A. (2001) *The Science of Sound*, 3rd edn. Addison-Wesley.

Santen, J., Sproat, R., Olive, J., and Hirschberg, J. (eds) (1997) *Progress in Speech Synthesis*. Springer.

Schroeder, M.R. (1993) A brief history of synthetic speech. *Speech Communication*, **13**(1), 231–237.

SIL, (2014) Ethnologue – languages of the world `http://www.ethnologue.com`.

Stevens, K.N. (1998) *Acoustic Phonetics*. MIT Press.

Sundberg, J. (1977) *The Acoustics of the Singing Voice*. Scientific American.

Titze, I.R. (1994) *Principles of Voice Production*. Prentice-Hall.

Uecker, M., Zhang, S., Voit, D., Karaus, A., Merboldt, K.D., and Frahm, J. (2010) Real-time MRI at a resolution of 20 ms. *NMR in Biomedicine*, **23**(8), 986–994.

W3C, (2004) Speech synthesis markup language (SSML) `http://www.w3.org/TR/speech-synthesis/`.

WALS, (2014) Ethnologue – Languages of the World. Technical report, The World Atlas of Language Structures Online. `http://wals.info/chapter/2`.

Wells, J. (1997) SAMPA computer readable phonetic alphabet. *Handbook of Standards and Resources for Spoken Language Systems*.

6

Musical Instruments and Sound Synthesis

Music is different from speech in that its role is not so much to convey linguistic and conceptual content – although it may have this role also – as it is to evoke an aesthetic and emotional experiences. Both speech and music are sort of 'utility' sounds that are experienced mostly with a 'positive' attitude, although in some circumstances they may become annoying or disturbing: noise.

In addition to speech, *music* is the second major form of acoustic communication between humans. The psychoacoustics of intervals and melodies in music will be discussed later in this book in Section 11.6. We begin with the discussion of the formation of sounds in acoustical and electric musical instruments in this section.

Acoustic instruments generate a large set of different sounds (Fletcher and Rossing, 1998). If electroacoustic and electric instruments are also accounted for, the discriminable set of different instrument sounds is, in practice, infinite. However, most music and speech is composed of a limited set of types and structures of sounds, where certain basic properties hold. For example, woodwind or plucked string instruments form distinguishable families of musical instruments, where the principle of sound generation and the structure of the sound signal are similar within each family. We will discuss shortly some basic properties of acoustic and electric instruments and their sounds.

6.1 Acoustic Instruments

6.1.1 Types of Musical Instruments

There exists a vast variety of more or less different musical instruments. In this book we are mainly interested in the sounds they produce and radiate to the environment. To deepen our interest, a basic understanding of their working principles is required.

von Hornbostel and Sachs (1961) classified musical instruments into four different categories based on how they produce sound:

Communication Acoustics: An Introduction to Speech, Audio, and Psychoacoustics, First Edition.
Ville Pulkki and Matti Karjalainen.

- *Idiophones* are instruments in which the body of the instrument is the main vibrating unit, and the instrument does not contain such elements as the string in a guitar, the membrane in a drum, or the column of air in a trumpet. Such instruments include, for example, the xylophone, the church bell, and a rattle.
- *Membranophones* have a membrane as their main vibrating unit and this also radiates the sound. Typical examples are different drums, which are often struck directly.
- *Chordophones* have a string as their main vibrating unit. In most cases the vibration is transduced into vibrations of the body of the instrument, which serves as the main source of acoustic radiation. Typical examples are the guitar, violin, and harp. The strings may be plucked, struck, or bowed.
- *Aerophones* are the class of instruments that do not have strings, membranes, or other vibrating bodies. However, the source of excitation may vibrate, like the reed in reed instruments or the player's lips in brass instruments. The vibration occurs typically in an air column. Examples are the clarinet, flute, church organ, and trumpet.

The classification of some instruments is the cause of some debate. For instance, the banjo has strings, but also a membrane, and typically it is classified as a membranophone.

6.1.2 Resonators in Instruments

Most instruments have one or more resonating bodies which modify and radiate the sound (Fletcher and Rossing, 1998). Let us use the body of the guitar as an example of this. The resonator of the guitar is formed by the *top plate* together with the *sound hole*, the *back plate*, and the air cavity between them, as shown in Figure 6.1.

The top and bottom plates and the air cavity have different resonances which all radiate sound together. Figure 6.2 shows an example of the acoustic response of a guitar measured with a microphone at a distance of 1 m in front of the sound hole in an anechoic chamber. The excitation was an impulse-like stroke with a sharp object to the *bridge*. The plot shows the lowest frequencies (<1 kHz) of the response, and the lowest modes are clearly visible. The lowest

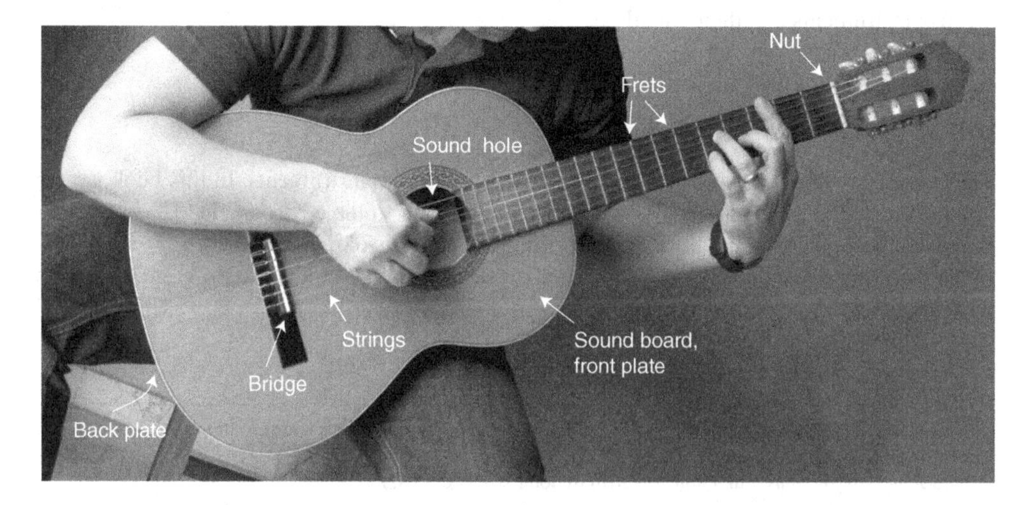

Figure 6.1 The classical acoustic guitar with the names of the most important parts labelled.

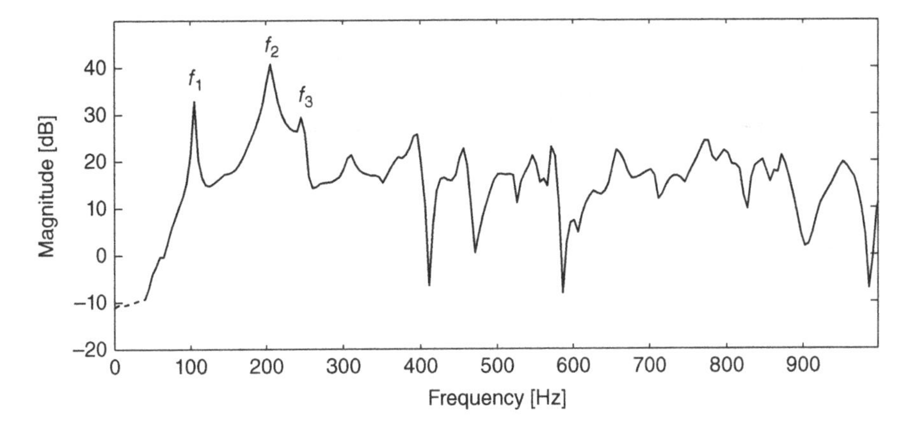

Figure 6.2 The magnitude spectrum of the guitar body at frequencies below 1 kHz. The lowest resonant frequencies are marked as f_1, f_2, and f_3.

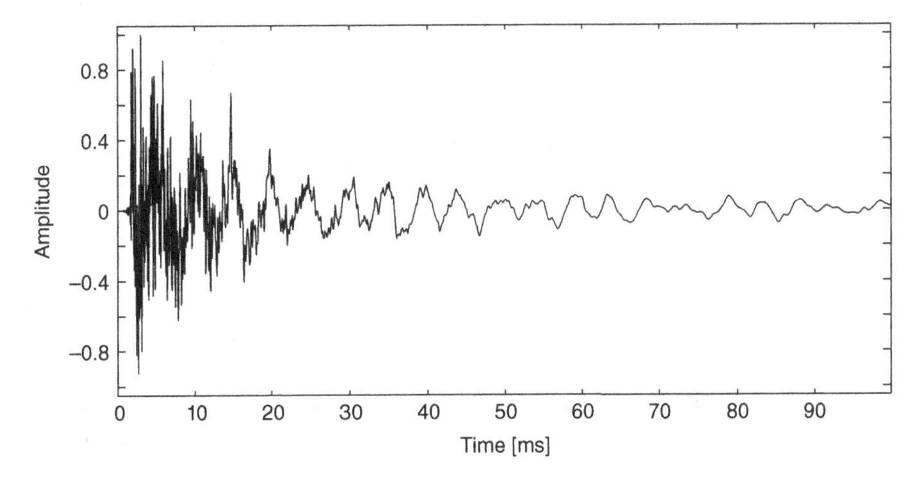

Figure 6.3 The impulse response of the guitar body.

resonance f_1 is the lowest resonance of the air cavity with a frequency of about 100 Hz. The next resonance f_2 is located just above 200 Hz, which corresponds to the in-phase movement of the front and back plates. When we get to the higher frequencies, the resonance structure gets more and more complex, due to the presence of more and more complex resonances in both plates and in the air cavity.

Each resonator colours the spectrum of the sound travelling through it, changing the spectral content of the sound. It is also important to know how the sound changes in the temporal domain. As will be shown in the following chapters, hearing can resolve temporal changes in the positions of the signal components if they change at least every few milliseconds, although with rapidly-changing signals, like a sawtooth wave, very small differences are audible. At the resonance frequencies, the build-up and decay time of the system response is usually larger than between the resonant frequencies. For example, the impulse response corresponding to the magnitude response of the body, shown in Figure 6.2, is so long that the ear perceives it as a sort of reverberation. The impulse response is shown in Figure 6.3. In principle, it is similar

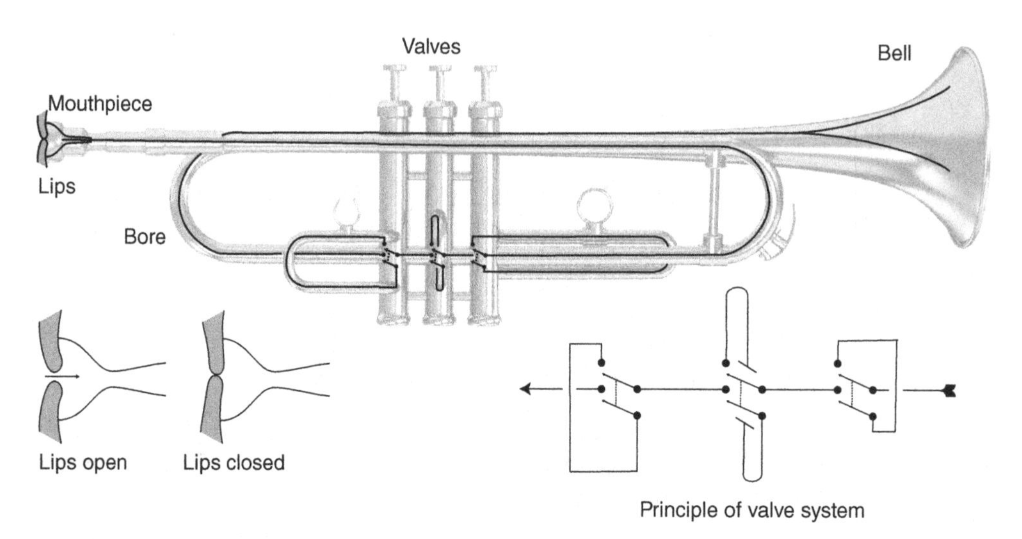

Figure 6.4 The trumpet with the names of its most important parts. The signal path through the valves is illustrated with lines. The movement of the lips of the player is shown schematically.

to the impulse response of a room, shown in Figure 2.20, but the modes are more sparsely distributed in frequency due to the smaller dimensions of the guitar body compared to those of the room.

In aerophones, the vibration occurs in an air column, which thus acts as a resonator in the system. An example of an aerophone is the trumpet, where the column is composed of the mouthpiece, the bore, and the bell, shown in Figure 6.4. The resonances of the trumpet remotely resemble the resonances of a tube closed at one end, as shown in Figure 2.12b on page 29. The acoustic effects of the mouthpiece and the bell, however, change the frequencies and amplitudes of the modes in a well-tuned manner to enable the musical use of the instrument. The interested reader is referred to Fletcher and Rossing (1998) for details on physics of the aerophones, as well as other musical instruments.

6.1.3 Sources of Excitation

In all cases, the vibrations in the instrument are caused by applying an external force to the system, such as by plucking, blowing, or striking (Fletcher and Rossing, 1998). Some examples are discussed here. When modelling musical instruments using signal processing, the excitation is thought to be the input signal to the system, and the rest of the system is assumed to act as a set of linear or non-linear filters.

Many aerophones resemble the human speech organs, where the glottal vibration is the source of excitation and the vocal and nasal tracts are the resonator. A basic difference exists though. In speech, the vibrations in the vocal folds are *not* coupled to the resonances of the tracts, in contrast to many instruments, where the vibrations in the excitation are strongly coupled and synchronized to the resonances in the air column. For example, in *brass instruments*, such as the trumpet, the lips of the player open and close in synchrony with one of the resonances of the bore (see the simplified illustration in Figure 6.4). In some *woodwind instruments*, such as the clarinet and the oboe, a reed or a pair of reeds function in a similar manner. In flutes, the player blows air sideways towards an opening, and the resonances of the air column cause the blowing stream to vibrate between the inside-hole direction and just-off-the-opening directions. These phenomena are non-linear from the viewpoint of signal processing, since the excitation is changed by the state of the system.

In plucked string instruments, the excitation is impulse-like. Ideally, the plucking imposes a pulse of acceleration on the position of plucking from where the wave starts to propagate in both directions of the string. Figure 6.1 shows the playing of the classical acoustic guitar, where the *strings* are plucked with fingers. When plucking with fingers, the temporal excitation of the string is smoother than an ideal impulse, causing a low-pass filtering effect in the sound. After the plucking, the string vibrates autonomously, the amplitude decaying with time. In this case, the modelling is relatively easy, because the vibrations of the string occurring after the excitation cannot affect the excitation itself.

In the piano, the strings are struck with a hammer covered with felt. The hammer gets its speed from pressing a key on the keyboard. The hammer hits the strings usually two to three times, depending on the force applied to the key. A similar effect is seen when the membrane of a drum is struck with a stick. This method of excitation is also non-linear, since the movements of the string or the membrane affect the excitation during the second and third hits.

In the violin, the bow and the string have a strong, non-linear interaction. When moving in one direction, the bow sticks to the string and moves it in the same direction until the tension of the string forces the string to retract and to stick to another position on the bow. The frequency of this repeated stick-and-slip effect depends on the resonances of the string. The movement of the bowing position of the string thus resembles the shape of the sawtooth wave (Equation (3.9) on page 46). From the point of view of modelling, this reminds us of the woodwind and brass instruments, where the excitation is continuous and non-linearly coupled and synchronized to the resonator.

6.1.4 Controlling the Frequency of Vibration

Many instruments are built such that a melody can be presented with them; that is to say, sounds with different frequencies can be produced. This section presents a very short overview of the basic methods to control the frequency, or the pitch, of the sound. The reader interested in more details is referred to Fletcher and Rossing (1998).

Chordophones typically include a number of strings tuned to different pitches. For example, the guitar has six strings, as shown in Figure 6.1. In some instruments, one string is used to produce only one pitch, and no variations are possible, as in the piano. In some other string instruments, the length of the vibrating part of the string can be changed by some means. In the guitar, the string is pressed with a finger against a *fret*, as shown in the figure, and the distance between the fret and the *bridge* defines the pitch of the sound. Yet another method to vary the pitch is the dynamic changing of the tension of the string.

In *aerophones*, the excitation and/or the acoustic properties of the air column define the pitch. In some cases, the effective length of the column is defined simply by the length of the bore and only one pitch is produced, such as in church organ pipes. In woodwinds, the effective length of the column is changed by opening and closing holes in the bore, which changes the pitch. In brass instruments, the vibration of the lips of the player is synchronized to one of the resonances of the instrument, which allows, in principle, the production of the harmonic series of frequencies above the lowest resonance. Additionally, in many of the brass instruments, the length of the column can be changed using valve mechanisms or sliding bores.

The trumpet utilizes valves, as shown in Figure 6.4. When the eight combinations of valve positions are combined with lip synchronization to different modes of the bore, a wide variety of pitches can be played with the instrument. Added to the control of the length of the air column, trained brass or woodwind players can also control the pitch slightly with the tension of their lips.

The pitch produced, if any, by *idiophones* is, in most cases, not controllable. In *membranophones* the pitch is defined by the properties of the membrane and the enclosure. In most cases it cannot be controlled, although exceptions exist.

6.1.5 Combining the Excitation and Resonant Structures

The final sound of the instrument is the sum of the excitation, the effect of the resonators and radiation properties. Also, the acoustic characteristics of the room where the instrument is played have a prominent effect on the sound.

For instruments that can be modelled as linear systems, the system can be expressed in the frequency domain to consist of the excitation $X(\mathrm{j}\omega)$ and the combined transfer function of the partial systems with transfer functions $H_i(\mathrm{j}\omega)$, which together produce the output $Y(\mathrm{j}\omega)$, as

$$Y(\mathrm{j}\omega) = X(\mathrm{j}\omega) \prod_i H_i(\mathrm{j}\omega), \tag{6.1}$$

if the partial systems are in *cascade*, that is the signal flows through them successively. In this manner, the guitar sound is formed by the excitation due to the plucking of the string and by the effects of the partial systems, which are the resonator formed by the string, the resonator formed by the body, and the radiation transfer function.

Unfortunately, many instruments are non-linear and such modelling is not possible. For example, in the trumpet, the air pressure affects the vibration of the lips in a complex manner that cannot be modelled by simple convolution. Also, the modelling of the dynamic control of the pitch brings an additional non-linear component to the system. This means that it is easier to model separated sounds with a constant pitch than an instrument whose physical state changes all the time, or, in technical terms, the frequencies of the resonances of vibrating structures change due to the playing of the instrument. Nevertheless, these effects can be numerically modelled in the time domain, which, however, may result in complicated models (Bilbao, 2009; Smith, 2010; Välimäki *et al.*, 2006).

6.2 Sound Synthesis in Music

The synthesis of musical sounds has been of interest for decades, and many musical instruments are based on such methods (Moore, 1990; Roads, 1996). The first systems utilized simple signal synthesis which did not have a counterpart in the world of acoustic instruments. One of the most influential early electronic synthesizers was manufactured by Robert Moog in the mid-1960s, and it included switchable oscillators, filters, and other modules for processing. Digital sampling systems became available in the late 1970s, and synthesizers with computational models of the physics of the instruments have been available commercially since the mid-1990s. The history of musical sound synthesis is thus different from the history of modelling of speech production, where the first models were based on the physics of speech organs, after which sampling methods were adopted.

6.2.1 Envelope of Sounds

The attributes of natural sounds typically change with time. Only periodic or noise-like sounds may sound the same for a longer time. Our hearing is very sensitive to temporal changes in sound, and a simple method to present the temporal variations in sound is to plot the instantaneous amplitude against time, which is called the amplitude envelope of sound. Another, slightly more complex method is to plot the spectrogram of the signal.

The envelopes of the sounds of musical instruments can often be divided into different phases in time. The *attack* represents the phase when the amplitude of sound increases. After this, the behaviour of the sound depends on the type of the source. In string instruments, the amplitude first decreases rapidly, after which a shallower decrease is reached. In electric instruments, the attack is often followed by *decay*, a decrease in amplitude; *sustain*, a constant amplitude; and *release*, a decrease of the amplitude until silence. This sequence is sometimes called the *ADSR-sequence*.

The amplitude envelope is not sufficient to represent the behaviour of different spectral parts of the sound produced by many musical instruments. The spectrogram (see Section 3.2.6) can be used to visualize instrument sounds. Figure 6.5 shows the magnitude envelopes of the harmonics of a kantele, a Finnish plucked string instrument. The plot shows the typical characteristics of the kantele sound; it has strong decay at higher harmonic partials, and it also has strong amplitude modulation at the harmonics, which appears with different modulation frequencies for different harmonics.

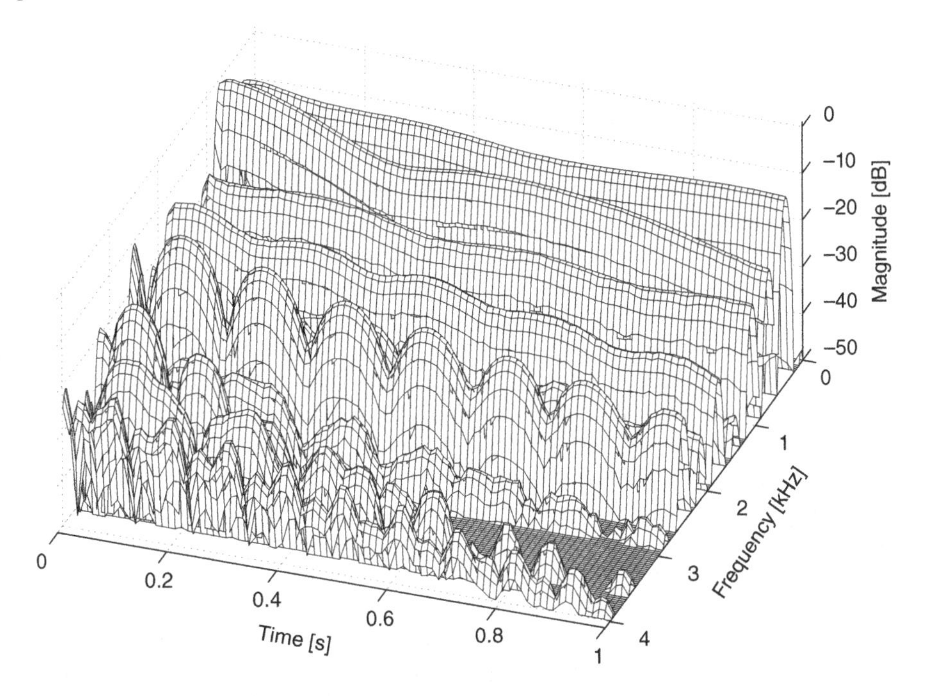

Figure 6.5 The amplitude envelopes of the harmonics of a single plucked string of the kantele.

Musical instruments with continuous excitation typically show slower attack times and steady sustain levels. For example, the sound of the flute starts from noise caused by blowing, and the actual vibration due to the resonance of the air column grows slower.

6.2.2 Synthesis Methods

The main methods used to synthesize musical sounds can be classified into the following categories (Rossing *et al.*, 2001):

- *Sampling*. In sampling, the sound of each note of an instrument is recorded and stored in the memory of a computer (see Section 3.3.1 on page 56) (Roads, 1996). After this, an interface, such as a musical keyboard, is used to control the playback of the notes. The system requires a considerable amount of memory, although the quality of sound can be very high. The memory requirement can be eased by repeating, or looping, certain parts of the sample using pitch shifting to reduce the number of samples in the memory and data compression algorithms (Borin *et al.*, 1997). However, modification of sampled sounds requires more sophisticated methods than the modification of sounds generated with other synthesis methods.
- *Additive synthesis*. In additive synthesis, each harmonic of a tone complex is synthesized separately and then added together (Grey and Moorer, 1977). In principle, the method can generate any sound desired. It is computationally relatively demanding, as musical sounds typically have a large number of harmonics. The system also allows the control of the level and phase of each harmonic arbitrarily, which makes possible the synthesis of any sound. The control data have to be obtained from some other process, for example using a set of heuristic rules, or they can also be analysed from sounds of musical instruments. In Serra and Smith (1990), the method was elaborated to *spectral modelling synthesis*, where a sound is partly represented by a set of sinusoids, and the remaining sound is modelled with white noise processed by a time-varying filter.
- *Subtractive synthesis* is computationally less demanding than additive synthesis. A spectrally rich sound is first synthesized, and filters are used to modify the spectrum to the desired form. The initial sound is typically a sound that can be easily synthesized with electronic circuits or digital computers, such as a sawtooth signal, a square or a triangular wave, impulses or pulses, or noise. The filtering can also be time-dependent. For example, in early synthesizers, the brass sounds were generated by synthesizing a sawtooth wave passed through a low-pass filter whose cutoff frequency and gain depended on the frequency of the sawtooth wave (Risset and Mathews, 1969). This system thus resembles the source–filter model, as discussed in speech synthesis. The early electronic synthesizers implementing subtractive synthesis produced a distinctive sound due to the many non-linear analogue effects, which can also be emulated with digital computers (Välimäki and Huovilainen, 2006).
- *Non-linear synthesis*. In non-linear synthesis, the system has an input–output relationship that produces new frequency components which are meaningful in musical sound synthesis.

  Perhaps the best-known example of non-linear synthesis is *FM synthesis* (Chowning, 1977), where a parameter of the carrier oscillator is modulated with another oscillator(s). Its basic form uses simple frequency modulation, such as $y(t) = \sin[\omega_c t + \sin(\omega_m)t]$, to form a spectrum with peaks at frequencies $\omega_c \pm n\omega_m$. If $\omega_c = \omega_m$, a harmonic spectrum is produced with the fundamental frequency $f_0 = 2\pi\omega_m$, which can be used in music sound synthesis. Different variations of FM synthesis have also been adopted, producing different sound colours.

- *Computational models of the physics of musical instruments*. In such systems, different parts of the physical instrument are modelled in the signal domain with functionally similar DSP blocks (Bilbao, 2009; Jaffe and Smith, 1983; Smith, 2010; Välimäki *et al.*, 2006). This approach is interesting in the context of this book, and an example of modelling plucked string instruments is given in the following section.
- *Computational models of the physics of natural sound sources*. 'Natural sources' refers here to non-musical and non-speech sources, such as sounds of walking in different terrains, the sound of a falling tree, thunder, and so on. Such synthesis methods have applications in computer games and other virtual environments. Different statistical and other methods are used to synthesize such sounds (Cook, 2007; Farnell, 2010).

6.2.3 Synthesis of Plucked String Instruments with a One-Dimensional Physical Model

The model of a string instrument is presented here as an example of a computational model of the physics of a musical instrument. The string in string instruments and the air column in brass and woodwind instruments are one-dimensional resonators, as already reviewed in Chapter 2. Figure 6.6 shows a computational model of such a vibrating string, called the *digital waveguide* (Smith, 2010), where each travelling direction is implemented as a digital delay line. When the transversal vibration arrives at the rigid end point of the line, in the ideal condition, it is fully reflected and the transversal polarity of vibration is changed. In real instruments, some damping of the vibration occurs both in the string and at the point of termination. In Figure 6.6, all of the damping is modelled as occurring at the ends, and the string itself is thought to be lossless. This allows the implementation of a computationally very effective method of synthesis. The excitation is brought to both lines, and the response of the string is captured from another position on the lines. The position may, for example, correspond to the position of the pick-up of an electric guitar or the bridge of an acoustic guitar.

Figure 6.7 shows the plot of a sound generated by the model of a string shown in Figure 6.6. The result is simply a decaying train of impulses. The parameters of the loop filter naturally affect the result; if the filter is set to be a low-pass filter, the decay of high frequencies is much faster than the low frequencies. With certain parameter values the model of the string sounds like a guitar not an acoustic guitar, but an electric guitar. This happens because the model does not include the model of the body of the guitar. The model of the body can be implemented simply by convolving the output of the string model with a measured impulse response of the guitar body shown in Figure 6.3. When the parameters of the model are tuned well, the resulting sounds are very realistic (Laurson *et al.*, 2001).

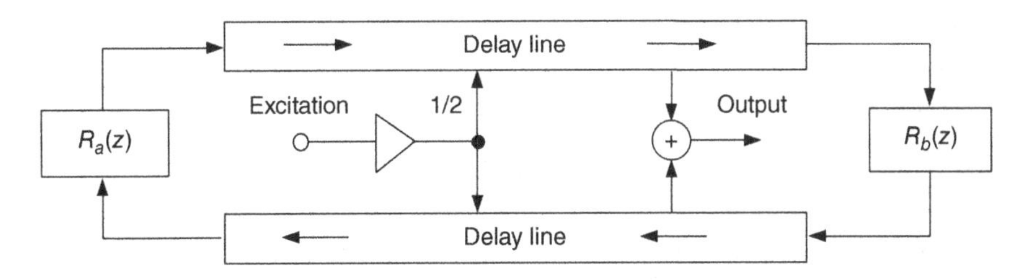

Figure 6.6 The digital waveguide, used here as a computational model of a string in the guitar.

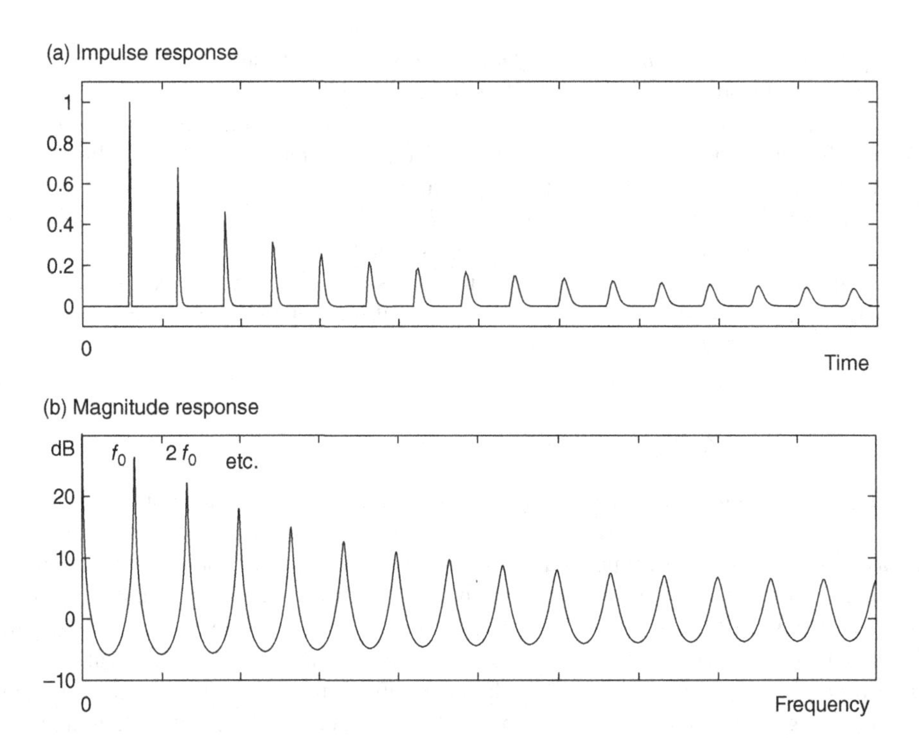

Figure 6.7 The output of the computational model of the guitar string implemented as the model shown in Figure 6.6.

Summary

This chapter has looked at music as a communication signal; how it is produced by humans, acoustic instruments, and technical systems. The modelling aspect has been taken to characterize the general principles that are, in most cases, common to human voice production, acoustic generation of sound, and to electronic and computer means in technical devices.

Further Reading

Musical acoustics has been studied widely, and interested readers should consider books such as (Bilbao, 2009; Fletcher and Rossing, 1998; Moore, 1990; Roads, 1996; Roederer, 1975; Rossing *et al.*, 2001; Smith, 2010).

References

Bilbao, S. (2009) *Numerical Sound Synthesis: Finite Difference Schemes and Simulation in Musical Acoustics*. John Wiley & Sons.

Borin, G., De Poli, G., and Sarti, A. (1997) Musical signal synthesis. In Roads, C., Pope, S.T., Piccialli, A., and De Poli, G. (eds) *Musical Signal Processing*. Routledge, pp. 5–30.

Chowning, J.M. (1977) The synthesis of complex audio spectra by means of frequency modulation. *Computer Mus. J.*, **1**(2), pp. 46–54.

Cook, P.R. (2007) *Real Sound Synthesis for Interactive Applications*. AK Peters.

Farnell, A. (2010) *Designing Sound*. MIT Press.

Fletcher, N.H. and Rossing, T.D. (1998) *The Physics of Musical Instruments*. Springer.

Grey, J.M. and Moorer, J.A. (1977) Perceptual evaluations of synthesized musical instrument tones. *J. Acoust. Soc. Am.*, **62**, 454–462.

Jaffe, D.A. and Smith, J.O. (1983) Extensions of the Karplus–Strong plucked-string algorithm. *Computer Mus. J.*, **7**(2), 56–69.

Laurson, M., Erkut, C., Välimäki, V., and Kuuskankare, M. (2001) Methods for modeling realistic playing in acoustic guitar synthesis. *Computer Mus. J.*, **25**(3), 38–49.

Moore, F.R. (1990) *Elements of Computer Music*. Prentice-Hall.

Risset, J.C. and Mathews, M.V. (1969) Analysis of musical-instrument tones. *Physics Today*, **22**, 23–30.

Roads, C. (1996) *The Computer Music Tutorial*. MIT Press.

Roederer, J.G. (1975) *The Physics and Psychophysics of Music: An Introduction*. Springer.

Rossing, T.D., Moore, F.R., and Wheeler, P.A. (2001) *The Science of Sound*, 3rd edn. Addison-Wesley.

Serra, X. and Smith, J. (1990) Spectral modeling synthesis: A sound analysis/synthesis system based on a deterministic plus stochastic decomposition. *Computer Mus. J.*, **14**(4), 12–24.

Smith, J.O. (2010) *Physical Audio Signal Processing: For Virtual Musical Instruments and Digital Audio Effects*. W3K Publishing.

Välimäki, V. and Huovilainen, A. (2006) Oscillator and filter algorithms for virtual analog synthesis. *Computer Mus. J.*, **30**(2), 19–31.

Välimäki, V., Pakarinen, J., Erkut, C., and Karjalainen, M. (2006) Discrete-time modelling of musical instruments. *Reports on Progress in Physics*, **69**(1), 1–78.

von Hornbostel, E.M. and Sachs, C. (1961) Classification of musical instruments: Translated from the original German by Anthony Baines and Klaus Wachsmann. *The Galpin Society Journal*, **14**, 3–29.

7

Physiology and Anatomy of Hearing

The purpose of hearing is to capture acoustic vibrations arriving at the ear and analyse the content of the signal to deliver information about the acoustic surroundings to the higher levels in the brain. The auditory system is the sensory system for the sense of hearing. According to current knowledge, the auditory system divides broadband ear-canal signals into frequency bands, and then conducts a sophisticated analysis on the bands in parallel and in sequence. The human auditory system largely resembles the hearing system in other mammals, but humans have one property that has developed much farther: the ability to analyse and recognize spoken language. As a consequence, the sensitivity to and resolution of some speech- and voice-related features of sound are very good. Thus, a major part of this book is devoted to discussing the different roles of hearing and the auditory perception in typical human communication.

The functional properties, or the *physiology of hearing*, are interesting in the context of understanding communication and engineering applications, especially from a basic research point of view. This topic includes the acoustic-to-mechanic and then to neural conversion occurring in the auditory periphery and the neural functions of the auditory pathway. From a communication point of view, however, a very detailed understanding of the physiology of hearing is not necessary. The *anatomical* structure of hearing is also somewhat interesting, though not very important as such, except in some special cases, such as audiology or spatial hearing. Thus, a brief introduction to both the anatomy and physiology of the auditory system is considered sufficient in this chapter.

7.1 Global Structure of the Ear

Humans, as well as most animals, have two sensors for sound – *the left and right ear* – and a complex neural system to analyse the sound signals received by them. The ear, more specifically the *peripheral auditory system*, consists of the *external ear* for capturing sound waves travelling in the air, the *middle ear* for mechanical conduction of the vibrations, and the *inner ear* for mechanical-to-neural transduction. Neural signals from the periphery are transmitted through the *auditory pathway*, where the neural signals are processed by different nuclei up to the auditory cortex where high-level analysis occurs.

Communication Acoustics: An Introduction to Speech, Audio, and Psychoacoustics, First Edition.
Ville Pulkki and Matti Karjalainen.

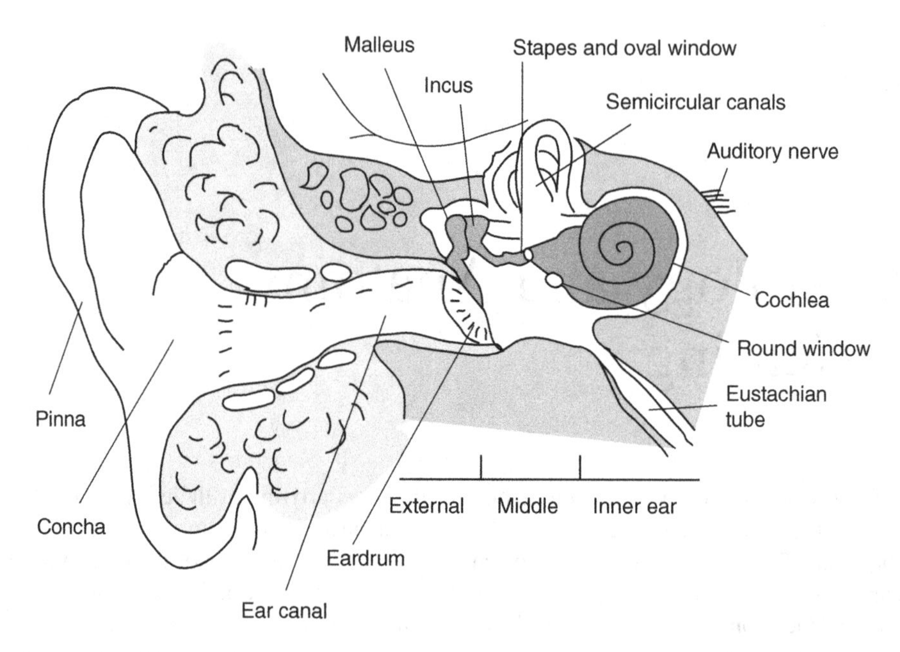

Figure 7.1 Cross-sectional diagram of one ear, showing the external, middle, and inner ear.

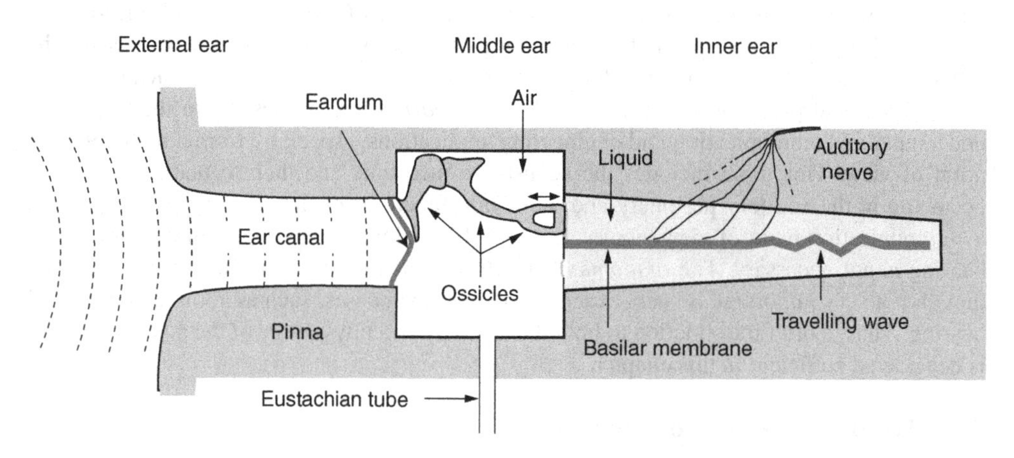

Figure 7.2 A simplified diagram of the ear.

Figure 7.1 depicts a cross-section of one ear, including the external (outer), the middle ear, and part of the inner ear. Figure 7.2 shows a more schematic diagram characterizing the most essential parts of the system and the path from acoustic wave to neural signal.

7.2 External Ear

The external ear (outer ear) consists of the *pinna* with the *concha*, the *ear canal* or *meatus*, and the *eardrum* or *tympanic membrane* as a borderline with the middle ear (Figure 7.1). The external ear is passive and linear, and its functioning is entirely based on the laws of acoustic

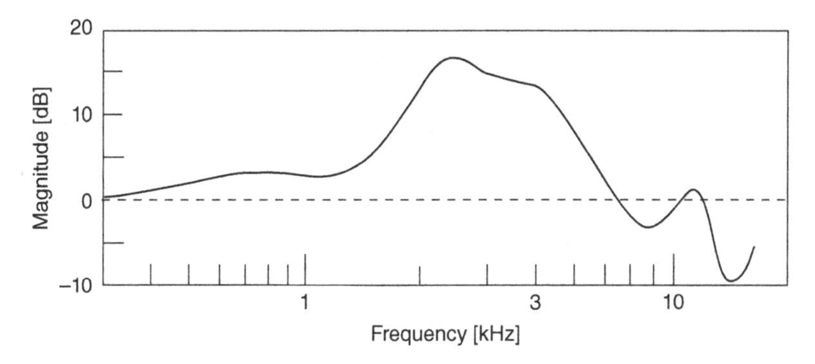

Figure 7.3 An approximate magnitude response of the outer ear at the eardrum to a frontal sound source in free-field conditions.

wave propagation. Passivity means that the external ear does not generate sound energy or 'react' to the sound, it only carries sound waves properly to the eardrum and the middle ear.

Acoustically, the entire head (and shoulders) makes part of the external ear. The distance between the two ears, located at opposite sides of the head, causes an arrival time difference of the sound wavefront that depends on the angle of incidence. A difference in sound level appears at high frequencies due to shadowing by the head to the side that is opposite to the sound source. These phenomena and related effects in spatial hearing are discussed in Chapter 12. In the present chapter, we concentrate on *monaural* phenomena; that is, hearing where interaural differences do not play a role.

The role of the *pinna* is to influence sound propagation to the ear at high frequencies by creating asymmetry, both front–back and top–down, that helps directional hearing. The cavity leading to the ear canal is called the *concha*. Together, the head and the external ear, up to the eardrum, emphasize frequencies of 1–5 kHz considerably, as depicted in Figure 7.3. This is measured for a sound source in front of a listener in free-field conditions. The dependency on the incidence angle is discussed in Chapter 12 based on the concept of the *head-related transfer function* (HRTF).

The *ear canal* (*external auditory meatus*) is a relatively hard-walled tube that is approximately 22.5 mm long and has a diameter of about 7.5 mm, so that its volume is about 1 cm^3. Acoustically, it is a short transmission line where sound waves propagate from the external environment to the eardrum. Because one end is open (low acoustic impedance) and the impedance of the eardrum is higher than for the tube itself, the ear canal acts as a quarter-wavelength resonator, emphasizing signals around frequencies of 3–4 kHz by about 10 dB, attenuating them at around 7–8 kHz, and showing the next (weak) resonance above 10 kHz.

The *eardrum* or the *tympanic membrane* is a membrane that converts sound waves arriving in the air through the ear canal into mechanical vibrations and passes these vibrations to the middle ear.

7.3 Middle Ear

The middle ear, which is located in a small, air-filled cavity between the eardrum and the inner ear, as shown in Figure 7.1, is a transmission system of mechanical vibration from the eardrum through little bones called *ossicles* to the *oval window* that leads to the cochlea of the inner ear. When the eardrum vibrates due to sound entering through the ear canal, the eardrum and

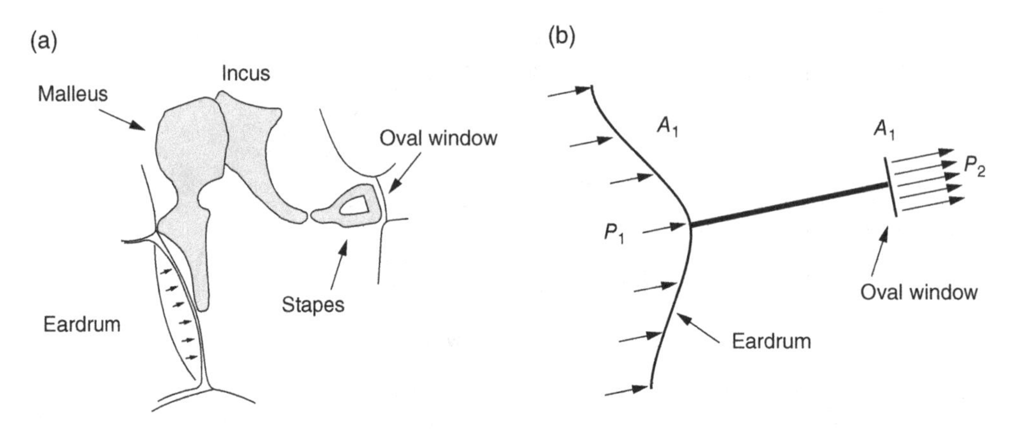

Figure 7.4 The ossicles in the middle ear: (a) the real structure, and (b) impedance matching due to the area ratio.

the ossicles vibrate along and transmit the vibration to the liquid medium in the inner ear. The ossicles consist of three very small bones called the *hammer* or *malleus*, the *anvil* or *incus*, and the *stirrup* or *stapes* (see Figure 7.4a).

The relatively complex construction of the middle ear is a result of evolution to provide sufficiently good impedance matching between different physical media. Since the characteristic impedances of air and water (in the cochlea) are very different (a ratio of about 1:3000), the vast majority of sound energy entering the ear canal would reflect back and only a minimal fraction could proceed to the inner ear (see Equations (2.25)–(2.27) and (2.29) for reflection, transmission, and absorption coefficients) if there were no special mechanism between the air in the ear canal and the liquid in the cochlea. A minor improvement is achieved by the chain of ossicles that works as a lever mechanism. Another, far more efficient effect is obtained by the area ratio of the eardrum and the oval window, as illustrated in Figure 7.4b, that works as an impedance transformer. This mechanism transforms a small pressure with large velocity in the air into a large pressure with small velocity in the liquid of the inner ear. Good efficiency of sound energy transfer is important in order to obtain the extremely high sensitivity that the auditory system has. The motion displacement of the eardrum at middle frequencies (1–4 kHz) at the threshold of hearing is only about 10^{-9} cm, which means just about a tenth of the diameter of a hydrogen atom!

The impedance-matching mechanism improves the pressure transfer by a factor of about 30; that is, by about 30 dB. The transfer function of the middle ear is a band-pass filter where mid-frequencies are emphasized, and the response rolls off at higher and lower frequencies, as characterized in Figure 7.5.

There is a narrow channel called the *Eustachian tube* connecting the middle ear to the oral cavity. The role of this tube is to balance the air pressure between the middle ear and the environment, for example during changes in altitude, such as in aeroplanes during takeoff or landing. Too high or too low a pressure in the middle ear displaces the eardrum and reduces the sensitivity of hearing, or may even cause pain.

Another detail of the middle ear that is related to the auditory function is the *acoustic reflex*. Loud sounds, approximately above 50–60 dB, trigger small muscles to contract so that conduction of sound by the ossicles is reduced (Geisler, 1998). This is understood as a mechanism

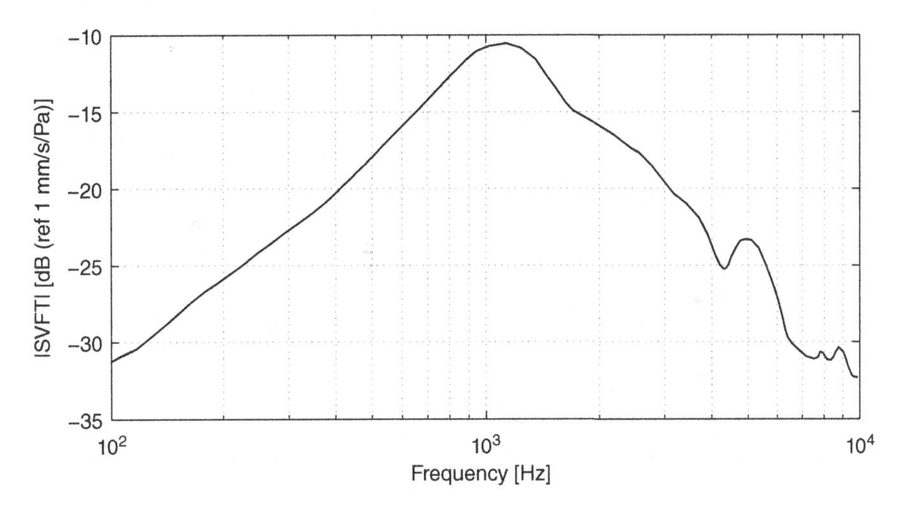

Figure 7.5 The middle ear pressure transfer function from the pressure at the eardrum to the velocity of the stapes window. Adapted from Aibara *et al.* (2001).

against overly loud sounds in order to protect the delicate mechanisms of the inner ear. This protection is, however, not very efficient, since it is slow (with a latency of tens or hundreds of milliseconds) against impulsive sounds and it attenuates prominently only low frequencies, 10–20 dB below 500 Hz and less above 2 kHz (Møller, 2006). However, the system may be effective against some impulsive sounds that the subjects generate themselves. It has been found that the reflex is activated subconsciously if a considerable vibration or sound is imminent, say when running on a hard surface, shouting, or using firearms. This reflex, when on, changes the acoustic impedance of the eardrum, and it can therefore be registered indirectly from the outer ear (Geisler, 1998).

7.4 Inner Ear

The inner ear consists of the cochlea and the semicircular canals. While the latter is a balance sensing organ and has no effect on monaural hearing, the cochlea is a marvellously complex system for transforming mechanically conducted sound from the middle ear first into the vibrations of the basilar membrane and then into neural impulses for higher-level analysis.

7.4.1 Structure of the Cochlea

The *cochlea* is a spiral-shaped and liquid-filled tube of about 2.7 turns with a length of 35 mm (see Figure 7.1) (Plack, 2013). The cochlea is in vibroacoustic connection with the middle ear via two ports called the *oval window* and the *round window*. The oval window connects the vibrations from the stapes to the liquid medium inside the cochlea. Mechanically, the most important part inside the cochlea is the *basilar membrane*, which is located between bony shelves along the length of the cochlea, as illustrated in the linearized model of Figure 7.6. There is an opening called the *helicotrema* at the apical end of the basilar membrane, connecting the liquid media on both sides of the membrane. The width of the membrane reaches about 0.5 mm at its widest point, which is at the apical end (Plack, 2013).

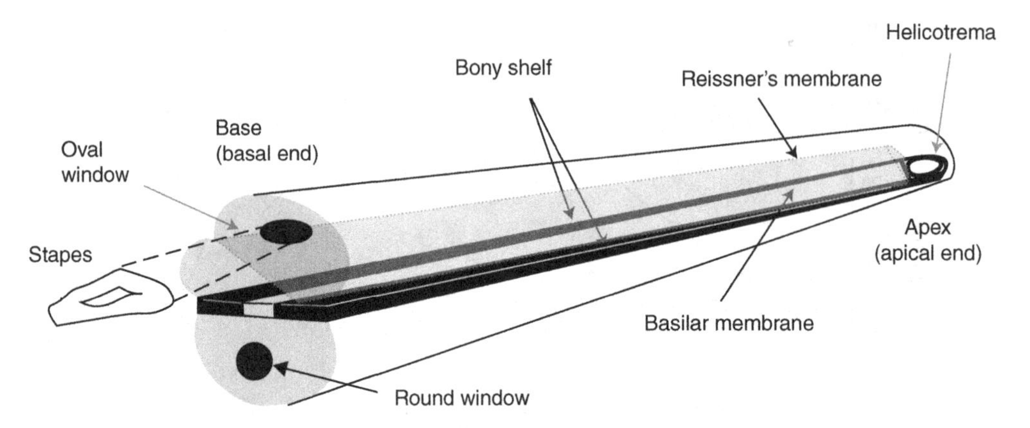

Figure 7.6 The structure of the cochlea depicted as a linear tube instead of its true, spiral form.

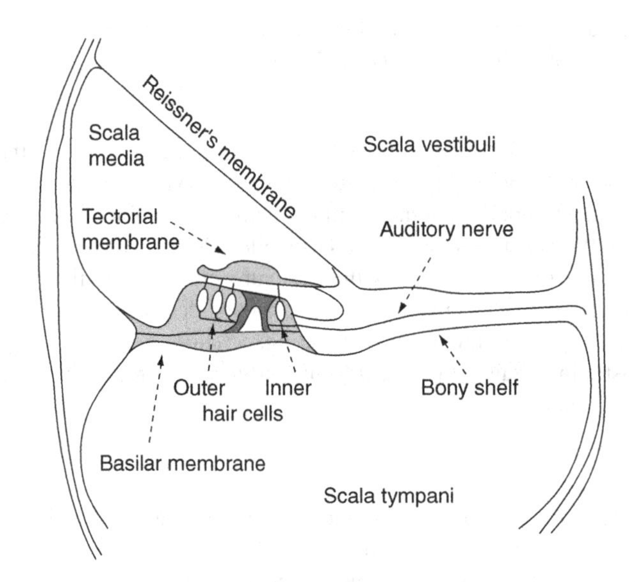

Figure 7.7 Cross-sectional structure of the cochlea.

A cross-section of the cochlea is depicted in Figure 7.7, showing other structural elements inside the tube. The basilar membrane and the bony shelf divide the tube into two halves. The liquid section above the basilar membrane is the *scala vestibuli* and that below the membrane the *scala tympani*. A thin membrane, *Reissner's membrane*, separates one more liquid section called the *scala media*. The scala media has a structure in its bony wall that leaks potassium ions (K^+) into it. Thus, since the ion concentrations of liquids in the scala media and the scala tympani are different, a small electric potential difference exists across the basilar membrane.

A vibration-sensitive structure called the *organ of Corti* resides on the basilar membrane (shaded in Figure 7.7). It has receptors called *hair cells*, which are specialized cells involved in the process of converting the vibration pattern of the basilar membrane into neural impulses in the *auditory nerve* fibres. Hair cells are organized in several rows: one row of *inner hair*

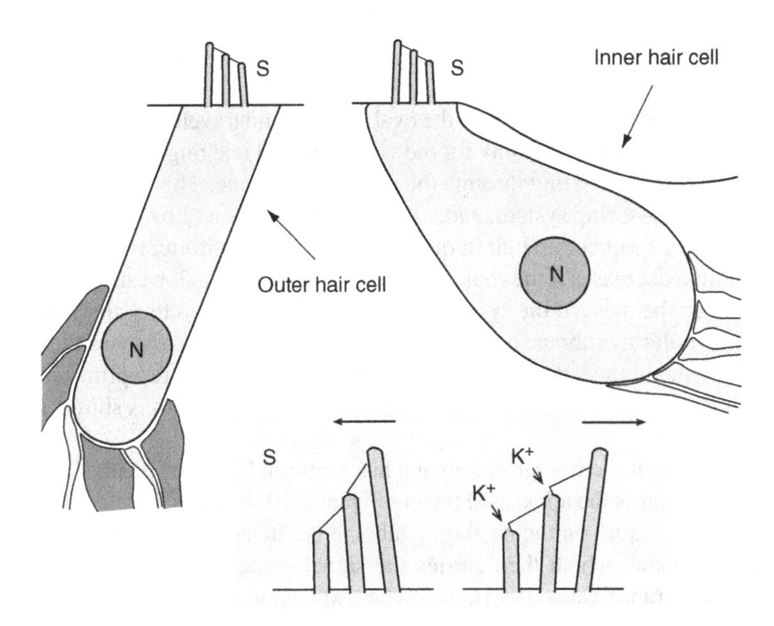

Figure 7.8 Top: A simplified schematic structure of the inner and outer hair cells. N = nucleus and S = stereocilia. Nerve endings at the bottom of a cell form the ascending path if shaded light grey and the descending path if dark grey. Below: Bending the stereocilia right opens the channels for K^+ ions on the tops of the hair cells.

cells and three to five rows of *outer hair cells*. There are about 3500 inner and 12 000 outer hair cells uniformly distributed along the organ of Corti. Each hair cell is equipped with a bunch of hair-like filaments on top of it, called *stereocilia*, structured in a U, V, or W shape. Figure 7.8 characterizes the structure of an inner and an outer hair cell. The lengths of the outer hair cells are about 12 μm at the basal end and about 90 μm at the apical end. Their positioning along the organ of Corti also varies systematically.

There exists one more membrane, the *tectorial membrane*, a gelatinous structure right above the hair cells and stereocilia. When the basilar membrane vibrates, the stereocilia are set into bending motion, and the hair cells are extremely sensitive in reacting to these movements. The inner hair cells are the primary receptors, while the outer hair cells have an active function to control the mechanical vibration state of the system, as discussed below.

7.4.2 Passive Cochlear Processing

The main role of cochlear processing is to act as a frequency-to-place mapping of signal components. Let us first assume that the basilar membrane is passive and time-invariant. The membrane is structured so that it is narrower, less massive, and stiffer at the basal end (oval window side) and gradually changes its characteristics to wider, more massive, and loosely moving at the opposite (apical) end. Notice that the widening of the basilar membrane towards its apical end is in contrast to the narrowing of the cochlear tube (see Figure 7.6). An inherent property of such a system is that each point along the line has a frequency of highest vibratory amplitude; that is, each point on the basilar membrane resonates at a specific frequency, which is called the *best frequency*, or *characteristic frequency*. This phenomenon was found and studied by von Békésy (1960), earning him the Nobel Prize in Physiology or Medicine in 1961 for his work.

The vibration enters the liquid through the oval window and travels at a very high speed in the duct. The round window acts as a sink for the vibration, and one might think that the vibration finds a route there by penetrating through the basilar membrane. This vibrating liquid column can be seen as a mass–spring system, and, as shown by Equation (2.6), the lower the mass, the higher the resonance frequency. High frequencies thus favour a shorter route, since the mass of the vibrating liquid decreases if the route is shorter. In contrast, the lowest frequencies favour a longer route, since the mass of the system is higher there. Added to this, the vibration favours penetrating the basilar membrane at a position where its frequency matches the characteristic frequency of the membrane. Thus, many different characteristics are optimized to make the frequency-to-place conversion of acoustic vibrations. This conversion is shown schematically in Figure 7.9.

The vibration seen in the basilar membrane has a certain behaviour with space and time, it seems to 'travel' towards the apex, as shown in Figure 7.10. It also has a clear resonance peak. The vibration of the liquid on the basilar membrane is often considered to cause a *travelling wave* to the membrane, which then carries the signal content with it and excites the resonances on the membrane (Yates, 1995). In another explanation, it is assumed that the vibratory

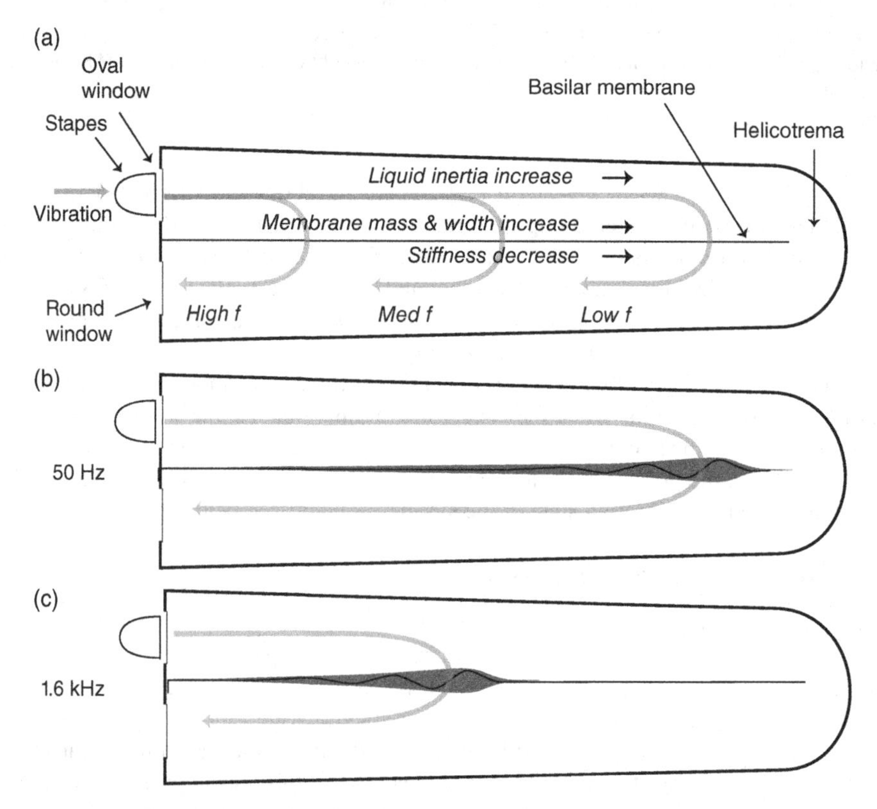

Figure 7.9 (a) A schematic illustration of the acoustic properties of the cochlear duct that change with distance from the stapes. The position at which a component of vibration entering the cochlea through the stapes penetrates the basilar membrane depends on its frequency. (b) and (c) The amplitude envelopes of the travelling waves on the basilar membrane caused by sinusoidal stimuli.

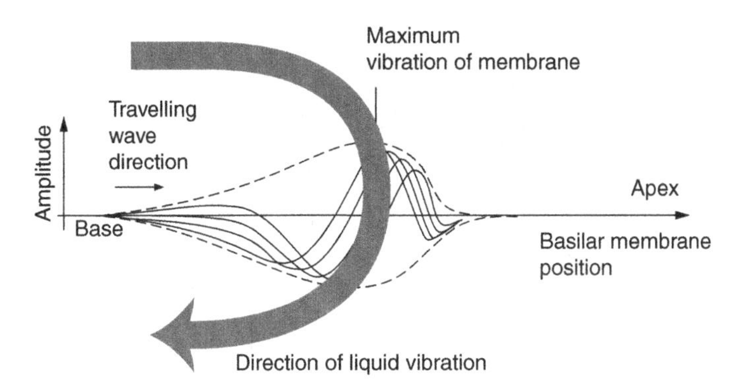

Figure 7.10 A travelling wave on the basilar membrane caused by a vibration in the liquid with the peak amplitude forming at a specific position for a specific stimulus frequency. The solid curves are waveforms at consecutive moments in time, while the dashed curve is the amplitude envelope of the vibration.

excitation from the liquid reaches each position on the membrane virtually at the same time, and that the formation of the wave is caused by the phase response characteristics of different positions of the basilar membrane. It is assumed that the positions with lower characteristic frequency respond a bit later, and such a wave phenomenon occurs (Schnupp *et al.*, 2011).

Figure 7.9 plots the amplitude envelope of a vibration on the basilar membrane for different frequencies. The lowest frequency signals resonate at the end of the membrane, while higher frequencies resonate earlier and then decay rapidly. The highest frequencies (near 20 kHz) have their maximum vibration at the very beginning of the membrane (not shown in Figure 7.9).

7.4.3 Active Function of the Cochlea

The explanation of the basilar membrane and the hair cell function above is qualitatively correct as a first approximation. Georg von Békésy and followers first measured the vibratory patterns of the membrane using ears that were not in prime physiological condition. They found that the frequency selectivity of mechanical vibration was far poorer than the selectivity measured from auditory nerve responses. This was explained by a hypothetical 'second filter' somewhere between these points.

Later studies have found that the collaboration of the basilar membrane and the hair cells makes a complex signal processing system – an *active* cochlea – where neural activity can control the vibratory pattern of the basilar membrane. The hair cells not only sense vibration and code this information to the auditory nerve, but they, particularly the outer hair cells, can actively amplify the mechanical motion. The term *cochlear amplifier* is used to refer to this phenomenon. The gain of the amplifier is shown in Figure 7.11a, where the velocity of the basilar membrane is measured at a single point in the cochlea of a cat, and the ratio between the velocity of the membrane and the pressure of the input signal is plotted as a function of input frequency. The gain function is high and has a high Q (good frequency selectivity; see Equation (3.30) on page 58) for low signal levels at frequencies near 9 kHz (Ruggero *et al.*, 1997). When the input sound pressure level is higher, the best frequency is somewhat lower, and the Q is significantly lower.

The active role of the cochlea and hair cells is based on non-linear positive feedback, whereby very weak signals are strongly emphasized in amplitude and the Q value of the cochlear resonance is improved. The outer hair cells have been shown to be able to modulate their length

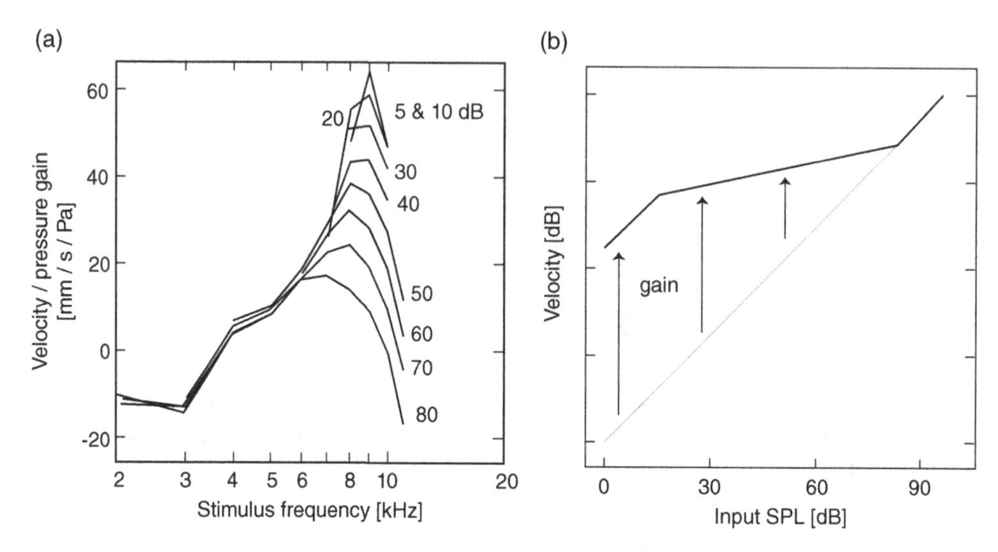

Figure 7.11 (a) The velocity of the basilar membrane recorded at a single point when excited with sinusoids of different frequencies and levels. The gain (velocity of the membrane divided by stimulus pressure) of the active cochlea is shown as a function of stimulus frequency. After Ruggero *et al.* (1997), with permission from The Acoustical Society of America. (b) A schematic plot of the gain function at a single position in the cochlea showing the dependence on the stimulus level.

when they are excited mechanically or by the descending neurons (Schnupp *et al.*, 2011). However, the exact mechanism of *how* the outer hair cells affect the cochlea movement is poorly understood and is not discussed here. A consequence of the active neural involvement in the cochlea is *otoacoustic emissions*, which are discussed more in Section 7.5.

When the stimulus level increases, active boosting will decrease until, for loud speech levels, the signal is no longer amplified. Cochlear compression is characterized in Figure 7.11b as a compressive input–output relation. A consequence of the outer hair cell contribution to the basilar membrane movement is that the system cannot be seen as a simple linear filter bank. It has many non-linear features in responses within frequency, time, and level.

To illustrate this phenomenon more clearly, Figure 7.12 shows the velocity patterns that arise at different positions of the membrane when two 500-Hz sinusoids with levels 40 dB SPL and 90 dB SPL are fed to a non-linear model of the cochlea. The high Q value at low levels of excitation results in considerable activation only near 500 Hz in the 40-dB case. When the SPL of the sinusoid is raised to 90 dB, the activation spreads over a large range towards higher frequencies. This behaviour matches the results shown in Figure 7.11a, where the position on the basilar membrane is sensitive to frequency very selectively only when the excitation level is low, and when the level is increased, it also responds to frequencies *lower* than its best frequency. Notice also in Figure 7.12 that for the 40-dB case the maximum velocity is obtained after 10 ms of stimulus, whereas in the 90-dB case it is reached at 4 ms. This is a result of the active amplification of the vibration, which takes some time to build up.

To show the functioning of the cochlea with time, the response of a non-linear cochlea model (Verhulst *et al.*, 2012) to an impulse train with a repetition rate of 10 ms is shown in Figure 7.13. The membrane responds at high frequencies after each impulse by damping the oscillation like a resonator. At the lowest frequencies the response follows the fundamental frequency of

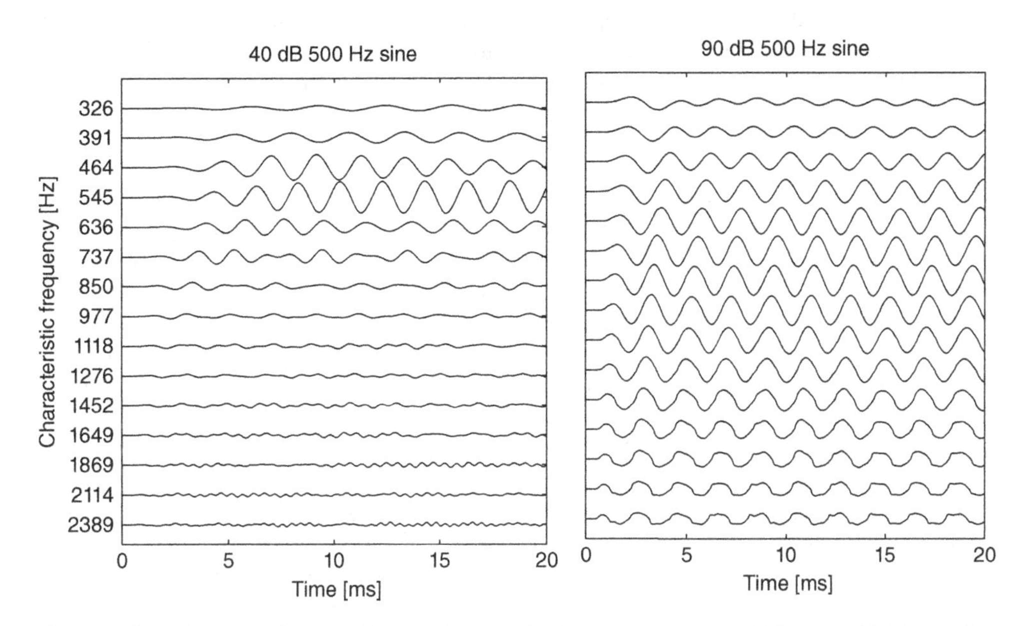

Figure 7.12 The normalized basilar membrane velocity response to a 500-Hz sinusoidal input a) at 40 dB SPL and b) at 90 dB SPL. Courtesy of Alessandro Altoè.

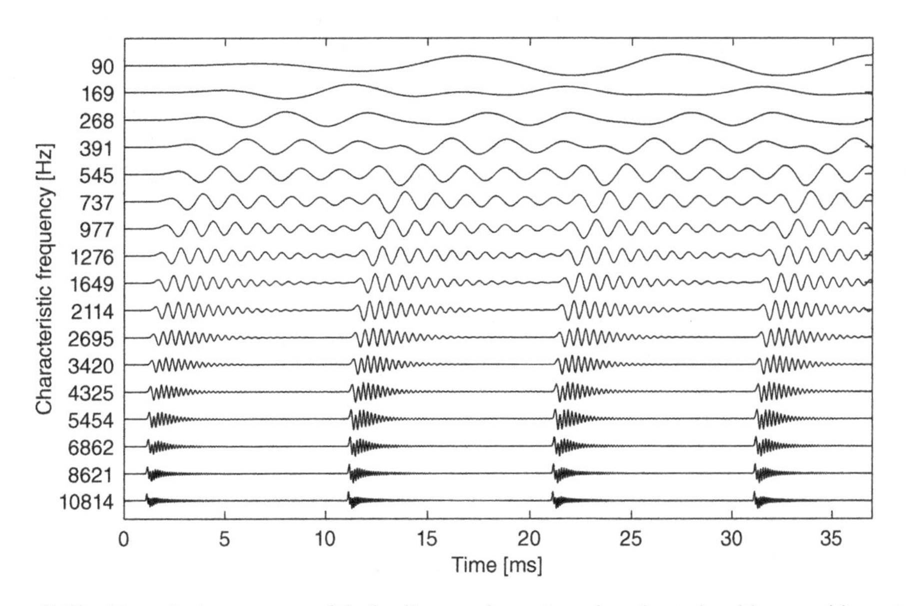

Figure 7.13 The velocity response of the basilar membrane to an impulse train with a repetition rate of 100 Hz as the excitation. The response is shown for different positions on the membrane corresponding to the indicated characteristic frequencies computed using a non-linear model of the cochlea (Verhulst *et al.*, 2012). Courtesy of Alessandro Altoè.

the excitation signal. A noticeable property is the delay of the response that increases towards lower frequencies due to the slower movement of the membrane at the apical end. The response resembles a travelling wave with a velocity that decreases with time.

7.4.4 The Inner Hair Cells

The conversion of the vibrations of the cochlea into the activity of neurons is performed by the inner hair cells. The vibration of the basilar membrane leads to the bending of the stereocilia which, in turn, causes the channels on the tops of the hair cells to open for K^+ ions, as shown in Figure 7.8. The movement of the K^+ ions modulates the potential difference across the membrane of the cell. The changes in the membrane potential trigger the release of *neurotransmitters* at *synaptic junctions* between the inner hair cells and neurons of the auditory nerve – the neurons that make up the auditory nerve are called the *spiral ganglion neurons*. Each inner hair cell is connected to about 20 spiral ganglion neurons, each of which receives an input from one inner hair cell only.

The frequency-dependent behaviour of the basilar membrane means that it acts, together with the signal detection by the hair cells, as a mechanical band-pass filter. A vibratory pattern on the basilar membrane is mediated by hair cells to the auditory nerve, thus coding frequency to place and then to neural fibre position. This position ordering of frequencies from high to low is called *tonotopical organization* (tonotopy), and the ordering is preserved throughout the neural processing stages up to the auditory cortex.

7.4.5 Cochlear Non-Linearities

There are other deviations from linear and time-invariant behaviour found in the neural responses in addition to the compression of dynamics shown in Figure 7.11. The two most prominent *cochlear non-linearities* are called two-tone suppression and combination tones. *Two-tone suppression* is the phenomenon where the addition of a second input tone at another frequency suppresses the activation caused by the test tone at its characteristic frequency (Delgutte, 1990). Figure 7.14 illustrates this effect in the form of a tuning curve where

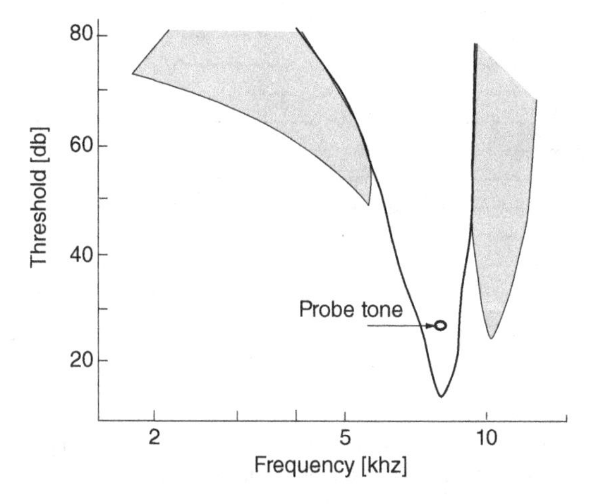

Figure 7.14 Two-tone suppression. A secondary tone with a frequency and level falling in the shaded area will suppress the response to the probe tone at its CF. Adapted from Arthur *et al.* (1971).

the shaded areas represent suppressed behaviour, in which if the second tone falls in these frequency–level areas, the neural activity due to the test tone at its CF will decrease. This phenomenon can be interpreted as a kind of masking, a means for stronger (and potentially more important) signal components to dominate over neighbouring weaker signal components that may be noise or other interfering sounds.

Another non-linearity of common interest is the generation of *combination tones*, a phenomenon common to all non-linear systems. Active collaboration of the basilar membrane and the outer hair cells generates harmonics for a single sinusoidal input, but due to the strong low-pass filtering along the basilar membrane beyond the point of the CF these harmonics do not play a prominent role. Combination tones of two (or more) sinusoids, however, are more audible. The difference tone of two frequencies, $f_{\mathrm{diff}} = f_2 - f_1$, can be perceived; for example, frequencies $f_2 = 1.1$ kHz and $f_2 = 1.0$ kHz produce the perception of a faint, low-frequency tone at 100 Hz.

An interesting kind of combination tone is the *cubic difference tone*, $f_{\mathrm{cubic}} = 2f_1 - f_2$. For example, frequencies $f_1 = 1.0$ kHz and $f_2 = 1.1$ kHz create $f_{\mathrm{cubic}} = (2 \cdot 1.0 - 1.1)$ kHz $=$ 0.9 kHz. Cubic difference tones are already generated at low levels of excitation, which is in contrast to the general behaviour of non-linear distortion. Therefore, studying non-linear cochlear mechanics with the cubic difference tone phenomenon has been of special interest.

7.5 Otoacoustic Emissions

When a sound is presented to the ear, an echo can be recorded. The echo can be delayed so much ($>$ 10 ms) that it cannot be explained by passive mechanics of the ear but neural systems must be involved. The resulting motion causes a faint sound, which is propagated back out of the ear and can be registered after stopping the acoustic stimulus.

The sounds are often weak, but they can be captured using sensitive microphones. The measurement of these emissions provides a means to research the functioning of the cochlea. These responses are called *otoacoustic emissions* or cochlear echoes. The echo can be related to the input signal frequency, it can be a combination tone, or a spontaneous generation of cochlear vibration. This is a behaviour characteristic of a healthy ear (Geisler, 1998). The subject is discussed in slightly more detail in Section 19.4.6.

In many hearing impairments, the active role of the cochlea is damaged and the echo cannot be detected. Such measurements are commonly used in the diagnostics of hearing. The first test of the hearing of newborn babies in some countries is done by measuring the level of otoacoustic emissions. A clear response is taken as evidence of a functional inner ear.

The ear also emits sound during silence. The level of emissions is typically lower than 5 dB SPL. Although such spontaneous emissions have no great importance in the research on hearing, some interesting details of hearing have been found to be connected to them. For example, subjects with strong sinusoidal acoustic emissions have higher sensitivity at frequencies close to emissions (Heise *et al.*, 2009).

7.6 Auditory Nerve

The basic building block of the nervous system is the neuron. Neurons are cells specialized in receiving, processing, and storing information via electrical and chemical means. A typical neuron consists of *dendrites*, the *soma*, and an *axon* (see Figure 7.15). The neuron receives input from other cells through its dendrites. This input converges in the soma, from which the output is sent to other neurons through the axon. The meeting point of the axon and the

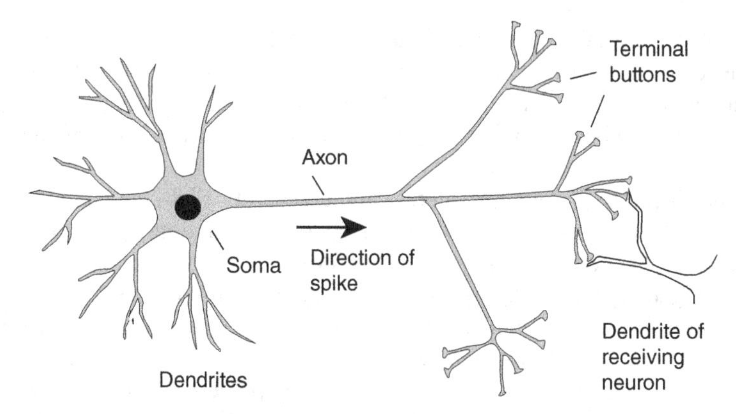

Figure 7.15 A neuron with a structure found typically in the brain. A multitude of dendrites conduct the action potential originating from other neurons to the soma. The soma processes the input and fires a spike through its axon, which then branches to reach the dendrites of receiving neurons.

dendrite of another neuron has a *synapse*, a tiny gap between the two neurons across which they communicate using chemicals called *neurotransmitters*. The axon releases these and they are received by the dendrite. The neurotransmitters modulate the membrane potential of the receiving neuron. These gradual fluctuations in the membrane potential progress to the soma. If the membrane potential reaches a threshold value, it gives rise to an *action potential*, which is an all-or-nothing cascade of events that includes a sudden large change in membrane potential followed by its return to the resting potential. This action potential then travels along the axon and triggers the release of neurotransmitters to the next neurons.

In the cochlea, the activity of the hair cells modulates the membrane potential of spiral ganglion neurons. The axons of the spiral ganglion neurons form the *auditory nerve* along which information travels from the cochlea to the brain in the form of action potentials. The vast majority of auditory nerve fibres transmit information from the inner hair cells, about 20 from each inner hair cell (see Figure 7.8). The rest of the fibres in the auditory nerve connect to the outer hair cells, and their role is not well understood (Plack, 2013).

Action potentials, often also called *spikes*, can be detected with the methods of neurophysiology. The rate at which the action potentials occur is often called the firing rate. A tiny recording electrode can be surgically placed in the proximity of the nerve. Neural responses in the auditory nerve fibres can be registered for different stimulus signals sent to the ear, and thus the whole chain of peripheral hearing can be investigated experimentally. Since the system is not linear and time-invariant (LTI), no single transfer function can represent it, but many points of view must be studied and characterized with different input–output relationships.

7.6.1 Information Transmission using the Firing Rate

The auditory nerve fibres show spontaneous firing even when no sound enters the ear. When the sound level is increased, the firing rate grows monotonically but starts to saturate above a specific level, as shown in Figure 7.16. The dynamic range is relatively narrow, about 20–40 dB for a single neuron (Moore, 2012). Auditory nerve fibres with a high spontaneous firing rate (about 20/s or more) are more sensitive but saturate first. The majority of neurons are of this type. Fewer medium spontaneous firing rate neurons also exist, and they are less sensitive but

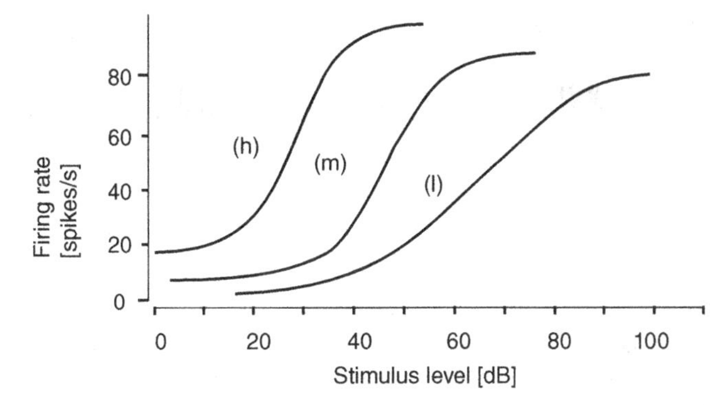

Figure 7.16 Level–rate functions for three neurons with different spontaneous firing rates: high (h), medium (m), and low (l). The level of stimulus is in dB SPL and the rate is in spikes per second. Adapted from Pickles (1988).

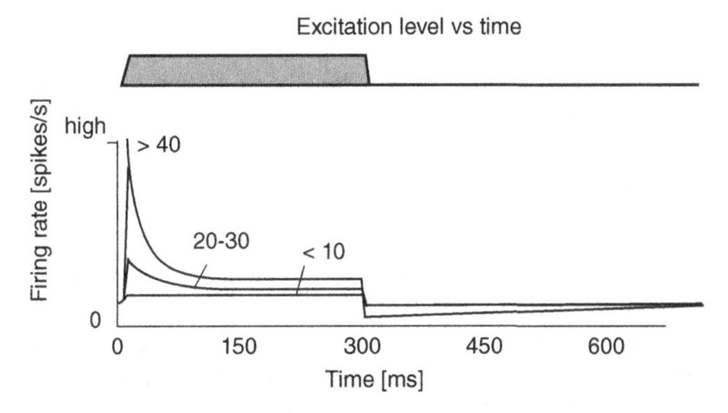

Figure 7.17 A schematic representation of a post-stimulus time histogram of the firing rate for a 300-ms long excitation at different levels. The curves exhibit overshoot after the onset and undershoot after the end of the excitation. The upper curve shows the time period of excitation. Lower curves show responses for different excitation levels (given as a parameter in dB).

work for higher levels of sound. The rare low-spontaneous-rate neurons remain unsaturated at high levels.

The output of the neuron to the auditory nerve is practically constant for signal levels above 50–60 dB, and the question then remains, how does our hearing analyse the loudness of sound at higher levels? A possible answer is that the cross-frequency spreading of the activation caused by loud sounds shown in Figure 7.12 is taken into account. With such an assumption, the auditory system analyses the auditory input across frequency and combines the excitation in several adjacent bands to form the loudness perception.

The static level–rate functions of Figure 7.16 exhibit the limited range and non-linearity of neural encoding, but this is just one viewpoint. Figure 7.17 shows a more dynamic view, called the *post-stimulus time histogram* (PST histogram), where the firing rate is plotted as a function of time for a constant excitation lasting 300 ms for different excitation levels. There is a significant *overshoot* after the onset of the excitation, but not for very low levels. This overshoot greatly exceeds a static saturation level, and it emphasizes attacks and onsets that

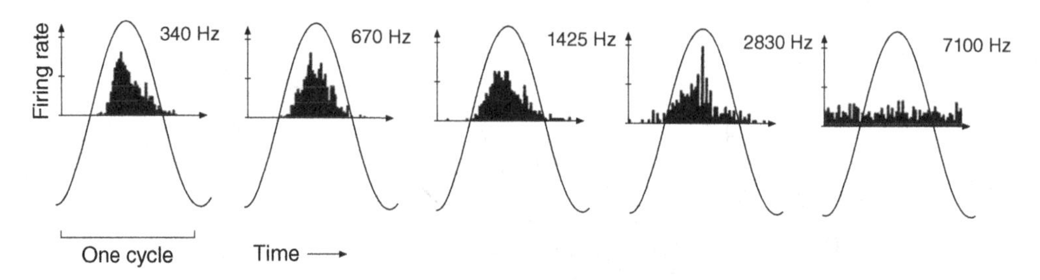

Figure 7.18 Post-stimulus time (PST) histograms for sinusoidal stimulation at different frequencies measured from a cat (data from Joris *et al.*, 1994). The neural firing in the auditory nerves at their best frequencies has been recorded for a continuous sinusoidal stimulus, and the results are summarized as PST histograms, which can be interpreted as the average activity of a neuron depending on the phase of the stimulus. The best frequencies of the nerves are specified. The sinusoidal stimuli are overlaid so that the phases match with the histogram. Reproduced with permission from The Acoustical Society of America.

are important from a perceptual point of view. When the excitation ends, there is an *undershoot* phenomenon before the spontaneous firing rate is recovered (Pickles, 1988).

7.6.2 Phase Locking

In some cases the neurons synchronize the firing with a certain phase of a periodic signal, a phenomenon called *phase locking*. This was observed in an experiment where the neurophysiological response from the auditory nerve of a cat to a periodic excitation was recorded, and the data were presented as a stimulus level versus synchronous firing rate plot, as shown in Figure 7.18 for different frequencies of a sinusoidal signal. The neural activity waveform is approximately a half-wave rectified (limited to positive values) form of the excitation waveform. In humans the synchronization clearly occurs at frequencies below 1500 Hz, degrades at frequencies above it, and is weak above 4 kHz. For such continuous signals, it is mainly the place and average rate that are mediated above this frequency, not the temporal details.

Most of neural processing is probabilistic and inaccurate when considering a single neuron, but neural signal processing is based on a large number of neurons working together, which can make the output of such neural networks very precise and robust.

The waveform of the signal also affects the synchronization, and phase locking does not necessarily occur at the same phase of each frequency component. With certain periodic sounds, the neurons in the cochlea synchronize to a specific phase of the fundamental frequency of the sound instead of the phase of the sinusoid(s) near the CF of the neuron. It has also been found that neurons with high characteristic frequency can phase lock to sinusoids with lower frequency.

So far we have not discussed the frequency selectivity of the neural responses. Since the basilar membrane (together with the outer hair cells) is able to do a selective-frequency analysis (frequency-to-place coding), it is reasonable to assume that frequency selectivity also exists in the auditory nerve coding. Figure 7.19 illustrates what happens when vowels are used as stimulus signals with varying levels. At low excitation levels, the neural firing in each characteristic frequency band shows a response that is approximately the vowel spectrum. The major formant ranges are easily seen (marked with arrows). At higher excitation levels, however, the

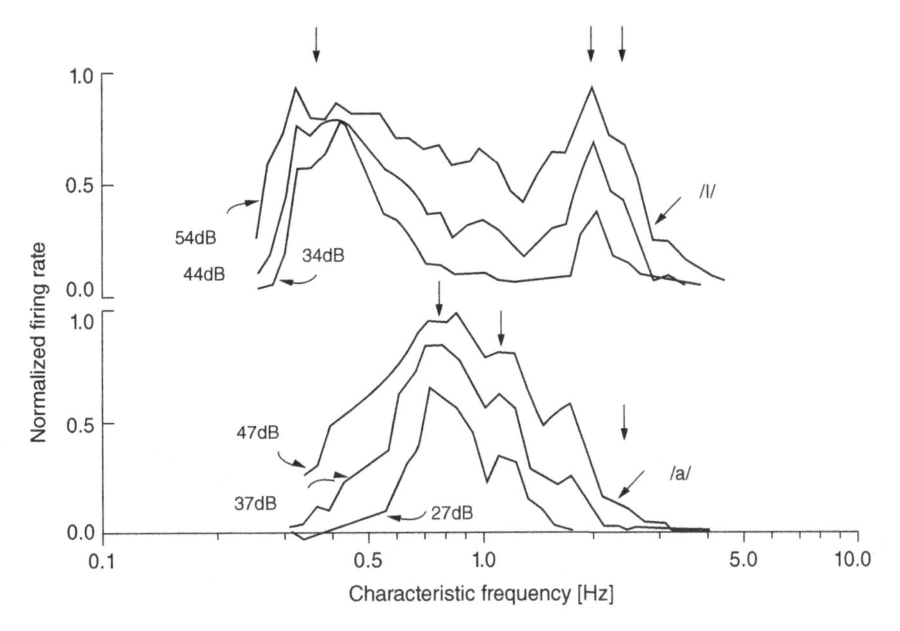

Figure 7.19 The averaged firing rate pattern produced for two vowels at different intensity levels measured from the auditory nerve of a cat. The higher stimulus level shows increasing saturation. Arrows indicate the main formant areas. Adapted from Sachs and Young (1979) and reprinted with permission from The Acoustical Society of America.

neural saturation effect distorts the spectral information, as seen in the frequency pattern of the firing rate (Sachs and Young, 1979).

From Figures 7.16 and 7.19 one may conclude that the frequency resolution of hearing might be lost at levels above 60 dB due to neural saturation. This is not true, as we know from our everyday experience that speech with levels of even 100 dB is still understandable. An evident source of information is the temporal pattern of neural activation, and indeed Young and Sachs (1979) show that a combination of spectral and temporal neural patterns contains the information of vocal formants even at high frequencies.

7.7 Auditory Nervous System

The auditory nerve carries information from the cochlea to the brain. The number of neurons in the human brain is humongous, of the order of 100 billion (100 000 000 000) (Goldstein, 2013). The great number of neurons, and the even greater number of connections between them, enables the complex functions the brain conducts all the time. These neurons and their connections are packed very tightly in the brain in an extremely complex organization. It is thus a real challenge for science to figure out what the individual tasks of the neurons are and what kind of overall function the individual neurons conduct together.

7.7.1 Structure of the Auditory Pathway

Figure 7.20 is a highly schematic organization diagram of the central auditory system showing only the main units of neural processing and only some of the most obvious ascending connections. The neural system is divided into the left and right *hemispheres*, which are roughly

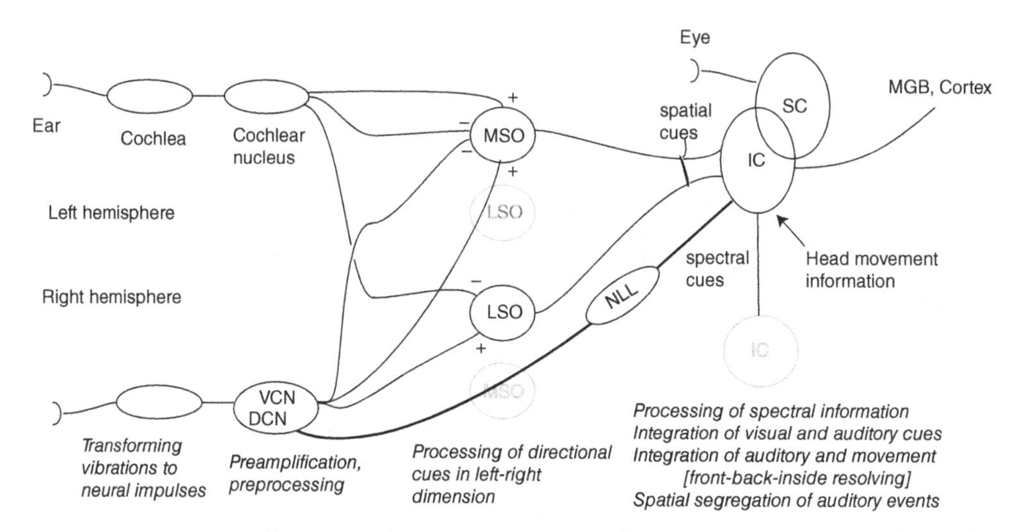

Figure 7.20 The most important nuclei and the auditory cortex as an organization of the central auditory nervous system. A simplified view of the most prominent ascending paths is shown.

symmetric with each other. All ascending fibres of the auditory nerve lead to the cochlear nucleus (CN), which is located in the brainstem. A plethora of neurons of different types exist, all performing different tasks. In engineering terms, the nucleus seems to act at least as a preamplifier, a signal router, and an initial periodicity and initial spectral analyser. The auditory pathway from the CN onwards to the inferior colliculus (IC) seems to be divided into two: one route leads to the IC directly or through the lateral lemniscus (LL) and the other through the superior olivary complex (SOC).

The neurons in the dorsal CN and in the postero-ventral CN give rise to the first pathway. The neurons have different forms, and they seem to perform a first-order analysis of the spectral content of sound, with some of them also phase locking to the repetitive structures in the time-domain signals, possibly affecting pitch perception (Schnupp *et al.*, 2011). The axons of these neurons lead directly, or through the LL, to the IC.

The antero-ventral CN, in turn, gives rise to the second route. It projects its output to the superior olivary complex (SOC). This connection is known for its accurate phase locking, exceeding the temporal accuracy found in the auditory nerve (Joris *et al.*, 1994). The enhancement is explained by the fact that CN neurons are innervated by multiple auditory nerves. The SOC is divided into the medial superior olive (MSO) and the lateral superior olive (LSO), both of which receive input from both hemispheres. Both the MSO and the LSO are known to be sensitive to interaural differences of ear canal signals (Grothe, 2003; Tollin, 2003). This processing thus enables the perception of left/right direction. The SOC projects to the IC directly or through the LL.

Numerous studies have measured the responses of IC neurons, but no single function has emerged for them. This may be related to the versatility of the IC: it is likely to perform a great variety of tasks. The IC transmits the auditory signals from the CN, LL, and SOC through the thalamus to the auditory cortex and also to the superior colliculus (SC) located next to it. The ICs in the two hemispheres are connected as well. A clear tonotopic organization can be found in the CN, SOC, LL, and in the IC.

The SC is also interesting in the context of hearing. It receives both visual and auditory signals, and it has been found to be one of the organs responsible for cross-modal interaction (Møller, 2006). One of its tasks is orienting the head and gaze towards the stimuli. Neurons in the SC form a topographical map of spatial locations; the neurons are organized according to the angle of incident of sound and/or light that they are most sensitive to. The SC includes a map of the auditory space that is aligned with the visual map in such a manner that the neurons responsive to auditory or visual stimuli from a certain direction are found close to one another. Some SC neurons also respond to multimodal stimulation originating from the same spatial location (Meredith and Stein, 1986). Interestingly, the topographical organization of the auditory space has been found only in the SC of all brain nuclei in the mammalian auditory system.

As the neuronal processes in the cochlea are only partly understood, it is quite understandable that the knowledge of neural processes beyond the cochlea is even more sparse. However, for acoustic engineering applications, understanding the functioning of the cochlea is usually enough, although certain spatial audio reproduction methods rely on knowledge of the function of the nuclei in the SOC.

From the IC, the auditory information is transmitted through the medial geniculate body of the thalamus to the auditory cortex. The tonotopical organization is maintained throughout the auditory pathway and is also seen in the core areas of the auditory cortex. The auditory cortex comprises the core areas with the tonotopical maps and surrounding areas in which neurons have increasingly complex response properties. Processing in the auditory cortex is also thought to be organized into two streams. The 'what' stream is thought to perform the analysis of the sound spectrum ('what is the source?'), whereas the 'where' stream is thought to analyse the spatial information of different sound events in the auditory scene ('where is the source?') (Rauschecker and Tian, 2000). A similar processing scheme is found in the neural analysis of the visual pathway.

There is no clear endpoint to the auditory pathway. The auditory cortex is interconnected to other areas of the cortex. Activity related to sound processing is also seen outside the auditory cortex, especially during listening tasks that involve attention, memory, and multimodal processing.

7.7.2 *Studying Brain Function*

The action potentials of single neurons can be recorded using tiny electrodes placed near the neuron or even inside it so that fluctuations in the membrane potential can also be monitored. These studies involve some invasive procedures and are therefore normally conducted only on animals. In such neurophysiological studies, sounds are presented to the animal, and the activity of single neurons or groups of neurons is then recorded. By carefully controlling the stimulus, and with a high number of repetitions, the representation of the auditory signals and analysis performed by the neurons can be figured out.

There is a variety of non-invasive brain-imaging techniques for studying human brain function, where the activity of the brain is recorded through the skull. These methods record the activity of entire brain areas, not single neurons. They are also best suited for recording the activity of the cortex. In *electroencephalography* (EEG), electrodes are placed on the scalp, and the electric potential in relation to a reference voltage measured from the subject is recorded with time. The neural activity during and after the stimulus causes small changes in the recorded voltages. The benefit of EEG is that it responds quickly to changes in brain activity. However, localizing the sources of activity and differentiating between the contributions of brain areas is difficult using EEG. Another, somewhat similar method is *magnetoencephalography* (MEG), where the magnetic fields caused by electrical currents in neurons are recorded with very sensitive magnetometers. It has better resolution in localizing the activity within the brain than EEG has, because it suffers less from distortions caused by the scalp and skull.

The anatomical structure of the living human brain can be mapped using magnetic resonance imaging (MRI). Functional MRI (fMRI) is also able to map the functional properties of the brain. fMRI records local changes in the metabolism of the brain that are related to neuronal activity. However, the temporal resolution of fMRI is in the order of only hundreds of milliseconds even at its best. fMRI scanners are also very noisy and therefore their use for studying the auditory system sets a special challenge. fMRI has, however, good spatial resolution and is well suited for identifying which brain areas are active.

7.8 Motivation for Building Computational Models of Hearing

As discussed above, studying the functioning of hearing is extremely difficult for many reasons, and all ethical direct or indirect approaches giving any clues on the functioning are valuable. An additional tool for this purpose is to build functional models of the auditory pathway called *auditory models*. The success or failure of the models then helps researchers understand the principles and details of the processing in the brain.

Human auditory perception has been modelled at many different levels. The principle is to build computational models of acoustic, physiological, or neural mechanisms that estimate the relation of perceived auditory attributes resulting from the presentation of stimuli with certain acoustic attributes. Ideally, the ear canal signals are fed into the system, and the auditory attributes that subjects perceive, on average, are estimated without bias or noise.

There are different motivations for building such models. First of all, being able to model the functionalities of different parts of the system provides basic knowledge of the auditory system. Often, more than one mechanism affects the perceived characteristics of an auditory object, making the analysis of the results of psychoacoustic tests complicated. The complications can be mitigated when the effects of the mechanisms can be isolated in the computational domain, making the interpretation of the results easier. Second, a very important application for such models is the evaluation of audio quality. Modern acoustic systems comprise various transfers in the acoustic, signal processing, and telecommunication domains. The measurement of quality of reproduced audio cannot be performed with simple linear measures, which calls for the usage of models of human hearing for such tasks. Ideally, the fidelity of audio systems could be estimated by computing the auditory cues reaching the conscious level. The third application of auditory models is in the diagnostics of hearing-impaired patients. The symptoms of the patient would be characterized with specific psychoacoustic tests, and the attributes of the model of human hearing were adjusted until the symptoms were explained. Optimally, this would track down the reason for the symptoms.

In principle, modelling could be performed by first modelling one neuron, followed by modelling the next neuron, until the entire auditory pathway was modelled. Unfortunately, this approach is doomed for two reasons: the current knowledge of the complex electrochemical functioning of the neurons and of the neuroanatomy of the pathway is far from the stage where all the details can be modelled. In addition, the enormous number of neurons devoted to hearing makes it computationally out of reach in the near future. The present status of auditory models is discussed in more detail in Chapter 13.

Summary

This chapter served the purpose of providing a brief introduction to the structure and function (physiology) of hearing: how sound waves from the surrounding air enter the auditory system through the external, middle, and inner ear. In the inner ear, the sound signals are divided into frequency channels and coded as neural spikes. The research methods used and the structure of the auditory pathway were also briefly introduced. Computational modelling of the mechanisms can then be used to supplement and test the knowledge obtained from neurophysiology.

Further Reading

Neural systems and their research is introduced in (Gelfand, 2004; Kalat, 2011). The methodology of investigating the physiology of the inner ear is described, for example, in Pickles (1988) and Yost (1994). There exist a lot of sources on the neuroanatomy and the neurophysiology of the auditory pathway. The reader might consider Popper and Fay (1992) or Webster *et al.* (1992) as an interesting starting point.

References

Aibara, R., Welsh, J.T., Puria, S., and Goode, R.L. (2001) Human middle-ear sound transfer function and cochlear input impedance. *Hearing Res.*, **152**(1), 100–109.

Arthur, R., Pfeiffer, R., and Suga, N. (1971) Properties of two-tone inhibition in primary auditory neurones. *J. Physiology*, **212**(3), 593–609.

Delgutte, B. (1990) Two-tone rate suppression in auditory-nerve fibers: Dependence on suppressor frequency and level. *Hearing Res.*, **49**(1), 225–246.

Geisler, C.D. (1998) *From Sound To Synapse: Physiology of the Mammalian Ear*. Oxford University Press.

Gelfand, S.A. (2004) *Hearing: An introduction to psychological and physiological acoustics*. Marcel Dekker.

Goldstein, E.B. (2013) *Sensation and Perception*, 9th edn. Cengage Learning.

Grothe, B. (2003) New roles for synaptic inhibition in sound localization. *Nature Reviews Neuroscience*, **4**(7), 540–550.

Heise, S.J., Mauermann, M., and Verhey, J.L. (2009) Investigating possible mechanisms behind the effect of threshold fine structure on amplitude modulation perception. *J. Acoust. Soc. Am.*, **126**(5), 2490–2500.

Joris, P.X., Carney, L.H., Smith, P.H., and Yin, T. (1994) Enhancement of neural synchronization in the anteroventral cochlear nucleus. I. responses to tones at the characteristic frequency. *J. Neurophys.*, **71**(3), 1022–1036.

Kalat, J.W. (2011) *Biological Psychology*. Cengage Learning.

Meredith, M.A. and Stein, B.E. (1986) Visual, auditory, and somatosensory convergence on cells in superior colliculus results in multisensory integration. *J. Neurophysiol.*, **56**(3), 640–662.

Møller, A.R. (2006) *Hearing: Anatomy, Physiology, and Disorders of the Auditory System*. Elsevier.

Moore, B.C.J. (2012) *An Introduction to the Psychology of Hearing*, 6th edn. Brill.

Pickles, J.O. (1988) *An Introduction to the Physiology of Hearing*. Academic Press.

Plack, C.J. (2013) *The Sense of Hearing*. Psychology Press.

Popper, A.N. and Fay, R.R. (1992) *The Mammalian Auditory Pathway: Neurophysiology*. Springer.

Rauschecker, J.P. and Tian, B. (2000) Mechanisms and streams for processing of "what" and "where" in auditory cortex. *Proc. Natl Acad. Sci.*, **97**(22), 11800–11806.

Ruggero, M.A., Rich, N.C., Recio, A., Narayan, S.S., and Robles, L. (1997) Basilar-membrane responses to tones at the base of the chinchilla cochlea. *J. Acoust. Soc. Am.*, **101**(4), 2151–2163.

Sachs, M.B. and Young, E.D. (1979) Encoding of steady-state vowels in the auditory nerve: Representation in terms of discharge rate. *J. Acoust. Soc. Am.*, **66**, 470–479.

Schnupp, J., Nelken, I., and King, A. (2011) *Auditory Neuroscience: Making Sense of Sound*. MIT Press.

Tollin, D.J. (2003) The lateral superior olive: A functional role in sound source localization. *Neuroscientist*, **9**(2), 127–143.

Verhulst, S., Dau, T., and Shera, C.A. (2012) Nonlinear time-domain cochlear model for transient stimulation and human otoacoustic emission. *J. Acoust. Soc. Am.*, **132**(6), 3842–3848.

von Békésy, G. (1960) *Experiments in Hearing*. McGraw-Hill and Acoust. Soc. Am.

Webster, D.B., Popper, A.N., and Fay, R.R. (1992) *The Mammalian Auditory Pathway: Neuroanatomy* volume 1. Springer.

Yates, G.K. (1995) Cochlear structure and function. *Hearing*, pp. 41–74.

Yost, W. (ed.) (1994) *Funfamentals of Hearing – An Introduction*. Academic Press.

Young, E.D. and Sachs, M.B. (1979) Representation of steady-state vowels in the temporal aspects of the discharge patterns of populations of auditory-nerve fibers. *J. Acoust. Soc. Am.*, **66**, 1381–1403.

8

The Approach and Methodology of Psychoacoustics

Auditory psychophysics, more often called *psychoacoustics*, is important in understanding the systemic and information processing properties of auditory functions (see Section 1.3). It is, in principle, independent of physiological research and knowledge, but it is always most fortunate if physiological and psychoacoustic facts and models support each other. As will become obvious below, many (but not all) psychoacoustic phenomena find a correlate in the physiology of hearing. Modern psychoacoustics, based on systematic experimentation, has been carried out for roughly a century. Its development has been greatly influenced by engineering sciences, especially by the challenges of communication technology.

Psychoacoustics has the advantage that experimentation in its basic form is easy and non-invasive, and thus the subject under study is not in danger of physical injury. However, this does not mean that making such behavioural experiments or interpreting their results is easy. Another advantage of the psychoacoustic approach is that higher level functions of the auditory system can be studied where physiological knowledge is missing or too weak to support our understanding.

Studies on auditory sensation and perception can be compared to measurements of a very complex physical system that is inherently non-linear, time-varying with both short-term and long-term effects, and it also shows minor or major variation due to innumerable other factors than those specifically being studied. It is like having an unreliable measurement device for measuring an unpredictably behaving system. With proper methodology and for properly specified problems, the task is, however, manageable and leads to useful theories and models.

8.1 Sound Events versus Auditory Events

Understanding psychophysical experimentation can be based on the system diagram shown in Figure 8.1. The outer box encloses the subject under study. Sound stimuli consist of *sound events* (or *sound objects*) s that enter the auditory system of the subject. They may be anything from simple tones to combinations of complex sound sources, deterministic sounds, or noises.

The equivalent of an external sound event, internal to the subject in Figure 8.1, is an *auditory event* (or *auditory object*) h_i. The subject has more direct access to this internal event than any

Communication Acoustics: An Introduction to Speech, Audio, and Psychoacoustics, First Edition.
Ville Pulkki and Matti Karjalainen.

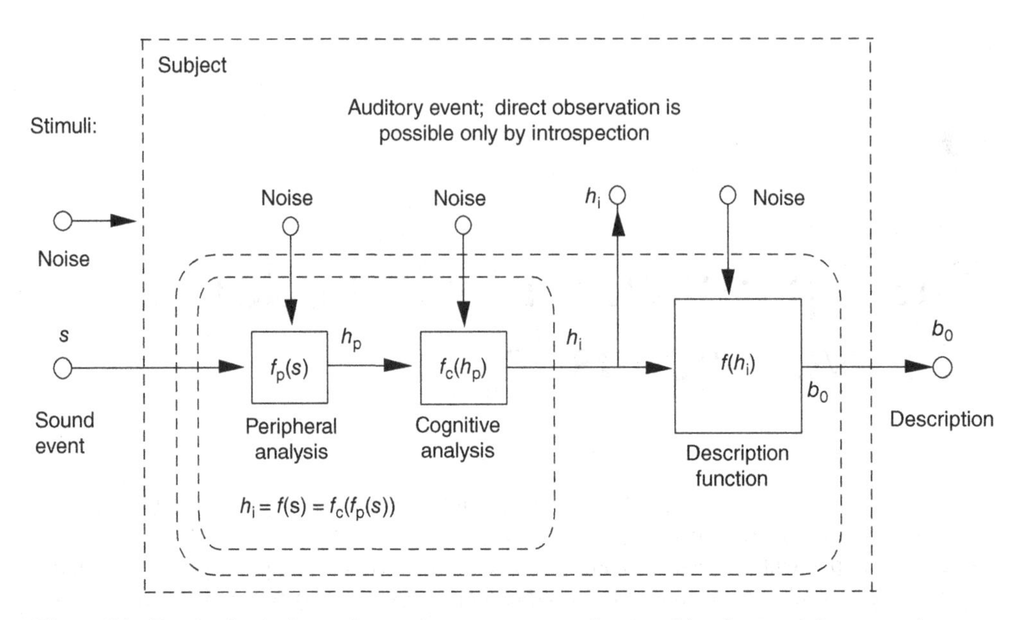

Figure 8.1 Psychophysical experimentation as a process where a subject is exposed to a sound event s (physical sound stimuli) and the external observer has only indirect access to internal auditory events h_i through an externalized description or reaction b_0.

external observer. This ability to observe one's own internal events is called *introspection*. This may be useful to a researcher in order to acquire a general picture of the auditory perception, but, on the other hand, introspective observation is seldom accepted as a scientific method as such. This is because of the subjective nature of such observations that may easily have a strong bias due to many disturbing factors and the fact that a single subject may not be representative of general behaviour. Therefore, in general, auditory events should be studied by an external observer, statistical methods should be applied (or at least one must be aware of the statistical nature of the process), and several subjects should be involved, unless the goal is to know the behaviour of a specific subject.

Figure 8.1 indicates that the peripheral analysis f_p produces result $h_p = f_p(s)$, which is available neither for the subject nor for the external observer. The output of the periphery is directly mapped to auditory event h_i with analysis f_c, as $h_i = f_c(h_p)$. h_i is available only to the subject through introspection. Both peripheral and cognitive transforms are disturbed by noise, which naturally has a different nature. The noise for peripheral analysis consists of such components as Brownian noise at the eardrum and tinnitus. The cognitive transformation may, in turn, be distracted by changes in concentration of the listener for various reasons.

The goal of an investigation may be to seek a relation $h_i = f(s)$, where $f(s) = f_c(f_p(s))$, which describes how the attribute(s) of the auditory event s is mapped onto the attributes of an auditory event. The external observer encounters another transform, $b_o = f(h_i)$, a mapping from the internal representation to the external reaction that can ultimately be registered objectively. This relation may be called the reaction function or the description function, depending on how the external observations are carried out. In a sense, it corresponds to the registration or display function of a measurement device used in physical or physiological experiments. Since the study of the mapping $h_i = f(s)$ is the focus of the investigation, the influence of $b_o = f(h_i)$

and noise should be eliminated or minimized. Statistical analysis is a powerful tool for this if the observed data are 'noisy'. If the effect of the reaction function $b_o = f(h_i)$ is not known well enough, different ways to study the same internal events may help to improve the reliability of interpreting $h_i = f(s)$. It is beneficial if there are physiological facts and knowledge that support the interpretation.

8.2 Psychophysical Functions

A *psychophysical function*, characterized by $h_i = f(s)$ in Figure 8.1, represents the relation between one or several properties of a sound event s and one or several properties of an auditory event h_i. Psychophysical functions may be mappings from one continuous scale to another continuous scale (such as sound pressure level → perceived loudness), from a continuous scale to a discrete scale (say, sound pressure level → audible or inaudible), and so on. A specific type of psychophysical function called the *psychometric function* refers to the mapping from a continuous scale to a yes/no scale expressed as a probability function of the detection of a signal.

In psychophysical functions from a continuous physical attribute to a continuous sensation variable, the auditory analysis does not typically make a linear mapping. The first studies on psychophysics were conducted in the early 1800s, when the Weber–Fechner law was derived. It was assumed that these mappings followed logarithmic characteristics

$$h = \mathrm{a} \log(s), \tag{8.1}$$

where h is, for example, the subjective loudness of a tone; s is a physical attribute, such as sound pressure; and a is a constant. In more careful studies, it turned out that psychophysical functions can have different forms. As will be discussed below, with SPL above about 40 dB subjective loudness follows the power law (rather than a logarithmic law)

$$h = \mathrm{c}\, s^{\mathrm{k}}, \tag{8.2}$$

where c and k are constants ($\mathrm{k} \approx 0.6$; see Section 10.2.3). On the other hand, the pitch (height) of a tone is almost a logarithmic function of the frequency of the tone, as will be shown in Section 10.1.

Each attribute of an auditory event depends typically on many properties of the sound event. For example, the loudness of an event depends not only on the sound pressure level of the sound event, but also on, for example, the frequency content and the temporal duration of sound. Often, one of the physical properties of a sound event is dominant, like the fundamental frequency of a tone complex that mostly defines the pitch.

8.3 Generation of Sound Events

In psychoacoustic tests, the sounds should be designed in such a way that the subject can report the characteristics of the auditory event reliably, which will eventually reveal properties of the peripheral or cognitive functions in hearing. This section describes the methods most commonly used to generate the sound events for different listening conditions.

8.3.1 Synthesis of Sound Signals

Relatively simple stimuli are often used in fundamental research on psychoacoustics. The stimuli are presented in more detail in Section 3.1.2 on page 45. These include:

- *Pure tone*
- *Amplitude- or frequency-modulated tone*
- *Tone burst*
- *Sine-wave sweep*
- *Chirp signal*
- Single *pulses*
- *White noise*
- *Pink noise*
- *Uniform masking noise*
- *Modulated noise*

The first five stimuli can also be realized using other simple signal waveforms such as a square wave, a sawtooth wave, an impulse train, and their filtered (low-pass, high-pass, band-pass) forms.

Since the auditory system is highly developed to receive complex sounds from the environment and other communicating subjects, simple stimuli are insufficient for psychoacoustic research. It is increasingly important to study the perception of complex sounds such as:

- *Harmonic tone complexes*;
- *Complex combination sounds* including inharmonic sounds;
- *Combinations of sinusoids, noises, and pulses*;
- *Speech sounds*: real speech and synthetic speech;
- *Musical sounds*: acoustic and electronic music;
- *Sounds from nature*: from animals and inanimate nature;
- *Noise*: harmful, loud, or annoying sounds.

Alternatively, the sound signals to be tested may originate from an engineering task, where the effect of processing an input signal with a system having different parameters is of interest. For example, in audio coding applications, the effect of the data rate on perceived quality of sound can be studied by processing a sound sample with codecs using different settings.

The non-linear processing in the ear causes sound with different sound pressure levels to be perceived differently. The loudness differences between the samples may cause undesired effects if the loudness itself is not studied. Thus, when conducting a listening test, the effect of the level of presentation of the signals should be taken into account. For example, if the audio quality provided by different loudspeakers is tested, and if one loudspeaker delivers slightly but noticeably higher SPL to the listener, it quite probably will be rated to provide the best quality of the tested items.

The effect of loudness should typically be avoided in the tests, and equal loudness should be produced by the listening test signals. Depending on the case, the task may be simple or complex. In some cases, the equalization of the signal energies may be sufficient, while in others no computational metric available is sufficient, and separate listening tests have to be organized to set the perceived loudness levels to be equal. Interested readers are referred to

Bech and Zacharov (2006) for a detailed discussion on *level calibration*, which is a process that aims to equalize the levels of the test items.

8.3.2 Listening Set-up and Conditions

Psychoacoustic experiments require that attention is paid not only to the sound signals but also to the acoustic environment and to how stimuli are presented. The sound source can, in principle, be any source that generates a desired and well-controlled sound. In practice, the stimuli are most conveniently generated by computers and played by electroacoustic reproduction means:

- *Loudspeakers* (Section 4.1.1), one or many, controlled by a single or several audio signals. The best control over the sound field is achieved in non-reverberant, free-field conditions in an *anechoic chamber*. If a loudspeaker is close to the listener (≤ 1 m), the sound field can be approximated by a spherical field. A plane wave can be approximated by placing a loudspeaker far enough away (≥ 2 m) in an anechoic chamber. Multiple loudspeakers are often needed when special effects of spatial hearing are being studied.
- *Headphones* (Section 4.1.1), which are in some cases an ideal source of sound, since the reverberant environment can be eliminated, and some headphones also attenuate external noise. Headphones are the only choice if very different sound stimuli are needed in each ear. On the other hand, spatial attributes may be difficult to reproduce using headphones unless very careful binaural reproduction techniques are applied (see Chapter 12).

It is important to pay enough attention to the acoustic environment of psychoacoustic experiments. If a very carefully controlled free field (anechoic chamber) is not needed, conducting experiments in a specially designed *listening test room* that resembles a living room with good acoustics may be better. As an example, conducting listening tests in an anechoic space when studying audio quality would give misleading results, since the reverberant field in normal rooms immensely affects the listening experience. Loudspeakers are meant to be used in normal rooms, and the response obtained in an anechoic chamber can be very different from the response in a normal room (Bech and Zacharov, 2006).

Sometimes a special room is necessary, for example a reverberation chamber, if real reverberation is being studied. Background noise should be minimized, unless it is an integral part of the study. Eliminating visual and other undesirable cues is also an important issue, since they may easily bias results or draw attention away from the focus of the study. (Of course there are also cases where just these effects are studied and therefore they have to be included.) Hearing is particularly influenced by vision, especially in the processing of information where vision is more reliable, such as in localization, object identification, and size. In these cases we 'often hear what we see' rather than what our ears receive.

Computers and digital signal processing have made psychoacoustic experiments easier and more precise. Computers with high-quality audio interfaces are ideal for generating practically any sound with high quality and repeatability. An inherent lack of ideality in loudspeakers and headphones may be compensated for by DSP, as was discussed in Section 4.3.

8.3.3 Steering Attention to Certain Details of An Auditory Event

An auditory event is often a comprehensive percept, where it is difficult or impossible to concentrate on specific parts or characteristics. For example, it is often impossible to concentrate on a single harmonic of a harmonic complex tone. Depending on the case, different methods can be used to route the attention to a specific aspect of sound. For example, a partial modulated by amplitude or frequency is easier to perceive. Presenting the sound first without the harmonic and then with it is another way of focusing attention on the harmonic. A third approach is to first play only the harmonic tone and then the whole complex with the harmonic.

8.4 Selection of Subjects for Listening Tests

When conducting the tests, it is also necessary to control the level of listening test experience of the subjects. The panel of test subjects may be trained to be as sensitive as possible to the researched attributes of sound, unexperienced listeners, or something in between. A trained panel is needed if one wants to conduct reproducible tests at the finest resolution of a specific property of hearing. Training makes subjects more 'analytic': they learn how to analyse the auditory input and to describe it in an objective manner. On the other hand, the opinion of trained listeners may not represent the opinion of the larger audience, as, for example, in product sound quality questions. In such cases, a relatively large set of listeners from the population segment of interest must be used. However, here, too, the ability of the subjects to report the properties of interest in the sound under study has to be taken into account (Bech and Zacharov, 2006).

8.5 What are We Measuring?

In psychoacoustic experiments, sound events are presented, and a question is asked of the subjects, which they should answer based on the properties of the auditory event created by the sound event. Ideally, they should respond to the question such that the attribute being investigated is revealed in the results of the tests. Typical properties which are measured are different thresholds. The relation between attributes of sound events and attributes of auditory events, that is, psychophysical functions in general, is also of interest. Auditory scales quantify these relations. Thresholds and scales used are briefly introduced below.

8.5.1 Thresholds

A basic question in the research on hearing is if any kind of auditory event is formed with a sound event, either in silence or in the presence of noise. There are two main types of threshold values:

- *Absolute threshold*: For example, the threshold of the sound pressure level for detecting an auditory event; in other words, the hearing threshold measured in audiometry. It quantifies the value of an attribute of the sound event and the respective psychoacoustic quantity above which the auditory event emerges. The corresponding task is called the 'detection task'.
- *Difference threshold*: The smallest change in one of the attributes of the sound event that is audible. A synonym is the term 'just noticeable difference' (JND). A special case of difference thresholds are different *modulation thresholds*, such as just noticeable amplitude or frequency modulation of a tone. The corresponding task is called the 'discrimination task'.

Note that absolute and difference thresholds can be quantified on both acoustic and auditory scales.

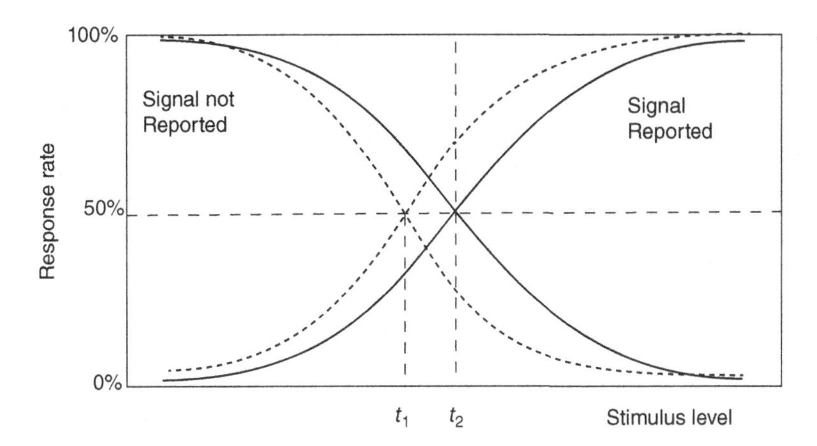

Figure 8.2 Psychophysical functions for the measurement of the absolute threshold of a sound event, where a subject is presented a signal with variable level and the rate of answers is measured. The dashed line shows the curve of the 'optimistic' type of subjects with threshold of t_1 and the solid line the curve for 'neutral' subjects with the threshold t_2.

The mapping of a physical quantity onto an auditory quantity is thus binary, either the value of h is above or below the threshold, as discussed in relation to Figure 8.1. However, the mapping does not change abruptly near the threshold, there is always a region where the auditory event may or may not be formed. This depends on some internal state of the subject, such as the level of inherent noise in sensory systems. When a listening test is organized and subjects are asked whether they perceive the auditory event, the description function $b_o = f(h_i)$ also comes in to play. For example, the personal character of the listener may have a large influence on the description, as typically some subjects tend to perceive auditory events although there are no sound events, and other subjects are more insecure and report an auditory event only when it is very clearly audible. This is illustrated conceptually in Figure 8.2. The rate of reporting that the signal is present is shown as a function of the signal level for two types of subjects: 'optimistic' and 'neutral'. Clearly, the test would produce very different threshold values if the stimulus level producing the rate of 'signal present' with a value of 50% was taken as the value of the threshold, as shown in the figure with vertical dashed lines.

Signal detection theory

This bias can be avoided if the subject is presented with an interval where the signal is or is not present. The interval is thus a period of time which can be indicated, for example, with a visual indicator on a computer screen. The length of the period is typically shorter than 10 seconds. The subject is asked whether the interval contains the signal, with the possibility of answering only 'yes' or 'no'. This gives information about the tendency of the subject to favour either 'signal present' or 'signal not present' cases. The analysis of the results from such testing is formalized in *signal detection theory*, which is a theory of perceptual mechanisms when measuring thresholds (Gescheider, 1997). It was developed during World War II to correctly detect planes in noisy radar measurements.

Signal detection theory can be applied to different cases, and we present it in its basic form. In the basic version, the subject is presented with a single interval containing either noise with the signal at a known low level near the threshold or just plain noise and asked to determine the presence or absence of the signal. This task is presented to the subject many times, and the rates for both 'signal present' and 'signal not present' are measured as percentages. The answers are then categorized as 'hit', the signal is present; 'miss', the signal is present but not perceived; 'correct rejection', the signal is not present and not reported; and 'false alarm', the signal is not present but reported. According to the theory, presenting the signal at a fixed level sets a value to the internal variable $h_s(t)$, which is not constant but varies due to the presence of external and internal noise. The mean value of $h_s(t)$ in this case is denoted as μ_s and the standard deviation as σ_s. When the signal is not present, the value of the internal variable $h_n(t)$ is assumed to have a lower mean value μ_n than μ_s, and the external and internal noise then causes the standard deviation μ_n.

The sensitivity d' is a statistic that shows the separation between the probability density functions of $h_s(t)$ and $h_n(t)$, which represent the distributions of the internal variable h in the presence of signal+noise and only noise, respectively. The sensitivity d' of the subject perceiving the signal at a given level can then be written as

$$d' = \frac{\mu_s - \mu_n}{\sqrt{(\sigma_s^2 + \sigma_n^2)/2}}. \tag{8.3}$$

In some cases we cannot directly measure the variables and their standard deviations, since h_s and h_n exist only as internal variables. In such cases, d' can still be estimated as

$$d' = Z(\text{hit rate}) - Z(\text{false alarm rate}), \tag{8.4}$$

where Z is the inverse of the cumulative Gaussian distribution. For example, if the hit and false-alarm rates are both 50%, $d' = 0$. For corresponding rate pairs (hit rate, false-alarm rate), (0.7, 0.4)→ $d' = 0.78$, (0.9, 0.1)→ $d' = 2.1$ and (0.9, 0.3)→ $d' = 1.8$. A more thorough description of the theory and further applications of it can be found in Gescheider (1997).

8.5.2 *Scales and Categorization of Percepts*

The absolute or difference thresholds can only be used to measure psychophysical functions that relate attributes of sound events to simple functions having only two values. Often, a psychophysical function which represents an auditory percept with a continuous scale is desired, for example the mapping of sound pressure levels between 0 dB and 120 dB to the perceived loudness of an auditory event. There are many methods to estimate such psychophysical functions.

The task is, naturally, very complicated, as the subjects cannot access the absolute measure of the auditory attribute. For example, the subject's ability to report, say, the loudness of a sound on an absolute scale or in relation to a sound heard more than a few seconds earlier is limited. As will be shown later, some clever systems have been found to define complex psychophysical functions, such as the relation of the SPL and loudness, or the sound spectrum and the loudness spectral density.

When estimating the magnitude of sensation on a scale, subjects are asked to describe a characteristic of an auditory event on a response scale either by itself or in comparison to other auditory events. In such tests the whole psychophysical function spanned by the attributes of the sound event may be researched, although a number of problems are inherent in such measurements, as already discussed at the beginning of this chapter. Different scales used in the tests are described below:

- The simplest response scale is the *nominal* scale, where the response (a number or symbol) implies that the auditory event belongs to a certain class. The classes may be, for example, *rough*, *reverberant*, *bright*, and so on. In nominal classification, the auditory events or their characteristics are not compared on any quantitative scale.
- When the auditory events, or some of their audible characteristics, can be ordered in an array, such a scale is called the *ordinal* scale. The position of an auditory event in the array is denoted with a positive integer, which does not mean that the differences between the ordinal numbers can be used as a measure of dissimilarity. Arithmetic operations between values are not applicable.
- The *interval scale*, in turn, is a numerical scale, which defines the differences of classes quantitatively. The zero point of the scale is not meaningful, because only the differences of subsequent values on the scale are defined. The valid arithmetical operations are thus based only on the differences.
- The *ratio scale* is defined similarly to the interval scale, but with the zero point defined. The valid analysis methods also include such operations where the position of zero on the scale has meaning (such as the geometric mean).

8.5.3 Numbering Scales in Listening Tests

The user interface normally contains a numerical scale, which is assumed to help the subject to describe the auditory attribute being studied. The numerical scaling of an auditory event is often performed on a scale with easily conceivable numbers, such as $p_i \in [1, 5]$, $p_i \in [1, 10]$ or $p_i \in [0, 100]$. A special case is the Mean Opinion Score (MOS) scale, MOS $\in [1, 5]$, which is commonly used to evaluate audio quality. A verbal description may correspond to different positions on the scale, for example, the value 5 may correspond to 'excellent' quality. MOS scales are discussed in more detail in Section 17.4.1.

A scale can also be set up using two concepts exhibiting opposite values, the *semantic differential*. Such pairs are, for example,

- soft $\leftrightarrow$ hard
- low $\leftrightarrow$ high
- distorted $\leftrightarrow$ clean

In the case of clearly opposite values, defining the scale symmetrically around zero can be meaningful, as in $p_i \in [-5, 5]$. The neutral case corresponds, in this case, to the value zero.

8.6 Tasks for Subjects

In formal listening tests, a task is given, sound events (or silence) are presented, and the subject responds to the auditory event according to the task. The subject typically has access to a human–computer user interface, such as a computer keyboard, a touch screen, or a specific response device. In some cases, speech- or movement-based reporting may also be used.

In any case, it is important to eliminate all sources of error and bias in psychoacoustic tests. The ideal situation is a blind test, where subjects have as little information as possible on the sounds they are hearing. The wording of the questions or tasks is very important in this respect.

Typical tasks used in listening tests are described below.

- *Detection*: The subject is presented with a single interval, which may contain a signal, noise or both, and a relevant question is asked, such as, 'did you hear a sound?'
- *Discrimination*: The subject is presented with a single interval containing two sound events with a small difference in an attribute of the sound event. The question may be, for example, 'do you hear a difference?'
- *Forced choice*: Subjects are presented with a number of temporal intervals from which they have to choose one based on the question asked. One of these intervals contains the signal while others are silent or contain some other sounds. The choice is thus based on the comparison of auditory events with a predefined criterion. Depending on the sounds presented to the subject, the task can be used to measure thresholds or can be scored. Such tests have different nomenclatures, such as *two-interval forced choice* (TIFC or 2IFC) or *two-alternative forced choice* (TAFC or 2AFC). The number of intervals can also be higher. Forced-choice methods are not sensitive to bias produced by subjects' tendencies, as discussed in Section 8.5.1.
- *Direct scaling*: Subjects are presented with the sound event being studied, which may be presented only once or it may be accessible many times using a user interface, after which they must report the magnitude of the sensation on a given auditory scale. This is also called *magnitude estimation*, or *grading*. The question asked may be, for example, 'how loud is the sound on a scale from zero to ten?'
- *Adjustment*: The task of the subject is to adjust the value of an attribute of a sound event until a desired attribute is obtained. This is more commonly called the *method of adjustment*, and it will be discussed more in detail in the next section.
- *Chronometric tasks*: Here, subjects are given a task where they must react to a specified auditory event as quickly as possible. An example task is 'press button A as quickly as possible when you hear a voice.' Different auditory events are then presented, and conclusions are drawn from the subjects' reaction times.
- *Verbal description*: Subjects are required to describe verbally the sounds they perceive. The description can be done using questionnaires, with free-form textual or oral descriptions, or by other means. The subject may also be asked to answer a formal question or to perform a task, and the verbal description complements the results. There are also methods that apply statistical tools to analyse verbal descriptions. Often, the target of such tests is to seek the perceptual dimensions in the background of a complex sound event, such as in product quality. In some cases, informal descriptions can also be reported, although they are seldom found sufficient for drawing conclusions in psychoacoustic experimenting. However, they may be a useful addition to other results.
- *Other tasks*: Different types of tasks can be utilized depending on the topic being studied and on the subjects. For example, a psychoacoustic test can be implemented as an application similar to a computer game, where the scoring of the subject defines the result. A typical use of such a task is to test hearing aids with children, who typically cannot concentrate on mechanical tasks for long periods. The use of such applications can produce less biased results than the simple use of formal tasks.

8.7 Basic Psychoacoustic Test Methods

So far, we have discussed the generation of sound events and auditory events, the scales (or psychophysical functions) to be researched with the tests, and also the tasks for the subjects. Another dimension in designing psychoacoustic tests is the procedures – how the tasks discussed in the previous section are presented to the subject in succession to ensure that meaningful results will be produced about the phenomenon being studied. The test method is here defined to be the logic underpinning how the attributes of sound events are chosen into the subsequent tasks presented to the subject. Psychoacoustic tests can be conducted using many different methods. The most important methods are described below.

8.7.1 Method of Constant Stimuli

The *method of constant stimuli* is used to quantify thresholds. To this end, the experimenter chooses a relatively large number of sound event attribute values around the assumed value of the absolute or difference threshold. The listening test is conducted for each attribute value with a detection or discrimination task, or preferably with a multiple-interval forced-choice task to avoid bias. The task must be repeated a considerable number of times, and a psychophysical function such as that shown in Figure 8.2 is obtained. The actual value for the threshold can then be selected to be, for example, the value of the abscissa where the function has the value 50%.

In some cases the value of the psychophysical function does not approach 0% at the lower end of the scale. This happens, for example, if the difference threshold is measured with the 2AFC method, where the chance of guessing correctly is 50% when the difference between the signal and the reference is below the threshold. In this case, the value of the threshold may be selected to be 71% or 75% of the maximum of the function.

In principle, the method of constant stimuli is the best method to measure the value of a threshold, as it avoids many subject-related sources of error. Additionally, the shape of a psychophysical function is also revealed with the method, whereas other methods reveal only the value of a certain threshold. Unfortunately, it is a relatively slow method to conduct. The number of presented stimuli required to obtain reliable data is, in many cases, relatively high, and often some adaptive methods are used instead.

8.7.2 Method of Limits

In the *method of limits*, an attribute of a sound event is changed automatically or by the experimenter, and the sound is presented to the subject during an interval. The subject can be given a *detection task*, reporting whether the stimulus was present in an interval. The gathered data are then used to measure absolute thresholds.

In an ascending series, the stimulus attribute value is first set well below the threshold and is then increased until the response changes. The attribute value where the response changes is called the 'limit'. In a descending series, the opposite is performed: the stimulus attribute is decreased from a level well above the threshold until the response changes. The experiment consists of many runs in both directions, possibly distributed randomly. The average of all the obtained limits is taken as the threshold.

Alternatively, in measurements of difference thresholds, two sound events are presented with a small difference in their attributes, and subjects perform a *discrimination task*, reporting if they perceive a difference in the auditory events. The method of limits, in general, is prone to bias, since the tendency to report the stimulus one way or another affects the results.

8.7.3 Method of Adjustment

The *method of adjustment* is actually a task, as already mentioned in the previous section. Here, the subject changes an acoustic attribute of the sound event until the auditory event corresponds to a reference value. For example, the level of a tone is adjusted to a level where it is just noticeable, or the frequency of a tone is adjusted to match the perceived pitch of another sound. The adjustments are conducted many times, and the results are averaged. The subject must be instructed to adjust the attribute value to one higher and lower than the reference before picking the final value. If possible, the adjustment is made in steps that are of the order of the JND to avoid the subject overestimating the change in the auditory event due to a minimal adjustment, the effect of which is actually not perceivable (Cardozo, 1965).

Besides finding thresholds, this method can be used in other tasks. For example, in *magnitude production*, the subject is asked to adjust a certain attribute of a sound event until the desired magnitude is reached. Similarly, in *ratio production*, the adjustment is made to obtain a ratio between the auditory attribute of each of two percepts. In early psychoacoustic experimenting, many of the tests were conducted by asking the subject to adjust an acoustic attribute of a test sound so as to produce the attribute that corresponded to, say, twice that of a reference value. By repeating this procedure for a new reference value every time, a relative psychophysical scale is derived, which can further be made into an absolute scale by choosing a single anchor point with a specified *anchor sound*. The subjective attribute value of the anchor sound corresponds to a certain value of the objective attribute value. This is used, for instance, for loudness and pitch scales defined later in this book. This technique has many variations.

8.7.4 Method of Tracking

In the *method of tracking*, the subject influences the direction of change of the studied attribute of sound. The historical example of this method is *Békésy audiometry*, where the task of the subject is to press a button whenever hearing a sound (von Békésy, 1960). The level of the tone whose frequency is swept gradually decreases as long as the button is pressed and increases when it is released. The local average of the level function with frequency can be taken as an absolute hearing threshold of the subject. The method is prone to bias, since the tendency to produce false positive and negative answers is not taken into account.

8.7.5 Direct Scaling Methods

Different methods are used to measure psychophysical functions with a direct scaling task. Measurements using a single auditory event only are prone to noise and bias, since the human senses have limited accuracy when evaluating the strength of a stimulus on an absolute scale. It is often beneficial to conduct the test so that the sound event being studied is compared to one or more known reference sound events, which are sometimes called *anchors*. The subject may or may not be aware which of the sound events are really the references. In the context of Figure 8.1, this means that the effect of the reaction function $b_{\mathrm{o}} = f(h_{\mathrm{i}})$ is minimized when the subject compares two auditory events with a minimal difference.

The task may also be to scale multiple sound events, which are compared with each other and potentially with some reference sound events. In audio quality measurements, such tests are often referred to as the multiple-stimulus-hidden-reference-with-anchors (MUSHRA) type of tests, which will be discussed in Section 17.4.2 in more detail.

8.7.6 Adaptive Staircase Methods

Adaptive staircase methods are similar to the method of tracking, except that, typically, forced-choice tasks are used. The value of the tested attribute is altered based on whether the subject's answer is correct or not. A wrong answer changes the attribute to make the task easier. Correct

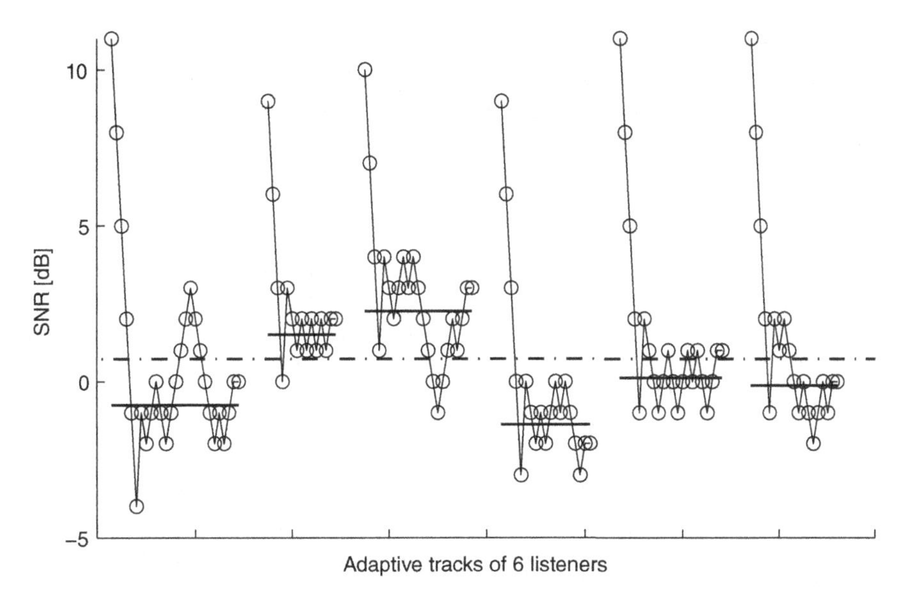

Figure 8.3 The detection threshold of a signal in noise measured with the adaptive procedure. The subject is presented with a signal+noise interval and two noise-alone intervals in random order with the forced choice approach. The SNR is reduced for correct answers and increased for wrong answers. The SNR step size is 3 dB first and is decreased to 1 dB after two reversals in the procedure. The results for six subjects are shown: the bold solid lines are the individual averages computed over the eight last turning points of the procedure. The bold dash-dotted line shows the average for 20 subjects.

answers, on the other hand, make the task harder. After a sufficient number of trials, the attribute ideally converges to the value of the threshold, and the average of last reversals in the tracking curve can be used as an estimate of the threshold. The value of the attribute can be plotted as a function of the number of trials, which often resembles a staircase, hence the name for the method. Staircases from such an experiment are shown in Figure 8.3 as an illustration.

The method is often designed with decreasing step size, starting with relatively large steps that are made smaller when convergence is thought to occur. There are variants of this procedure, where the level of convergence in the psychophysical function of a threshold is changed. The variants either apply different step sizes for the up and down movement or they require a different number of correct or wrong answers before changing the level of the attribute (García-Pérez, 2011; Levitt, 1971). Adaptive procedures are quite common, since they avoid the subject-related bias effects, and since the threshold value can be found with a smaller number of tests than with the method of constant stimuli. However, the down side of this method is that it does not guarantee convergence, and the experimenter must carefully select the parameters used and verify that the results obtained are meaningful.

8.8 Descriptive Sensory Analysis

The previous sections discussed listening tests, where it is often implicitly assumed that the difference between auditory events is in a single auditory attribute, such as in loudness or pitch. As will be discussed later in the context of sound quality, the auditory events studied may differ from each other in multiple dimensions, for example in both loudness and pitch. The situation becomes even more challenging if the attributes are not known *a priori*. For example, when perceptually motivated audio codecs are developed, and different versions of the codecs are tested, the listeners may perceive changes in 'crispness', 'noisiness', and 'loudness'. *Descriptive sensory analysis* is a family of methods which targets revealing the palette of attributes of a given set of stimuli, and in some cases also scaling (or grading) the stimuli in the attribute dimensions.

In the tests, a set of sounds is defined by the experimenter, and the perceptual properties that differentiate them are to be measured. Descriptive sensory analysis aims to identify, describe, and quantify the sensory attributes of stimuli using naive or trained human subjects (Piggott *et al.*, 1998), and it is often described as the most sophisticated tool in sensory science (Lawless and Heymann, 1998). An overview of the techniques in the context of audio is given by (Lorho, 2010), and is summarized here.

A number of techniques have been specifically designed for this purpose, mainly in food science but also in speech and audio. The term *elicitation* is often used in this context, which means 'the process of getting information from someone'. In this case, the elicited information is how the auditory events differ from each other, specifying all the attributes in which the differences are found, and also how much they differ in each of the dimensions specified by the attributes. This means that the researched sounds are projected to an N-dimensional space spanned by the elicited attribute dimensions.

The set or palette of attributes is often thought of as a *vocabulary*, which means a set of meaningful words that can be associated with the attributes. As shown in Figure 8.4, the techniques can be divided as follows:

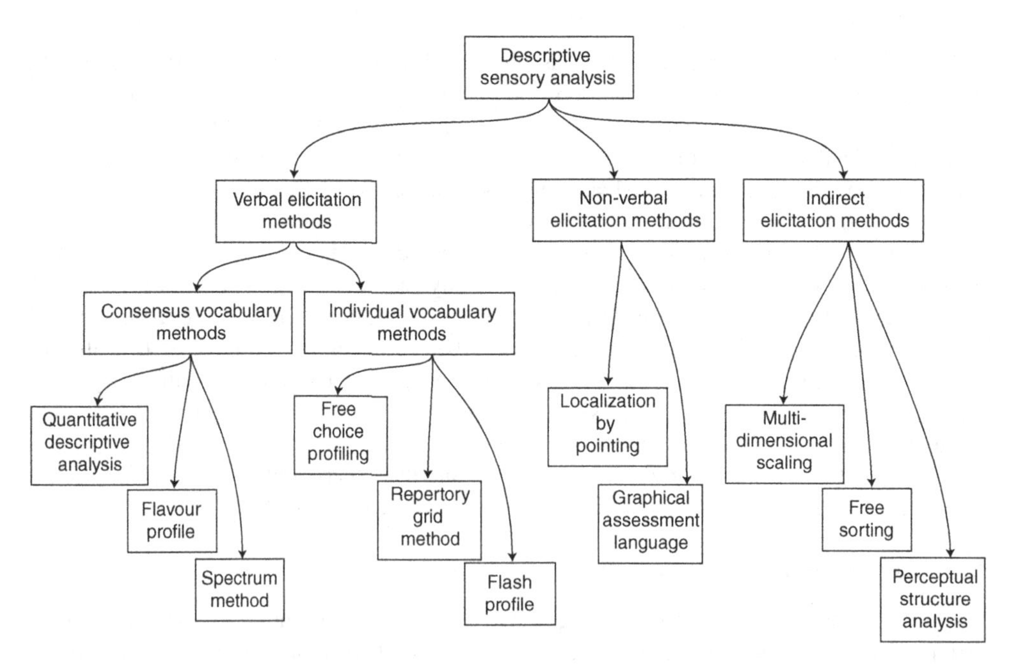

Figure 8.4 Descriptive sensory analysis methods commonly encountered in the field of sensory science. Adapted from Lorho (2010).

- *Verbal elicitation*: Methods relying on a verbal description of perceived sensations, for instance, techniques employing a vocabulary development process with a group of subjects.
- *Non-verbal elicitation methods*: These techniques are based on bodily gestures.
- *Indirect elicitation*: Covers those methods working without direct sensation labelling.

8.8.1 Verbal Elicitation

Techniques based on verbal elicitation are extensively utilized in the field of sensory science, and they form the largest group of descriptive analysis methods. Two distinct categories of techniques exist for establishing the sensory descriptors, consensus vocabulary (CV) methods and individual vocabulary methods. CV methods use a panel of assessors to develop a common set of descriptors, or dimensions, characterizing the sensory properties of the stimuli being investigated. Examples of CV methods are:

- *Flavour profile method* (Cairncross and Sjöström, 1950): A major component of the flavour profile method is a highly trained panel of four to six members, who individually evaluate the stimuli and then work in discussion as a group to determine a consensus profile. This consensus leads to data that act as a representative value; this it is not an average of the panellists; scores, it is a single score agreed upon by all panel members. This component of the profile method was criticized in the 1960s and 1970s as offering too much potential for the panel leader or an opinionated panellist to introduce bias. It is also claimed that the appropriate selection of panellists, extensive training, and the blind nature of the testing can protect against bias.
- *Quantitative descriptive analysis* (Stone *et al.*, 2004): During training, a representative set of stimuli is used for the consensus language development. The panel leader facilitates communication without involvement and interference in panel discussions. Known reference stimuli can be used to generate sensory terminologies, especially when panellists disagree with each other on some sensory attributes. The subjects then conduct the actual analysis of the stimuli separately using the descriptors found in the training period.
- *Spectrum method* (Meilgaard *et al.*, 2006): In spectrum descriptive analysis, the panel consists of 10 to 15 screened subjects who develop technical expertise through a comprehensive training procedure. A descriptive terminology is built covering all the perceptual attributes using a set of absolute category scales calibrated to have equal intensity across the attributes. For example, grade 5 on a sweetness scale is defined to have equal intensity with grade 5 on a saltiness scale. The scales are based on the systematic use of reference points with corresponding reference samples, but the magnitude estimation of the attributes is made individually by the assessors.

Individual vocabulary methods let subjects in turn develop their own, individual set of sensory descriptors. Examples of such techniques are:

- *Free-choice profiling* (Williams and Langron, 1984): In free-choice profiling, subjects are assumed to differ mainly in the way they describe sensory characteristics and not so much in the way they perceive them. This allows assessors to first elicit the dimensions in the stimulus set and then to quantify the stimuli with the attributes following their own vocabulary. The effort used in panel training is thus considerably reduced because the difficult and time-consuming step of agreeing on the descriptors is side-stepped. The output of the analysis is thus a set of grades on individual scales. For example, when analysing a specific sample, subject 1 may give the value 6 to the dimension 'reverberance' and value 2 to 'bass', while subject 2 may give the value 7 to 'hall-sound' and 1 to 'warmth'. A data analysis procedure known as generalized procrustes analysis (Gower and Hand, 1995) is then used to project the results to a common set of dimensions representing the sensory attributes.

- *Repertory grid technique* (Kelly, 1955): The basic idea of this technique is to get subjects to define their own constructs by asking them to describe the ways in which elements and their associated meanings vary. This is done, for example, by presenting three samples, where each assessor states the characteristic for which two of the samples are similar to each other and different from the third. After a number of trials, an individual set of descriptors of the dimensions is obtained and can then be applied to evaluate all the stimuli. Different types of data analysis, such as principal component analysis, can be exploited to study individual and multiple aspects of the experiments.
- *Flash profile* (Delarue and Sieffermann, 2004): The individual elicitation approach of free-choice profiling and the pair-wise comparative evaluation technique are combined in flash profile. During the descriptive analysis process, all stimuli are compared in pairs, which apparently removes the need for a phase of familiarization and a phase of individual training with the attributes. In addition, flash profiling assumes that assessors are familiar with descriptive analysis, which ensures that discriminant attributes can be generated in a short time. Generalized procrustes analysis or a similar method has to be used to reveal the main dimensions in the data. As a result, a relative sensory grading of the stimuli on the scales found in the test can be obtained in just one to three sessions with this technique.

8.8.2 Non-Verbal Elicitation

In *non-verbal elicitation techniques*, which form the second group of descriptive analysis methods, the aim is to achieve a direct elicitation of perceived sensations, but without using a formal set of verbal descriptors (Mason *et al.*, 2001). Several techniques based on bodily gestures, such as *localization by pointing* in the direction of the tested or reference object, have been used (Choisel and Zimmer, 2003). The rationale is that verbal elicitation is not always appropriate to describe the complexity of an auditory space. Drawing techniques have also been exploited in the *graphical assessment language* to quantify the auditory perception created by spatial sound reproduction systems (Ford *et al.*, 2002).

8.8.3 Indirect Elicitation

The third group of descriptive sensory analysis methods comprises techniques based on *indirect elicitation*, as shown in Figure 8.4. The test methods included in this group are significantly different from the verbal and non-verbal elicitation methods discussed earlier, since the subjects do not elicit directly the perceived sensory characteristics of the stimuli. *Multidimensional scaling* (Carroll, 1972) is an example of an indirect elicitation method commonly utilized. A number of samples are produced, and the target is to find the main auditory attributes responsible for the dissimilarities between the samples. The listener rates the perceived dissimilarity pairwise between all combinations of the samples, thus forming distance matrices between the samples. The matrices are scaled to a lower-dimensional space for easier interpretation. The perceptual attributes are assumed to be present in the resulting space. However, the distance matrices alone do not offer a way to interpret the perceptual dimensions associated with the spatial map, because no labelling of the sensation is asked of the subjects.

Free sorting requires subjects to create groups containing stimuli that are perceived similar, based on their own criteria. In addition, they can be asked after the sorting task to describe each group of stimuli with verbal descriptors. This *a posteriori* labelling is assumed to facilitate the interpretation of perceptual dimensions during the analysis (Cartier *et al.*, 2006). The interview data may be analysed by means of the *grounded theory* (Corbin and Strauss, 2008), where a theory is systematically developed beginning from the collected data. The key points in the interview notes are labelled with codes, which are further organized into categories to form the basis of the theory, explaining, for example, sound quality.

Perceptual structure analysis is an example of an indirect elicitation technique recently used in the field of audio by Wickelmaier and Ellermeier (2007). This approach is based on Heller's theory of semantic structures (Heller, 2000), where the processes of identification and labelling of perceived characteristics are separated. In the test, the subjects are presented with three stimuli and are asked to indicate if the first two stimuli share a common feature with the third stimulus or not. After verifying that the subjects really use consistently the features that the data indicate, a representation of the individual perceptual structure can be derived indirectly. The method has been applied by Choisel and Wickelmaier (2007) to develop a set of auditory attributes that describe the differences in perception of multi-channel sound reproduction.

8.9 Psychoacoustic Tests from the Point of View of Statistics

An important part of the research on psychoacoustics is the statistical analysis of the results from listening tests. Actually, in some cases, the statistical considerations should be taken into account in the design of the experiments. Tests should be conducted with the proper number of subjects and a meaningful selection of stimuli and tasks, listening conditions, and repetitions. In many cases, such designs should be conducted carefully so that the results prove or disprove the existence of the phenomenon that is hypothesized based on informal listening before the test.

The attributes of sound events are frequently called independent variables in experiment design and statistical analysis. In subjective tests, the independent variables are manipulated to produce different stimuli for testing, and when the selected test method is applied to the subjects, their responses (possibly after some post-processing) then yield the 'dependent' variable(s). The results should then reveal the relation of the 'independent' variable(s) to the 'dependent' variable(s), or in general the psychophysical function $h_{\mathrm{i}} = f(s)$, as discussed in Section 8.1.

Testing typically produces a large data set to which proper statistical methods must be applied. In simple cases, some basic descriptors, such as means, variances, and 95% confidence intervals can be used. However, often the influence of attributes in the tests on the data obtained should be examined using appropriate parametric or non-parametric statistical tests. Quite commonly *analysis of variance* (ANOVA) is used, which answers the question, 'do any of the independent variables have an effect on the dependent variable?' If an independent variable is found to have an effect, *posthoc tests* can then be used to measure how the independent variable affects the result.

Summary

This chapter discussed various methods used to study the functionality of hearing mechanisms by psychoacoustic means; that is, by presenting sound events to subjects and asking them to perform some tasks in a formal listening test method. The field is quite mature: if a test is designed carefully, the results indeed provide valid information on the attributes of an auditory event, generated by acoustic attributes of a sound event. In other words, psychoacoustic test methods can be used to measure the psychophysical functions that transfer acoustic attributes into auditory attributes. Descriptive sensory analysis involves methods of finding, in a formal way, the attributes of auditory events perceivable by subjects.

Further Reading

An introduction to psychophysical research methods of perception from all senses is found in Goldstein (2013). In Bech and Zacharov (2006), various considerations of listening-test design are made in the context of audio quality, and Gelfand (2004) discusses the psychoacoustical methods in general in more detail. A good source regarding early investigations into the quantitative formulation of auditory sensation and perception is Fletcher (1995). The use of descriptive sensory analysis techniques in the field of audio is reviewed by Bech and Zacharov (2006); Neher *et al.* (2006). The statistical analysis of quantitative attributes resulting from descriptive sensory analysis is reviewed by Næs and Risvik (1996).

References

Bech, S. and Zacharov, N. (2006) *Perceptual Audio Evaluation – Theory, Method and Application*. John Wiley & Sons.

Cairncross, S. and Sjöström, L. (1950) Flavor profiles – a new approach to flavor problems. *Food Technology*, **54**(4), 308–311.

Cardozo, B. (1965) Adjusting the method of adjustment: SD vs DL. *J. Acoust. Soc. Am.*, **37**, 786–792.

Carroll, J.D. (1972) Individual differences and multidimensional scaling. *Multidimensional Scaling: Theory and Applications in the Behavioral Sciences*, **1**, 105–155.

Cartier, R., Rytz, A., Lecomte, A., Poblete, F., Krystlik, J., Belin, E., and Martin, N. (2006) Sorting procedure as an alternative to quantitative descriptive analysis to obtain a product sensory map. *Food Quality and Preference*, **17**(7), 562–571.

Choisel, S. and Wickelmaier, F. (2007) Evaluation of multichannel reproduced sound: Scaling auditory attributes underlying listener preference. *J. Acoust. Soc. Am.*, **121**(1), 388–400.

Choisel, S. and Zimmer, K. (2003) A pointing technique with visual feedback for sound source localization experiments *Audio Eng. Soc. Convention 115* AES.

Corbin, J. and Strauss, A. (2008) *Basics of Qualitative Research: Techniques and Procedures for Developing Grounded Theory*. Sage.

Delarue, J. and Sieffermann, J.M. (2004) Sensory mapping using Flash profile. Comparison with a conventional descriptive method for the evaluation of the flavour of fruit dairy products. *Food Quality and Preference*, **15**(4), 383–392.

Fletcher, H. (ed.) (1995) *Speech and Hearing in Communication*. Acoustical Society of America.

Ford, N., Rumsey, F.J., and Nind, T. (2002) Subjective evaluation of perceived spatial differences in car audio systems using a graphical assessment language *Audio Eng. Soc. Convention 112* AES.

García-Pérez, M.A. (2011) A cautionary note on the use of the adaptive up–down method. *J. Acoust. Soc. Am.*, **130**, 2098–2107.

Gelfand, S.A. (2004) *Hearing: An introduction to psychological and physiological acoustics*. Marcel Dekker.

Gescheider, G.A. (1997) *Psychophysics: The Fundamentals*. Psychology Press.

Goldstein, E.B. (2013) *Sensation and Perception*, 9th edn. Cengage Learning.

Gower, J.C. and Hand, D.J. (1995) *Biplots* volume 54. CRC Press.

Heller, J. (2000) Representation and assessment of individual semantic knowledge. *Methods of Psychological Research*, **5**(2), 1–37.

Kelly, G. (1955) The Psychology of Personal Constructs. Norton.

Lawless, H.T. and Heymann, H. (1998) *Sensory evaluation of food. Principles and practices*, Chapmann & Hall.

Levitt, H. (1971) Transformed up–down methods in psychoacoustics. *J. Acoust. Soc. Am.*, **49**(2B), 467–477.

Lorho, G. (2010) *Perceived quality evaluation: an application to sound reproduction over headphones*. Ph.D thesis, Aalto University.

Mason, R., Ford, N., Rumsey, F., and De Bruyn, B. (2001) Verbal and nonverbal elicitation techniques in the subjective assessment of spatial sound reproduction. *J. Audio Eng. Soc.*, **49**(5), 366–384.

Meilgaard, M.C., Carr, B.T., and Civille, G.V. (2006) *Sensory Evaluation Techniques*. CRC Press.

Næs, T. and Risvik, E. (1996) *Multivariate Analysis of Data In Sensory Science* volume 16. Elsevier.

Neher, T., Brookes, T., and Rumsey, F. (2006) A hybrid technique for validating unidimensionality of perceived variation in a spatial auditory stimulus set. *J. Audio Eng. Soc.*, **4**, 259–275.

Piggott, J.R., Simpson, S.J., and Williams, S.A. (1998) Sensory analysis. *Int. J. Food Sci. & Technol.*, **33**(1), 7–12.

Stone, H., Sidel, J., Oliver, S., Woolsey, A., and Singleton, R.C. (2004) Sensory evaluation by quantitative descriptive analysis. In Gacula, M.C. (ed.) *Descriptive Sensory Analysis in Practice*. John Wiley & Sons, pp. 23–34.

von Békésy, G. (1960) *Experiments in hearing*. McGraw-Hill and Acoustical Society of America.

Wickelmaier, F. and Ellermeier, W. (2007) Deriving auditory features from triadic comparisons. *Perception & Psychophysics*, **69**(2), 287–297.

Williams, A.A. and Langron, S.P. (1984) The use of free-choice profiling for the evaluation of commercial ports. *J. Sci. Food Agri.*, **35**(5), 558–568.

9

Basic Function of Hearing

This chapter starts by discussing the most fundamental of questions regarding an auditory object: under what conditions does it exist? Two physical attributes limit the audibility of a frequency component of sound: the sound pressure level (SPL) and frequency. If the SPL is too low, nothing is heard. If it is too high, the sensation of sound is accompanied by the sensation of pain, and beyond some limit in the SPL the ears are destroyed. The threshold of pain is commonly taken as the upper limit of the SPL to generate auditory events in sound and voice techniques. On the other hand, sound components with too low frequencies are not perceived as sound, but more as a sensation of vibration, and sound components with too high frequency content are not perceived at all. The attributes also interact with tonal signals; the SPL threshold of audibility depends in a complicated manner on frequency. The first section in this chapter will discuss these issues.

When multiple sounds are presented to the subject, they influence each other's audibility. In spectral masking, sounds in different frequency regions make each other inaudible, and in temporal masking, the same happens in the temporal dimension. The basics of masking are discussed in the second and third sections of this chapter.

The last section in this chapter discusses the first steps of spectral analysis conducted in hearing; that is, the characteristics of the frequency bands in hearing. The processing of sound begins in the cochlea, where the sound is divided into narrowband time-domain signals, which are then processed in the brain more or less individually. When a spectrally broad sound is presented to a subject, a relevant question is what is the frequency resolution of the neural presentation? Two methods to measure the widths of the bands are described, and different estimates of human 'critical bandwidths' are reviewed.

9.1 Effective Hearing Area

The auditory system is able to receive and process a very wide range of different sounds. The first characterization of the abilities of hearing is to describe the working range of possible signal frequencies and amplitudes. Figure 9.2 defines the upper and lower limits of normal working hearing as well as equal loudness curves as functions of frequency. The useful range

---

Communication Acoustics: An Introduction to Speech, Audio, and Psychoacoustics, First Edition.
Ville Pulkki and Matti Karjalainen.

of frequency is from about 20 Hz to 20 kHz, although higher and lower frequencies can be perceived if they are intense enough. Sounds below the lower frequency limit are called *infrasound* and those above the upper limit are called *ultrasound*.

The smallest amplitude of a tone causing an auditory event is called the *hearing threshold*. This threshold curve is measured as the *minimum audible field* (MAF), the sound pressure level of the weakest binaurally perceived sound from the front. The MAF level is measured in the free field after the subject is removed, in the position where the centre of the subject's head was.

The decibel scale of sound pressure (see Section 2.2.3) is defined so that 0 dB is close to the hearing threshold for a pure tone at 1 kHz, which corresponds to the reference pressure p_0 in Equation (2.14), $2 \cdot 10^{-5}$ Pa. The sensitivity of hearing is best at about 3–4 kHz, and the threshold increases at low (below 250 Hz) and high frequencies (above 5 kHz).

The upper amplitude limit is about 130 dB, beyond which a hearing percept changes to pain. Just above this level the hearing system is in danger of being injured, even for a short impulse of sound. Even at considerably lower levels, at about 85 dB, exposure for eight hours daily over a long period results in hearing loss (see Section 19.3.2).

Figure 9.1 illustrates sound pressure levels in dB for typical sound sources that are useful for ear-based level estimation. A good reference in the middle frequency range is to remember that typical speech measured from a distance of one metre is about 60–70 dB. This is the optimal level for communication, since it is not too loud to be harmful for the functioning of hearing but is loud enough to produce a good signal-to-noise ratio in most environments, and it also provides good conditions for temporal and spectral analysis to the auditory system.

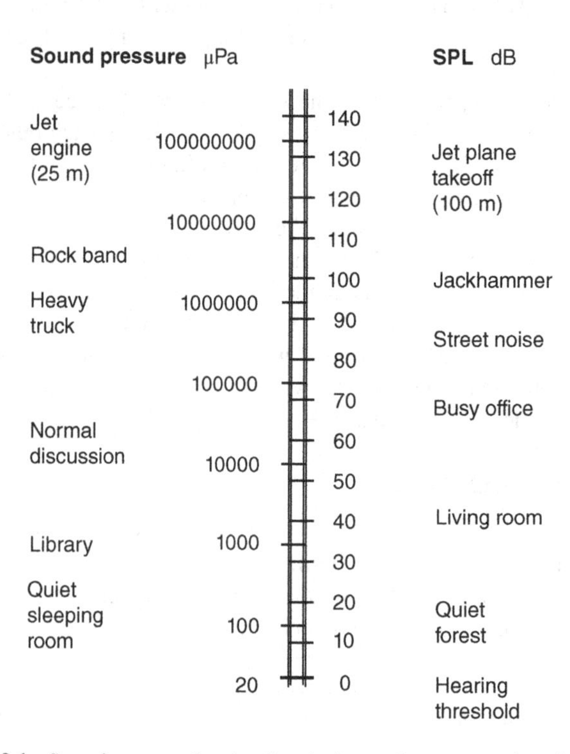

Figure 9.1 Sound pressure levels of typical sound sources and environments.

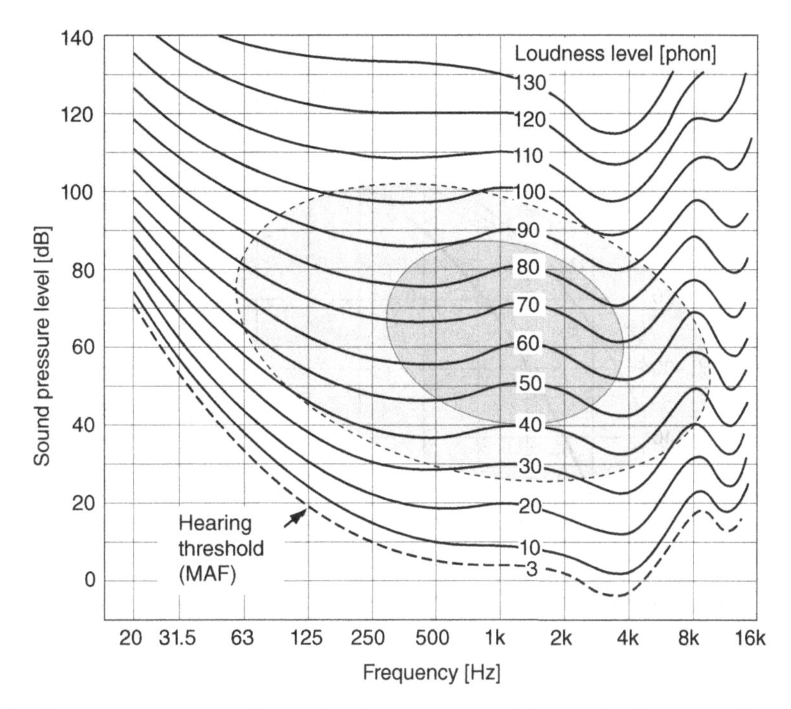

Figure 9.2 The working range and equal loudness contours of human hearing on the frequency–SPL plane. The lowest curve specifies the minimum audible field (MAF), the field producing a just audible sound to the subject, as measured in free-field conditions for a sound source in front of the subject. The uppermost curve specifies the approximate threshold of pain. Each separate curve specifies a constant of the perceived loudness for a pure tone as a function of frequency.

Although the full functional range of hearing is vast, about 130 dB for the level and 20 Hz to 20 kHz for frequency, a much smaller area of the total range is utilized in practical communication situations. For acoustic music the effective range is about that shaded light grey in Figure 9.2. In speech communication the most basic range is still more limited. For example, the frequency range for understandable speech communication over a telephone connection is from about 300 Hz to about 3.4 kHz and the amplitude range is typically 40–70 dB (the darker shade in Figure 9.2) (Fletcher, 1995).

9.1.1 Equal Loudness Curves

Sounds between the threshold of hearing and threshold of pain are perceived with increasing 'strength' or 'volume'. This subjective feature of sound is called *loudness*. The *loudness level* has been defined such that the sound pressure level of a 1-kHz pure tone in dB has the same loudness level in phon units (SPL, Equation (2.14)).

Based on this definition, it is possible to measure the equal loudness curves using subjects who compare a 1-kHz reference tone at a given level and a test tone of another frequency by adjusting the latter to have the same perceived loudness. Curves in Figure 9.2 are averaged for a large number of subjects under conditions standardized in ISO-226 (2003). Loudness perception will be discussed in more detail in Section 10.2.

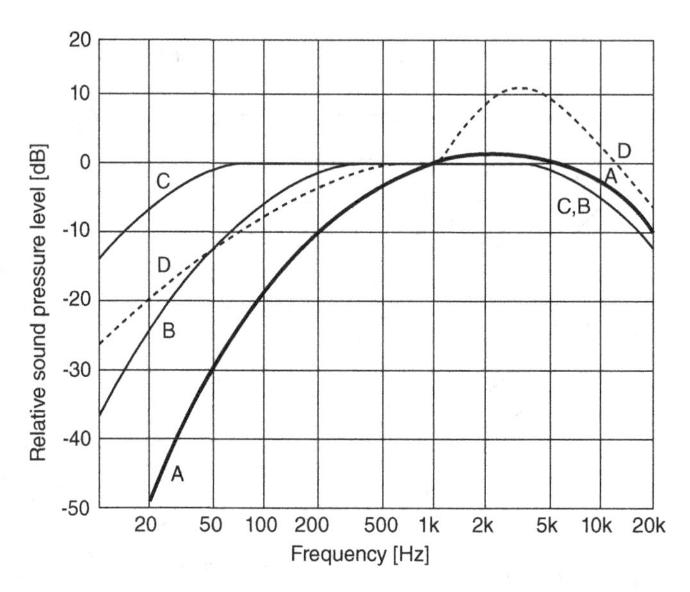

Figure 9.3 Weighting curves A, B, C, and D for sound level measurement.

9.1.2 *Sound Level and its Measurement*

Sound pressure was introduced in Section 2.2.2 as one primary physical measure of sound waves. The sound pressure level (SPL) in decibels, defined in Equation (2.14), is a logarithmic measure of amplitude, which is more convenient because of its large range and better corresponds to subjective perception. Even after this mapping to the new scale, sound pressure does not work well as a perceptual measure since the subjective level is strongly frequency-dependent, as can be seen easily from Figure 9.2.

In order to have a better, yet relatively simple, measure of the perceived level of sound, the concept of *sound level* is defined as a measure that is weighted with frequency so that the level roughly approximates the frequency sensitivity of hearing. Four different weighting curves are defined: A, B, C, and D, as shown in Figure 9.3. The *A-weighted* sound level is commonly used in noise measurements to characterize the perceived level and the risk of hearing loss. The A weighting slightly emphasizes the levels at mid frequencies and attenuates them at low and high frequencies. When the weighting curve is compared with the inverted equal-loudness curves, it is easy to see that the A weighting is just a very rough estimate. Technical simplicity and extensive use in practice are its advantages.

The other curves (B, C, and D) are rarely used. The unit of all sound levels is the decibel [dB], the same as for SPL, but it is common to also denote the weighting curve, for example, dB(A) for the A-weighted curve.

9.2 Spectral Masking

The inaudibility of soft sounds in the presence of louder sounds is very common in everyday listening scenarios. The phenomenon itself is present in all technical devices aiming to detect a weak signal in the presence of an interfering strong signal. The masking effect is an important characteristic of human hearing, which is commonly exploited in sound and voice technologies, as in the design of lossy audio codecs (see Section 15.3.1). The effect of spectral masking occurs when a sound with certain spectral content makes the detection of another sound with different spectral content harder, even though the spectra do not necessarily overlap. Spectral masking, that is, how the *masker* sound affects the detection threshold of the test sound, can be best described by plotting the masking threshold as a function of frequency.

9.2.1 Masking by Noise

Let us first investigate the effect of broadband noise on the masking threshold of a tone. The threshold is presented in Figure 9.4 when the masker sound is white noise with different power spectral densities. The test sound is a tone whose detection threshold has been measured in psychoacoustic experiments (Fastl and Zwicker, 2007). The dashed line shows the hearing threshold of the tone without the presence of the masker.

The masking threshold plots have a constant value up to about 500 Hz, after which they increase with a slope of 3 dB/octave (10 dB/decade). If spectral masking with a constant effect at all frequencies is desired, a *uniform masking noise* must be used, where the spectral density decreases correspondingly by 3 dB/octave at frequencies above 500 Hz.

The nature of spectral masking can be revealed by using narrowband noise as the masker. The masking thresholds then have the form shown in Figure 9.5. For example, a 160-Hz-wide noise masker with centre frequency 1 kHz and sound pressure level 60 dB generates a slightly asymmetric curve with its peak at 1 kHz, a few decibels below the level of the masker. The other curves shown in the figure for maskers with different centre frequencies f_c produce similar masking threshold curves.

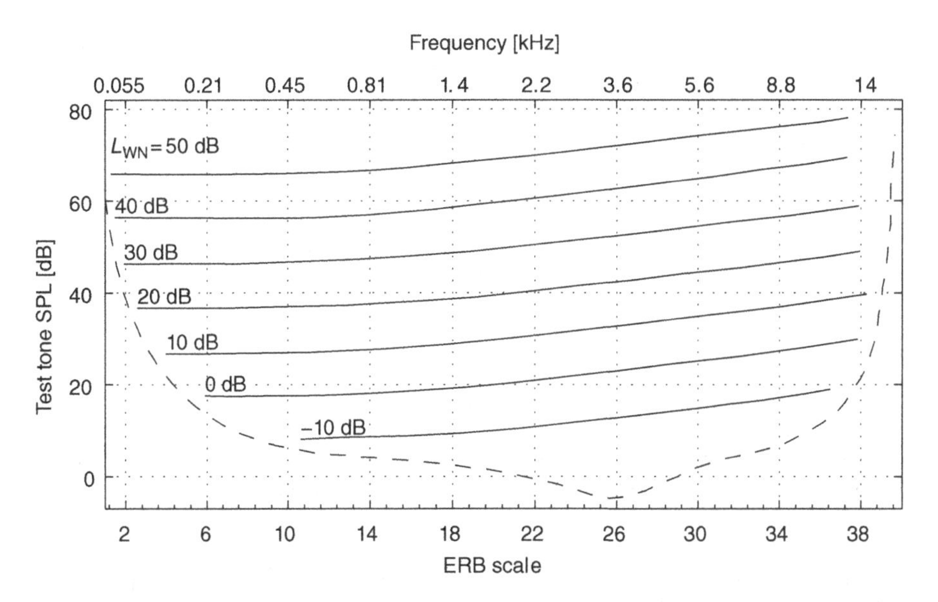

Figure 9.4 Masking thresholds (solid lines) in the presence of white noise with different power spectral density values L_{WN}. The ERB scale is an auditory frequency scale, and it will be described in Section 9.4.3 on page 167. The dashed line represents the threshold of audibility. Adapted from Fastl and Zwicker (2007).

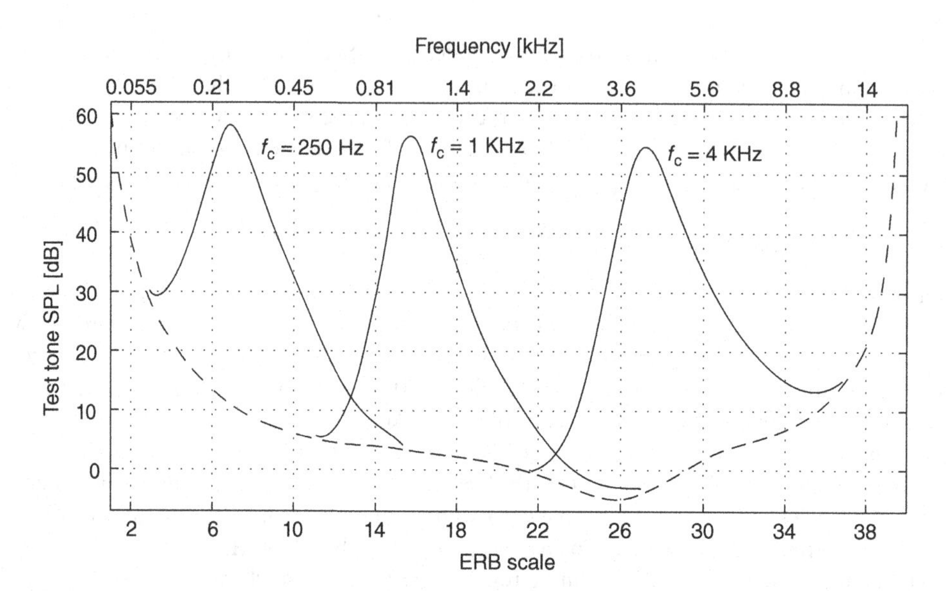

Figure 9.5 Masking threshold curves caused by narrowband noise with centre frequencies 250 Hz, 1 kHz, and 4 kHz with a sound pressure level of 60 dB. Adapted from Fastl and Zwicker (2007).

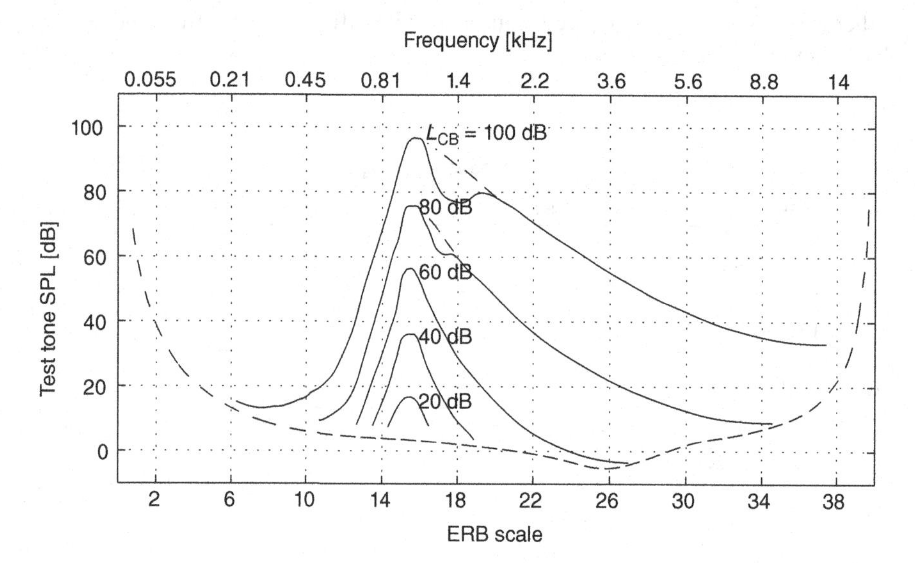

Figure 9.6 Masking threshold curves caused by narrowband noise with centre frequency 1 kHz at different sound pressure levels L_{CB}. Adapted from Fastl and Zwicker (2007).

When narrowband noise with a centre frequency of 1 kHz at different sound pressure levels is used as a masker, the masking curves shown in Figure 9.6 are obtained. When the level of the masker is increased, the curvature of the masking thresholds changes. With an increase in level, the drop in the threshold due to masking is shallower at frequencies above the centre frequency of the masker. However, the slope of the masking threshold below the centre frequency of

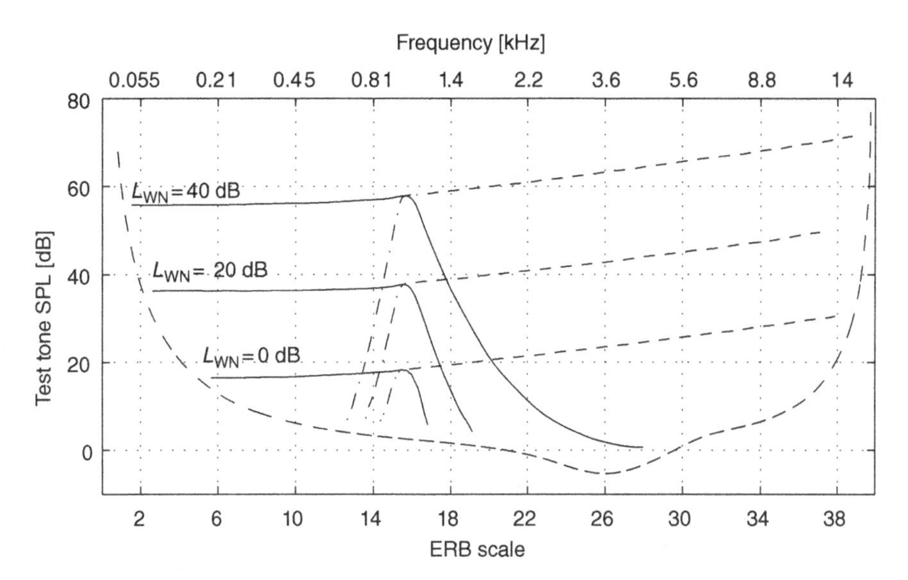

Figure 9.7 Masking threshold curves obtained with low-pass (solid line) or high-pass (dashed line) filtered white noise by a filter with a cutoff frequency of 1 kHz. Adapted from Fastl and Zwicker (2007).

the masker is not affected. The dips in the masking threshold curves with L = 80 and L = 100 dB at frequencies above the centre frequency of the masker are caused by the audibility of combination tones caused by non-linearities in hearing. The test sound itself becomes audible only when the level shown with a dashed line is exceeded.

If low-pass or high-pass filtered noise filtered at a cutoff frequency of 1 kHz is used as the masker, the masking threshold curves shown in Figure 9.7 are obtained, where the curves follow the results shown in Figures 9.5 and 9.6, but only on the half of the spectrum that has no signal.

9.2.2 *Masking by Pure Tones*

A pure tone – a sinusoid with fixed frequency – causes a masking effect similar to that of narrowband noise. However, due to the non-linear processing in hearing, in the presence of a second test tone, called the probe tone, close to the first tone (the masker tone) some beating and combination tones are also perceived. Beats are the periodic changes in the amplitude of the signal, which may be audible as roughness or fluctuations, as will be explained in Chapter 11. Thus, the presence of a probe tone with a tone masker can be detected with lower amplitudes than with a narrow bandnoise masker, as shown in Figure 9.8. In many cases, the probe tone itself is not audible, the listeners perceive only the beating or the combination tones (Fastl and Zwicker, 2007; Wegel and Lane, 1924).

9.2.3 *Masking by Complex Tones*

The masking effect caused by other sounds follows the same principle as with the special cases described above. The masking threshold caused by a harmonic tone complex consisting of the ten lowest partials is shown in Figure 9.9. Each harmonic can be thought to result in a partial masking pattern, and the total masking pattern is the sum of the individual patterns

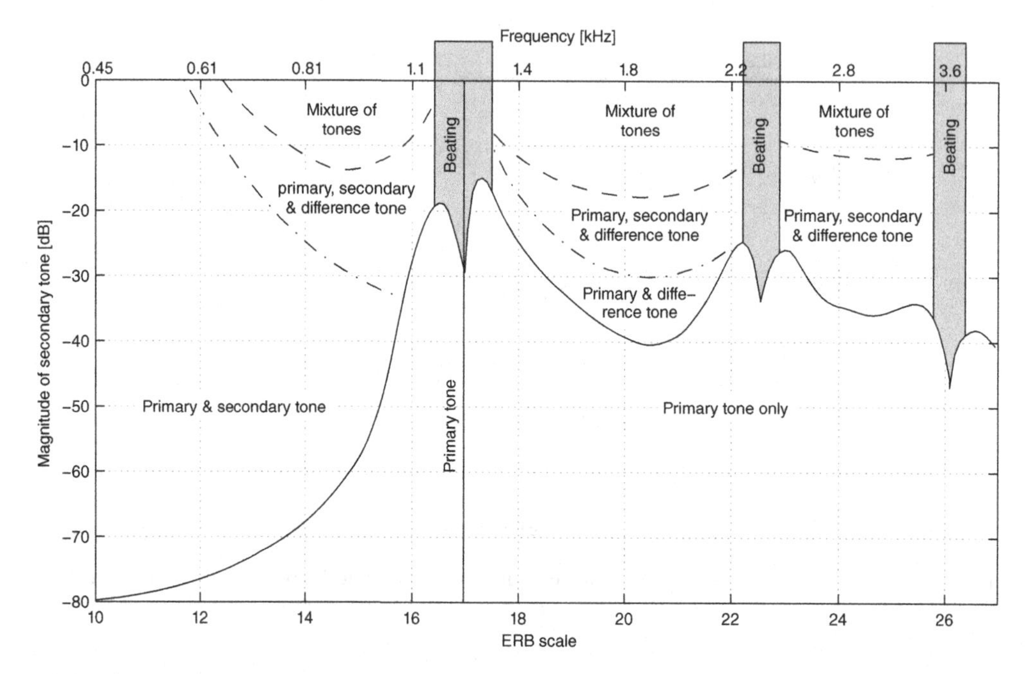

Figure 9.8 Masking threshold patterns of a probe tone in the presence of a tone masker (1.2 kHz). The probe and the masker tones interfere with hearing mechanisms at the frequencies shaded grey. The audibility regions of the masker tone, the test tone, and the difference tones are marked. Adapted from Wegel and Lane (1924).

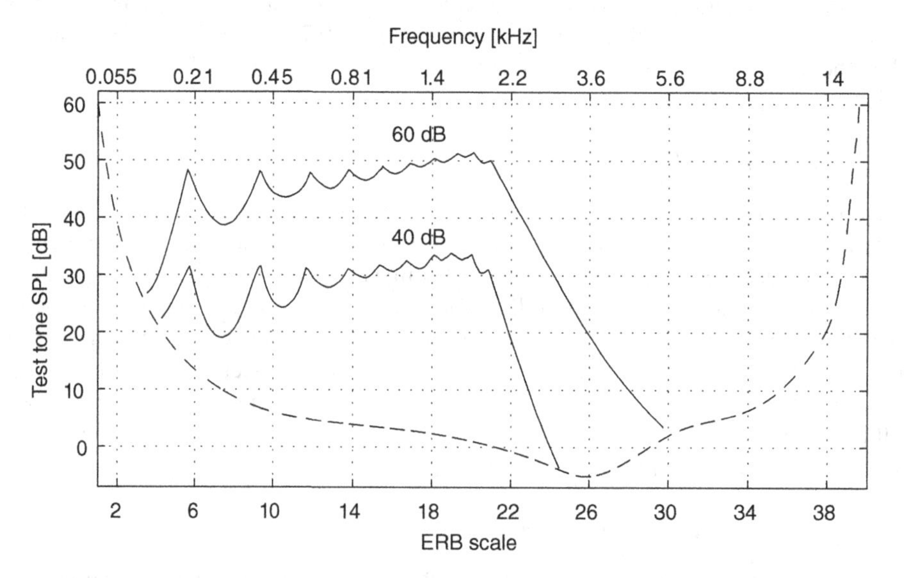

Figure 9.9 The masking patterns caused by a harmonic complex tone consisting of the ten first partials of equal level. The fundamental frequency of the masker is 200 Hz. The effect of the masker is shown for two different sound pressure levels. Adapted from Fastl and Zwicker (2007).

(Fastl and Zwicker, 2007). Similar results are obtained with instrument sounds similar to this spectrum. Overall, spectral masking is of great importance in music and audio reproduction. For example, the audibility of one instrument sound is limited by the masking effects caused by other, simultaneously playing instrument sounds.

9.2.4 *Other Masking Phenomena*

Co-modulation masking release is the decrease in the masking effect when the masker is amplitude-modulated with the same modulating function at all frequencies. Let us consider the case where the signal is a single sinusoid and the masker is a noise band around the frequency of the sinusoid. If the noise band is not modulated, the masking effect grows stronger when the noise bandwidth is extended. Interestingly, if the masker is amplitude-modulated coherently in all frequencies of the masker at the rate of 10–20 Hz, the masking effect deviates from the non-modulated case when the bandwidth of the masker exceeds 100 Hz (Hall *et al.*, 1984). This can be seen to imply that the hearing system groups different frequency bands together based on the similarity of the variation of the level in the bands.

In some cases the masking effect is reduced and cannot be explained with the masking effects discussed in the previous sections, which have been measured using simple sinusoids, noise, or transient sounds – types of masking called *energetic masking* (Watson, 2005). When the maskers and signals are more complex, reduced masking effects may be obtained, such as in speech-in-speech masking (Brungart, 2001). Such masking effects are called *informational masking* and can be thought to be caused by the ability of the human listener to segregate the signal and masker at a higher processing level.

Dau *et al.* (1997) suggest that the outputs of the auditory filters are processed by a bank of overlapping 'modulation filters' (analogous to auditory filters), each tuned to a different modulation frequency – *a modulation filter bank*. This helps the system to group signals sharing the same source, since different frequency bands originating from the same source are often modulated coherently. Although the model explains many aspects of modulation perception, the existence of the bank has not been proven, and is controversial (Plack, 2013).

9.3 Temporal Masking

The masking effect has been discussed so far only for continuous sounds. Sounds mask each other in time as well; that is, a sound affects the audibility of a preceding or following sound. A conceptual illustration of temporal masking is shown in Figure 9.10, both for a sound occurring before the masker, called *backward masking*, or *pre-masking*, and after the masker, called *forward masking*, or *post-masking*. Such thresholds of audibility are measured using psychoacoustic tests, where the listener hears a masker sound long enough (>200 ms), which is preceded or followed by a short burst of sound. Forward masking is, in general, more significant and more consistent a phenomenon than backward masking.

Backward masking has an effect only 5–10 ms before the onset of the masker, and relatively low-level sounds are masked. The effect seems to appear only in inexperienced listeners. Thus, backward masking is not interesting in the context of acoustic communication, and it is not discussed further in this book.

The forward masking effect, in turn, has an effect over a much longer period of time, about 150–200 ms after the offset of the masker. Relatively high-level sounds are also masked, thus it is relevant in the context of this book. The forward masking effect of a noise masker on a

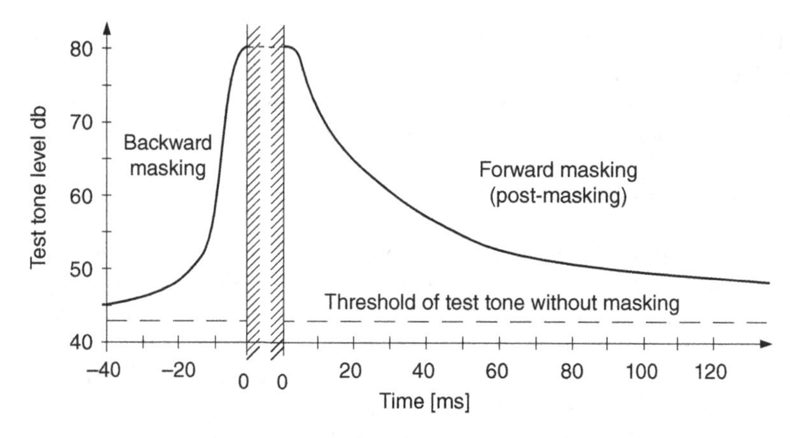

Figure 9.10 The temporal masking effects in hearing. A probe sound at a sufficiently low level arriving after a masker sound is not audible due to forward masking, and a probe sound at a low enough level arriving just before the masker sound is not audible due to backward masking. The vertical hatched lines represent the start and end times of the masker sound lasting at least 200 ms.

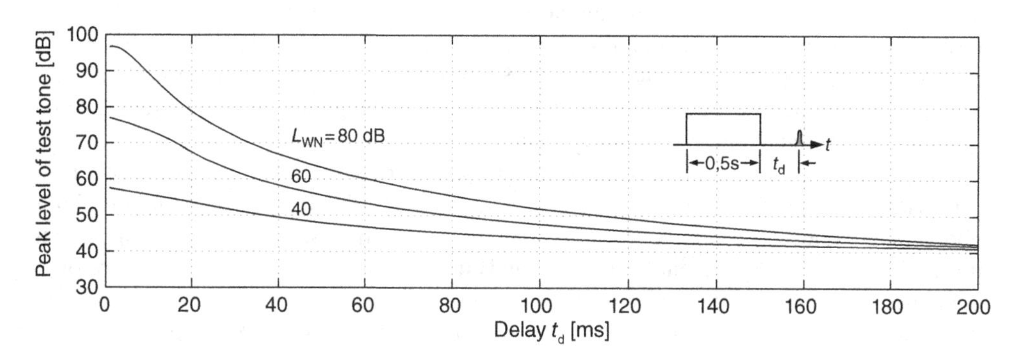

Figure 9.11 The forward masking effect at three different masker levels. Adapted from Fastl and Zwicker (2007).

probe tone is shown for three different levels of the masker on a linear time scale in Figure 9.11. The masking level decreases linearly after 5–10 ms from the offset, and after about 200 ms the threshold of hearing in silence is reached (Fastl and Zwicker, 2007).

The forward masking effect is a complex phenomenon, which also depends on the length of the masker sound. Figure 9.12 shows the effect for maskers lasting 5 ms and 200 ms. The effect is significantly milder with the shorter sound.

The masking effect has also been studied with periodically repeating bursts, modulated signals, pulses, and impulses (Duifhuis, 2005). For example, impulses or pulses can be used to measure the resolution of hearing of the temporal fine structure of sound, as shown in Figure 9.13. Using a masker composed of an impulse, the masking threshold has been measured for a probe impulse as a function of the temporal position with respect to the position of the masker impulse. A distinct pattern is seen, where the maximum value of the masking threshold pattern of about −10 to −5 dB is within a distance of 1 ms from the position

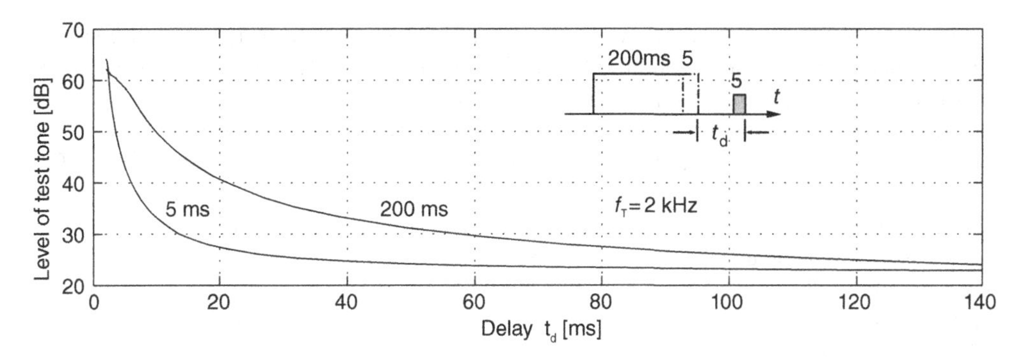

Figure 9.12 The forward masking effect for two different temporal lengths of the masker sound. Adapted from Fastl and Zwicker (2007)

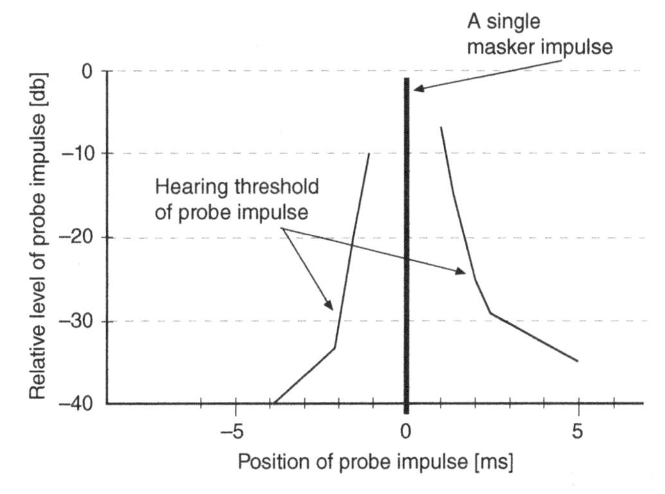

Figure 9.13 The temporal masking threshold pattern caused by an impulse. The probe signal is a single impulse at different positions in time with respect to the masker impulse. Adapted from Feth and O'Malley (1977).

of the masker impulse. The threshold decreases quickly at longer temporal distances from the masking impulse, reaching −40 dB after 5 ms. This shows that a relatively weak impulse is still perceived as an individual auditory event if it is not preceded or followed immediately by other impulses. If the impulses arrive within a time window of about 1–2 ms, they are merged into one auditory event, especially if one of them has a higher level. The value of 1–2 ms at which the impulses are just perceived as separate sounds can be treated as the best time resolution of hearing.

9.4 Frequency Selectivity of Hearing

The frequency masking patterns shown in Figure 9.5 imply that a narrowband sound presented to the listener affects the perception of other sounds in nearby frequencies. If two narrowband sounds at similar levels have different enough frequencies, our hearing resolves them into two separate auditory events. If they are located sufficiently close to each other in frequency, they are perceived as a single auditory event. The ability of our hearing to segregate sounds separated in frequency is called *frequency resolution.*

Frequency resolution and selectivity are important properties in terms of understanding the functioning of human hearing. Frequency resolution has been studied a lot, and slightly different results have been obtained with different approaches. Frequency selectivity is commonly thought to stem from cochlear processing; a broadband stimulus arriving at the cochlea is converted using mechanical and neural processing into a neural output. Each of the inner hair cells has a different best frequency to which it responds, but they also respond strongly to frequencies near the best frequency. This frequency region is called the *critical band* or the *auditory filter*. In practice, the width of this band, which depends on frequency, is of interest in sound and voice technologies.

9.4.1 Psychoacoustic Tuning Curves

The frequency selectivity of hearing measured from a single point on the basilar membrane was shown in Figure 7.11. A related measurement can be conducted applying psychoacoustic methods. In the measurement, the masking sound is usually narrowband noise, whose centre frequency is a parameter, and the test signal is often a tone, whose frequency and level (say, 10 dB) are kept constant when measuring a single curve. The task of the listener is to adjust the level of noise so that the test tone is just audible. The resulting functions obtained with this set-up are shown in Figure 9.14. The cochlear gains for the lowest signal levels in Figure 7.11 and psychoacoustic tuning curves in Figure 9.14 are clearly approximately reciprocal to each other, which verifies the validity of the approaches.

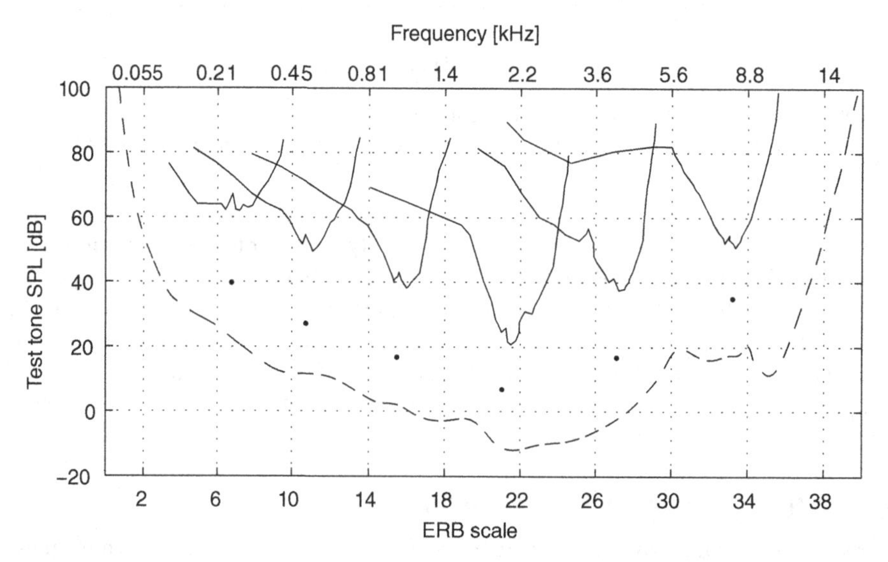

Figure 9.14 Psychoacoustic tuning curves measured using a low-level sinusoid as the signal and narrowband noise as the masker. The level of the sinusoid was 10 dB, and the frequencies of the tones were 0.25, 0.5, 1, 5, and 10kHz. Adapted from Vogten (1974).

Bark bandwidths

A classic method to measure the frequency resolution of human hearing is outlined in this section. A narrowband noise with a fixed centre frequency is used as the reference sound, to which the subject compares the test sound, band-limited noise whose sound pressure level and centre frequency are equal to the reference sound (see Figure 9.15). The change in perceived loudness is measured with some psychoacoustic test, and a schematic plot of such a test result is shown in Figure 9.16. Interestingly, the loudness is constant up to a certain value of bandwidth (Fastl and Zwicker, 2007). This bandwidth is 160 Hz for the center frequency 1 kHz, as shown in the figure, beyond which the loudness increases steadily. The knee point in the plot is thought to be the position where the test sound's spectrum spreads over more than a single critical band

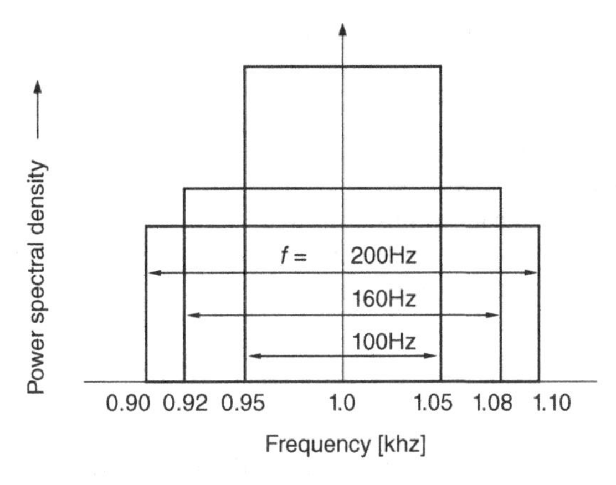

Figure 9.15 Measuring the critical bandwidth with band-limited noise having equal centre frequency and sound pressure level.

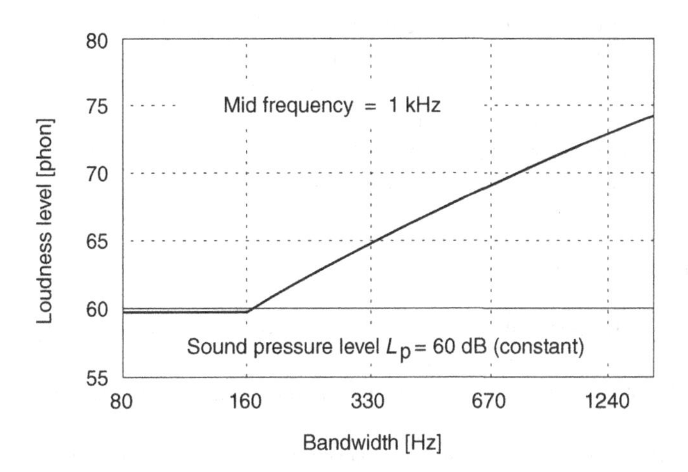

Figure 9.16 The perceived loudness as a function of the frequency bandwidth of the noise stimulus. The critical bandwidth is defined as the point on the curve above which the perceived loudness starts to increase. The centre frequency in this example is 1 kHz, and the width of the critical band is measured to be 160 Hz. Rossing *et al.* (2001).

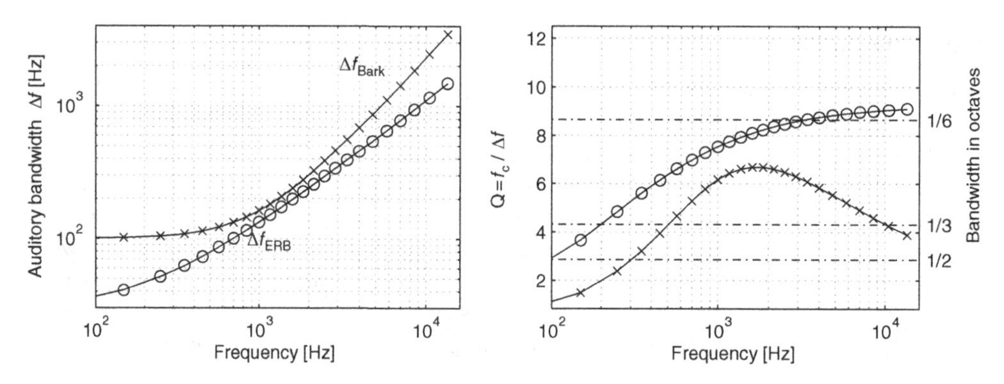

Figure 9.17 (a) Two estimates of the critical bandwidths Δf in Bark bandwidths (crosses) and ERB bandwidths (circles) as a function of the centre frequency f_c. (b) The same data plotted as Q values.

and so is processed inside more than one band. The knee point is regarded as the measure of the bandwidth of the critical band.

Zwicker named the critical bandwidths measured using this approach *Bark bandwidths* after Heinrich Barkhausen, who proposed the first subjective measure of loudness. The bandwidths Δf_{Bark} [Hz] measured via listening tests are estimated as

$$\Delta f_{\text{Bark}} = 25 + 75[1 + 1.4(f_c/1000)^2]^{0.69} \tag{9.1}$$

and are shown in Figure 9.17a as a function of the centre frequency f_c. The bandwidth is 100 Hz at low frequencies, and above 500 Hz it increases with frequency, being a bit less than 1/3 octave wide. Near the upper end of the audible frequency range the width is several kHz.

The fact that the perceived loudness increases (instead of, say, decreasing or staying constant) is an interesting phenomenon. The perceived loudness can be assumed to be related to the broadening of the excitation pattern on the basilar membrane with the level of the signal, as shown in Figure 7.12. The increase in loudness with signals having a broader spectrum can be explained by the mechanism whereby the auditory bands analysed to originate from the same source are integrated, and the total loudness is computed from the sum of the signals from each auditory band.

9.4.2 ERB Bandwidths

Another method to measure the bandwidth of the auditory filters in our hearing uses the concept of *ERB* (equivalent rectangular bandwidth) bands. ERB gives an approximation to the bandwidths of the filters in human hearing, using the unrealistic but convenient simplification of modelling the auditory filters as rectangular band-pass filters. The frequency widths of filters can be measured with different methods. In a commonly used listening test, bands of masking noise applied above and below the test signal, as shown in Figure 9.18, eliminate the possibility of hearing responding in the cochlea to the test tone outside the frequency region being tested, a phenomenon called *off-frequency listening*. The detection threshold of the tone is measured as a function of the width of the notch $2\Delta f$ to give an estimate of the frequency selectivity of

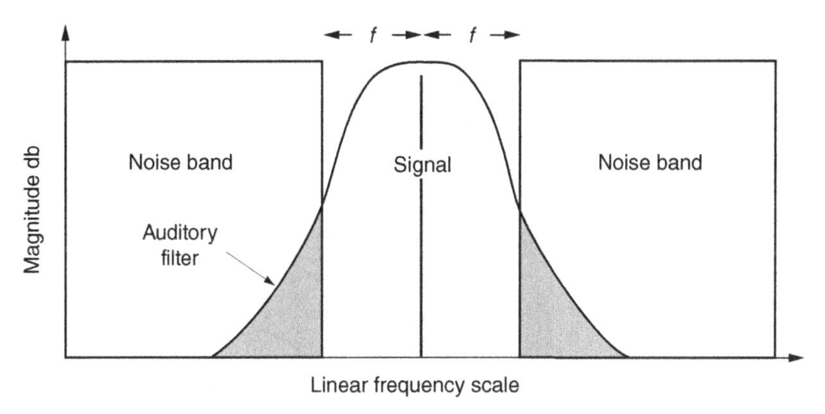

Figure 9.18 Estimation of the critical bandwidth by measuring the detection threshold of a tone masked by notched noise. The threshold is measured as a function of the width of the notch, which is further used to estimate the width of the critical band at different frequencies as ERB (equivalent rectangular bandwidth) bands.

human hearing. The obtained widths of critical bands (ERB) are typically 11–17% of the value of the centre frequency f_c, and the widths can be estimated as

$$\Delta f_{\text{ERB}} = 24.7 + 0.108 f_c \tag{9.2}$$

(Glasberg and Moore, 1990). The bandwidth here also follows a logarithmic relationship with the centre frequency but over a larger range than the Bark bandwidth. This is seen from the fact that the relation between the ERB bandwidth and centre frequency in Figure 9.17b changes less with frequency.

The bandwidth of hearing also changes with level: the higher the level, the poorer the frequency resolution, as discussed previously. However, for acoustic communication technologies, it seems to be a fair assumption that the frequency window in human hearing used in the spectral analysis of complex sound signals is around 10–15% of the centre frequency. This is not to be confused with the other frequency scale in humans, the pitch scale, which will be discussed in the next chapter.

9.4.3 *Bark, ERB, and Greenwood Scales*

Both Bark and ERB bandwidths define a frequency scale. If the bandwidths are computed and stacked on top of each other starting from a very low frequency, a scale is obtained for both methods. With Bark bands, this scale is called the *Bark scale*, z_{Bark}, and it can be estimated from the frequency f as

$$z_{\text{Bark}} = 13 \arctan(0.76 f/1000) + 3.5 \arctan(f/7500)^2. \tag{9.3}$$

Correspondingly, the frequency scale based on ERB bandwiths is derived by:

$$z_{\text{ERB}} = 21.3 \log_{10}(1 + f/228.7). \tag{9.4}$$

Both scales are presented as logarithmic functions of frequency in Figure 9.19. The Bark scale is linear up to about 500 Hz, as was seen in Figure 9.15. Above 500 Hz it is roughly logarithmic.

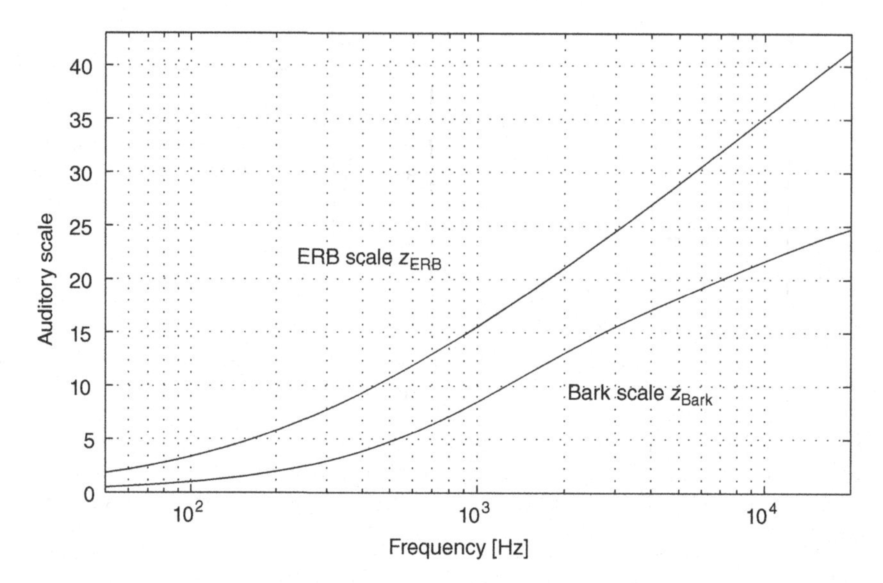

Figure 9.19 Bark and ERB frequency scales as functions of frequency.

The slope of the ERB scale is relatively straight at all frequencies with the logarithm of the frequency. Thus, it is logarithmic over a larger frequency region than the Bark scale.

The auditory scale also has an interesting relation to the anatomy of the inner ear. Each of the scales has been interpreted to have a linear relation to the position of resonance on the basilar membrane. This means that a constant change in the auditory frequency scale corresponds to a constant change in the resonance position on the basilar membrane at all frequencies. It seems that the scale based on ERB bands is closest to this relation, where a bandwidth of 1 ERB corresponds to 0.9 mm on the basilar membrane, which has about 90 inner hair cells.

To conclude, the ERB bandwidths and the ERB scale are commonly used in hearing sciences to approximate the auditory frequency resolution and scale, respectively. However, the Bark scale has not been abandoned. Many technical applications utilize it or related scales, for example the *mel* pitch scale is relatively commonly used in certain speech technologies, as will be discussed in Section 10.1.4.

A related auditory scale is the scale proposed by Greenwood (1990), where the resonance position on the basilar membrane x [mm] and the frequency f [Hz] have been found in mammals to have the relation

$$f = A(10^{ax} - k), \tag{9.5}$$

if x is known. If f is known, x is computed as

$$x = (1/a)\ \log_{10} \frac{(f + kA)}{A}, \tag{9.6}$$

where, for humans, $A = 165.4$, $a = 0.06$ and $k = 1$.

Summary

This chapter has drawn a picture of the first processing steps conducted in our hearing. The ear has a certain working range limited in frequency and in SPL. The range is actually very impressive, as the 120-dB range in SPL with the lowest levels very low and the frequency range starting from 20 Hz and ending at 20 kHz are hard to achieve with man-made devices.

Frequency masking effects and the measurement of critical bandwidths indicate that the frequency resolution of hearing corresponds to a bandwidth of about 1/6 octave at best. The best resolution is obtained at frequencies above 1 kHz with an SPL level below about 60 dB. Temporal masking effects are strong after the dying out of a loud sound. However, the temporal resolution is about 1–2 ms, which is quite remarkable in accuracy.

Further Reading

Due to a relatively long period of research and an active interest in auditory mechanisms, there exists a rich literature on this topic. For general introductions to the topic, some being more specific than in this book, the reader is referred to, for example, (Fastl and Zwicker, 2007; Moore, 2012; Plack, 2013; Schnupp *et al.*, 2011) and the appropriate chapters in Bregman (1990); Crocker (1997); Moore (1995); Yost (1994). A good source on the early investigations towards a quantitative formulation of auditory sensation and perception is Fletcher (1995).

References

Bregman, A. (1990) *Auditory Scene Analysis*. MIT Press.

Brungart, D.S. (2001) Informational and energetic masking effects in the perception of two simultaneous talkers. *J. Acoust. Soc. Am.*, **109**, 1101–1109.

Crocker, M.J. (ed.) (1997) *Encyclopedia of Acoustics*, Volume 3. John Wiley & Sons.

Dau, T., Kollmeier, B., and Kohlrausch, A. (1997) Modeling auditory processing of amplitude modulation. I. detection and masking with narrow-band carriers. *J. Acoust. Soc. Am.*, **102**(5), 2892–2905.

Duifhuis, H. (2005) Consequences of peripheral frequency selectivity for nonsimultaneous masking. *J. Acoust. Soc. Am.*, **54**(6), 1471–1488.

Fastl, H. and Zwicker, E. (2007) *Psychoacoustics – Facts and Models*. Springer.

Feth, L.L. and O'Malley, H. (1977) Influence of temporal masking on click-pair discriminability. *Percep. Psychophys.*, **22**(5), 497–505.

Fletcher, H. (ed.) (1995) *Speech and Hearing in Communication*. Acoustical Society of America.

Glasberg, B.R. and Moore, B.C.J. (1990) Derivation of auditory filter shapes from notched-noise data. *Hear. Res.*, **47**(1–2), 103–138.

Greenwood, D.D. (1990) A cochlear frequency-position function for several species – 29 years later. *J. Acoust. Soc. Am.*, **87**(6), 2592–2605.

Hall, J.W., Haggard, M.P., and Fernandes, M.A. (1984) Detection in noise by spectro-temporal pattern analysis. *J. Acoust. Soc. Am.*, **76**(1), 50–56.

ISO-226 (2003) Acoustics – normal equal-loudness-level contours. International Organization for Standardization.

Moore, B.C.J. (ed.) (1995) *Hearing*. Academic Press.

Moore, B.C.J. (2012) *An Introduction to the Psychology of Hearing*, 6th edn. Brill.

Plack, C.J. (2013) *The Sense of Hearing*. Psychology Press.

Rossing, T.D., Moore, F.R., and Wheeler, P.A. (2001) *The Science of Sound*, 3rd edn. Addison-Wesley.

Schnupp, J., Nelken, I., and King, A. (2011) *Auditory Neuroscience: Making Sense of Sound*. MIT Press.

Vogten, L. (1974) Pure-tone masking: A new result from a new method *Facts and Models In Hearing*, Springer. pp. 142–155.

Watson, C.S. (2005) Some comments on informational masking. *Acta Acustica United with Acustica*, **91**(3), 502–512.

Wegel, R. and Lane, C. (1924) The auditory masking of one pure tone by another and its probable relation to the dynamics of the inner ear. *Phys. Rev.* **23**(2), 266–285.

Yost, W.A. (ed.) (1994) *Fundamentals of Hearing – An Introduction.* Academic Press.

10

Basic Psychoacoustic Quantities

As seen in the earlier chapters, the basic functioning of hearing can be characterized as a kind of time–frequency analysis of the ear canal pressure signals resembling a bank of band-pass filters. This chapter describes the psychoacoustic quantities at the lowest level of analysis: pitch, loudness, timbre, and duration, which are more or less related to the physical quantities frequency, level, magnitude spectrum, and time.

10.1 Pitch

Pitch is defined by the American National Standards Institute as 'that auditory attribute of sound according to which sounds can be ordered on a scale from low to high' (ANSI-S1.1, 2013). Pitch is perceived from many types of sounds, such as sinusoids, vocals, instrument sounds, and noisy sounds (Fastl and Zwicker, 2007; Hartmann, 1996). However, the definition is problematic in the sense that not all sounds have clear pitch, and some of them have more than one pitch. However, in the context of the basic properties of hearing, it is meaningful to restrict the discussion to relatively simple sounds producing a single, more or less salient pitch. The nearest counterpart of pitch in the physical world is the frequency of repetition of a signal portion, even though pitch depends on some other parameters as well.

Pitch is a relevant cue in speech communication. It is the primary cue that distinguishes between male, female, and child speakers. The prosody of speech is also perceived as pitch changes during spoken sentences. Of course, pitch plays a very central role in music, as melody is composed of the perception of successive changes in pitch. The pitch of sound produced by a physical object also conveys information about its geometry, physical properties, and size, for example, large barrels produce a different pitch from small cans.

10.1.1 Pitch Strength and Frequency Range

The clarity and salience of pitch perception, or the *pitch strength*, depends significantly on the nature of the sound. Pitch strength has been measured using psychoacoustic tests with different types of signal, as shown in Figure 10.1 (Fastl and Stoll, 1979; Fastl and Zwicker, 2007).

Communication Acoustics: An Introduction to Speech, Audio, and Psychoacoustics, First Edition.
Ville Pulkki and Matti Karjalainen.

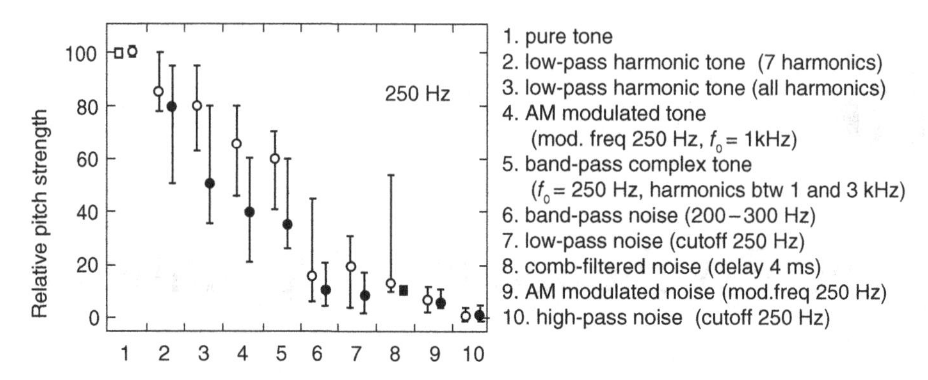

Figure 10.1 Pitch strength with different types of sounds producing a pitch of 250 Hz. The pitch strength is scaled so that 100% corresponds to a 250-Hz sinusoid. The unfilled markers show data measured using a pure tone as the reference, and filled markers denote results using comb-filtered noise (type 9) as the reference. The pre-assigned values for references (100 and 10, respectively) are marked with squares. Adapted from Fastl and Stoll (1979), and reprinted with permission from Elsevier.

The pitch strength is compared to the pitch of a 250-Hz sinusoid that serves as the reference on the scale. Periodic signals are clearly perceived to have higher pitch strength than aperiodic ones. Sinusoids produce the clearest pitch perception, closely followed by low-pass filtered harmonic complexes, both of which are perfectly periodic signals. Quite interestingly, an aperiodic sound, narrowband noise, is reported to have a pitch strength just below that of a low-pass filtered harmonic complex and a higher strength than an AM tone and band-pass filtered complex tones, both of which are periodic signals. The perception of some of the signals is analysed in detail below.

The frequency range of pitch perception has been studied by testing where musical melodies are still recognized correctly. The range covers repetition frequencies from about 30 Hz (Pressnitzer *et al.*, 2001) to 4 kHz (Plack, 2013). At lower frequencies, vibration is still perceived, but the musical tone is no longer recognized. At frequencies above 4 kHz the perception of pitch is also weak. A musical melody at these frequencies may just sound peculiar.

10.1.2 JND of Pitch

We begin our exploration of pitch perception by looking at the basic results of JND – just noticeable difference – measurements. As shown in the previous section, tones presented alone evoke the highest pitch salience, where the frequency of the tone defines the pitch in general. Thus, to get an idea of the best performance of hearing with pitch, measuring the smallest perceived change in the frequency of the tone $\Delta f = |f_1 - f_2|$, where f_1 and f_2 are the frequencies of two tones presented in succession to the subject, is of relevance. Δf then characterizes the JND of pitch. The result of such a measurement is shown in Figure 10.2. At the frequencies 250 Hz and 500 Hz, the JND is about 1 Hz, and it rises quickly to 200 Hz for 8-kHz tones (Sek and Moore, 1995). The same result plotted as a ratio with tone frequency is shown in the right panel. This shows that the ratio is at best about 0.3% and clearly rises at frequencies above 2 kHz; it also seems to rise at frequencies below 500 Hz. The poorer relative JND at higher frequencies can be explained simply by the neural loss of phase synchronization, while at low frequencies the decreasing sensitivity of hearing explains the degrading results.

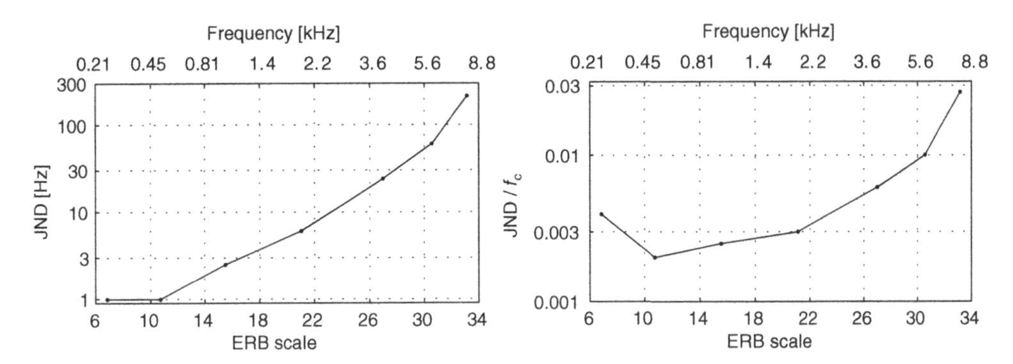

Figure 10.2 The left panel shows the JND of frequency for two successively presented tones, with the frequency difference on the y-axis, as a function of the mean frequency of the tones. The same data are plotted as a ratio between the JND and the mean frequency in the right panel. Adapted from Sek and Moore (1995).

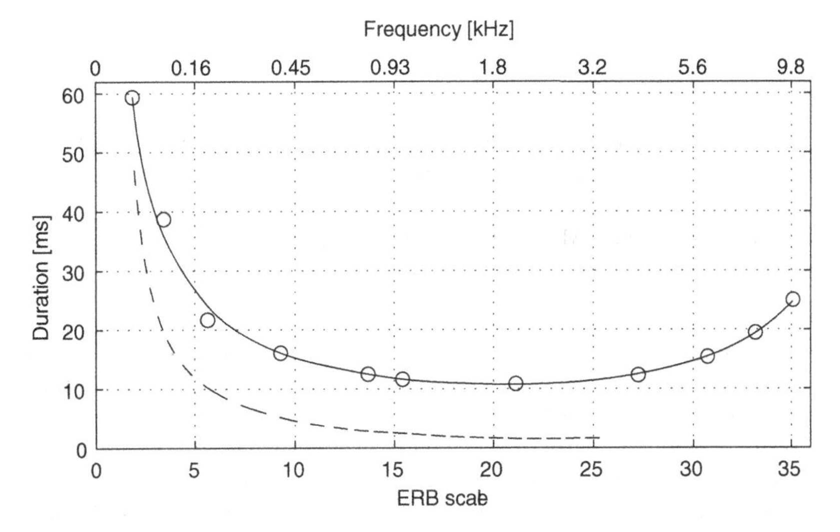

Figure 10.3 The minimum length of a tone burst required for pitch perception. The dashed line shows the length of two periods depending on frequency. Adapted from Burck *et al.* (1935).

10.1.3 Pitch Perception versus Duration of Sound

The formation of the percept of pitch and improving its precision requires a sample of a signal with non-zero length. In theory, at least two periods of a periodic signal are necessary before the pitch percept can be formed. The minimum length of a tone burst after which a pitch is perceived is shown in Figure 10.3. The dashed line shows the length of two periods of sine signals. The shortest time window required for pitch perception is seen to be between 400 Hz and 6 kHz, where the length is less than 20 ms. The performance of human hearing in pitch perception is remarkable at low frequencies; the required length for a tone burst is only slightly longer than two periods.

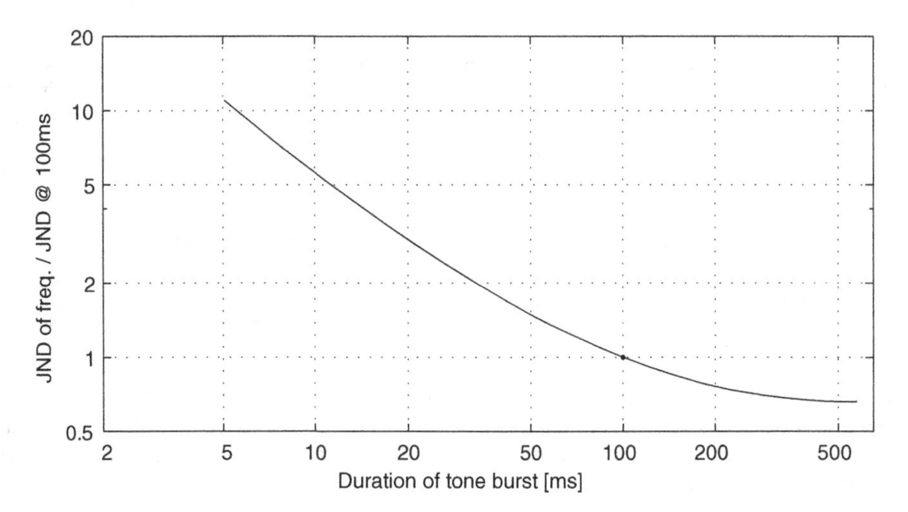

Figure 10.4 The normalized JND of pitch as a function of the duration of a 2-kHz tone. Adapted from Fastl and Zwicker (2007).

The best accuracy in pitch perception is achieved 100–200 ms after the onset of the sound. The normalized JND of pitch as a function of 2-kHz tone length is plotted in Figure 10.4. It can be seen that the JND decreases only slightly when increasing the duration to over 100 ms, and the JND is practically constant after 200 ms. The variation in the JND with duration differs somewhat across frequencies (Moore, 1973).

10.1.4 Mel Scale

A frequency scale was created using psychoacoustic tests to quantify pitch. In the tests, the task of the subject was to adjust a tone or a harmonic complex to a specified ratio of the reference sound for a range of ratios to obtain a subjective scale. An example task would be 'adjust the pitch of the test tone to be two times higher than the reference tone.' The phrase 'two times higher' may seem arbitrary, and different people may interpret the task differently. For example, raising the pitch of a musical note by one octave is commonly considered to 'double the height of the note'.

The scale thus obtained is called the *mel scale*, and the name of the unit is the *mel*, which comes from the English word *melody*. It turns out that when subjects are asked to double the pitch they double the frequency, on average, for frequencies below about 500–1000 Hz (Fastl and Zwicker, 2007). Above this frequency limit, subjects adjust the frequency to increase in considerably larger steps than the pitch (see Figure 10.5). It can be seen that when doubling the pitch of a 2-kHz tone, subjects have adjusted the frequency to about 15 kHz. This can be explained by the low sensitivity of hearing to pitch at high frequencies due to the inability of the cochlea to lock onto phases of tones there.

The mel scale can be approximated at low frequencies with relatively simple equations, such as

$$m = 2595 \log_{10}(1 + f/700), \tag{10.1}$$

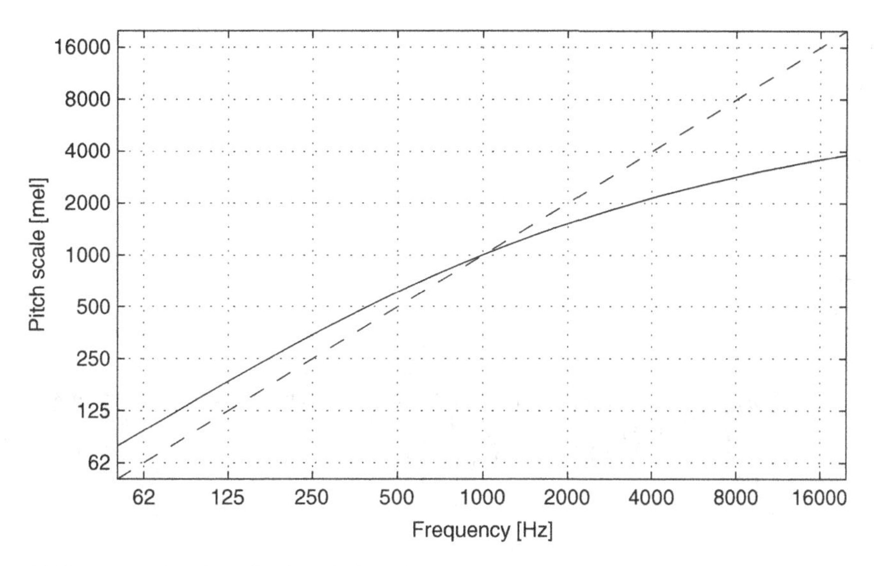

Figure 10.5 The mel scale defined with Equation (10.1) plotted as a function of frequency in Hz. The anchor point is 1000 mel ↔ 1000 Hz. The dashed line shows, for reference, the linear relationship with a slope of one.

where m is the mel value and f is frequency in Hz. The anchor point in the equation is 1000 Hz ↔ 1000 mel. 1 mel equals to there are also other equations proposed for this purpose.

10.1.5 Logarithmic Pitch Scale and Musical Scale

As will be discussed later in this book, musical notes can, in most cases, be interpreted simply as harmonic complexes. The pitch scales form sets of notes with different fundamental frequencies. The scales are logarithmic universally, since the distances between musical notes are defined as ratios between their fundamental frequencies. In simpler terms, when the fundamental frequencies of two musical notes have the ratio 2:1, their distance in frequency, or *interval*, is called an octave. Octaves are perceived to have similar width in pitch in different frequency areas, especially between about 100 and 1000 Hz. There are a number of methods to define the fundamental frequencies of notes within an octave, which will be discussed in Section 11.6.2. A simple method is equal temperament, where an octave is logarithmically divided into 12 semitones, and the intervals between successive notes are set as $\sqrt[12]{2}$.

The logarithmic frequency scale plotted against the piano keyboard illustrating the diatonic scale used in music is shown in Figure 10.6. The widths of intervals – octave, fifth, major third, and semitone – are also shown. The reference frequency commonly used in Western music is 440 Hz for the note A. The reference has changed with time, and the overall tendency has been upwards in pitch to obtain a brighter sound for classical orchestras (Rossing *et al.*, 2001). The figure also shows the relative frequency of occurrence of different tones in five randomly selected piano sonatas by Beethoven. It is clear that most of the notes reside in the frequency region between 50 Hz and 2000 Hz, although outliers also exist. Since the piano is an instrument with one of the largest ranges of notes, this frequency region can be taken in general as an example of the most commonly found range of pitches in tonal music.

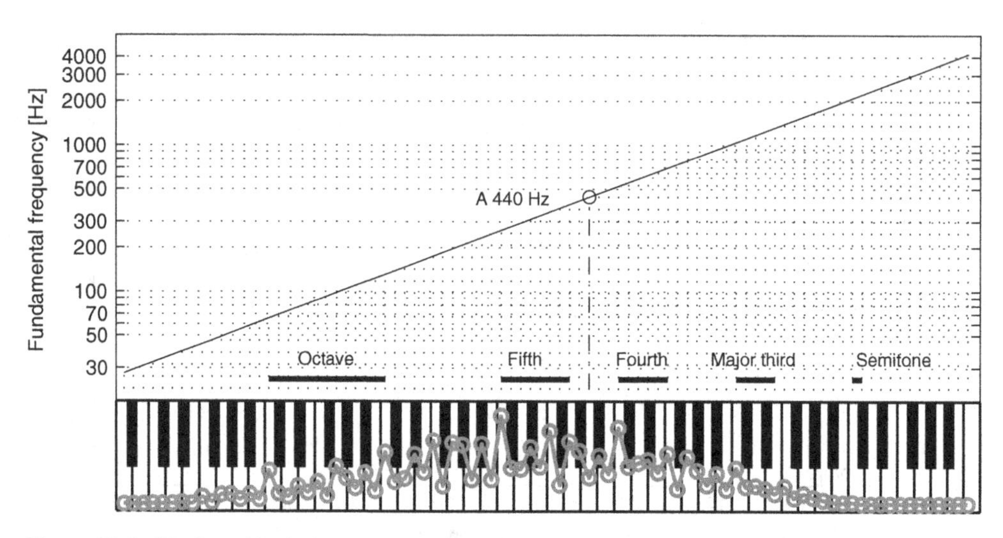

Figure 10.6 The logarithmic frequency scale and the diatonic scale illustrated as a piano keyboard. The grey bold line imposed on the keyboard illustrates the relative frequency of appearance of each note in five randomly selected Beethoven piano sonatas.

The mel scale defined in the previous section is essentially a linear scale up to about 1000–2000 Hz. It is thus in conflict with the logarithmic definition of musical scales. In addition, the JND of the frequency of a sinusoid is about 0.3% from 250 Hz to 2 kHz instead of being a constant value expressed in Hz, which bolsters use of the logarithmic pitch scale in this frequency region. Although the mel scale does not correspond to the musical pitch scale, it is still useful in some technical applications, such as speech recognition.

10.1.6 Detection Threshold of Pitch Change and Frequency Modulation

The JND of frequency depending on the frequency of the tone was shown earlier in Figure 10.2. Often, the frequency of continuous sound changes with time and varies around a centre frequency. The changing of frequency may or may not be audible as a change in pitch or some other perception. This change in frequency can be studied using frequency modulation. The JND of frequency modulation of carrier tones at different frequencies is shown as a function of modulation frequency in Figure 10.7.

A practical example of the perception of frequency modulation is present in mechanical audio devices such as LP players caused by the fluctuations in speed of the player. Figure 10.7 shows that humans are most sensitive to modulations at 4 Hz (Demany and Semal, 1989; Fastl and Zwicker, 2007).

10.1.7 Pitch of Coloured Noise

In addition to periodic signals, non-periodic signals may also give rise to a perception of pitch in many situations. A band-pass filtered noise signal with a single spectral peak is perceived to have pitch corresponding to the centre frequency of the peak. The salience of pitch perception is better the narrower the peak is.

A low-pass or high-pass filtered noise signal with steep roll-off causes a perception of pitch which matches the cutoff frequency. In the case of band-pass filtered noise, one or

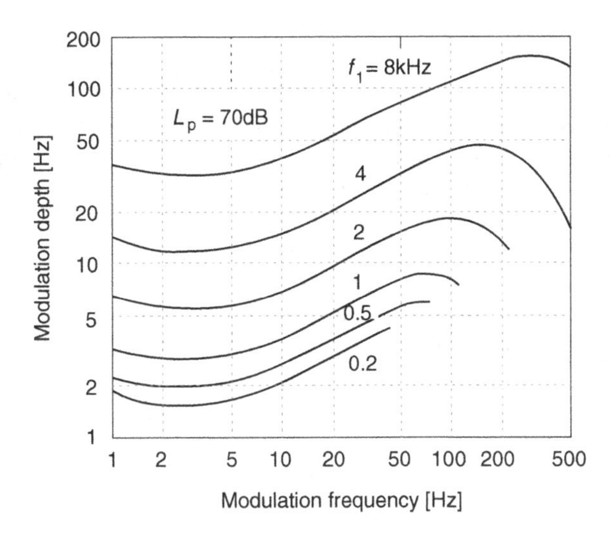

Figure 10.7 A schematic presentation of the detection threshold of frequency modulation as a function of the modulation frequency for different carrier frequencies. Adapted from Demany and Semal (1989).

two pitches may appear depending on the frequency characteristics. If the cutoff frequencies are far enough from each other and the roll-off curves are steep enough, two pitches may appear. Otherwise, for sufficiently shallow roll-off curves, a single pitch emerges, corresponding either to the centre of gravity of the sound energy in frequency or to one of the edge frequencies.

The preceding examples pertained to cases where the frequency spectrum of sound contained peaks or other variations. A weak pitch percept is also generated by modulating sinusoidally the amplitude of white noise (sinusoidally amplitude-modulated noise, or SAM noise). The pitch is perceived at the frequency of amplitude modulation, although the long-term spectrum of sound is flat.

10.1.8 Repetition Pitch

An example of pitch generated by coloured noise is *repetition pitch*. In repetition pitch, a broadband signal such as noise or a rapid transient is delayed and summed to the non-delayed signal. The perceived pitch corresponds to the frequency that is the reciprocal of the delay. In signal processing this is known as the *comb-filter effect*, where the spectrum of the signal has evenly spaced peaks and valleys. The frequency of the first peak in the comb-filter spectrum corresponds to the repetition pitch (Bilsen and Ritsma, 1969).

Repetition pitch often occurs outdoors, where a noisy sound is reflected from a wall or from hard ground and is summed to the direct sound in the ear canals. Relatively often it is simply perceived as colouring of sound and not as pitch. However, if the delay changes with time, as when the source or the receiver moves, the pitch is perceived more prominently.

The frequency of a single, sharp, moving spectral peak can be perceived as pitch, where the signal itself can also be a harmonic complex. An example of this phenomenon is the 'wah wah' effect pedal commonly used with guitars, which effectively performs peak filtering with a high-Q filter, the centre frequency of which is controlled. Another example is overtone singing, as described in Section 5.4.

10.1.9 *Virtual Pitch*

Pitch perception in all the cases presented so far, except for the SAM noise case, can be explained using the place theory of pitch perception, where the location of the maxima in the place-dependent response of the basilar membrane is responsible for the percept. However, as shown in Figure 7.12, the activation of the basilar membrane spreads to a larger range with an increase in level of the stimulus, and one could expect that the perceived pitch would change noticeably if only the maximum of the activation in frequency defined the perception of pitch. Such a noticeable change in pitch depending on the level does not occur, and another process is likely to be involved in pitch perception. There exist many other cases where the place theory does not at all explain the perception. A simple case is a harmonic sound, where the partials at higher frequencies are dominant. The perceived pitch follows the fundamental frequency in many cases, although the centre of gravity of the spectrum is elsewhere (Plack *et al.*, 2005).

An important special case of this is the phenomenon of perceiving the *missing fundamental*. If the lowest partial(s) of a harmonic tone complex are missing, the pitch of the sound still corresponds to the fundamental frequency of the tone complex. The sound is perceived somewhat differently, as if the colour of sound is a bit thinner, and the pitch strength may be weaker, but in many cases the pitch does not change. The perception of pitch as being the frequency of the missing fundamental is an example of *virtual pitch*, also known as *residue pitch*.

The mechanism leading to the perception of the missing fundamental can, at least partially, be understood by considering the half-wave rectification performed by the hair cells in the cochlea. If more than one partial is present in one critical band, non-linearity in the cochlea creates the presence of sum and difference signals in the output of the band. In practice, the fundamental frequency is present as amplitude modulation in the output. This notion calls for some sort of periodicity analysis that is applied to the auditory band outputs. The result of the analysis is then perceived as pitch.

10.1.10 *Pitch of Non-Harmonic Complex Sounds*

Pitch perception becomes an even more complex phenomenon if the sound consists of partials that are not necessarily whole number multiples of a common fundamental frequency. For example, church bells and other vibrating bars and plates generate tone complexes with partials in more or less inharmonic succession. There may be one or more pitches perceived from the bell sound, and the pitches may correspond to some high-level partials in the spectrum, or they may be virtual pitches generated by some partials.

An interesting experiment has been conducted to test the validity of place and timing theories of pitch perception. In the experiment, a set of harmonic partials is generated; for example, three to ten harmonics along with the fundamental frequency of, say, 100 Hz. Thus, the spectrum spans from 300 Hz to 1000 Hz in intervals of 100 Hz. If all partials are shifted upwards by 30 Hz, the output envelopes of hair cells should still have a strong modulation with 100 Hz frequency. However, the pitch perception rises a bit, even though the modulation in each auditory band remains the same (Plack *et al.*, 2005). It seems that the more complex the spectral relationships of the partials are, the harder it is to explain the perception using simple models.

10.1.11 *Pitch Theories*

Pitch is a remarkably stable percept. The SPL of sound and the direction of sound source do not have a noticeable effect on it. The only factors that seem to affect pitch perception are the magnitude spectrum and repetitive structures in the ear canal signals.

The mechanism leading to pitch perception has been explained by two theories: the *place theory* and the *timing theory* (Plack *et al.*, 2005).The place theory assumes that the frequency–place mapping occurring in the cochlea explains pitch perception. Unfortunately, this does not explain all pitch phenomena, and in some cases it is clear that a time-domain analysis of periodicity is also performed in hearing. The current assumption is that hearing analyses pitch using both mechanisms, but, unfortunately, they are not known well enough to be able to construct a functional model of the pitch perception mechanisms for all cases, although some quite advanced models already exist (Plack *et al.*, 2005).

10.1.12 Absolute Pitch

Human hearing is very accurate when comparing the characteristics of two sounds, but quite inaccurate in making absolute assessments of the characteristics of sounds in isolation. This is also true in the perception of pitch. An interesting exception is the ability of some people to evaluate pitch on an absolute scale without any reference, this ability is called *absolute pitch* (Rossing *et al.*, 2001).

Although musicians can improve their ability to assess pitch, the ability of a person with absolute pitch is about ten times more precise in assessing pitch than a person who is trained but does not naturally have this ability. It is as if people with this ability have an inner reference frequency. It has been known for this reference to change with increasing age. Some people also claim that the pitch of recorded music has changed with time. Absolute pitch is not necessarily connected to musical giftedness, though undoubtedly it may help an otherwise musical person in their musical training.

10.2 Loudness

Loudness is 'that attribute of auditory sensation in terms of which sounds can be ordered on a scale extending from quiet to loud' (ANSI-S1.1, 2013). Loudness perception is a relatively complex, yet consistent phenomenon. The theory describing loudness perception is one of the central theories of psychoacoustics. The theory is used below to explain different phenomena revealed by psychoacoustic experiments using different kinds of stimuli, starting with simple sinusoids and continuing gradually to more complex cases.

10.2.1 Loudness Determination Experiments

As with the concept of pitch, the concept of loudness is easiest to understand by first looking at how we perceive sinusoidal signals. Subjects are given the task of adjusting a test tone with the same frequency as the reference tone to two times (or half) the loudness of the reference tone. The adjusted tone then becomes the reference tone, and the process is continued. There may be individual differences in terms of how the phrase 'twice as loud' is interpreted, and the adjusted level may be different in each trial (Fastl and Zwicker, 2007). However, when the results from a large set of subjects and from many repetitions of the same task are averaged, and when an anchor point is selected, a psychophysical function is obtained which describes the relation between the level of a tone and perceived loudness. Such a function is presented in Figure 10.8 for a 1-kHz tone. The result is also presented for the same test made with white noise.

The unit of loudness, the *sone*, is defined by stating that a loudness of 1 sone is equivalent to the loudness of a 1-kHz tone at 40 dB SPL.

Figure 10.8 shows that the sound pressure level does not estimate perceived loudness directly, since different signal types at equal sound pressure levels do not generate an equal loudness

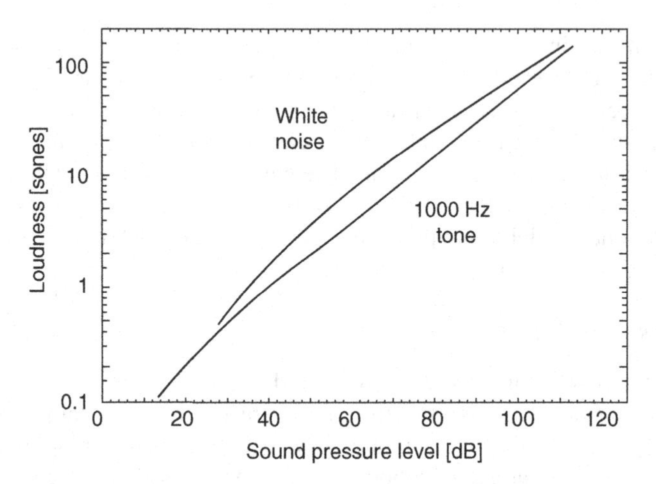

Figure 10.8 The loudness of a 1-kHz tone and white noise as a function of the sound pressure level. Reprinted from Canteretta and Friedman (1978) with permission from Academic Press.

perception. The figure also shows that the loudness of white noise is higher, although its sound pressure level is equal. This effect is actually a manifestation of the same phenomenon that was used to measure Bark bands in Section 9.4.1; the perceived loudness increases when the band-pass spectrum is made wider while keeping the sound pressure constant.

10.2.2 Loudness Level

The equal loudness curves presented in Figure 9.2 on page 155 show that the perception of loudness depends on the level and frequency of sinusoidal signals in a manner which cannot be easily expressed in mathematical terms. However, the curves can themselves be used to define the *loudness level.* The curves represent sound pressure levels of tones that are perceived to have equal loudness with reference values located at 1 kHz with 10 dB spacing in the sound pressure level. The unit of loudness level is the *phon.* It is defined such that at 1 kHz the sound pressure level in dB and the loudness level in phons have the same magnitudes.

The linear dependence between the SPL in dB and the loudness level in phons is different from the dependence between SPL and loudness in sones introduced in the previous section. The SPL in Figure 10.8 is expressed in dB and the loudness is expressed in sones on a logarithmic scale. A 1-dB change in SPL thus produces very different changes in the value of loudness in sones at low and high levels of the SPL. Technical applications often use the loudness level, since the JND of the SPL of sound events is constant at about 1 dB (or 1 phon) over a wide range of SPLs (Fastl and Zwicker, 2007). A change in a certain number of phons thus corresponds to a similar change in the perceived 'strength' of an auditory event, which is beneficial in auditory models and sound and voice techniques.

10.2.3 Loudness of a Pure Tone

The result of the 1-kHz loudness listening test discussed in Section 10.2.1 is shown in Figure 10.9, where one can see that at levels higher than 40 dB the loudness is doubled with an increase in level of 10 dB. This can be expressed as

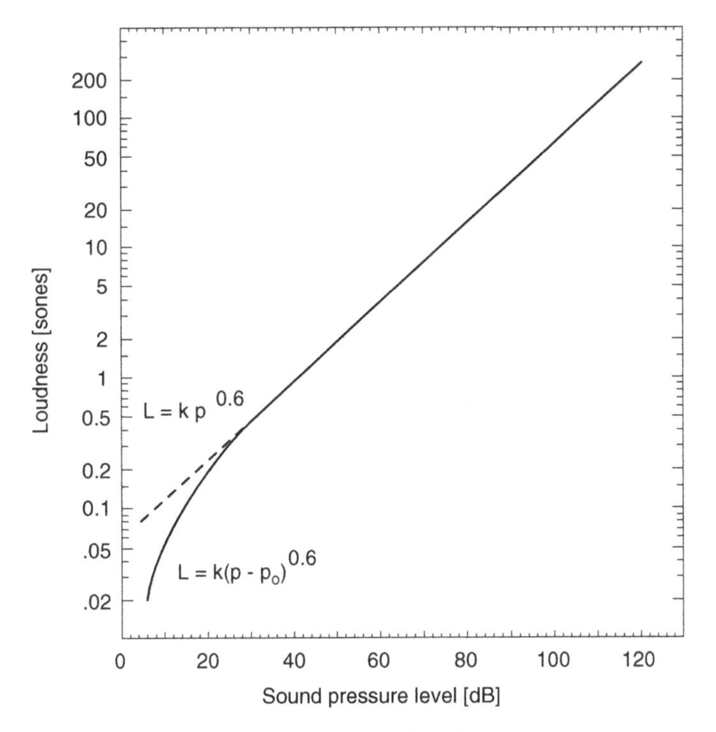

Figure 10.9 The loudness of a 1-kHz tone in sones as a function of the sound pressure level. Reprinted from Canteretta and Friedman (1978) with permission from Academic Press.

$$N = 2^{(L_L - 40)/10}, \tag{10.2}$$

where N is the loudness in sones and L_L is the loudness level of a tone at 1 kHz. Note that at 1 kHz, $L_L = L_P$ by definition, where L_P is the sound pressure level in dB. At levels below 40 phons, the curve is steeper. A theoretical relation between sound pressure and loudness can thus be derived as

$$N = \mathrm{k} \cdot p^{0.6}. \tag{10.3}$$

The curve can also be fitted to the results at low levels by introducing p_0, the pressure level near the threshold of hearing, into the equation:

$$N = \mathrm{k} \cdot (p - p_0)^{0.6}. \tag{10.4}$$

The constant k must be selected so that the loudness at a 40-dB sound pressure level matches 1 sone.

When the test sound is partially masked due to the presence of white noise, the perceived loudness of the signal is also affected, as shown in Figure 10.10. The threshold above which Equation (10.2) holds rises with the level of masking noise. When the level of the masker is increased, the loudness level of the tone decreases gradually before becoming inaudible.

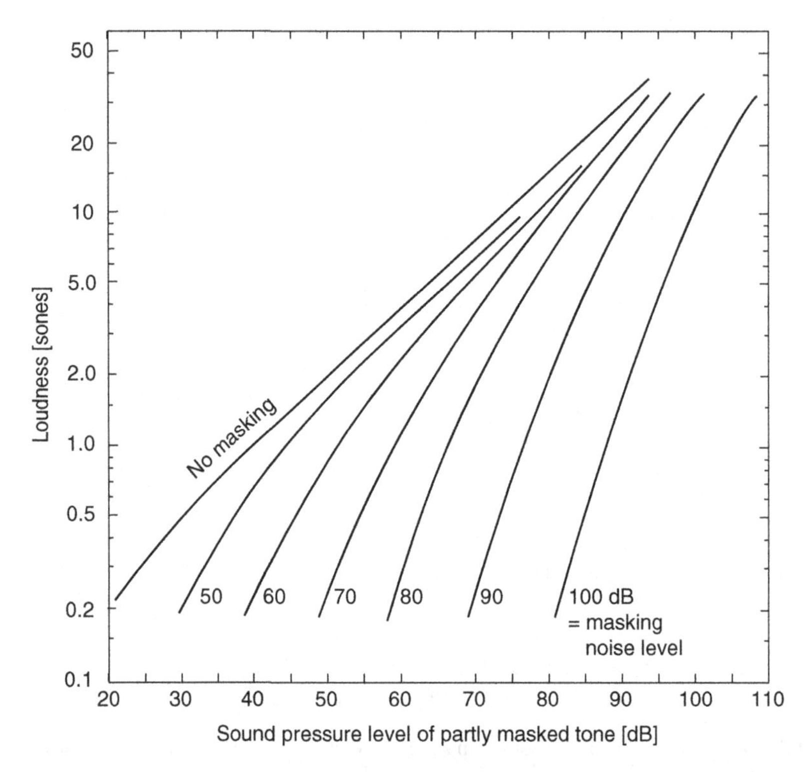

Figure 10.10 The loudness of a tone as a function of its SPL in the presence of masking white noise with tabulated SPL. Reprinted from Canteretta and Friedman (1978) with permission from Academic Press.

10.2.4 Loudness of Broadband Signals

A broadband sound is perceived to be louder than sound with a narrower band of equal sound pressure level, as shown by both the phenomenon used to define Bark bands in Section 9.4.1 and the higher loudness generated by white noise than a sinusoid shown in Figure 10.8. Our hearing thus analyses the loudness of different auditory bands separately and adds the obtained results to form a total loudness for the system, which turns out to be higher than it should be in principle. This implies that the mechanism detecting loudness accounts for the effect of the inherent broadening of the excitation of narrowband signals in the basilar membrane. It can be assumed that the mechanism uses the broadening as a loudness cue, and higher loudness is analysed with broader frequency content. This makes the system measuring the loudness dependent on the signal bandwidth itself, which could be thought to be an issue. However, very possibly this does not cause any deficiencies in the perception of the surrounding world, since no large evolutionary advantage can be seen to occur if the loudness of sound was not dependent on the spectral content of sound. On the other hand, an important task of hearing is to estimate if a sound source is approaching or if its distance is changing in general. In such cases, the loudness of the auditory event is compared to earlier similar events, and the relative loudness gives the cue of distance. Such a comparison does not require an accurate estimate of the sound pressure level, and thus a relative difference is already very usable.

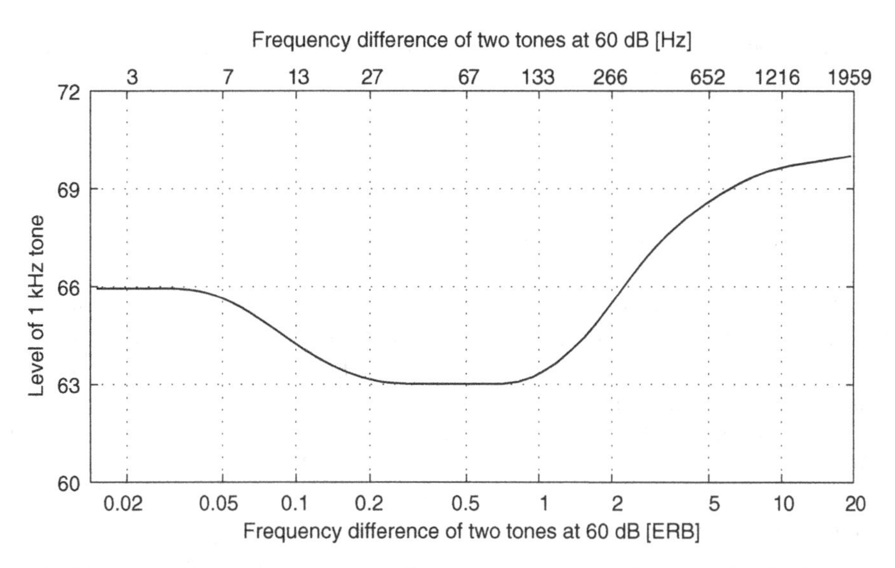

Figure 10.11 The level of a reference tone adjusted to match the loudness with a pair of tones having the frequency difference shown in the abscissa. The level of each of the tones in the pair is 60 dB. Adapted from Fastl and Zwicker (2007).

The formation of the loudness percept on the frequency axis can be investigated with different psychoacoustic tests. Figure 10.11 shows a result where the level of a 1-kHz tone was adjusted to match the perceived loudness with the loudness generated by a pair of concurrent tones. When the frequency of a pair of sinusoids differed less than a few hertz, the level of the reference tone was raised by 6 dB from the level of the non-summed tones. This effect can be explained by assuming that hearing reacts to the instantaneous maximum levels in the input and, due to beating, the maximum corresponds to a direct sum of the amplitudes (6 dB equals double the amplitude). When the frequency difference is from 20 Hz to 160 Hz, the adjusted level of the reference tone is only 3 dB higher than the non-summed test tones, which corresponds to a doubling of the power of the sound. The hearing can then no longer follow the maxima of beating, but only the average of the signal, corresponding to the sum of the powers. At frequency differences above 160 Hz, the tones appear in different critical bands, and the adjusted level of the reference tone approaches 10 dB, which corresponds to the sum of the loudnesses (Fastl and Zwicker, 2007). However, some aspects of this explanation are not supported by data shown by (Moore *et al.*, 1999).

10.2.5 Excitation Pattern, Specific Loudness, and Loudness

As already discussed, broadband sound with SPL equal to a narrowband sound is perceived to be louder and that sound in one frequency region affects the perception at other frequencies. Thus, we have to differentiate between the *excitation pattern*, *specific loudness*, and *loudness* (Fastl and Zwicker, 2007).

Let us assume that a sound stimulus entering the hearing mechanism has a power spectral density $S(f)$ on a linear scale. As the cochlea processes $S(f)$ on the auditory frequency scale, a change of frequency axis is needed. The resulting spectrum is $S'(z)$, and the change can be computed on an arbitrary frequency scale z as

$$S'(z) = S[f(z)]\frac{\mathrm{d}f}{\mathrm{d}z}, \quad (10.5)$$

$f(z)$ is a function that transforms the linear frequency scale to an auditory one, for example, the inverse of Equation (9.4). The derivative $\mathrm{d}f/\mathrm{d}z$ represents the change in spectral density, as the density grows at higher frequencies.

The effect of each narrowband component of the input signal spreads in frequency in a manner similar to the effect of the narrowband masker in Figure 9.6. This spread can be characterized by a *spreading function* $B(z)$, which depends both on frequency and level, like the masking curves. For average levels of speech (about 60 dB SPL), a single-peak function has an approximate peak (at −3 dB) of about 1 ERB (or Bark), and the slopes are about 8 dB/ERB (10 dB/Bark) towards high frequencies, and about 20 dB/ERB (25 dB/Bark) towards low frequencies. The closest physiological correlate of the spreading function is the envelope of the vibration on the basilar membrane when excited by a single sinusoid.

The *excitation pattern* $E(z)$, that is, the effect of $S'(z)$ spread on the basilar membrane, can be thought to result from a convolution between the input signal power spectrum S' and the spreading function B,

$$E(z) = S'(z) \star B(z). \quad (10.6)$$

Note that the convolution, or the filtering operation, is computed here between two functions depending on auditory frequency.

The *specific loudness* N' can then be computed from the excitation pattern by scaling:

$$N'(z) = \mathrm{c}\,E(z)^{0.23}. \quad (10.7)$$

The constant c is chosen such that the SPL of 40 dB with a 1-kHz sinusoid results in 1 sone of loudness, and the exponent 0.23 is chosen so as to double the loudness when the SPL of a sinusoid increases by 10 dB.

The last concept, *loudness*, expressed in sones, is computed by integrating specific loudness over all the critical bands:

$$N = \int_0^M N'(z)\mathrm{d}z, \quad (10.8)$$

where M is the number of critical bands. If desired, the loudness level can be set to 0 sones at the hearing threshold with the following computation:

$$N = \mathrm{c}\int_0^M \max\{[E(z) - E_0(z)]^{0.23}; 0\}\mathrm{d}z, \quad (10.9)$$

where $E_0(z)$ is the excitation pattern corresponding to the hearing threshold, which can be interpreted as the background noise level of the hearing system.

Figure 10.12a shows the excitation patterns caused by a sinusoid (1 kHz, dashed line), and by uniform masking noise (solid line). Figure 10.12b, in turn, shows the corresponding specific loudness patterns in sones/ERB. The excitation pattern produced by the uniform masking noise follows the sensitivity function of hearing. The figures also show, at least in principle, how much the sounds mask each other, shown as the area with the cross hatching in Figure 10.12b.

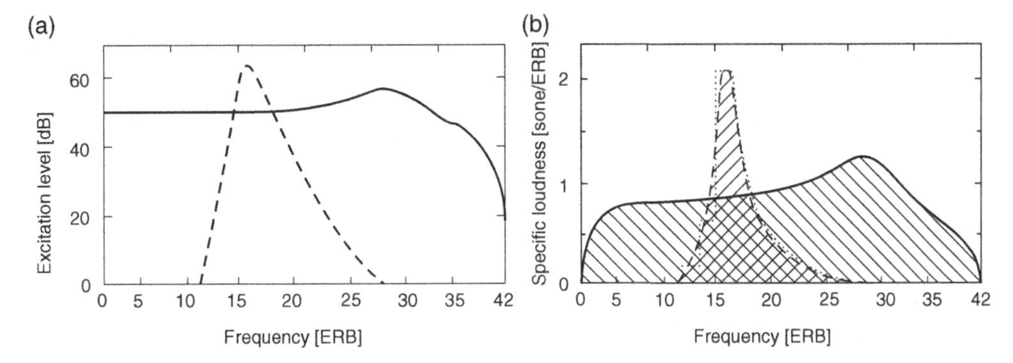

Figure 10.12 (a) Excitation patterns created by a sinusoid (dashed line) and by uniform masking noise (solid line) and (b) the corresponding specific loudness functions. Adapted from Fastl and Zwicker (2007).

These concepts are central to the understanding of the formation of the perception of loudness, and they are also widely used in computational models of loudness perception. Such models are discussed in more detail in Chapter 13.

10.2.6 Difference Threshold of Loudness

As discussed above, loudness depends on many acoustic parameters of a sound event, such as level, spectral content, and duration. It is also interesting to quantify the JND of loudness to see how small a change is detected at all. This has been researched as the JND of level, which has a very strong influence on loudness.

The detection threshold of amplitude modulation indicates some properties of human ability to perceive changes in loudness. As will be shown below, the greatest sensitivity is obtained when the modulation frequency is about 4 Hz. The detection threshold of the level of modulation of a 1-kHz tone and white noise is shown in Figure 10.13 as a function of the level of the signal. When the level of the carrier signal is increased from silence, the threshold decreases monotonically. With relatively low levels (20–50 dB) the detection threshold is about 1 dB, and at higher levels the threshold keeps decreasing, reaching about 0.2 dB at 100 dB. Interestingly, when the same test is conducted with white noise, the detection threshold is constant at all levels above 25 dB.

Interestingly, the detection threshold of an amplitude-modulated tone depends on its level, but such dependence is not found with noise. This can be understood when the excitation pattern of the tone on the basilar membrane is investigated in the cases of amplitude and frequency modulation. Schematic diagrams of these cases are shown in Figure 10.14. In the case of amplitude modulation, the shape of the excitation pattern depends on the level such that at maximum levels of the signal a much broader frequency range is covered than with the lowest levels. This implies that when the level is modulated, the frequency range is also modulated, which provides an additional cue to detect the modulation. The frequency coverage is, on the other hand, constant in the case of frequency-modulated tones, and such an additional cue is not available.

The dependence of the detection threshold of the modulation index as a function of the modulation frequency is shown in Figure 10.15. When the carrier signal is a tone, the result is similar to that obtained with frequency modulation in Figure 10.7. The threshold level

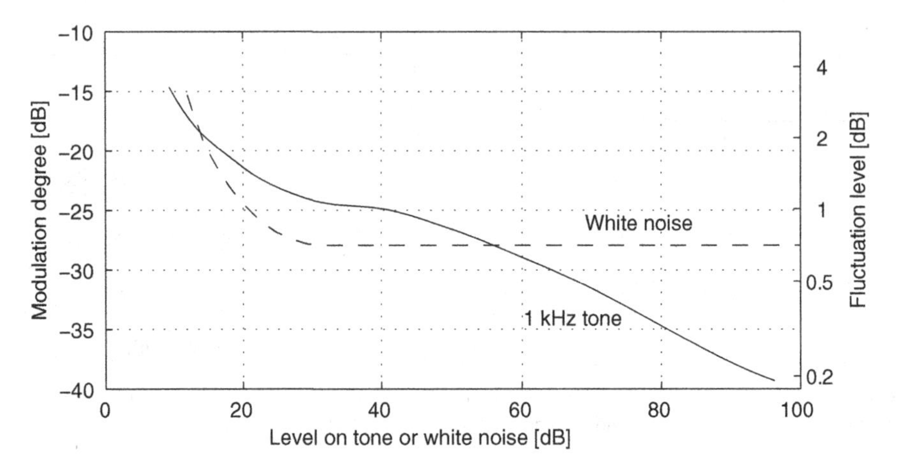

Figure 10.13 The just noticeable level of amplitude modulation of a 1-kHz tone (solid line) and white noise (dashed line). The modulation degree m (see Equation (3.4)) is expressed in dB on the left-hand y-axis, and the resulting fluctuation of level is shown on the right-hand y-axis. The modulation frequency is 4 Hz. Adapted from Fastl and Zwicker (2007).

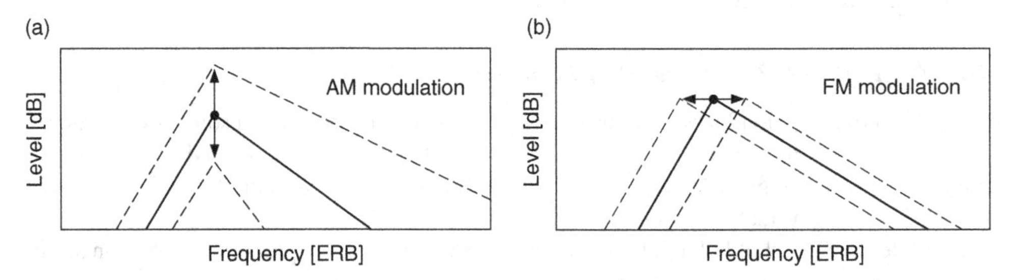

Figure 10.14 Excitation patterns in the cochlea for (a) an amplitude-modulated and (b) a frequency-modulated tone. Adapted from Fastl and Zwicker (2007).

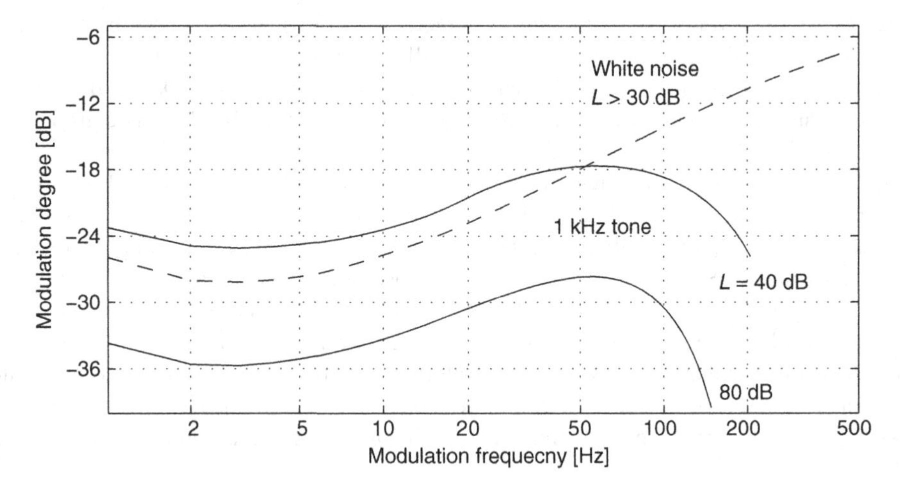

Figure 10.15 A schematic presentation of just noticeable amplitude modulation of a 1-kHz tone (solid line) at two different levels, and of white noise (dashed line) as a function of modulation frequency. Adapted from Fastl and Zwicker (2007).

is at its lowest with a modulation frequency of 4 Hz, and increases for other frequencies (Fastl and Zwicker, 2007). In the case of a sinusoidal carrier, when the modulation frequency reaches about half the width of the critical band, the threshold starts to decrease again. As the spectrum of the modulated signal spreads to an area wider than one critical band, hearing detects the change more easily. For a white noise carrier, the signal is broadband to begin with, and the spectral effect of modulation is not detectable.

10.2.7 Loudness versus Duration of Sound

The loudness percept requires some time to build up. The loudness level caused by a 2-kHz tone burst is shown in Figure 10.16. When the length of the burst decreases from 100 ms, the loudness decreases 10 phones each time the length is decreased to one tenth, indicating that the loudness percept is formed by integrating sound energy over time. This operation performed by our hearing is called *temporal integration*. For a sound burst lasting longer than 200 ms, loudness no longer increases (Fastl and Zwicker, 2007).

A similar temporal effect of a change in the level of sound is seen in measurements of the JND, as shown in Figure 10.17. When the burst length is decreased from 200 ms, the detection

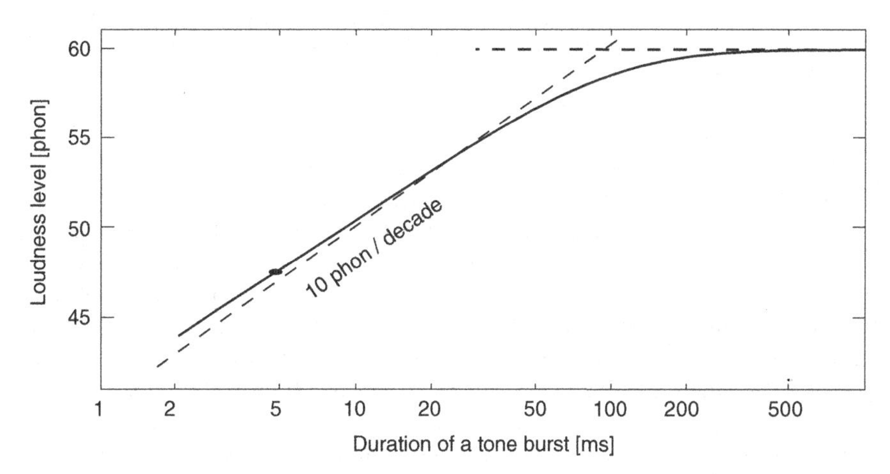

Figure 10.16 The dependence of loudness level on the duration of a tone burst with a frequency of 2 kHz and a sound pressure level of 57 dB. Fastl and Zwicke 2007.

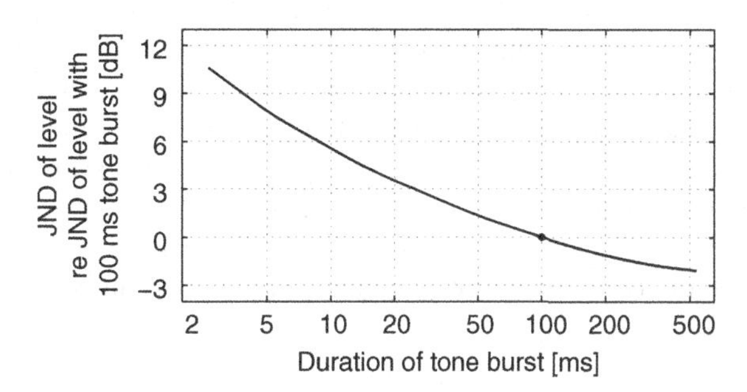

Figure 10.17 Just noticeable difference of the level of a 1-kHz tone as a function of the temporal length. The ordinate value is normalized with the JND of a 100-ms-long tone burst. Adapted from Fastl and Zwicker (2007).

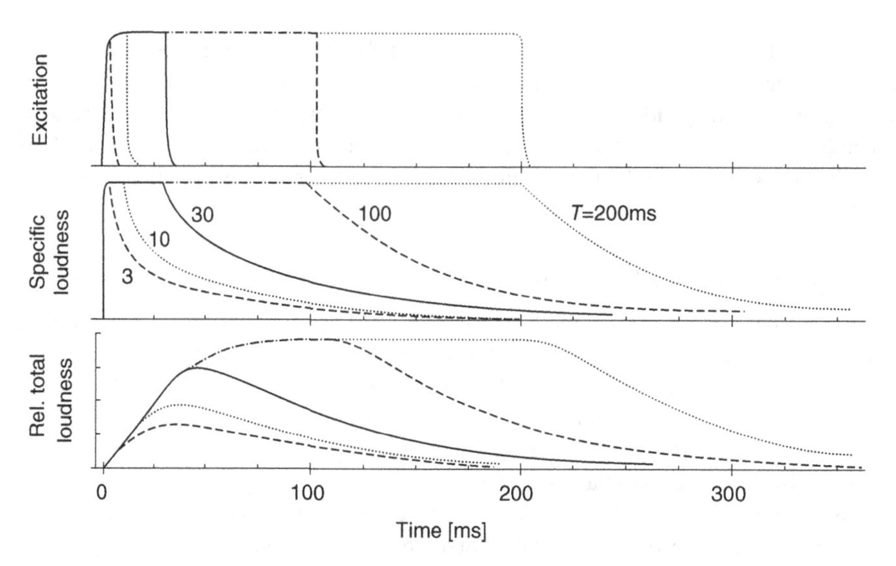

Figure 10.18 The change in loudness depending on time. Upper panel: sound pressure level. Middle panel: specific loudness. Lowest panel: relative total loudness. Tone bursts with length 3, 10, 30, 100, and 200 ms are used as stimuli in the simulations. Adapted from Zwicker (1984), and reprinted with permission from The Acoustical Society of America.

threshold increases monotonically (Fastl and Zwicker, 2007). The loudness of temporally shorter sounds is thus subject to added uncertainty in the hearing mechanism.

So far we have reviewed the perception of loudness caused by tone bursts. Speech and music are composed of sounds whose spectrum changes all the time. The perceived loudness can thus be expected to change all the time. However, it has been found that the loudness of such time-varying sounds seems to be the result of a short-time integration of sound power followed by a processing of the peak values by another mechanism. The perceived loudness is thus defined by loudness peak values. For example, a sound, in which 200-ms sound bursts alternate with 200-ms pauses is perceived to have loudness similar to a corresponding continuous sound.

Fastl and Zwicker (2007) present a model to estimate the loudness of sound with a dynamically changing level. The time-varying level of a tone-burst excitation with different lengths is shown in Figure 10.18 (topmost panel). The resulting loudness densities varying with time are shown in the middle panel, and the relative total loudness is shown in the lowest panel.

The specific loudness shown in the middle panel increases rapidly, matching the stimulus level, but decreases with a time constant that depends on the length of the corresponding stimulus. With short excitation signals, the specific loudness decreases rapidly, and with longer signals, the decrease is notably slower. The specific loudness represents the excitation in the inner ear, and the time integrated total loudness is assumed to be formed at higher stages of processing. The curves in the lowest panel represent the result after integration over time and frequency. The total loudness perception then corresponds to the highest peak in the curve.

10.3 Timbre

The tonal colour, or *timbre*, is a multidimensional psychoacoustic measure. When two sounds have the same pitch, loudness, and duration, timbre is what makes one particular musical sound different from another. For example, the same musical notes played by a piano and a trumpet are easily distinguished by listeners as being different. The best physical explanation for this difference comes from the spectrum and its variation with time. Certain factors that affect the formation of timbre are discussed next.

10.3.1 Timbre of Steady-State Sounds

If the short-term amplitude spectrum of the sound is constant with time, the timbre is also constant in most cases. In some cases with periodic sounds, some alterations in the phase spectrum of sound but no changes in the amplitude spectrum may also change the perceived timbre. This effect is discussed more thoroughly in Section 11.5. However, in such steady-state conditions, the specific loudness (or the corresponding excitation pattern) represents the perceived timbre quite accurately if the sound is broadband noise.

The number of possible timbres is enormous. Theoretically, since the frequency resolution of hearing is 1 ERB and the entire auditory frequency range consists of 42 bands, and since the level resolution is 1 dB for a dynamic range of about 100 dB, there are about 100^{42} possible timbres. However, when the masking effect is taken into account, this number is smaller. Nevertheless, even for a dynamic range of only 6 dB, the number of timbres is $6^{42} > 4 \cdot 10^{32}$, which is still a colossal number.

10.3.2 Timbre of Sound Including Modulations

In many musical instruments, such as the piano or the human singing voice, the partials are modulated in amplitude, in frequency, or in both. These modulations show up as modulations in the auditory band signals, which give a new dimension to the perceived timbre. These fluctuations are visible in the *amplitude envelope* of the auditory bands. Hearing is especially sensitive to modulations at 4 Hz. Faster modulations (>10 Hz) lead to the sound being perceived as ‘rough’, and still faster modulations are not perceived, since the signal components spread over different critical bands (Fastl and Zwicker, 2007, Chapter 10). In general, the presence of modulation in the signal of an auditory channel makes that channel more detectable, and the rate of modulation can also be perceived as beating.

The *onset* of sound is also important in timbre perception. Sounds from many instruments have a distinctive onset, where the partials of the sound build up and possibly contain some transient-like, noisy components. The timbre during the onset defines the individual characteristics of the instrument in many cases. For example, some brass instrument sounds cannot be distinguished from each other if the onset is digitally removed.

10.4 Subjective Duration of Sound

Quantifying the duration of sound, the attribute used to order a set of sounds from ‘short’ to ‘long’, can be done in a similar manner to quantifying the loudness and pitch of sound. The subjective duration of a burst of tone is shown in Figure 10.19. The curves originate from a test conducted in a fashion similar to that for loudness tests; the task for the subjects was to find sounds with half or twice the duration of the reference sound. The unit of subjective duration is the *dura*, which corresponds to the perceived duration of 1 s of a 1-kHz tone (Fastl and Zwicker, 2007, Chapter 10). As shown in the figure, the subjective and physical duration of sound correspond well with each other for sounds longer than about 200 ms. For physical durations shorter than 200 ms, the subjective duration is perceived to be longer than the physical one.

The subjective duration of a relatively long sound surrounded by silence is thus quite accurately perceived. It is also interesting how accurately humans perceive the duration of silence.

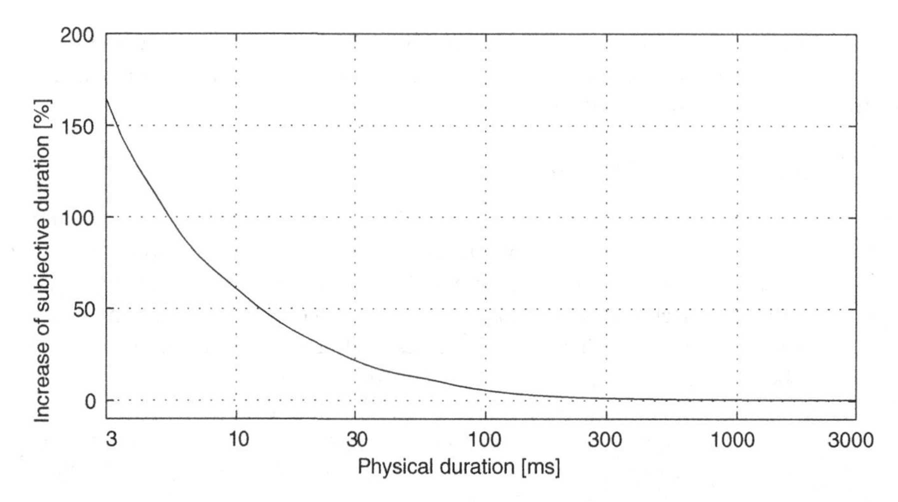

Figure 10.19 The relative subjective duration of sound as a function of the physical duration of a 1-kHz tone at an SPL of 60 dB. A similar result is obtained with noise. Adapted from Fastl and Zwicker (2007).

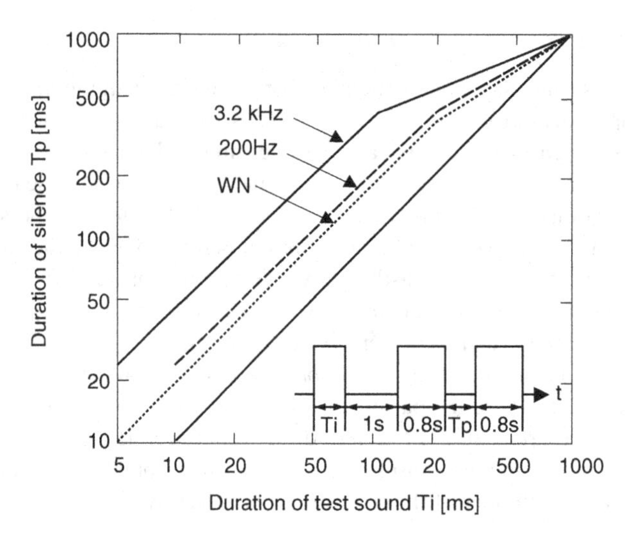

Figure 10.20 The physical duration of silence T_p between two bursts is adjusted to match the perceived length of a test sound. The physical duration of the test sound is T_i. Fastl and Zwicke 2007.

This has been studied by asking subjects to adjust the length of silence between two bursts so that it is equal to the length of a test sound. The result of this experiment is shown in Figure 10.20, which indicates the physical duration of the silence T_p when it has been adjusted to match the corresponding perceptual durations of the test sound. The physical duration of the test sound is T_i.

The physical duration of test sounds interpreted as durations of silence by subjects match well when the duration is near 1 s. The difference between the values, however, increases with a shortening test sound, with durations less than 100 ms being perceived as much longer than the length of the physical test signal (Fastl and Zwicker, 2007, Chapter 10). It is clear that this phenomenon has an effect on the perception of speech and music. For example, short silences existing in speech are hard to perceive, at least partly due to this effect. It is also clear that the longer perceptual duration of the burst tone relative to silence is related to the time-masking phenomenon. The silence immediately after the burst is masked, as shown in Figure 10.18, which makes the silence appear shorter than it actually is.

Summary

The basic quantities of sound measured by the hearing system, namely pitch, loudness, timbre, and subjective duration, were introduced in this chapter. Based on how our hearing perceives these quantities, a general view of the functioning of hearing, which is coherent though rich in details, was drawn. In general, our hearing shows remarkable capabilities in analysing sounds arriving at the ear canals. Identifying these basic quantities is the first step taken to identify different sounds, such as speech, natural sounds, and sounds from musical instruments. The next chapter discusses some of the further analysis mechanisms used by our hearing, which then lead to our abilities to generate and enjoy music, and to organize complex sound scenarios into meaningful streams.

Further Reading

The basic quantities in psychoacoustics are also discussed in the corresponding chapters of the books by Fastl and Zwicker (2007), Moore (1995), and Yost (1994). Recent advances in theories of loudness and pitch perception are reviewed by Florentine *et al.* (2005) and Plack *et al.* (2005), respectively.

References

ANSI-S1.1 (2013) Acoustical terminology. Standards Secretariat, Acoustical Society of America.

Bilsen, F. and Ritsma, R. (1969) Repetition pitch and its implication for hearing theory. *Acta Acustica United with Acustica*, **22**(2), 63–73.

Burck, W., Kotowski, P., and Lichte, H. (1935) Die horbarkeit von laufzeitdifferenzen. *Elek. Nachr.-Techn.*, **12**, 355–362.

Canteretta, E.C. and Fridman, M.P. (eds)(1978) *Handbook of Perception*. Academic Press.

Demany, L. and Semal, C. (1989) Detection thresholds for sinusoidal frequency modulation. *J. Acoust. Soc. Am.*, **85**(3), 1295–1301.

Fastl, H. and Stoll, G. (1979) Scaling of pitch strength. *Hearing Res.*, **1**(4), 293–301.

Fastl, H. and Zwicker, E. (2007) *Psychoacoustics – Facts and Models*. Springer.

Florentine, M., Popper, A., and Fay, R.R. (eds) (2005) *Loudness*, volume 37. Springer.

Hartmann, W.M. (1996) Pitch, periodicity, and auditory organization. *J. Acoust. Soc. Am.*, **100**(6), 3491–3502.

Moore, B. (1973) Frequency difference limens for short-duration tones. *J. Acoust. Soc. Am.*, **54**(3), 610–619.

Moore, B.C.J. (ed.) (1995) *Hearing*. Academic Press.

Moore, B.C., Vickers, D.A., Baer, T., and Launer, S. (1999) Factors affecting the loudness of modulated sounds. *J. Acoust. Soc. Am.*, **105**(5), 2757–2772.

Plack, C.J. (2013) *The Sense of Hearing*. Psychology Press.

Plack, C.J., Oxenham, A.J., Fay, R.R., and Popper, A.N. (2005) *Pitch: Neural Coding and Perception*, volume 24. Springer.

Pressnitzer, D., Patterson, R.D., and Krumbholz, K. (2001) The lower limit of melodic pitch. *J. Acoust. Soc. Am.*, **109**, 2074–2084.

Rossing, T.D., Moore, F.R., and Wheeler, P.A. (2001) *The Science of Sound*, 3rd edn. Addison-Wesley.

Sek, A. and Moore, B.C. (1995) Frequency discrimination as a function of frequency, measured in several ways. *J. Acoust. Soc. Am.*, **97**, 2479–2486.

Yost, W.A. (ed.) (1994) *Fundamentals of Hearing – An Introduction*. Academic Press.

Zwicker, E. (1984) Dependence of post-masking on masker duration and its relation to temporal effects in loudness. *J. Acoust. Soc. Am.*, **75**, 219–223.

11

Further Analysis in Hearing

Chapter 9 presented two important concepts in psychoacoustic experimentation – masking and critical bands. Chapter 10 presented the four central quantities or dimensions of psychoacoustics – pitch, loudness, timbre, and subjective duration, all of which are relatively well defined and orthogonal to each other, except perhaps timbre. Timbre is a multidimensional and complex measure, and a number of quantities that describe the perception of timbre have been defined in the literature; these can be regarded as subcategories of timbre.

This chapter describes a few of these quantities which are useful in the research on psychoacoustics or in technical applications. These quantities are sharpness, roughness, fluctuation strength, tonality, consonance, and dissonance. These quantities also provide an interesting connection to the intervals in music scales.

Sound signals are primarily one-dimensional, time-dependent functions. However, our hearing distils them into features and characteristics that depend both on time and frequency, which makes sound a multidimensional entity for neural processing. The psychoacoustic foundations of our perception of sound and music, which is an art form that takes full advantage of the time–frequency structures of sound, are discussed in this chapter. In natural listening conditions, our hearing attempts to represent sound scenarios as objects, just as the other senses do. Certain complex sound scenarios, such as sound in environments containing many sources or music presented by an ensemble of instruments, are often structured into auditory streams in our hearing.

11.1 Sharpness

When subjects are asked 'how sharp is the sound?' on being presented with sounds having different spectral content, a relatively stable, repeatable, and subject-independent response can be measured. The resulting response is the psychoacoustic quantity called *sharpness* (Fastl and Zwicker, 2007, Chapter 9). The higher in frequency the centre of gravity of the amplitude spectrum is located, the higher is the sharpness measured. Figure 11.1 shows the sharpness of narrowband noise with a width of 1 Bark as a function of frequency. The figure also shows the sharpness for low-pass and high-pass filtered noise as a function of cut-off frequency.

Communication Acoustics: An Introduction to Speech, Audio, and Psychoacoustics, First Edition.
Ville Pulkki and Matti Karjalainen.

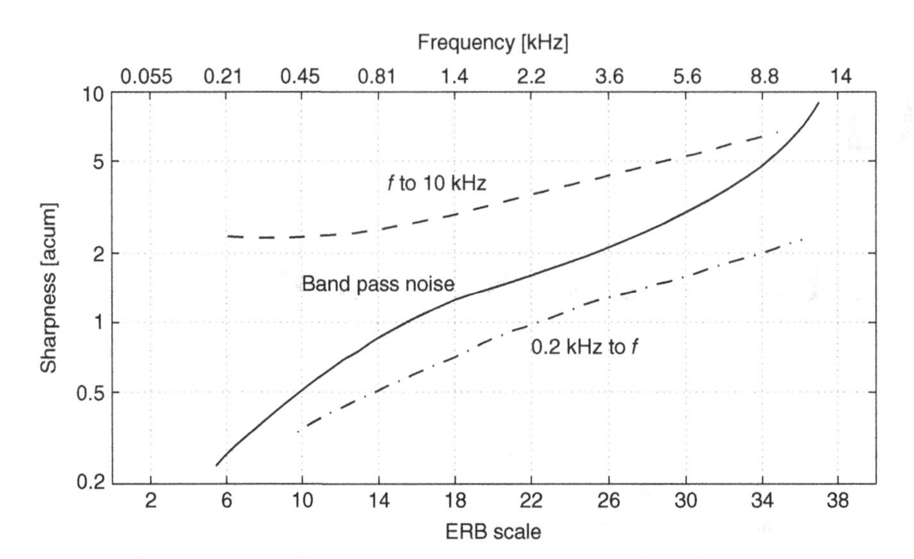

Figure 11.1 The sharpness of narrowband noise (solid line), high-pass filtered noise (upper cutoff is at 10 kHz), and low-pass filtered noise (lower cutoff is at 200 Hz) as a function of the centre frequency or cutoff frequency. Adapted from Fastl and Zwicker (2007).

The unit of sharpness, the *acum*, is defined such that narrowband noise of width 1 Bark and centre frequency 1 kHz at an SPL of 60 dB produces a sharpness value of 1 acum. Psychoacoustic testing reveals that the level of sound also affects the perceived sharpness, but not very prominently. Increasing the level from 30 dB to 90 dB only doubles the sharpness value (Fastl and Zwicker, 2007, Chapter 9).

For narrowband noise, the sharpness rises monotonically on the frequency scale. Below 1 kHz and above 4 kHz the rise is steeper. The steady rise of sharpness S in the middle frequencies and the steep rise at high frequencies can be modelled simply by defining the gain $g(z)$ on the Bark scale z, as in Figure 11.2, and by writing

$$S \sim g(z)\, z. \tag{11.1}$$

Figure 11.1 shows that the sharpness of broadband sounds depends on where the 'centre of gravity' of specific loudness is located, and when it is situated at higher frequencies, particularly high sharpness is obtained. This observation can be used to derive a simple computational model for sharpness, which can be written as

$$S = 0.11\, \frac{\int_0^{24\,\mathrm{Bark}} N'(z)\, g(z)\, z\, \mathrm{d}z}{\int_0^{24\,\mathrm{Bark}} N'(z) \mathrm{d}z}, \tag{11.2}$$

where $N'(z)$ is the specific loudness. The equation does not represent the relationship between the level of the sound and sharpness. Although the equation is relatively simple, it provides a fairly accurate estimation of sharpness for different sounds.

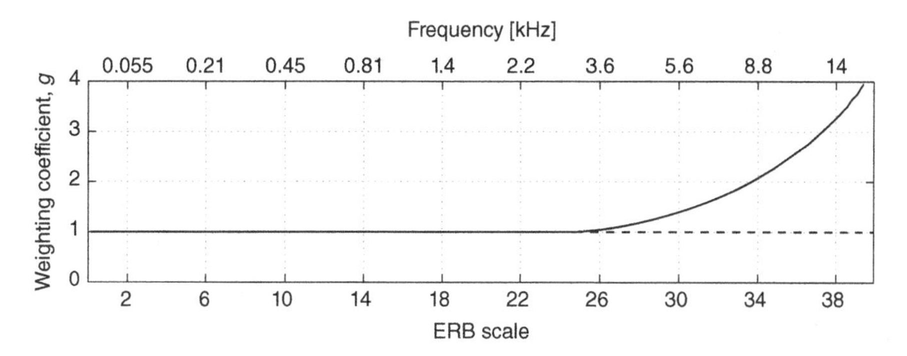

Figure 11.2 The gain factor $g(z)$ as a function of the Bark scale z used to compute sharpness. Adapted from Fastl and Zwicker (2007).

11.2 Detection of Modulation and Sound Onset

Considerable changes in the acoustic characteristics of sound with time are perceived as beating, warbling, impulsiveness, fluctuation, or just changing in general. If the level of sound fluctuates relatively slowly over a certain range of rates, between about 1 Hz and 16 Hz, a significant *fluctuation strength* may be perceived. When the modulation rate far exceeds 16 Hz, our hearing is unable to follow the level of sound, and the sound is perceived as *rough*, associated with the psychoacoustic quantity *roughness*. A fast-rising sound is perceived as *impulsive*, and the corresponding psychoacoustic quantity for this perception is *impulsiveness*.

11.2.1 Fluctuation Strength

The simplest examples of fluctuation in sound are amplitude and frequency modulation. Such modulations give rise to the perception of the psychoacoustic quantity called *fluctuation strength* (Fastl and Zwicker, 2007). If the modulation has a relatively low frequency, below about 0.5 Hz, our hearing does not detect it anymore, since the short-term auditory memory does not have a very good reference to the past. If the frequency of modulation is higher than about 16 Hz, the sluggishness of the hearing mechanisms starts to limit the resolution of fluctuation, and the modulations are perceived to produce *roughness*, as discussed later in Section 11.3. The hearing is most sensitive to fluctuations and modulation near the frequency 4 Hz.

The unit of fluctuation strength is the *vacil*, where the reference point is selected by prescribing that a 1-kHz sinusoid at a level of 60 dB with 100% amplitude modulation at 4 Hz produces a fluctuation strength of magnitude 1 vacil. The fluctuation strength is shown in Figure 11.3 as a function of modulation frequency for three cases: amplitude-modulated broadband noise, an amplitude-modulated tone, and a frequency-modulated tone. The curves are similar, and their peak is found at about 4 Hz (Fastl and Zwicker, 2007, Chapter 10).

The dependence of fluctuation strength on modulation depth is shown in Figure 11.4 (AM BBN) for broadband noise, where it can be seen that modulation of a few dB is needed to produce notable fluctuation strength, and that the phenomenon saturates at a modulation depth of 20–30 dB. The corresponding dependence of a frequency-modulated tone on frequency deviation is shown in the same figure in the panel labelled FM tone. Increasing the frequency deviation is seen to cause increasing fluctuation strength, and no saturation effects

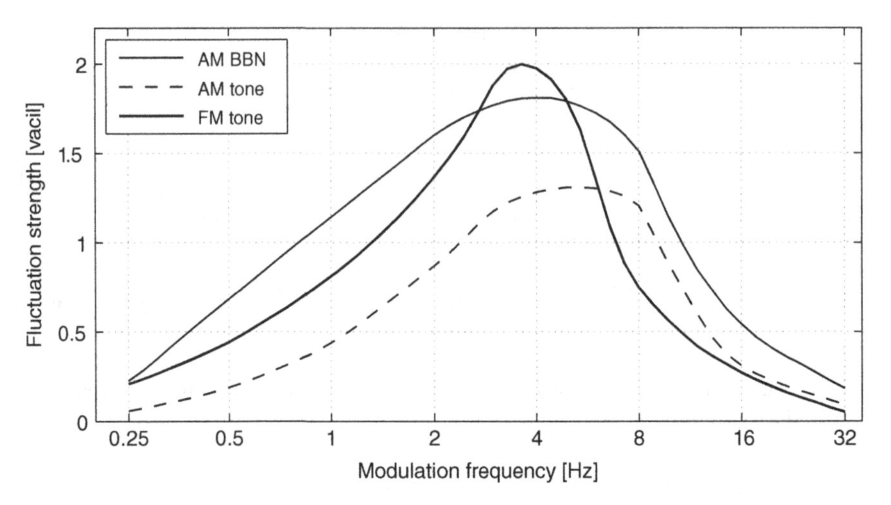

Figure 11.3 Fluctuation strength as a function of modulation frequency for three cases: (AM BBN) amplitude-modulated white noise with 40-dB modulation depth, (AM tone) amplitude-modulated tone with 40-dB modulation depth, and (FM tone) frequency-modulated 1-kHz tone with ± 700 Hz frequency variation. Adapted from Fastl and Zwicker (2007).

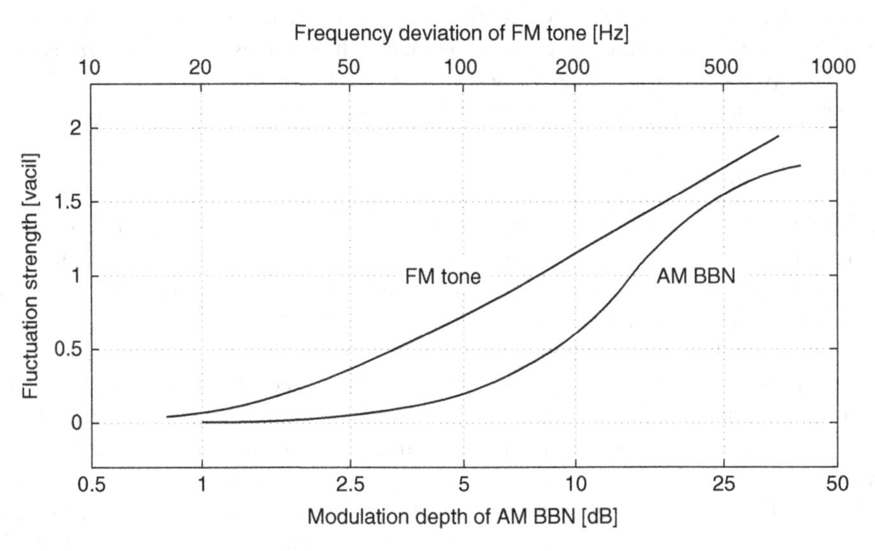

Figure 11.4 The fluctuation strength of a 4-Hz amplitude-modulated broadband noise as a function of modulation depth in dB (AM BBN). The fluctuation strength of a 1.5-kHz tone at a level of 70 dB and frequency-modulated at the rate of 4 Hz as a function of the modulation deviation in Hz (FM tone). Adapted from Fastl and Zwicker (2007).

are seen. The fluctuation strength also depends on loudness. For example, a change in the level of amplitude-modulated sound from 50 dB to 90 dB causes a fivefold increase in fluctuation strength.

The information presented so far suggests that our hearing forms the fluctuation strength percept by summing the modulation in the level of the signal in all of the auditory bands. The

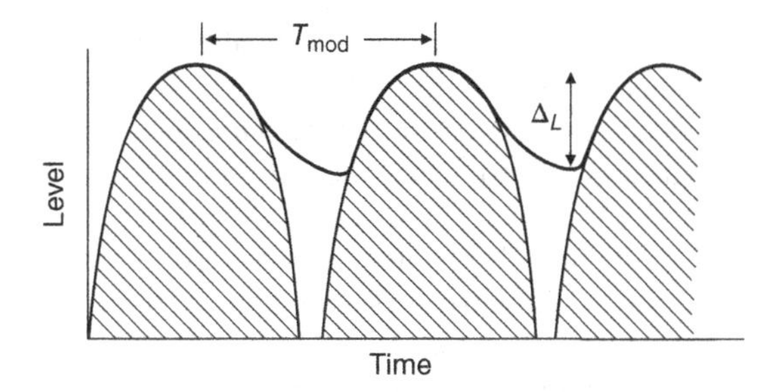

Figure 11.5 The temporal masking pattern caused by a sinusoidally amplitude-modulated masker, and the depth of the pattern ΔL and modulation period $T_{mod} = 1/f_{mod}$. Adapted from Fastl and Zwicker (2007).

principles illustrated in Figure 10.14 on page 186 can also be applied in this context, since they also explain the formation of the fluctuation strength with both amplitude and frequency modulation. Although the modulated signal components reside only in a narrow frequency region, the spreading function in the basilar membrane causes the effect of modulation to be present over a much wider region.

A schematic diagram of the phenomenon leading to the perception of fluctuation strength is shown in Figure 11.5. In the illustration, the input signal is an amplitude-modulated sinusoid with a modulation frequency of f_{mod}. The level of excitation in a single auditory band is shown as the hatched area, which exhibits strong modulation with time. The thick line shows the specific loudness schematically, which does not follow fast modulation. The quantity ΔL represents the depth of modulation of the temporal masking pattern, and the fluctuation strength F can be estimated as

$$F \sim \frac{\Delta L}{(f_{mod}/4\,[\text{Hz}]) + (4\,[\text{Hz}]/f_{mod})} \tag{11.3}$$

A slightly more general dependence between amplitude- and frequency-modulated signals can be written (Fastl and Zwicker, 2007, Chapter 10), if $\Delta L(z)$ is known:

$$F[\text{vacil}] = \frac{0.008 \int_0^{24\,\text{Bark}} (\Delta L(z))\text{d}z}{(f_{mod}/4[\text{Hz}]) + (4[\text{Hz}]/f_{mod})}. \tag{11.4}$$

This relation can be implemented if $\Delta L(z)$ is measured with a suitable time-dependent auditory model. The denominator may also be implemented, if f_{mod} is known, or is measured by band-pass filtering L.

11.2.2 Impulsiveness

A sound is perceived to be 'impulsive', if it contains transients where the level of sound increases rapidly. As shown in Figure 3.13, impulses produce a very strong response in the auditory nerve, while on the other hand, as shown in Figure 10.18, very short sound events produce lower responses than their level would imply. However, transients are often perceived as an increase in annoyance caused by noise, and they very effectively grab a subject's attention. This attention steering may be seen as a method to make one aware of one's surroundings, since it may carry valid information about active processes that are often caused by a change of state in physical objects. A strongly impulsive sound also contains high frequencies, since the sound cannot be very short without them.

There is no simple psychoacoustic measure for the *impulsiveness* of sound. In noise measurements there are some measures used, where the levels of measured noise are characterized as long-term mean levels. For example, the sound pressure level can be measured so that it exceeds the average 10% of the time. When the measured level is compared to the equivalent level of the noise, an estimate of impulsiveness is obtained. It has also been proposed that onsets which define impulsiveness can be identified by finding the regions where the positive slope of the instantaneous sound pressure level exceeds 10 dB/s (Pedersen, 2001).

11.3 Roughness

When two tones close to each other in frequency are added, the resulting broadband signal contains amplitude fluctuation at frequencies corresponding to the difference in the frequencies. This fluctuation is called *beating*. The resulting perception of the auditory event is also interesting, as it depends relatively systematically on the difference in frequency. This is shown conceptually in Figure 11.6.

When Δf is small, a slight and often pleasant beating is perceived, which can be measured as a fluctuation in the strength of sound. When Δf exceeds 10–15 Hz, the percept turns more 'rough', which is generally perceived to be unpleasant. After a certain value of Δf, the tones are not anymore fused, but instead are resolved and their frequencies are perceived independently of each other, however, with some roughness still present. When Δf is increased further, and the width of a critical band is exceeded, the roughness vanishes, and the presence of the two tones no longer has any influence on perception.

Roughness (Daniel and Weber, 1997; Fastl and Zwicker, 2007) is a psychoacoustic quantity which is caused by relatively fast amplitude modulations (15–300 Hz) that take place for stimuli within the range of a critical band. For example, narrowband noise always sounds slightly rough, since there are random amplitude fluctuations in it.

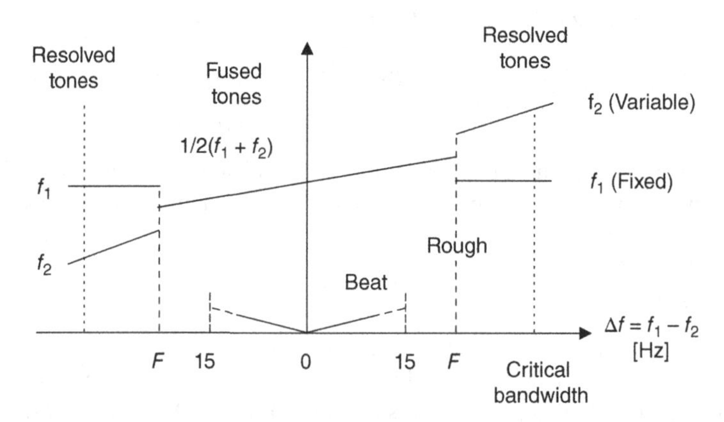

Figure 11.6 The nature of auditory perception of two sinusoids, one with a fixed frequency f_1 and another with a variable frequency f_2. The frequency difference is shown on the abscissa and the ordinate shows the frequency corresponding to the fused or resolved tones. Adapted from Roederer (1975) with permission from Academic Press.

The unit of roughness is the *asper*, and 1 asper is obtained with a 1-kHz tone which is 100% amplitude-modulated at a rate of 70 Hz. The roughness of such a sound is presented in Figure 11.7 for different modulation depths (Fastl and Zwicker, 2007, Chapter 11). Figure 11.8 presents roughness as a function of modulation frequency for different carrier frequencies. Most of the curves shown in Figure 11.8 have a coinciding low-pass shape with their maximum near 70 Hz, but the results with low-frequency sinusoids are different. At low frequencies, the width of the critical band (in hertz) is narrower than at high frequencies, and the maximum roughness is obtained with lower values of modulation frequency than at higher frequencies. Figure 11.8 shows that roughness can be explained as a kind of band-pass operation. The frequency of amplitude modulation present in the critical band can be thought to be band-pass filtered by a filter with the frequency response shown in Figure 11.8, and the level of the output of the filter is related to the value of roughness.

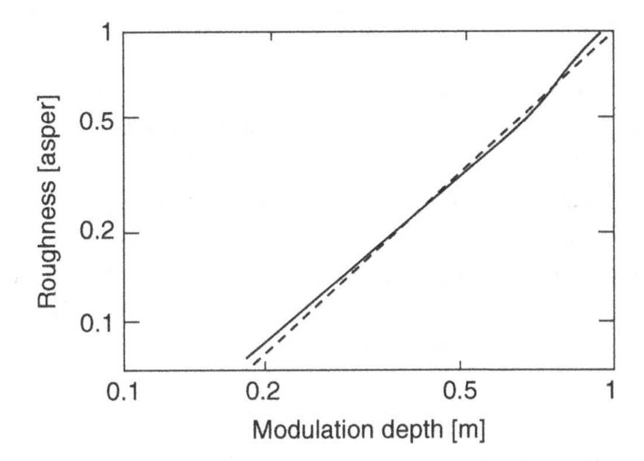

Figure 11.7 The solid line presents the roughness of an amplitude-modulated tone as a function of modulation depth, when the frequency of the tone is 1 kHz and the modulation frequency is 70 Hz. The dashed line is a linear fit of the roughness for reference. Fastl and Zwicke 2007.

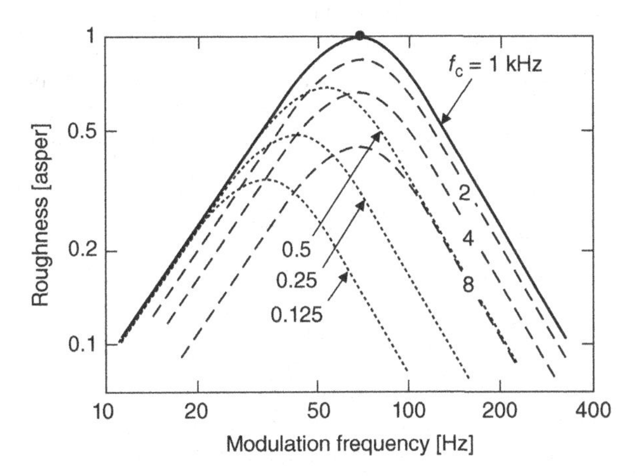

Figure 11.8 The roughness of a 100%-amplitude-modulated tone as a function of the modulation frequency. The different curves show the roughness for different carrier frequencies. Fastl and Zwicke 2007.

Roughness is also perceived to be high when broadband noise is amplitude modulated. Also, when a sinusoid is frequency modulated, the auditory event is perceived to be rough. In both cases the signal produces roughness in many critical bands. In auditory modelling, the roughness perception caused by an arbitrary sound can be estimated by summing the partial roughness created in each critical band.

11.4 Tonality

Tonality (or *tonalness*) is similar to the voicing character of speech: tonality is low with noisy sounds and high with sounds containing prominent frequency components, or tones. The tonality of a sound also increases with one or many high-amplitude and narrowband components, where the listener resolves either the fundamental frequency or a tonal component of the sound. Tonality is also related to pitch salience. When pitch salience is high, such as with pure tones or with narrowband noises, the tonality is high as well. However, with some signals, such as bell sounds, pitch is ambiguous, producing low pitch salience but yet high tonality. Note that tonality in psychoacoustics is a quantity which reflects how much the sound contains the characteristics of a *tone*. It is not to be confused with the terms from music, *tonality*, *tonal*, *atonal*, and *microtonal*, which are descriptors of musical scales and styles of musical composition.

There is no really clear method to measure tonality, although there are several starting points from which it can be quantified. Some methods are described below. A simple measure for tonality is the *tone-to-noise ratio, T/N,* which has also been standardized in ISO-7779 (2010). It measures the level of a tone compared to the level of the surrounding frequency band. Unfortunately, if there are multiple tones present in a critical band, the T/N method does not match the perceived tonality. The *prominence ratio, PR* (Bienvenue and Nobile, 1991) is a better approach in such cases; here, the power of the critical band containing the prominent tone is compared to the mean power in adjacent critical bands.

Tonality is also discussed and modelled by (Terhardt *et al.*, 1982, 1996). In the model, the local maxima are sought from the spectra. If the level difference between the maximum and its surroundings is more than 7 dB, the signal component corresponding to the maximum is thought to be tonal.

Tonality has also been studied in the context of audio technologies and speech transmission, since knowing how well a partial of a harmonic or noise masks another partial or noise is relevant in audio coding (Johnston, 1985). It has been found that noise is masked easier by a partial than vice versa. The *spectral flatness measure* (SFM) is first computed as the ratio of the geometric mean G_{m} to the arithmetic mean A_{m} of the power spectrum of the signal in dB,

$$\mathrm{SFM_{dB}} = 10\,\log_{10}\frac{G_{\mathrm{m}}}{A_{\mathrm{m}}}, \tag{11.5}$$

and the tonality α is then computed as

$$\alpha = \min\left(\frac{\mathrm{SFM_{dB}}}{\mathrm{SFM_{dBmax}}}, 1\right), \tag{11.6}$$

where $\mathrm{SFM_{dBmax}} = -60\,\mathrm{dB}$ is a reference value. If $\mathrm{SFM_{dB}} = 0\,\mathrm{dB}$, the signal is set to have a tonality value of 0 and thus to have a noise-like character. If $\mathrm{SFM_{dB}} < -60\,\mathrm{dB}$, the tonality is 1, and it is considered to have a tone-like character. Johnston (1985) also explains how the estimated tonality α is further used to estimate the audibility of some component of sound, so that it is retained or discarded during audio coding.

11.5 Discrimination of Changes in Signal Magnitude and Phase Spectra

The previous and the present chapters have discussed many properties of hearing, starting from simple signals and considering only their amplitude spectra. The analysis of basic psychoacoustic quantities with such signals helps to explain many phenomena in hearing. On the other hand, audio and speech signals can also be analysed by objective means, as shown in Chapter 3, for example by using spectral analysis, time–frequency plots, and spectrograms. This section focuses on the characteristics of hearing in relation to signal analysis, especially the ability to hear different characteristics in the magnitude and phase spectra. This helps develop an understanding of quality evaluation of audio techniques, which will be discussed further in Chapter 17.

11.5.1 Adaptation to the Magnitude Spectrum

A remarkable property of hearing is its ability to adapt to the acoustics of different listening conditions. To help in identifying a sound event across a transmission channel, it appears that listeners try to remove spectral distortion caused by the channel. Listeners seem to compensate for, or adapt to, the channel (Olive *et al.*, 1995; Pike *et al.*, 2013). In other words, it seems that listeners are able, at least partially, to inverse-filter the spectral effect of the channel to more accurately estimate the signal emitted by the source (Toole, 2006; Watkins, 1991).

The ability to do this is most probably a result of neural processes within the hearing system. For example, a study by Summerfield *et al.* (1984) shows that exposure for just 1 second to a sound can result in spectral adaptation, possibly via neural adaptation in the auditory periphery. Summerfield *et al.* (1984) played a sequence of sounds to the listener the first of which consisted of a harmonic complex containing the first 50 harmonics of a tone, with a number of those 50 harmonics reduced in amplitude. The complex was then heard again but with the harmonics which had previously been reduced in level increased to the level of the remaining harmonics. When listening to the second sound, the components that had changed in level were heard to stand out perceptually from the rest of the tone. This enhancement effect appears to demonstrate the auditory system's heightened sensitivity to temporal changes in spectrum and may show a mechanism which helps listeners to perceive differences in magnitude spectra between different sounds, as well as a system that can remove spectral distortion.

The temporal dynamics of the enhancement effect suggest that it is the result of rapid adaptation of neural responses in the peripheral hearing system (Summerfield *et al.*, 1984, 1987). However, the results of onset and recovery measurements are somewhat controversial (Cardozo, 1967; Viemeister, 1980; Wilson, 1970). Therefore, the extent to which the measured temporal dynamics of adaptation are compatible with processing only in the periphery is not clear.

Spectral adaptation may also be caused, at least partly, by central brain processes. Adaptation in the central hearing system appears to be stronger and to have a longer time constant (Holt, 2006; Ulanovsky *et al.*, 2003, 2004; Watkins, 1991). This effect is assumed to occur at the level of the auditory cortex, and it is called *stimulus-specific adaptation*, which is the specific decrease in the response to a frequent (standard) stimulus (Holt, 2006). In addition, such adaptation may also occur at even higher stages in the brain in the cognitive processes. This would mean subconscious calculation of the average spectrum and its inverse filtering during listening. Adaptation may also be inextricably linked to auditory memory, whereby adaptation occurs by a process of forgetting (McKeown and Wellsted, 2009). These adaptation effects are relatively poorly understood. The time constants in peripheral adaptation and central adaptation are not yet confirmed, and the mechanisms behind adaptation are largely unknown.

11.5.2 Perception of Phase and Time Differences

The Fourier analysis in Equation (3.17a) produces a complex spectrum, which can be presented in polar coordinates using the magnitude and phase

$$\mathcal{F}\{x(t)\} = \mathrm{Re}\,\{X(\omega)\} + \mathrm{j}\,\mathrm{Im}\,\{X(\omega)\} = |X(\omega)|e^{\mathrm{j}\varphi(\omega)}. \tag{11.7}$$

The phase $\varphi(\omega)$ is mathematically problematic, since if it is computed from

$$\varphi(\omega) = \arctan[\mathrm{Im}\,\{X(\omega)\} \,/\, \mathrm{Re}\,\{X(\omega)\}] \tag{11.8}$$

the result is wrapped discontinuously between 0 and 2π. To obtain a continuous function of phase, it has to be *unwrapped*, which is numerically a critical operation and can lead to errors with a magnitude of 2π or its multiples.

The capability of the phase spectrum of sound to transmit information between humans in typical acoustic conditions is now discussed. As an example, the signals in Figure 11.9 have equal magnitude spectra but different phase spectra. In principle, such a prominent change in the temporal structure of the signal could be used to encode some meaning into communicated sounds, such as speech sounds. However, real listening conditions with a relatively distant source contain at least some reflections from nearby surfaces and potentially some reverberation. The impulse response of such a space corresponds to a very complex transfer function,

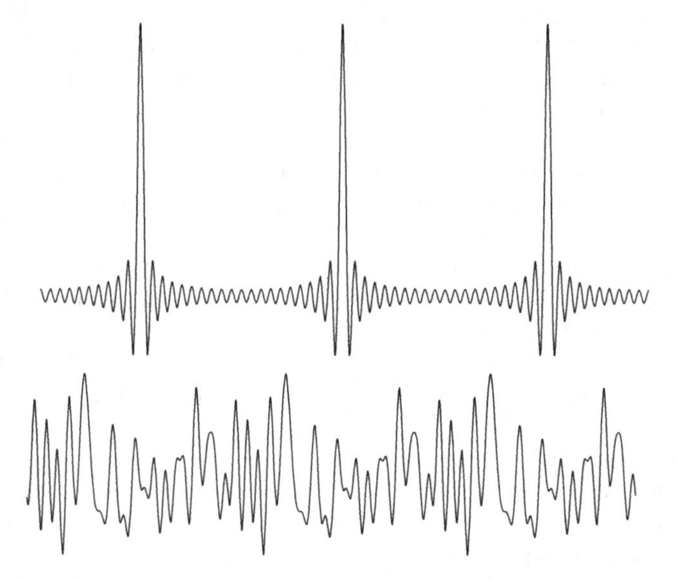

Figure 11.9 Two signals with 40 harmonic partials with identical magnitude spectra. In the upper signal, the phases of the harmonics are synchronized so that the partials have their maxima at the same temporal position. Correspondingly, the phases of the partials are randomly set in the lower signal.

which skews the phase spectrum to almost a random form. Thus, in practice, if the source emitted a signal such as in the upper panel of Figure 11.9, the signal reaching the receiver in a room outside the reverberation radius of the source would look something like the signal in the lower panel. It is thus natural that hearing has developed to be almost immune to phase, although some sensitivity to it exists. In 1843, Georg Ohm proposed that hearing measures the characteristics of a sound as the strengths of its partials and disregards the phase relations between them, which is known as Ohm's acoustic law. This is not completely true, as will be discussed below.

Such insensitivity to phase can often be assumed, although it has been shown that in some cases the phase and a modification of the phase have a prominent effect on perception. In fact, if the signals in Figure 11.9 are listened to with headphones or with loudspeakers closeby (without the effect of the room), a clear difference is perceived. In the upper panel, all partials of the harmonic complex have their maxima aligned at some temporal position, resulting in a prominent, repetitive peak at corresponding positions. In the lower panel, the phase relationships of the harmonics are random, and no such 'peakiness' is seen. The first signal is sharper; that is, it is perceived to have more energy at high frequencies. In addition, the tonal character of such sound has been called 'buzzy' (Moore, 2002), or as having high 'buzzyness' (Laitinen *et al.*, 2013), referring to a low vibrating sound like that of a bee.

Factors leading to a perception of buzzyness

In many cases, when a pair of signals with equal amplitude spectra but different buzzyness is found, a clear difference in the 'peakiness' is seen in the time domain that is, the location and height of peaks occurring in the two signals in the time domain differ. Such a change in peakiness is obvious in the broadband sound signal plotted in Figure 11.9. The peakiness, in this case, is present in the response of the signal at the basilar membrane, as shown in the highest ERB band of the signal, which is plotted in Figure 11.10.

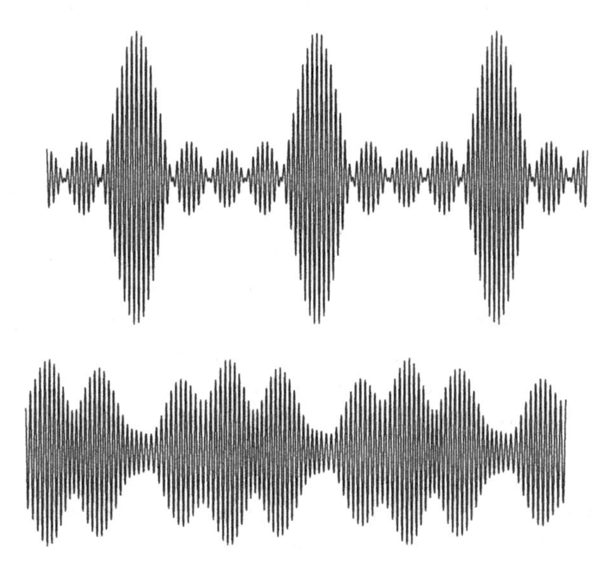

Figure 11.10 The responses of the signals in Figure 11.9 in the ERB band with frequency corresponding to the high end of their spectra.

This difference in peakiness has been studied further by Laitinen *et al.* (2013), where the timbral characteristics produced by various phase alterations are addressed. The study shows that phase randomization of the partials of a sawtooth signal perceptually creates a slightly larger change in timbral characteristics than when the amplitudes of the harmonics are randomized within a standard deviation of 4 dB. In other tests it was found that hearing is sensitive to the peakiness of the signal integrated within bandwidths of about two octaves. This can be simulated using an auditory model which estimates the neural firing rate at different critical bands depending on time. If the firing rate is synchronized in time within neighbouring critical bands, a buzzy character is perceived in the sound.

For an example, signals with equal amplitude spectra but different phase spectra, which have different timbres in headphone listening, can be visualized using the auditory model as shown in Figure 11.11. The signal in the top-left corner is a sawtooth signal with f_0 = 100 Hz, and the firing rate is shown below as a spectrogram. The excitation pattern shows vertical stripes, which means that the neurons of the cochlea at all frequencies fire simultaneously, leading to a very buzzy character of sound. The time-domain signal in the top-right corner is otherwise the same as in the top-left corner signal, but the 31st harmonic (at 3100 Hz) is inverted in polarity, making that frequency clearly audible, corresponding to about 6–9 dB amplification of that harmonic with the sawtooth signal (Laitinen *et al.*, 2013). The corresponding firing rate spectrogram shows an added horizontal stripe at that frequency, implying a time-smeared response, which seems to cause added loudness in this case.

The bottom-left corner shows a time-inversed (equivalently, having reversed polarity) signal of the one shown in the top-left corner. The signal creates an auditory event with a differently perceived level of bass, which can be explained by the time scattering of the firing rate there (Laitinen *et al.*, 2013). Correspondingly, the bottom-right corner of the figure shows the same signal, but now with a random phase spectrum. This results in a sound with low buzzyness, which is seen through the lack of vertical alignment of the firing rate peaks in time.

The reader is encouraged to listen to the effects using headphones; however, note that the group delay of the headphone has a dramatic effect on the phenomenon, and the results are not easily reproducible with all headphone models.

Perception of frequency-dependent group delay of audio devices

A technical measure commonly used to describe the temporal-domain characteristics of a signal or a transfer function is the group delay $\tau_g(\omega)$, defined in Equations (3.23a) and (3.23ab) on page 51. It is related to the delay of the components of the envelope (or modulation) of a signal. For example, the phase characteristics of a loudspeaker can be described by group delay. The mean value of the delay has no importance in one-way communication, such as listening to music records. However, if two-way communication is utilized, the group delay should be minimized to avoid latencies. If the group delay changes significantly with frequency, audible degradations of quality may occur.

The JND of group delay has been measured to be about 4–5 ms across the entire audible frequency range, when the group delay changes smoothly with frequency (Patterson, 1987). Such changes in group delay, depending on the frequency, are most audible with impulsive sounds. For example, in perceptual coding of audio, the coding artefacts are often first heard as time-smearing of impulses. The impulsive sounds may be degraded to sounds resembling chirps or noise bursts. The JND is smaller when the group delay changes less smoothly with frequency. In Blauert and Laws (1978), the detection threshold of a change in group delay was

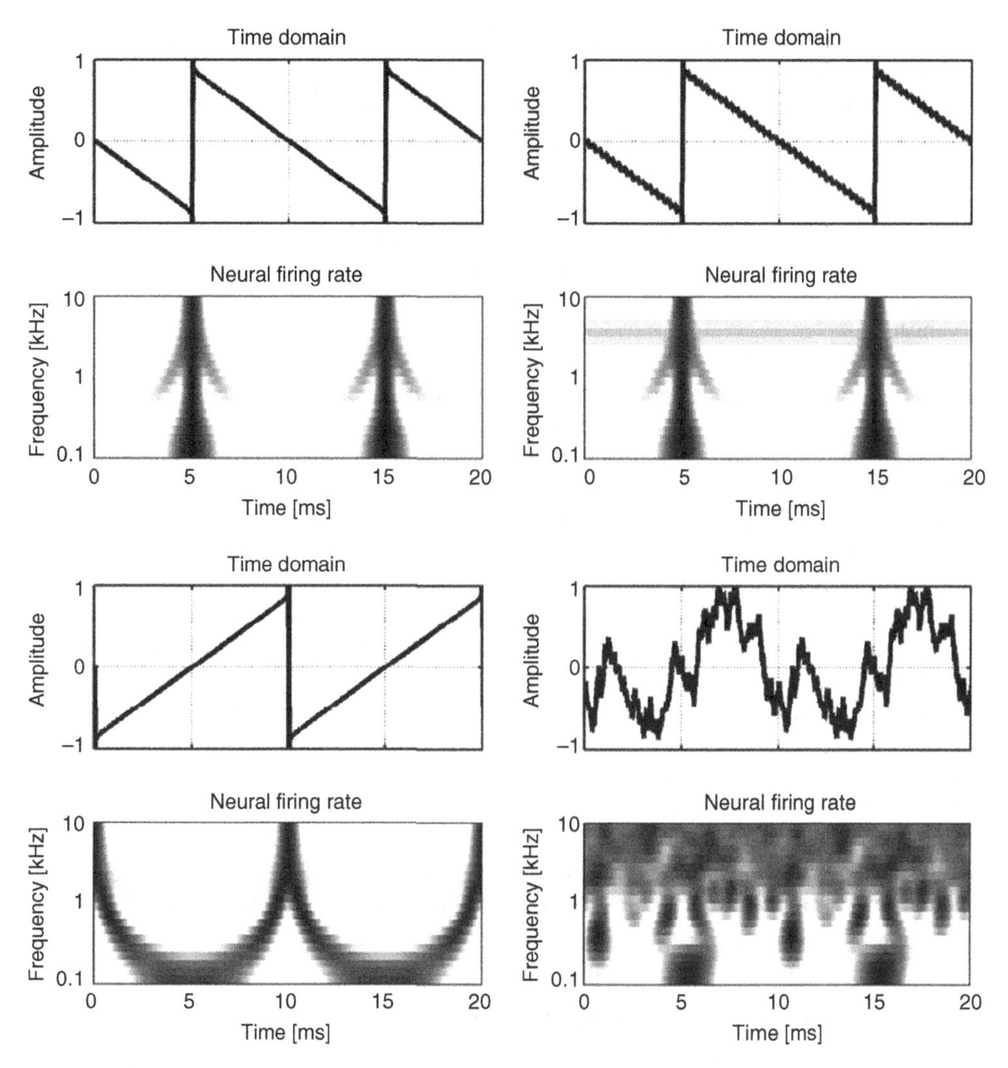

Figure 11.11 Four signals with identical magnitude spectra. The firing-rate panels show a spectrogram type of plot of the instantaneous neural firing rate in the cochlea. See the text for details. Courtesy of Mikko-Ville Laitinen.

as low as 0.4 ms when the frequency dependency of group delay was otherwise smooth, and a local group delay peak about half an octave wide was situated at a certain frequency.

The perceptibility of smoothly frequency-dependent group delays can be described as follows by a set of heuristic rules (Karjalainen, 2008):

- At best, our hearing can detect group delay changes, which are smooth in time, of 1 ms within a critical band. For example, sound signals from the elements of a multi-way loudspeaker at different distances from the listener arrive at different times. In laboratory conditions, delay differences of a little below 1 ms can be detected when directly compared to the non-delayed sound. This means that the distances from the elements of a loudspeaker to the listener should be similar, not deviating by more than 20–30 cm, which is normally not a problem in loudspeaker design.

- Group delay changes have to be about 3–5 ms with speech and music before they can be perceived. If the changes are of the order of 5–10 ms, they begin to be detectable if the effect of the listening room is minimal. With some phase-insensitive signals, such as noise or diffuse sound, changes in group delay may not be perceived even with longer delays.
- Although the changes in group delay may be perceived as changes in the timbre of sound, the effect on communication, or on speech intelligibility, can be small. For example, even 100-ms group delay changes can be introduced into speech signals before intelligibility starts to vanish.

If the group delay introduced by a communication channel is not smooth with frequency, and has abrupt phase delay differences in adjacent frequencies, the above-mentioned rules no longer apply. One such typical response is reverberation, where, at frequencies over the critical frequency (see Section 2.4.4), the phase response behaves randomly. In practice, most of the phase-related effects are not audible at all in reverberant conditions. When a buzzy signal is filtered with such a random phase response, the buzzyness vanishes, and the firing rate spectrogram resembles the lower-right plot in Figure 11.11. If the phase delay is changed at one frequency in such a case, almost invariably, no audible changes are introduced. In other words, a phase-sensitive signal can be made phase-insensitive by adding the effect of room reverberation to it.

In anechoic listening to buzzy signals, minimal changes in group delays may be audible. This has already been demonstrated with a sawtooth wave, modified by a group delay smooth with frequency, but containing local abrupt change. In the top-left and top-right cases in Figure 11.11, the amplitude spectra are otherwise equal, but the phase of one harmonic has been reversed. This corresponds to applying a group delay of only 0.32 ms at 3100 Hz and none at other frequencies, making that frequency much louder. Such frequency-dependent variations in an otherwise smooth group delay can thus have a large effect during anechoic or headphone listening.

11.6 Psychoacoustic Concepts and Music

Music is a form of art which is composed of sound and silence. It is an art communicated through hearing, and the relations of musical sounds to psychoacoustics are interesting here. The sounds used in music have diverse dimensions, and many of these dimensions are related to psychoacoustics and to the theory of hearing. Some of the phenomena in music can be directly explained by results obtained from psychoacoustics.

This section contains a discussion on two psychoacoustic views of music in two specific perspectives: melody and harmony, and rhythm. Note that in music, terms like consonance, dissonance, and rhythm are very complex and depend on the style of music, and the discussion in this book is therefore limited to only a 'sensory' perspective.

11.6.1 Sensory Consonance and Dissonance

A musical note, simply a 'note' in the following, is defined as a sound that evokes pitch and has a defined duration. This is a somewhat restrictive definition, since it excludes, for example, drum sounds. However, for simplicity, a note is here defined to be a complex harmonic tone

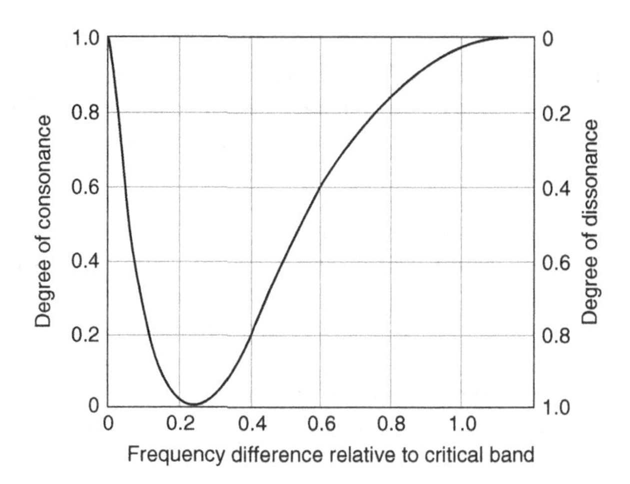

Figure 11.12 The consonance and dissonance levels created by two tones as a function of their frequency difference on a linear scale from zero to the width of the critical band. Adapted from Plomp and Levelt (1965), and reproduced with permission from the Acoustical Society of America.

with a fixed fundamental frequency. Each note with a certain frequency is also often named, for example, C, D, E, F, and so on. The notes may also have a defined duration. Most musical instruments produce a harmonic spectrum when a single note is played on them. Thus, for simplicity, the term 'note' refers to a sound with a harmonic spectrum in this section.

Two notes of different frequencies presented at the same time create a pleasant or an unpleasant sound. If the resulting sound is pleasant, the original notes are said to be in consonance, and correspondingly, for an unpleasant result they are said to be in dissonance. These concepts are central to music, especially Western music, and to the structures to which the discussion in this book is limited.

The level of consonance and dissonance has been measured with listening tests for two sinusoids as a function of the difference in frequency, and the result is presented in Figure 11.12 (Rossing *et al.*, 2001). When the frequencies are the same, consonance has a maximum value of 1.0 and dissonance 0.0. When the frequency difference is about a quarter of the critical band, dissonance has its maximum and consonance is at its minimum. After this, consonance approaches asymptotically the value 1.0 when the frequency difference is increased.

In principle, a single note can produce considerable roughness or dissonance, since, with certain pitches, some of the harmonics at high frequencies may be at a distance of a quarter of a critical band. This may be understood better with the results shown in Figure 11.8, which shows that the separation between partials should be less than 100 Hz to create a noticeable perception of roughness. With harmonic complexes, or notes, this can be achieved only with frequencies below 100 Hz and when at least some of the higher frequency partials have a large amplitude. Typically, instrument sounds have an amplitude that decreases with frequency, and such roughness issues are not encountered. On the other hand, in some synthetic bass sounds such roughness may exist.

The roughness phenomenon is, on the other hand, much more easily obtained when two notes are played at the same time. If the fundamental frequencies are different, the distances of their harmonics on the frequency scale define the roughness. If some of the harmonics are located at distances relative to each other that create roughness, a dissonant sound is perceived, as shown for three intervals in Figure 11.13. When the frequency ratio is near 3:2, as shown in the uppermost case in the figure, most of the partials of the spectra are either located very near to each other or far enough away from each other to cause no roughness. Only partials with number 10 or higher are at frequencies influencing each other. When the frequency ratio is about 5:4, a larger number of partials are seen to have such a ratio with each other that higher roughness is perceived. With a frequency ratio of $\sqrt{2}$, most partials interfere with each other, and a high level of roughness, and also a high degree of dissonance, is obtained.

Dissonance has been measured for different ratios between the fundamental frequencies of two notes. Figure 11.14 shows the degree of dissonance and consonance when the fundamental frequency of one of the sounds stays at 250 Hz, and the frequency of the other sound changes from 250 Hz to 530 Hz (Plomp and Levelt, 1965). Each sound consists of the fundamental frequency and five harmonic components. Perfect consonance is obtained with frequency ratios 1:1 and 1:2. Consonance has a local maximum with all simple integer ratios, such as 2:3, 3:4, 3:5, and so on, and a minimum in between.

11.6.2 Intervals, Scales, and Tuning in Music

The notes in a *musical scale* differ from each other by fundamental frequencies, which are specified by the tuning system. The notes are at distances on the scale called *intervals*, and they are quite often specified as a ratio between the frequencies. In most scales, an octave corresponds to doubling or halving the frequency. A *melody*, by its simplest definition, is a succession of notes with varying frequencies and temporal lengths. A *chord* is composed of multiple notes presented simultaneously. *Harmony* can then be defined simply as the tonal character of the chords and possibly also their succession in time.

Sensory consonance and dissonance provide the foundation to understanding the structure of melodies and harmonies in music. Different musical styles can be characterized based on their tendency to involve dissonance. For example, choral works from the Renaissance era hardly ever have any dissonant chords; classical music uses dissonance to build tension, which is released in consonant chords; and Jazz music basically avoids consonant chords completely.

The plot in Figure 11.14 helps us understand why the intervals of notes with simple integer ratios are the basis of melodies and harmonies. The *octave* with ratio 2:1; the *fifth* with ratio 3:2; the *fourth*, 4:3; the *sixth*, 5:3; and the *third*, 5:4 are examples of simple intervals. Almost all scales in music utilize such intervals. As a basic rule, the larger the integers in the ratio are, the lower the consonance is.

The diatonic scale divides an octave into seven notes and into repeated octaves. The intervals between adjacent pitches are five whole tones and two semitones. The order of the intervals in the Western major scale has two whole tones followed by one semitone, three whole tones and one semitone, ending at the octave of the first note. The notes in C major are called C, D, E, F, G, A, B, and the one-octave higher C. Each of the notes may also be 'raised' (or in some cases 'augmented') or 'lowered' (or in some cases 'diminished') by a semitone. The notation is then, for example, D♭ (D flat) or D♯ (D sharp), respectively. The number of notes is thus 12, assuming that a whole tone is divided into two semitones, resulting in one note in between. This means that C♯ equals D♭; D♯ equals E♭, and so on. The whole tones in the scale do not necessarily correspond to an equal frequency ratio, and the same holds for semitones.

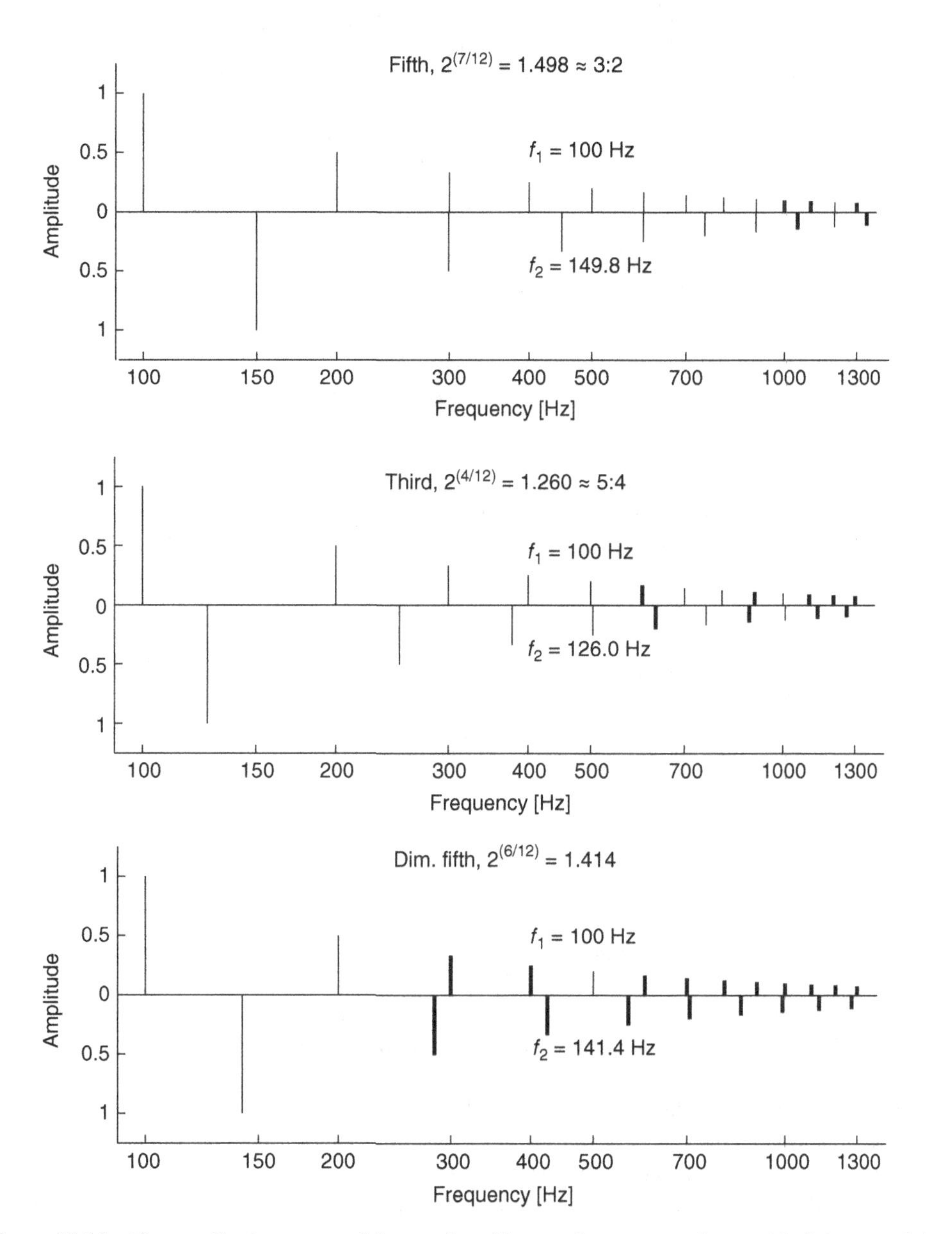

Figure 11.13 The amplitude spectra of three pairs of harmonic tone complexes with the name of the corresponding interval and frequency ratio given in the figures. The abscissa shows the frequency on a logarithmic scale, and the ordinate is the amplitude of each harmonic. The spectrum of a sound with lower pitch is shown with spectral peaks upwards, and correspondingly, a higher pitch is shown with spectral peaks downwards. The bold spectral peaks represent those harmonic components that interact with the components of the other sound, causing roughness and a higher degree of dissonance.

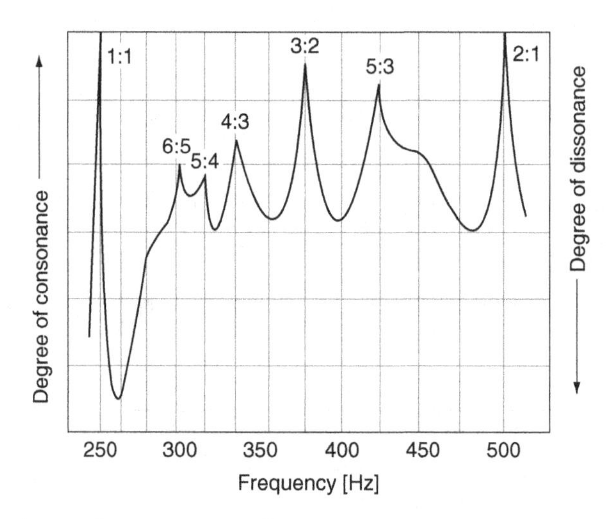

Figure 11.14 The degree of consonance and dissonance created by two harmonic complexes plotted as a function of the frequencies of one of the sounds when the fundamental frequency of the other sound is a constant 250 Hz. Reproduced from Plomp and Levelt (1965) with permission from The Acoustical Society of America.

The frequencies of the notes in the scales then have to be specified, or 'tuned'. The following tuning methods are of theoretical and possibly practical interest:

- *just intonation*;
- *Pythagorean scale*;
- *mean-tone temperament*;
- *equal temperament.*

Just intonation is based purely on *triads*, where the frequencies of the triads C-E-G, G-B-D, and F-A-C have the ratio 4:5:6 and octaves are always tuned to 2:1. Unfortunately, with this method, many intervals are not in tune. For example, the interval D-A should have the ratio 1.5, but in this tuning it is 1.48, which is audibly mistuned.

The Pythagorean tuning system has the greatest number of pure fifths and octaves. Unfortunately, the thirds are then mistuned.

We can thus conclude that no single method exists that tunes the seven notes of the diatonic scale so that all the intervals between the notes have intervals with simple integer ratios. Some methods have been developed to overcome this problem. For instance, the mean-tone temperament alters the Pythagorean tuning so that the fifths are tuned a bit narrower, which overcomes some problems of the system.

Many musical instruments, such as pianos and synthesizers, are nowadays tuned according to the equal temperament system, where a pure octave is divided into 12 semitones having an equal frequency ratio $\sqrt[12]{2} \approx 1.05946$. A whole tone thus corresponds to two semitones $\sqrt[6]{2} \approx 1.225$. The intervals shown in Figure 11.13 use the equal temperament system, where, for example, the interval 'third' should have a frequency ratio of 5:4 (1.25), but the equal temperament system defines the ratio to be 1.26. If the interval ratio was 5:4, the fifth partial of f_1 = 100 Hz note would have a frequency of 500 Hz, and the fourth partial of f_2 = 125 Hz note

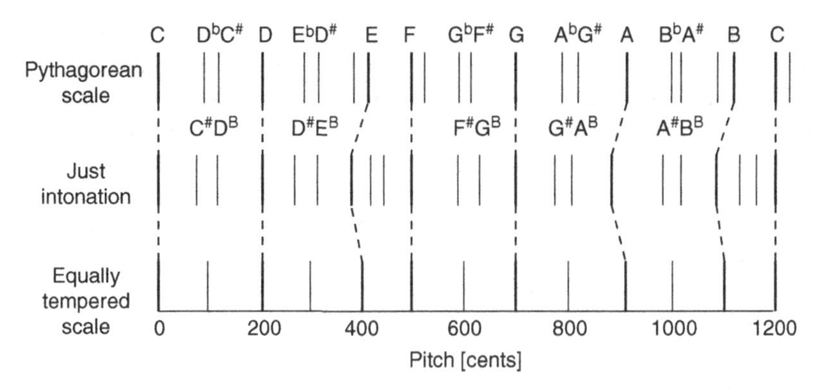

Figure 11.15 The pitches in the Pythagorean scale just intonationand the equally tempered of one octave as a function of logarithmic pitch expressed in cents. Rossing *et al.*, 2001.

would also be 500 Hz, producing no beating or roughness. However, in equal temperament, the fourth partial of f_2 = 126 Hz note is located at 504 Hz (as shown in the figure), which interferes with the partial of f_1 note at 500 Hz. However, the difference is so small that roughness is not generated, and only some beating is perceived. A semitone is further divided into 100 *cents*, each cent corresponding to the ratio $\sqrt[1200]{2} \approx 1.0005778$.

The Pythagorean scale, just intonation, and the equally tempered scale are compared in Figure 11.15. The scales clearly exhibit irregular differences. Note that only with equal temperament is the number of individual frequencies 12, since the sharp notes have equal counterparts in flat notes. With the other scales, the sharp and flat notes do not have the same frequencies in any case, i.e., for example, C♯ does not equal D♭.

11.6.3 Rhythm, Tempo, Bar, and Measure

Probably an even more important musical structure than melody and harmony is *rhythm* and its related features. Rhythm is a complex concept which refers to different temporal structures in music. The existence of rhythm is based on natural repetitions in time, such as walking, running, the heartbeat, and breathing. In general, a systematic description of rhythmic structures is harder than one for the structures in melody and harmony. The concepts related to rhythm are also slightly imprecise.

Some other concepts related to rhythm are:

- *Note value*: The relative temporal length of a note. The basic values are the full note 𝅝 ; the half note 𝅗𝅥 ; the quarter note ♩ ; the eighth note ♪ ; and the sixteenth note 𝅘𝅥𝅯. The duration of the notes is halved each time when stepping forwards in the presented list of note values. The lengths of pauses, or silences, between notes are also defined accordingly.
- *Measure* or *bar*: A rhythmic 'placeholder' which indicates a prototype repeated rhythm in music. The number of 'prototype' notes and their duration is shown in the time signature N/M, where N is the number of notes with the value of $1/M$ that can be included in each measure. For example, the time signature $3/4$ means that the temporal length of each measure is such that three quarter notes can be played during the measure as | ♩ ♩ ♩ |. In actual

music, the prototype notes in a bar can be replaced by rests, divided into shorter notes, or joined into longer ones without the temporal length of the measure changing.

- *Tempo*: The speed of presentation. In the notation of written music, ♩ = 80 means that the speed of presentation is 80 quarter notes per minute. The tempo also typically varies during the performance, and the tempo notation is simply a recommendation of the tempo.
- *Beat*: The accenting of specific temporal positions in a bar. In music with a 4/4 measures, the downbeat denotes that the first note should be emphasized with a milder accent on the third note. The upbeat, in turn, denotes that the second and fourth prototype notes should be emphasized. There are also other meanings for 'beat', but they are beyond the scope of this book.

Performed music quite seldom exactly follows in rhythm such mathematically defined temporal lengths. Musical notation must be taken as a simplified way of communicating and storing musical melodies, rhythms, and harmonies, and musicians must know the style of playing to realize the musical piece in the manner intended by the composer.

In a simple case, a single-voiced melody with a static tempo and a static time signature can easily be described and analysed. Unfortunately, there are plenty of examples of music with more complicated rhythmic patterns, whereby a description cannot be made easily (Fastl and Zwicker, 2007). Sadakata *et al.* (2006) use Bayesian theory to relate the rhythmic patterns of music production to the perception of it.

11.7 Perceptual Organization of Sound

The perception of the auditory environment can either be comprehensive or focus on certain details in it. There appear to be two processes running in parallel during perception. A primitive, bottom-up mechanism automatically orders incoming acoustic stimuli via certain acoustic features. The second, top-down mechanism allows us selectively to attend to whichever features we desire (for instance, a certain pitch, location, time interval, or frequency range). In general, we can focus our attention only on one detail at any one time, although by switching rapidly and by using our short-term auditory memory we can attend to a few targets at the same time. Simultaneously, the top-down processing parses the details and organizes the auditory environment as a whole and tries to focus on some new details for active monitoring.

The hearing mechanism involves certain inborn capabilities to analyse the summed sounds of the auditory environment arriving from multiple sources with or without room reflections and reverberation. The capability to discern the source signals is also partially learned and based on experience of similar situations. Some general principles on how human hearing performs this task will be discussed in the following and are discussed extensively by Bregman (1990).

The organization of sensations and perceptions has been a long-standing topic in experimental psychology. *Sensation* refers to the representation of a real-world object by a sensory organ, whereas *perception* means the higher-level interpretation of the real-world object formed by the brain based on a set of sensations. Perceptions are complete interpretations of objects or events, whose organization does not require exhaustive sensory information about the object being inspected.

Universal laws of pattern formation have been found, particularly in the domain of visual perception. For example, the school of Gestalt psychology has focused on finding the principles of grouping in order to explain the emergence of organized patterns. The most common Gestalt laws of grouping are:

- *Principle of proximity*. When two sensory elements are close to one another both in time and space, they tend to be grouped perceptually.
- *Principle of similarity*. Sensory elements resembling each other tend to be grouped together, while differing elements are thought to belong to another object.
- *Principle of closure*. Complete forms and figures tend to be perceived even if part of the figure is hidden. In the case of a pure tone being interrupted sequentially by bursts of white noise, the human auditory system assumes the pure tone continues uninterrupted during the noise bursts.
- *Principle of continuity*. Continuation of a pattern in time, space, or in some other dimension is a strong assumption, unless contradicting sensory information is presented. For example, the smooth pitch variations and smooth formant changes in speech imply to the listener that the speech originates from the same speaker and is organized into a single stream. If too-rapid changes in pitch or formants are introduced, multiple streams may be created (Plack, 2013).
- *Principle of common motion*. If sensory elements move in the same direction at the same rate, they tend to be grouped as parts of a single stimulus.
- *Principle of belongingness*. Each sensory element can (usually) only belong to a single perceptual object.

Gestalt psychology has many principles that are common to the sensory modalities (vision, auditory, somatosensory, olfactory, and gustatory). Naturally, each of the modalities can have individual implementations of these principles.

The theory and techniques of *pattern recognition* have developed through research into machine vision and machine hearing. In pattern recognition, information captured from the real world (for example, a visual or auditory signal) is pre-processed and transformed into a vector of relevant features. The acquired feature vector is then compared to a set of known feature vectors using a suitable algorithm, with the goal of recognizing (classifying) the unknown pattern. Pattern recognition often requires the formation of quantitative metrics for distance or similarity (compare with the principles of grouping).

11.7.1 Segregation of Sound Sources

We know by experience that an auditory event and the corresponding sound event can usually be related to a specific sound-producing object, the *sound source*. Thus, to a human listener, grouping sound events or their constituents based on the sound sources is more meaningful than basing a grouping on the acoustic properties of the sound events (although they are often interrelated). For example, an approaching car produces engine noise, tyre noise, or a horn signal that are distinctively different sound events. However, when properly aligned, we experience their combination foremost as a unified object, a sound source to which we can attribute a functional meaning thanks to our previous encounters with similar sound sources.

One of the most remarkable capabilities of the human auditory system is *sound segregation*, or *source separation*. A traditional example is the so-called *cocktail party effect*, referring to our ability to focus on a single speaker in a multi-speaker situation, or in strong background noise. Psychophysical tests have found that subjects use many cues in stream segregation in such cases (Bronkhorst, 2000).

Another example is listening to music. An experienced musical listener can distinguish details from complex sound masses with amazing accuracy. However, not everybody gets to practise their analytical listening skills, because music does not have a communication function similar to speech.

A third example is street noise. We can be somewhat aware of our surroundings by using auditory information only (the blind have to learn this to an even greater degree). Spatial hearing and hearing the space are important components in orienting in the sonic environment.

11.7.2 Sound Streaming and Auditory Scene Analysis

As an example of the perceptual organization of auditory events, we will inspect the formation of a melodic line, which is an important structural factor in Western musical tradition. If we listen to sound events placed properly both in time and frequency, depending on the case, we will perceive one or more *auditory streams* (Bregman, 1990; McAdams and Bregman, 1979). A broader term for the perceptual organization of sound is *auditory scene analysis* (Bregman, 1990).

Figure 11.16 presents a simple example of a repeating sequence of six notes with a relatively wide frequency separation. If we listen to this sequence in a slow tempo, we hear only a single auditory stream (melodic line). When the rate of notes is increased, the higher and lower notes will be separated into two distinct auditory stream, or melodic lines.

Following the principle of belongingness, each note is organized into one of the competing auditory streams (or it can be perceived as a separate stream), but no note is part of more than one streams. Depending on the case, in Figure 11.17 the final note (F) may be organized either into the upper or lower auditory stream. This also changes the perceived rhythmic pattern accordingly.

As the rate of notes being presented increases (and/or the frequency separation becomes wider), the auditory perception is organized into more and more auditory streams, until finally only an ensemble timbre, without any melodic lines, is perceived (Figure 11.18).

It can be stated that the streaming of auditory events, or stream segregation, depends mainly on the rate of notes presented and the width of their frequency separation. Figure 11.19 presents thresholds for coherent auditory streaming, stream segregation, and for the uncertain area

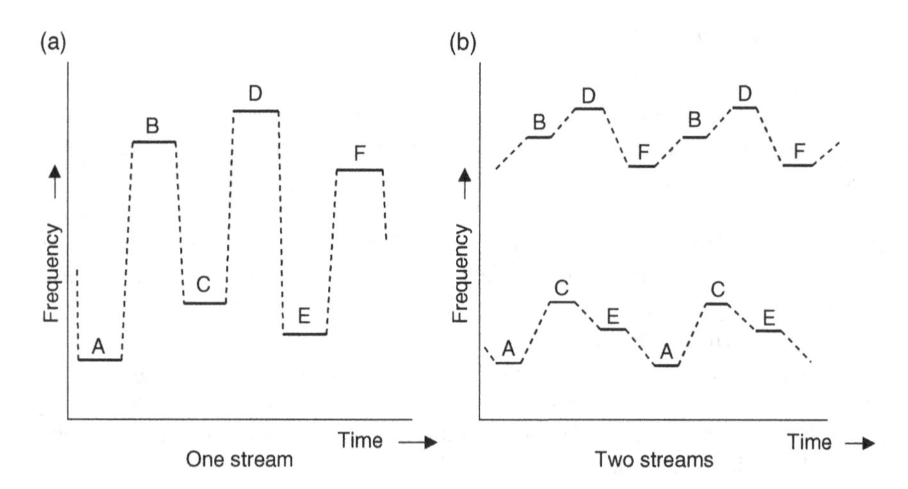

Figure 11.16 The organization of auditory events, depending on the tempo, (a) into one auditory stream for 5 notes per second and (b) into two auditory streams for 10 notes per second.

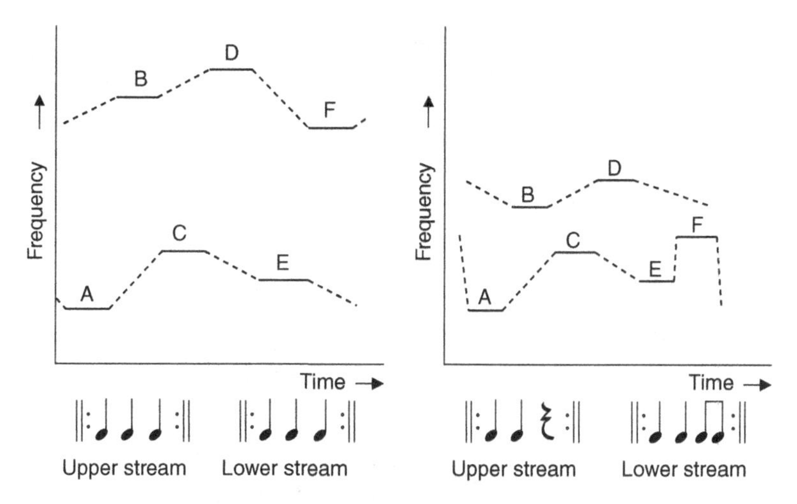

Figure 11.17 A single note (F) may be organized into either of the auditory streams, depending on the case, which also affects the perceived rhythmic pattern.

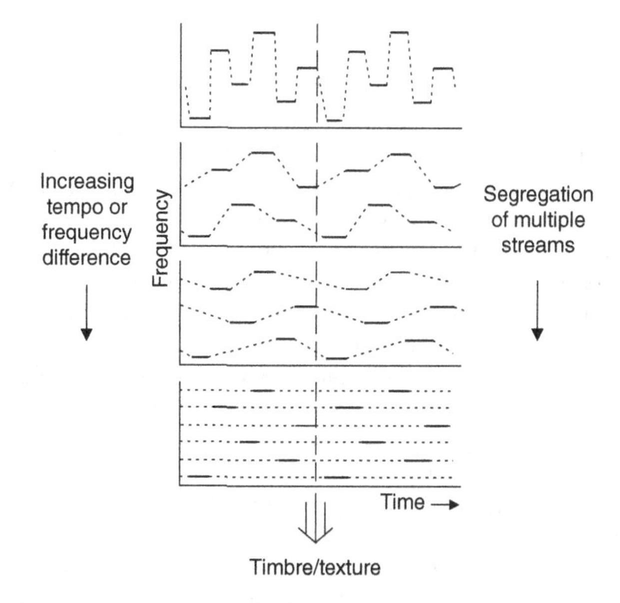

Figure 11.18 As the tempo of the sequence is increased, it begins to segregate into multiple auditory streams, until finally an ensemble timbre, without melodic lines, is perceived.

between them. If the frequency separation of notes is approximately one semitone, auditory streaming occurs regardless of the rate of presentation of notes. Correspondingly, if the temporal interval between notes is more than 150 ms, the probability of the tones being grouped into a single stream is also increased for wide frequency separation between notes. However, this happens in the uncertain area where listeners may perceive either a single coherent auditory stream or segregated streams (the hatched area).

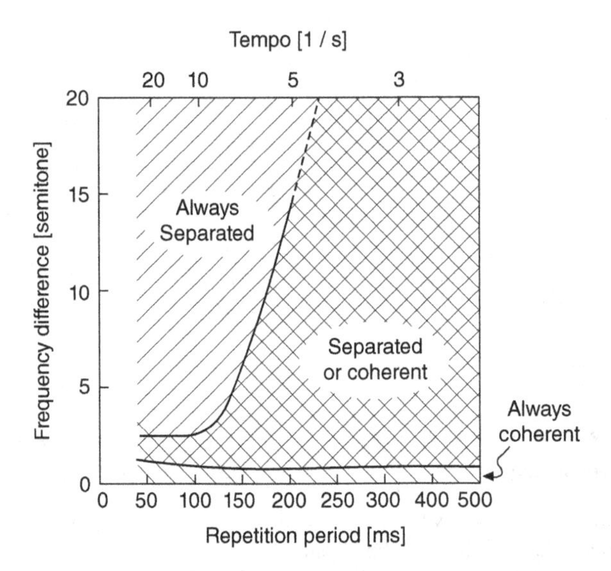

Figure 11.19 Auditory event streaming or segregation as a function of the rate of presentation and frequency separation. Adapted from van Noorden (1975).

Furthermore, factors other than the tempo and frequency separation affect the auditory streaming too. Similarity in the timbre or in some other factor tends to aid stream integration. Additionally, continuation factors such as smooth changes in frequency at the beginning or end of a note may cue that specific note for stream integration.

So far, we have studied only the *sequential streaming* of notes, or, in other words, streaming of auditory events occurring in succession, not simultaneously, in time. Perhaps an even more fundamental concept is the integration of simultaneous auditory components; that is, organizing a set of concurrent components into subsets to represent distinct sources of sound. This is closely related to, for example, perceiving the fundamental frequency of a sound with a harmonic or near-harmonic structure as a coherent whole. Other factors facilitating simultaneous grouping are, for instance, the synchronization of onsets, synchronized amplitude or frequency modulation, and uniform spatial information.

In a sense, auditory scene analysis is a continuation of the fundamentals of psychoacoustics, and it offers future research challenges in understanding and modelling the higher-level properties of the human auditory system. *Computational auditory scene analysis* (CASA) (Wang and Brown, 2006) is a promising approach to solving the puzzle.

Summary

This chapter complemented the discussion on psychoacoustic quantities by introducing sharpness, fluctuation strength, impulsiveness, tonality, and roughness, which can all be seen to be subcategories of timbre. The quantities are interesting when trying to understand what kind of signals draw the attention of the listener. Many of the quantities have a connection to human sensitivity to modulations in sound. We seem to be tuned to pay attention to amplitude and frequency modulations in the range from about 1 Hz to 16 Hz. In turn, modulations at 30–100 Hz result in the perception of rough and unpleasant sound.

Perception of the magnitude response and phase spectrum variations has also been described; this clearly shows that humans are sensitive, in some cases, to the phase spectrum of a signal. Time-domain peaks seem to cause the perception of buzzyness, but if the sound source is in a room, the response smooths out such peaks in many cases.

A clear connection between psychoacoustics and music has been shown. Roughness is closely related to consonance and dissonance perceived between two or more musical notes; the more the partials of the notes generate roughness, the higher the perceived dissonance is. This leads directly to basic melodic and harmonic structures in music.

The perceptual organization of complex auditory scenes into streams was also described in this chapter. The laws of pattern formation describing how the sensory information is grouped into auditory objects were discussed, and some illustrative cases in music shown.

Further Reading

The fundamentals of the psychoacoustic quantities discussed in this chapter are outlined by Fastl and Zwicker (2007). A good introduction to music acoustics is provided by Rossing *et al.* (2001), and the primary source for auditory scene analysis is (Bregman, 1990).

References

Bienvenue, G.R. and Nobile, M.A. (1991) Prominence ratio for noise spectra with discrete tones: A procedure based on Zwicker's critical band research *Proc. of Inter-Noise*, **1**, 53–55.

Blauert, J. and Laws, P. (1978) Group delay distortions in electroacoustical systems. *J. Acoust. Soc. Am.*, **63**, 1478–1483.

Bregman, A. (1990) *Auditory Scene Analysis*. MIT Press.

Bronkhorst, A.W. (2000) The cocktail party phenomenon: A review of research on speech intelligibility in multiple-talker conditions. *Acta Acustica United with Acustica*, **86**(1), 117–128.

Cardozo, B.L. (1967) Ohm's law and masking. *IPO Annual Progress Report*. Institute of Perception Research, Eindhoven The Netherlands, **2**, 59–64.

Daniel, P. and Weber, R. (1997) Psychoacoustical roughness: Implementation of an optimized model. *Acta Acustica United with Acustica*, **83**, 113–123.

Fastl, H. and Zwicker, E. (2007) *Psychoacoustics – Facts and Models*. Springer.

Holt, L.L. (2006) The mean matters: Effects of statistically defined nonspeech spectral distributions on speech categorization. *J. Acoust. Soc. Am.*, **120**(5), 2801–2817.

ISO-7779 (2010) Acoustics – measurement of airborne noise emitted by information technology and telecommunications equipment. International Organization for Standardization.

Johnston, J.D. (1985) Transform coding of audio signals using perceptual noise criteria. *IEEE J. Selected Areas in Commun.* **6**(2), 314–323.

Karjalainen, M. (2008) *Kommunikaatioakustiikka* (in Finnish), *Communication Acoustics*. Teknillinen korkeakoulu.

Laitinen, M-V., Disch, S., and Pulkki, V. (2013) Sensitivity of human hearing to changes in phase spectrum. *J. Audio Eng. Soc.*, **61**(11), 860–877.

McAdams, S. and Bregman, A. (1979) Hearing musical streams. *Computer Music J.*, **3**(4), 26–60.

McKeown, D. and Wellsted, D. (2009) Auditory memory for timbre. *J. Experimental Psych.: Hum. Percep. Perform.* **35**(3), 855.

Moore, B.C. (2002) Interference effects and phase sensitivity in hearing. *Philosoph. Trans. Royal Soc. London. Series A: Mathematical, Physical and Engineering Sciences*, **360**(1794), 833–858.

Olive, S.E., Schuck, P.L., Sally, S.L., and Bonneville, M. (1995) The variability of loudspeaker sound quality among four domestic-sized rooms *Audio Engineering Society Convention 99*.

Patterson, R.D. (1987) A pulse ribbon model of monaural phase perception. *J. Acoust. Soc. Am.*, **82**, 1560–1586.

Pedersen, T.H. (2001) Objective method for measuring the prominence of impulsive sounds and for adjustment of LAeq. *Proc. Int. Congr. and Exhibition on Noise Control*.

Pike, C., Brookes, T., and Mason, R. (2013) Auditory adaptation to loudspeakers and listening room acoustics *Audio Engineering Society Convention 135* AES.

Plack, C.J. (2013) *The Sense of Hearing*. Psychology Press.

Plomp, R. and Levelt, W.J. (1965) Tonal consonance and critical bandwidth. *J. Acoust. Soc. Am.*, **38**, 548–560.

Roederer, J.G. (1975) *The Physics and Psychophysics of Music: An Introduction*. Springer.

Rossing, T.D., Moore, F.R., and Wheeler, P.A. (2001) *The Science of Sound*, 3rd edn. Addison-Wesley.

Sadakata, M., Desain, P., and Honing, H. (2006) The hearing-aid speech quality index (hasqi) version 2. *Music Perception: Interdisc. J.*, **23**(3), 269–288.

Summerfield, Q., Haggard, M., Foster, J., and Gray, S. (1984) Perceiving vowels from uniform spectra: Phonetic exploration of an auditory aftereffect. *Percep. Psychophys.* **35**(3), 203–213.

Summerfield, Q., Sidwell, A., and Nelson, T. (1987) Auditory enhancement of changes in spectral amplitude. *J. Acoust. Soc. Am.*, **81**(3), 700–708.

Terhardt, E., Stoll, G., and Seewann, M. (1982) Algorithm for extraction of pitch and pitch salience for complex tonal signals. *J. Acoust. Soc. Am.*, **71**(3), 679–688.

Terhardt, E., Stoll, G., and Seewann, M. (1996) Pitch of complex signals according to virtual pitch theory: Examples and predictions. *J. Acoust. Soc. Am.*, **55**, 671–678.

Toole, F.E. (2006) Loudspeakers and rooms for sound reproduction - A scientific review. *J. Audio Eng. Soc.*, **54**(6), 451–476.

Ulanovsky, N., Las, L., and Nelken, I. (2003) Processing of low-probability sounds by cortical neurons. *Nature Neurosci.*, **6**(4), 391–398.

Ulanovsky, N., Las, L., Farkas, D., and Nelken, I. (2004) Multiple time scales of adaptation in auditory cortex neurons. *J. Neurosc.*, **24**(46), 10440–10453.

van Noorden, L. (1975) *Temporal Coherence In the Perception of Tone Sequences*. Institute for Perceptual Research.

Viemeister, N.F. (1980) Adaption of masking. In van den Brink, G. and Bilsen, F.A. (eds) *Psychophysical, physiological and Behavioural Studies in Hearing*. Delft University Press, pp. 190–199.

Wang, D. and Brown, G.J. (2006) *Computational Auditory Scene Analysis: Principles, Algorithms, and Applications*. Wiley-IEEE Press.

Watkins, A.J. (1991) Central, auditory mechanisms of perceptual compensation for spectral-envelope distortion. *J. Acoust. Soc. Am.*, **90**(6), 2942–2955.

Wilson, J.P. (1970) *Frequency Analysis and Periodicity Detection in Hearing*. Sijthoff, Leiden, The Netherlands, 303–318.

12

Spatial Hearing

During our evolution, the ability to locate the source of a sound has been critical for our survival. Since the appearance of mammals as primarily nocturnal animals more than 200 million years ago, they have relied heavily on their sound localization abilities for locating friends, foes, and food. Locating the source of a sound remains an important sensory ability for prey and predator alike. Even for humans in modern times, spatial hearing is extremely useful and important for orientation in one's environment. Sound sources are often localized with relatively good accuracy by our hearing, even in cases when the sources are not visible.

The mechanisms of how localization is performed are, however, not generally understood by the layperson. Many of them have been uncovered largely by science. There is actually a plethora of complex, robust, and accurate mechanisms for spatial hearing that are based on signal analysis of either binaural or monaural inputs. For example, the JND in the detection of delays between binaural signals is of the order of 20 μs, which is amazingly accurate when one remembers that it is obtained using neurons whose latency times and output spike lengths are of the order of 1 ms.

Spatial hearing develops substantially through learning and adaptation to gain more accuracy and better performance in complex environments. The fundamental role of learning is easy to understand because spatial hearing is dependent on individual factors, such as the size and form of the head and geometries of the pinnae. The auditory system learns to analyse sound environments by utilizing the properties of direct sound, reflections from surfaces and objects, and reverberant sound arriving at the two ear canals of the subject.

12.1 Concepts and Definitions for Spatial Hearing

12.1.1 Basic Concepts

We begin the discussion by defining some basic concepts related to spatial hearing. The term *localization* is the process by which the location of an auditory event in the auditory space is associated with the attributes of a sound event in an acoustic environment. In general, the human auditory system represents the external sound environment by an internal auditory

---

Communication Acoustics: An Introduction to Speech, Audio, and Psychoacoustics, First Edition.
Ville Pulkki and Matti Karjalainen.

image or scene (Bregman, 1990). The localization of a sound source can be described by the perception of its direction, distance, and spatial extent. Basically, all auditory events are 'localized', implying that a more or less precisely defined position in the surrounding world or inside the listener's head is associated with them. Reflected and reverberated sound can also be localized, which may lead to perception of the attributes of the geometry of the room and the acoustic properties of the surfaces in it.

The two following concepts are useful in spatial hearing:

- *Monaural hearing* refers to listening in conditions where there is no interaural difference information, or where such differences are ignored. The simplest case of monaural hearing is when the same signal enters both ears (see the definition of diotic listening below).
- *Binaural hearing* means listening to sounds where information due to interaural differences exists and is taken into account.

Spatial hearing is often studied using arrangements enabling the signals entering the two ears to be controlled separately. This is easily possible by using headphones. For such experimental studies, it is useful to define the following concepts:

- *Monotic listening* is a situation where a signal is fed to one ear only.
- *Diotic listening* is the case where a signal is fed equally to both ears.
- *Dichotic listening* is the arrangement where the two ears receive different sounds (which can originate from the same sound source but are processed differently for each ear.)

Note that these concepts do not characterize explicitly the sound-generation techniques used – monophonic, stereophonic, binaural, or multi-channel techniques.

Headphones offer the opportunity for dichotic listening, where the stimulus to each ear can be controlled separately. This is why there have been many studies of dichotic listening with the binaural cues controlled independently. Figure 12.1 illustrates the set-ups for ITD and ILD adjustment using a) controllable delays and b) level attenuators.

Unless special techniques are used, headphone listening typically results in *inside-the-head localization*. In *lateralization*, inside-the-head, localized auditory events are controlled in the left–right direction by modifying interaural differences. Lateralization experiments with

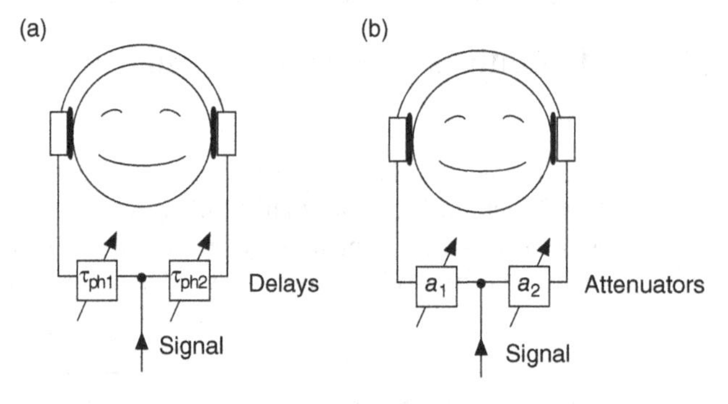

Figure 12.1 Headphone listening set-ups for auditory event lateralization experiments: (a) control of time-delay difference and (b) control of signal-level difference.

headphones are interesting from a theoretical point of view due to the possibility of processing the ear signals independently. On the other hand, such lateralized reproduction often sounds very artificial and does not necessarily reveal much about spatial perception in the real world.

12.1.2 Coordinate Systems for Spatial Hearing

In general, spatial hearing is a three-dimensional phenomenon and therefore requires a three-dimensional coordinate system to describe the acoustic environment. The same spatial information can be mathematically expressed in several different coordinate systems. In this context we define three coordinate systems, each of which has its specific strengths:

1. *Rectangular coordinates* or *Cartesian coordinates* $\{x, y, z\}$. This is a natural way to describe three-dimensional information when observing a sound space and a listener from an external point of view. The listening subject can be positioned at the origin, looking, for example, along the x-axis. From the perceptual point of view of the listener, the rectangular coordinate system is not the most intuitive one.
2. *Spherical coordinates* $\{\varphi, \delta, r\}$ from the most natural 'head-related' or 'listener-centric' coordinate system for subjects when orienting themselves in an acoustic environment. Figure 12.2 illustrates this case using the azimuth angle φ of a direction, $-180° \leq \varphi < +180°$; the elevation angle δ of a direction, $-90° \leq \delta < +90°$; and the distance r of an object.
3. *Cone of confusion coordinates* $\{\varphi_{cc}, \delta_{cc}, r\}$. The geometry of the head and external ear implies a concept called the cone of confusion, discussed below, which is that set of directions where direction discrimination is relatively difficult due to symmetry. This set of directions forms the surface of a cone, hence the name. The cone of confusion coordinate system, characterized in Figure 12.3, is suitable for describing sound source positions from this point of view. The angle φ_{cc} covers the range $-90° \leq \varphi_{cc} < +90°$ and δ_{cc} the range $-180° \leq \delta_{cc} < +180°$.

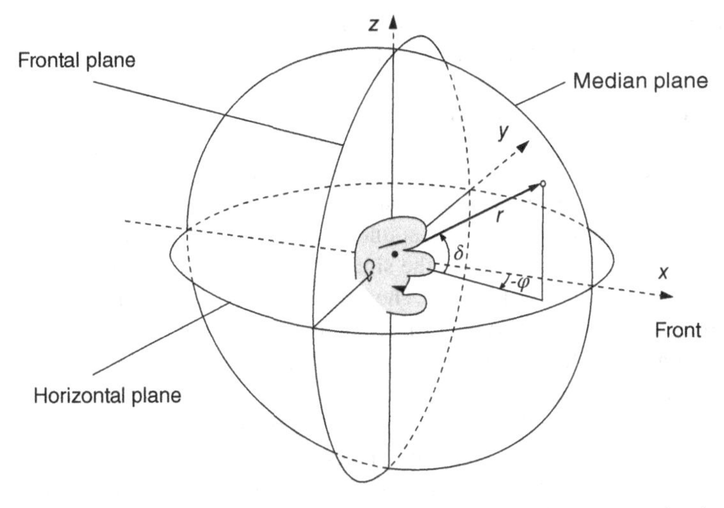

Figure 12.2 The spherical coordinate system, which is natural from the perceptual point of view of a listener. The angle φ is the azimuth and δ the elevation. The distance to the source is r. The three main planes characterized by the circles are the median plane, the horizontal plane, and the frontal plane.

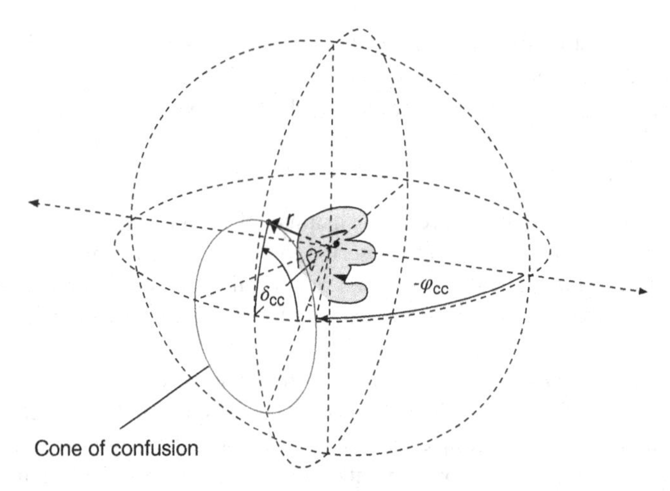

Figure 12.3 The cone of confusion coordinate system, which is motivated by the approximate acoustic symmetry of the head. The angle φ_{cc} is the azimuth angle to the cone of confusion in the horizontal plane and δ_{cc} is the angle on the cone of confusion. The distance to the source is r.

Three planes of symmetry defined below are indicated in Figure 12.2:

- *Frontal plane* $x \equiv 0$ divides the space in the front–back direction.
- *Median plane* (or *median sagittal plane*) $y \equiv 0$ where the distance to the ears is equal.
- *Horizontal plane* where $z \equiv 0$ and $\delta \equiv 0$.

Sometimes, reporting the results from binaural hearing studies requires that the side relative to the ear is specified. The two following terms are often used for this purpose:

- *Ipsilateral* refers to the side of the head that is being discussed.
- *Contralateral* refers to the side of the head opposite to that being discussed.

12.2 Head-Related Acoustics

As discussed in Section 7.1, the transfer function of sound from the source to the ear canal entrance depends on the position of the sound source and the acoustic properties of the listener's head. The azimuth angle φ of the sound source direction, if different from zero, causes interaural time and level differences in the signals entering the ear canals. The torso, particularly the shoulders, also adds reflections. The asymmetry of the head in the front–back and the up–down directions contributes to the difference in signals from different directions.

The *pinna* (Figure 12.4a) makes important contributions to the front–back and up–down differences at frequencies above 4 kHz (Blauert, 1996). The cavities resonate at specific frequencies and affect signals in the ear canal depending on the angle of incidence of sound from a sound source. In particular, the perception of elevation of a source in the median plane, where the ear canal signals are practically equivalent, is aided by the direction-dependent filtering of sound by the pinna.

The combined effect of the torso, the head, and the external ear acoustics is compactly represented in *head-related transfer functions* (HRTFs) or the corresponding *head-related impulse*

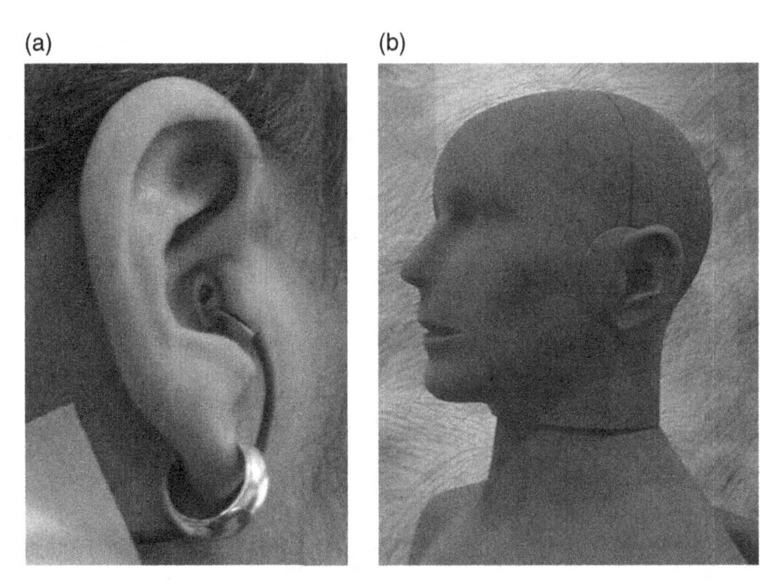

Figure 12.4 (a) The pinna of a subject. The ear canal is blocked by a miniature microphone to measure the head-related transfer function. (b) A dummy head (Cortex MK2) used for binaural measurements and recordings.

responses (HRIRs). Note that the term 'head-related transfer function' is often used generically even when referring to an impulse response.

The head-related transfer function H_{HRTF} is defined by the response to a sound source at a specified position in the ear canal H_{ec} normalized to the response H_{ff} in the middle position of the head when the head is absent:

$$H_{\mathrm{HRTF}}(\omega) = H_{\mathrm{ec}}(\omega) \,/\, H_{\mathrm{ff}}(\omega) \tag{12.1}$$

HRTFs or HRIRs are measured in free-field conditions with the sound source (a loudspeaker) placed in desired positions, for example, 2 metres from a subject at each azimuth and elevation angle of interest (Møller, 1992). An anechoic room can be used for the measurement, but the reflections from walls can also be eliminated by windowing the response to before the arrival of the first reflection. Small microphones are used to capture the response close to the entrance or inside the ear canal. In the former case, the ear canal is typically blocked by the microphone set-up, as illustrated in Figure 12.4a, or it can be kept open. In the ear canal, the measurement position can be any specified point, including the immediate vicinity of the tympanic membrane, in which case the microphone probe must be inserted carefully to avoid damage to the membrane.

In some cases it is practical to use *dummy heads* in binaural response measurements and recordings (Møller, 1992). The dummy heads are also known as *head and torso simulators*, *artificial heads*, or *binaural microphones*. They are designed to approximate the head-related acoustics of a typical human subject. Figure 12.4b shows a dummy head with a removable pinna unit, where condenser microphones are used in place of the ear drums.

A set of HRTFs measured from both ears for three directions is shown in Figure 12.5, and the corresponding HRIRs are shown in Figure 12.6. It can be seen that when the source is in

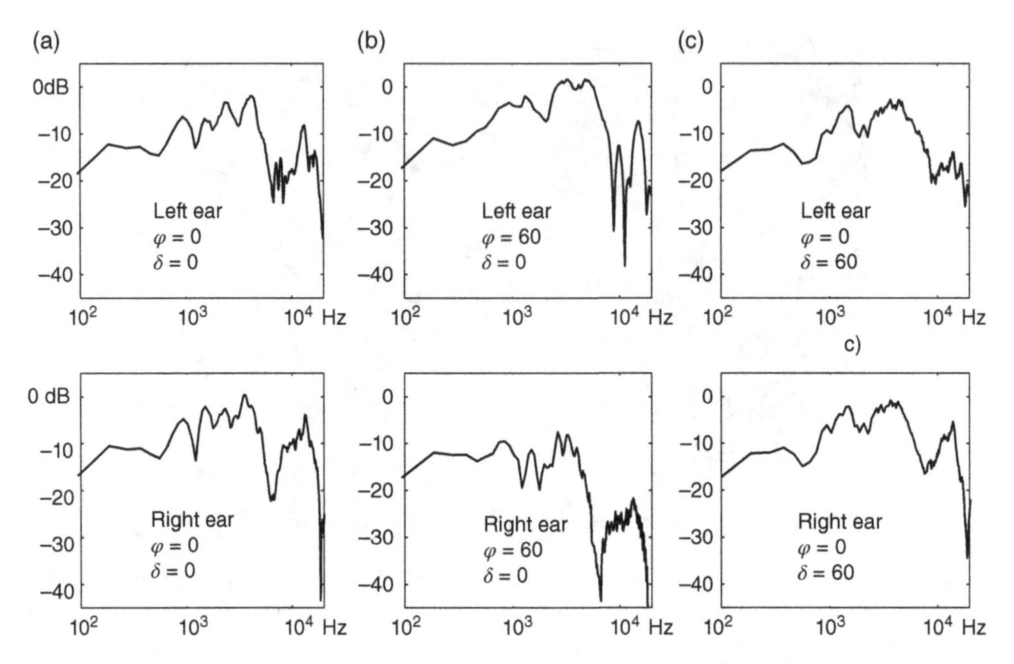

Figure 12.5 A set of HRTF magnitude responses measured from a subject at the blocked ear canal entrance for different directions of sound incidence: (a) $\varphi = \delta = 0°$ (b) $\varphi = 60°, \delta = 0°$, (c) $\varphi = 0°, \delta = 60°$. x-axis: frequency. y-axis: magnitude in dB.

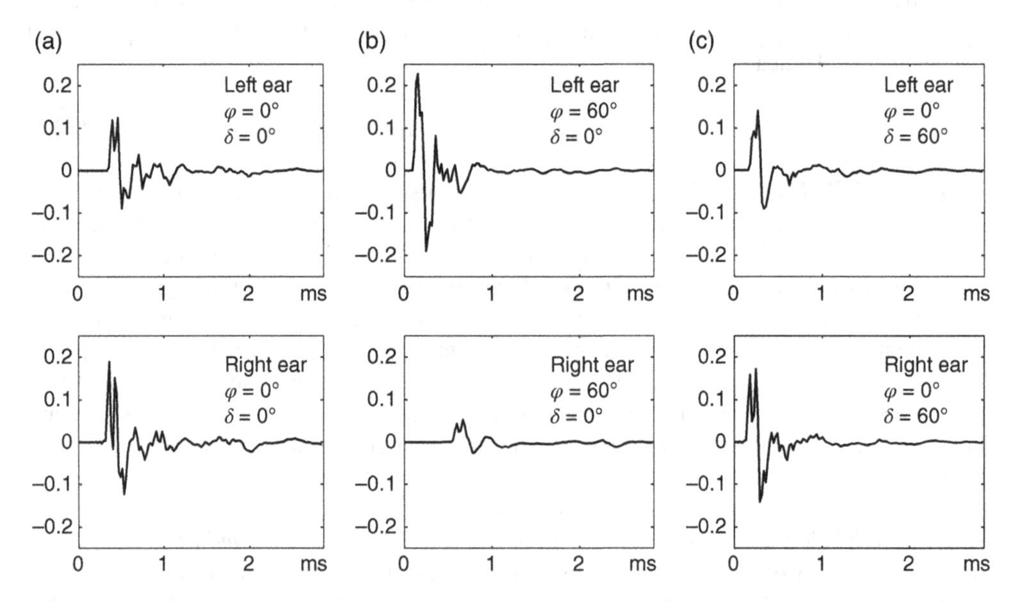

Figure 12.6 A set of HRIRs measured from a subject at the blocked ear canal entrance for different directions of sound incidence (a) $\varphi = \delta = 0°$ (b) $\varphi = 60°, \delta = 0°$, (c) $\varphi = 0°, \delta = 60°$. x-axis: time. y-axis: sound pressure (relative amplitude).

front, the HRTFs and HRIRs are similar in both ears, which is due to left–right symmetry in humans. Only at high frequencies are there some irregular differences in the responses. The HRTFs are relatively flat up to about 1 kHz, after which there is a broad hump up to about 5 kHz and a dip around 8 kHz, as indicated in Figure 7.3 on page 113.

When the source is moved to the azimuth direction, $\varphi = 60°$, the ipsilateral HRTF shows a boost at high frequencies when compared to the contralateral HRTF. Correspondingly, the direct sound in the ipsilateral HRIR arrives earlier and has a higher amplitude, while the sound in the contralateral HRIR is delayed due to the longer distance it must travel and has lower amplitude due to head shadowing. When the source is in the median plane with elevation $\delta = 60°$, the difference between the contralateral and ipsilateral HRIRs and HRTFs is negligible due to symmetry. An interesting question is how localization is performed, since the frontal responses ($\varphi = 0°, \delta = 0°$) appear to be similar to the elevated case ($\varphi = 0°, \delta = 60°$). This is analysed further below with a larger set of measured responses.

To demonstrate better their dependency on direction, a large number of HRTFs and HRIRs are plotted as 2D surfaces for different azimuths and elevations. Figure 12.7a is a plot of HRIRs measured with a source at 72 directions in the horizontal plane from the left ear of a subject. The time of arrival of the direct sound varies with direction, which is because the centre position of the head of the subject is at the centre of a circle along which the source moves, leaving the ear out of that position by about 8 cm. Thus, the distance between the source and the ear canal changes when the source moves. The source is closest to the ear when in the direction of about 90°, where the earliest start of the response is seen. When the source is on the contralateral side, the response has peaks with lower amplitudes which appear later in time than on the ipsilateral side. An interesting phenomenon is seen at about 1.3 ms for the source direction −100°. The crests of the wavefronts cross, forming a 'bright' spot at the crossing in the 2D visualization of the response. This is because the sound arrives at the ear not only along the shortest route, but also by diffracting around the head along all other paths. This causes multiple peaks in the response, and when the most prominent contributions meet at the ear canal, a higher response occurs.

The corresponding horizontal-plane HRTFs are plotted in Figure 12.7b as a function of the azimuth angle φ. When the source is on the contralateral side, high frequencies are clearly attenuated, which is shown as the black area in the plot. When the source is on the contralateral side in directions from −100° to −110°, the 'bright spot' phenomenon discussed above causes

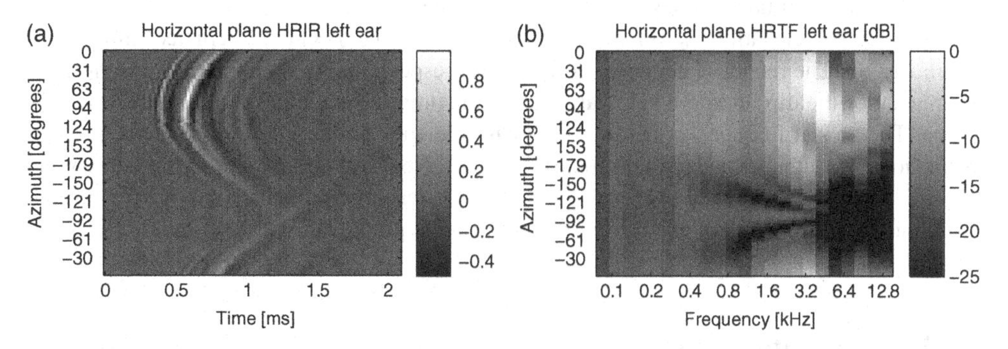

Figure 12.7 (a) The head-related impulse responses measured from the left ear of the subject. (b) The head-related transfer function from the same measurement. The HRTF data originate from Gómez Bolaños and Pulkki (2012).

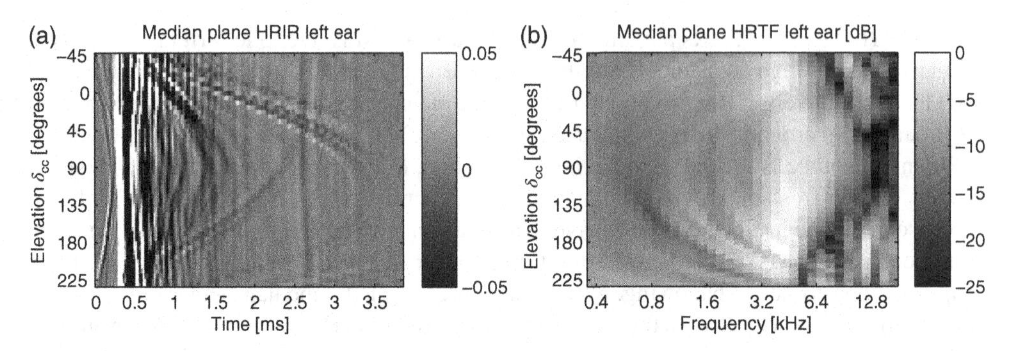

Figure 12.8 (a) HRIRs and (b) HRTFs measured from the left ear of the subject for 72 directions in the median plane. The HRTF data originate from Algazi *et al.* (2001). To emphasize the structure of the HRIR responses after 1 ms, the colour code saturates at ±0.05, although the maximum value of the HRIRs has been scaled to unity.

an amplified response compared to adjacent source directions. The effect is most pronounced at frequencies 1–4 kHz.

The change of HRIR in the median plane is shown in Figure 12.8a as a function of the cone of confusion elevation δ_{cc}. The peaks caused by the sound arriving along the direct path are located at temporal positions 0.3–0.5 ms, which means that the ear canal is relatively well at the centre of the rotation of the source. Perhaps the most interesting details in the figure are the faint arcs or half-arcs which coincide at 0.5 ms with source directions −45° and 225°. The arcs or half-arcs seem to have their apexes at temporal positions of about 1.5 ms, 2.5 ms, and 3.5 ms. The first arc, with apex at 1.5 ms, is caused by the reflection from the shoulders. When the source is above the subject, the sound reaching the shoulders is reflected back to the ears, and thus it travels a distance about 30 cm longer than the direct sound, which causes the 1-ms delay seen in the figure. When the source is lower in the median plane, the extra travel time is shorter, and the shoulder contribution arrives earlier. The reason for the appearance of the other arcs in the figure is not so evident. Very probably they are due to similar systematically changing reflections, for instance from the measurement devices, the chair, or from the feet of the sitting subject. Although the reflections seem quite faint in the figure, they have a noticeable effect on the magnitude response, as will be shown below.

The effect of elevation δ_{cc} in the median plane on the HRTF is depicted similarly in Figure 12.8b. The response depends on the angles, especially above 1 kHz. For example, the frequency of the dip around 8 kHz varies as a function of the elevation angle in the median plane. The arcs at frequencies below 4 kHz with the apex pointing left are indeed caused by reflections from the subject, which manifest themselves as the faint arcs in the corresponding HRIR response. At frequencies above 4 kHz, the response contains a lot of irregularities, which are due to the complex spectral filtering by the pinna and will be discussed below in more detail.

12.3 Localization Cues

The ability of humans to localize sound sources is surprisingly good, especially when you consider that it is based on real-time analysis of two signals entering the ear canals.

Localization is determined or guided by *localization cues* (Grantham, 1995). These are properties of auditory stimuli that are relevant to spatial perception. The combined effect of all cues, weighted by their dominance and relevance, then forms the spatial attributes of subjective auditory events. This section describes the cues that localize the direction of the sound source, the directional cues.

Any physical aspect of the acoustic waveforms reaching a listener's ears that is altered by changes in the position of the sound source may be considered a potential cue in localization. The sound signal undergoes linear distortion, or changes in spectral or temporal content, when propagating from a source to the listener's ears. This distortion forms the basis of most acoustic cues. The cues are commonly grouped as binaural (or interaural) and monaural cues (Blauert, 1996).

Binaural cues are derived from the differences in the signals between the two ears. The two binaural cues, which are also the main acoustic cues in general, are the *interaural time difference* (ITD) and the *interaural level difference* (ILD). The ILD is also often called the *interaural intensity difference* (IID). Particularly in the horizontal plane, these cues have been found to be dominant in localization.

12.3.1 Interaural Time Difference

The *Interaural time difference* (ITD) occurs due to the finite velocity of sound and differing distances from the source to the ears. Figure 12.9 illustrates the principle of ITD formation for a spherical head model and the dependence of the ITD value on the azimuth angle φ in the horizontal plane, computed from a simple, yet relatively accurate approximation rule:

$$\tau = \frac{D}{2c}(\varphi + \sin \varphi), \tag{12.2}$$

where φ is the azimuth direction of the source in radians, D is the diameter of the head, and c is the velocity of sound. This rule is easily derived from the spherical geometry of the head.

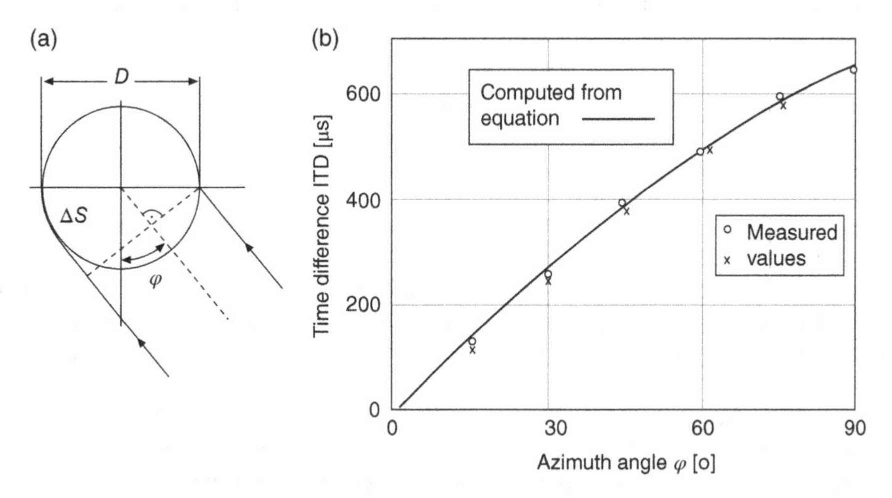

Figure 12.9 (a) The interaural time difference due to propagation path difference ΔS. (b) The ITD in microseconds plotted as a function of azimuth angle φ in the horizontal plane, computed using Equation (12.2) (solid line) and measured values (markers o and ×).

The ITD varies from 0 μs to approximately 600–700 μs when the azimuth angle varies from 0° to 90°. The values derived by the spherical model match the measurements from real heads well, as shown in the figure. Figure 12.10a shows the dependency of the ITD on frequency and on direction as a surface. The values are computed using a cross-correlation type of binaural auditory model. The ITD does not depend significantly on frequency. Only at low frequencies below about 700 Hz are the maxima of the ITDs a bit higher. Some irregularities in ITD functions at frequencies above 2 kHz also exist, and these are due to the complex multi-path wave propagation on the contralateral side. The coherence of the binaural signals is typically lower there, which causes the irregularities in cross-correlation computation.

The hearing mechanisms are sensitive to the ITD between the critical bands of the left and right ears sharing the same frequency. For low-frequency signals up to about 1.5 kHz, the auditory system is sensitive to the phase difference between the narrowband signals in each auditory band. Note that the cue is still often called the ITD, although in some cases the slightly more precise term interaural phase difference (IPD) is used. For frequencies higher than about 800 Hz, the size of the head is comparable to or larger than half the wavelength of the sound, making phase differences ambiguous. Nevertheless, humans are sensitive to the IPD at frequencies up to about 1.6 kHz (Blauert, 1996). ITD cues are still extracted at higher frequencies by the auditory system from the delays between temporal envelopes of the signals.

The perceived lateral position of an auditory event with varied ITD cues while keeping the ILD constant has been measured in headphone experiments, and the results are shown in Figure 12.11a. The position is measured with impulse-like stimuli having a time difference τ_{ph}. In principle, the perceived lateral position shifts in the direction of the ear at which the signal arrives first. For time differences from −600 μs to +600 μs, the dependence on lateral position is linear, and for time differences in the range of about 1–20 ms, the sound is perceived in the ear corresponding to the preceding impulse (Blodgett *et al.*, 1956). With even larger ITDs and for continuous signals, the subjects can no longer tell on which side the auditory event occurs; they may perceive one on each side or perceive a diffuse source in all directions.

Humans are relatively sensitive to small changes in ITD. The value of the JND of the ITD depends on the ITD itself, on the level of the signal, and on the temporal and spectral content of the signal. In the best case, with relatively broadband stimuli which have strong temporal envelopes, and with base ITD corresponding to directions near the median plane, the JND of ITD is of the order of the 10 μs to 20 μs (Blauert, 1996; Hafter and De Maio, 1975).

12.3.2 Interaural Level Difference

The *interaural level difference*, ILD, is another major binaural cue. It is commonly known as the level difference between the ear canal signals. Plane waves produce an ILD due to the scattering – the shadowing and reflection – of sound waves by the head. When a sound wave from a distant source arrives at the head, the pressure level increases due to reflection at the ipsilateral ear and the pressure level decreases at the contralateral ear, as shown in the centre panel in Figure 12.5.

Scattering is a frequency-dependent phenomenon, since wavefronts arriving from distant sources – plane waves – produce ILD only where the head is large enough compared to the wavelength to affect the acoustic field around it. As can be seen from Figure 12.7, the level of the HRTF magnitude responses is constant for frequencies below about 400 Hz, and the dependence on the azimuth angle increases for higher frequencies, reaching more than 20 dB at frequencies above 4 kHz. Therefore, it is a natural consequence that the ILD is a strong

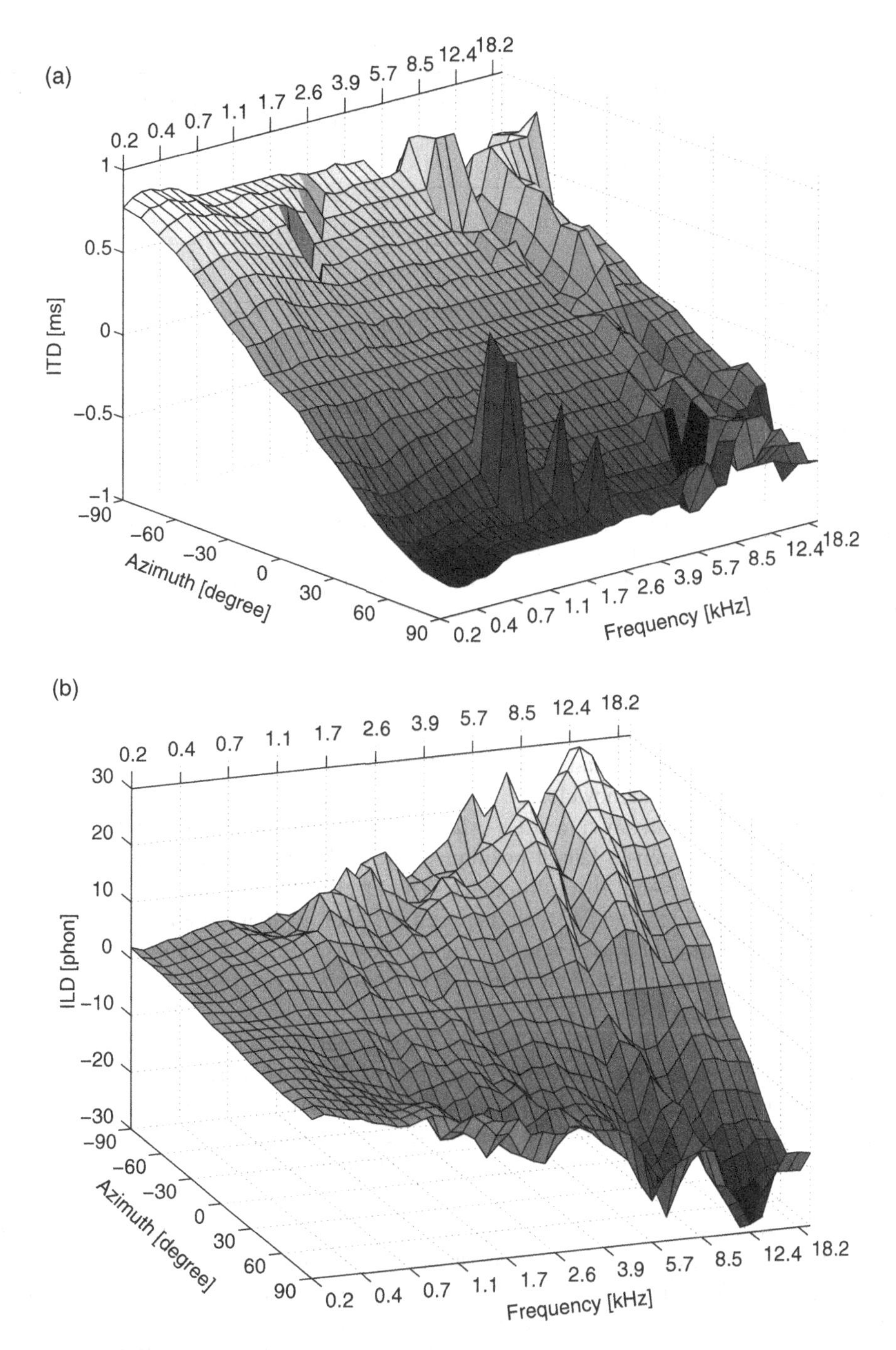

Figure 12.10 (a) The interaural time difference (ITD) functions created by a sound source emitting pink noise in different azimuth directions and in different frequency regions. The ITD is computed using a Jeffress-type auditory model based on the cross-correlation between ear canal signals (see Section 13.5.1). (b) The interaural level difference (ILD) functions expressed as the loudness-level difference in phons computed using a simple auditory model from the same measurement as in a).

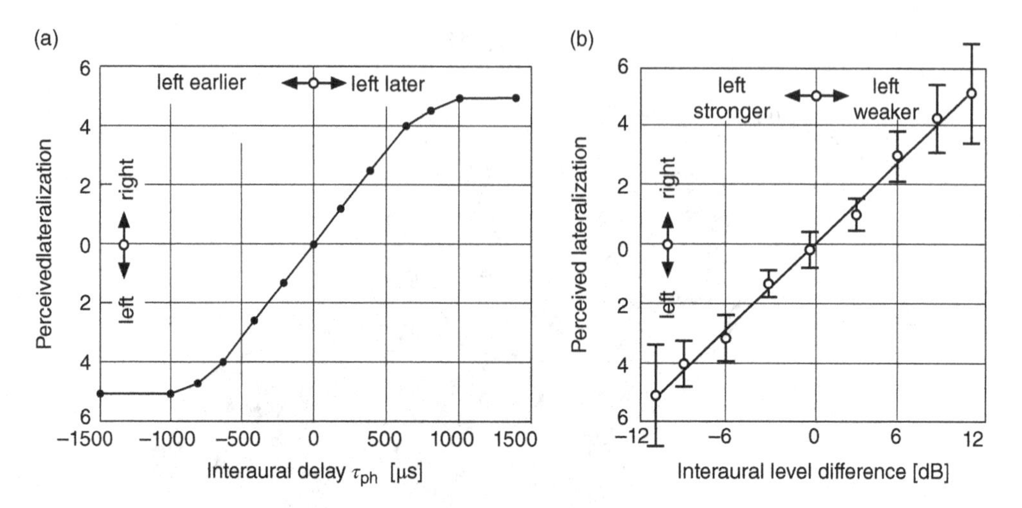

Figure 12.11 (a) The perceived lateralization position as a function of the interaural time difference and (b) the level difference. The ITD experiment is based on impulse-like stimuli (Toole and Sayers, 1965) and the ILD experiment on a 600-Hz sinusoid as the stimulus (Sayers, 1964). Reproduced with permission from The Acoustical Society of America.

directional cue for plane waves (or distant sources) at high frequencies. The magnitude of the ILD as a function of source sound direction and frequency is shown in Figure 12.10b, which shows the frequency dependency clearly.

The sensitivity of lateralization to ILDs using sinusoidal stimuli is shown in Figure 12.11b, which indicates the perceived lateral position of the auditory event created with a 600-Hz sinusoid having an interaural level difference. The level difference range of approximately $-12 \ldots +12$ dB covers the lateral scale between the ears, so that the auditory event is displaced towards the direction of the louder signal (Sayers, 1964). A complete saturation of the localization of the auditory event in the left or right ear is obtained when the ILD has an absolute value of 15–20 dB.

An apparent inconsistency in spatial hearing is the fact that although plane waves cause an ILD only at high frequencies, humans are sensitive to ILDs at all frequencies. As shown in Figure 12.12, the JND of the ILD is less than 3 dB with sinusoids at frequencies below 2 kHz for absolute ILD values less than 22 dB. With larger absolute values and at higher frequencies, the JND is 3–7 dB. The lowest JNDs are obtained when the ILD value is near zero (Weiping *et al.*, 2010).

As noted earlier, plane waves do not generate ILDs when their wavelengths are long compared to the size of the human head. The question then remains, *why are we sensitive to ILD at low frequencies?* There are at least two reasons for this. At low frequencies, the ILD is clearly a distance cue, since when the source reaches the listener, much higher ILDs are measured between the ear canals. The other use of the ILD seems to be in the perception of coherence of ear canal signals. It has been shown that humans are sensitive to very fast changes of ILD in time; the time resolution of ILD changes is of the order of a few milliseconds. If the ear canal signals are incoherent, as may happen in the case of diffuse sound, large instantaneous fluctuations in ILD are analysed (Goupell and Hartmann, 2007).

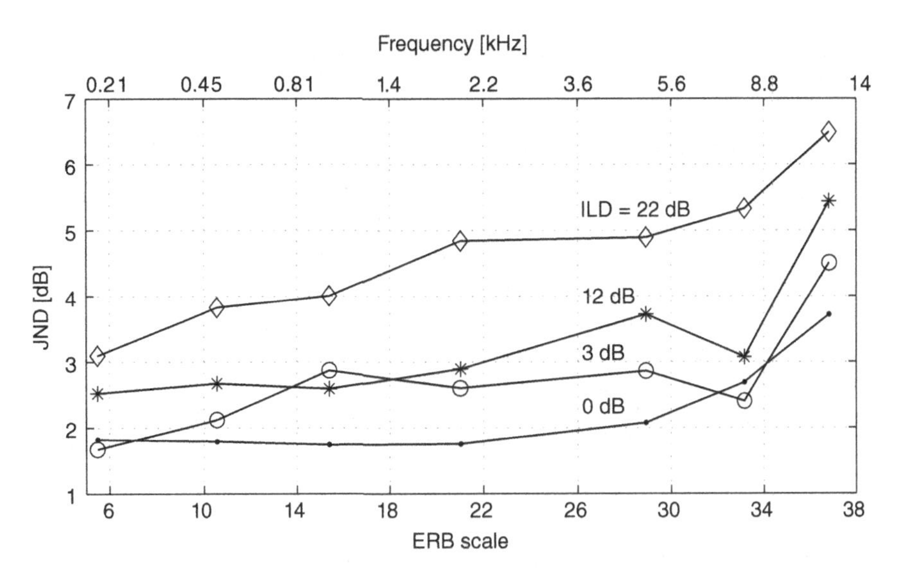

Figure 12.12 The JND of the ILD as a function of frequency and the base ILD. Adapted from Weiping *et al.* (2010)

12.3.3 Interaural Coherence

Interaural coherence (IC) is a measure of the similarity of the signals at the ear canals (Faller and Merimaa, 2004). Humans are sensitive to IC, but it is not clear if the auditory system computes it as an independent cue or if it is perceived due to the effect it has on the values of other directional cues. Some binaural models, which will be discussed further in Section 13.5.1, assume that IC is computed in the auditory system. In this section we define the coherence simply by stating that the coherence of ear canal signals is close to unity when listening to a single sound source in a free field. Low coherence is obtained in a diffuse field, or when sound from multiple sources arrives from different directions at the listener. The value of IC also depends on the frequency in the diffuse field. Low coherence is obtained only at frequencies above about 400 Hz, where the distance between a subject's ears is comparable to the wavelength (Borss and Martin, 2009).

When IC is high, as when in an anechoic chamber, the auditory events are localized to be point-like. When a broadband sound signal, such as pink noise, is presented with low IC, either using headphones or multiple loudspeakers, subjects perceive an auditory event that is located in surrounding directions or in many directions at the same time (Blauert, 1996). It has been found that humans are especially sensitive to coherence at low frequencies with narrow-band signals (Culling *et al.*, 2001). However, the perception of broad sources is not limited to low-frequency signals only. When multiple loudspeakers are used to reproduce incoherent high-frequency content, a spatially broad auditory event is perceived (Santala and Pulkki, 2011).

In normal rooms, the IC is high when the onset of an impulsive sound arrives at the ear canals, and the IC is lower when reflections and the room effect in general arrive at the ear. Faller and Merimaa (2004) suggest that localization occurs only if the IC has a high enough value. Alternatively, as discussed in the previous section, low coherence also causes ILD and ITD cues to fluctuate randomly with time and frequency (Goupell and Hartmann, 2007). These fluctuations, caused by the low IC, are assumed to lead directly to a perception of surrounding auditory events, and no separate IC cue is computed in the auditory system.

12.3.4 Cues to Resolve the Direction on the Cone of Confusion

At least in principle, the ITD and ILD do not change when changing the position of a sound source on a cone of confusion. This means that decoded values of the ITD and the ILD do not resolve the direction from which the sound arrives at the listener, since the same cue values are produced by sound arriving from any direction specified by the surface of the cone. The fact that we can perceive the front–back dimension quite well and the elevation angle correctly indicates that there must be effective mechanisms for resolving the direction inside the cones of confusion. There are two mechanisms by which we perceive the elevation correctly, the analysis of spectral cues and the utilization of dynamic cues, which will now be discussed.

Spectral cues

Monaural cues, based on information from a single ear or common to both ears, are important in localization. Because no interaural differences are used, the information for monaural cues is derived from the ear canal signal itself, particularly from the properties of the sound source and its variation due to the effect of the head, but also due to the acoustic environment. The main monaural cues, when the direct sound from a source is dominant, are *spectral cues*. Since the monaural time resolution of hearing is about 1–2 ms, and as most of the details in HRIR are located within a 1-ms time window, monaural temporal cues are considered unimportant.

Due to scattering and reflection caused by the head and the pinna, the magnitude spectrum of the sound entering the ear(s) is dependent on the direction of arrival. These cues are denoted as *spectral cues*. Spectral cues are important when localizing sound sources in the median plane and its vicinity. To illustrate the spectral cues, the left ear HRTFs measured with a source in different directions in the median plane are shown in Figure 12.13 for four subjects. It can be seen that the HRTFs indeed carry information on direction; there are vast changes in the spectra when the elevation changes. The pinnae of subjects are different, and so are the spectral cues. The figures have similar structures, the arcs have similar shape, but they are located at different frequencies. The notches and humps at 4–10 kHz are also located at different frequency-elevation positions, although some similarities between subjects are evident. The overall black–white contrast in the plots is also different. For example, subject 17 has less contrast than subject 12, which implies that the spectral notches are deeper with subject 12 than with subject 17.

Broadband signals are generally needed to make spectral cues efficient, since otherwise the cues cannot be decoded. This is a fair assumption, since most natural sounds have a relatively flat spectrum at frequencies where the pinna cues are effective, namely at 4–10 kHz. On the other hand, it has been shown that when a sinusoid is presented in that frequency region, using a static loudspeaker in the median plane, the perceived direction depends significantly on the frequency of the sinusoid. The listener perceives the auditory event as moving in space when the frequency is changed (Blauert, 1996), although the sound source does not move.

Learning also has an effect on the utilization of spectral cues. This is natural, since the ears of humans develop steadily throughout life, and the hearing system has to continuously adapt to the acoustic effect of the pinnae. In an experiment by Hofman *et al.* (1998), the cavities of subjects were made smaller using putty, thus destroying the natural spectral cues. The subjects

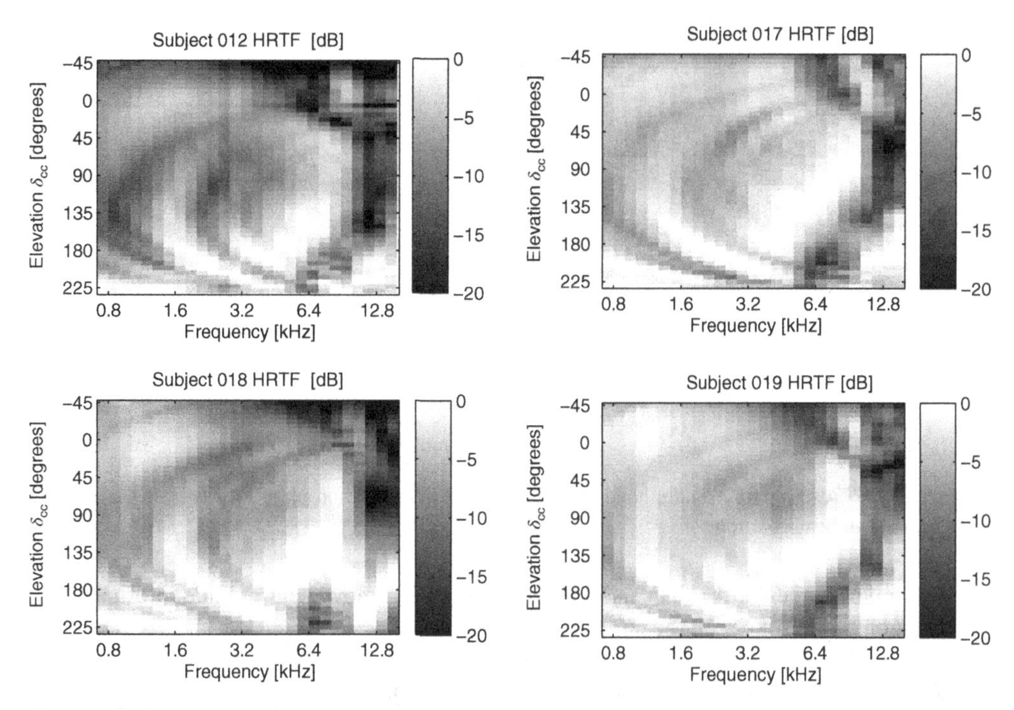

Figure 12.13 The HRTF magnitude spectra measured for 50 elevations in the median plane and smoothed over 1/6th-octave bands. The smoothed magnitudes have been normalized with the maximum of the magnitude in each frequency band. Different panels show the results of different subjects. The HRTF data for the illustrations originate from Algazi *et al.* (2001).

immediately lost their ability to perceive elevation, however this returned in a few weeks to normal if the putty was not removed from the ears during that time. The listeners thus learned the acoustic properties of the new ears during the period they had putty in their ears. Very interestingly, after removing the putty, the listeners regained their elevation perception ability immediately. For some reason, no after-effect was present.

Another case where monaural (spectral) cues are of the utmost importance is when a listener has only one ear in effective use. Localization of sound sources is still possible, although the accuracy is strongly degraded (Blauert, 1996).

Dynamic cues

The other mechanism humans use in resolving the direction inside the cone of confusion is the utilization of dynamic cues (Blauert, 1996). When the listener moves his or her head by rotating, tilting, or moving, the binaural cues change accordingly. Let us assume that a sound source is in front of the listener. If the listener rotates his or her head right, the left ear comes closer to the source and the right ear moves farther away. This changes the ITD and ILD, favouring the left ear. If the source was located behind the listener, with the same movement the binaural cues would change to favouring the right ear. The localization mechanisms implicitly assume that the sound source locations are static in the global coordinates, and thus the sources do not move in synchrony with the listener's head movements, which most often is true. Thus,

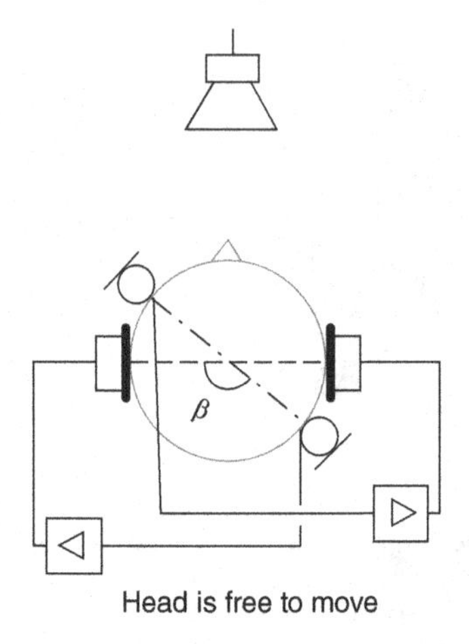

Figure 12.14 A set-up for pseudophone localization experiments. Microphones away from the listener's ears move with head rotation.

the dynamic changes of the ITD and the ILD during head movement provide information on the position of the source on the cone of confusion.

An interesting experimental set-up called *pseudophone localization* is presented in Figure 12.14. Signals from microphones are reproduced through headphones so that the subject perceives sounds as if having ears at the positions of the microphones. A particularly confusing situation is achieved if the angle β is 180°, so that the ears are effectively interchanged, resulting frontal sound sources generating auditory events behind the listener, and vice versa (Blauert, 1996). One of us (Ville Pulkki) had a striking experience with such a device. A person was talking in front of me at a distance of about one metre. When I rotated my head with the pseudophone in action, the auditory object jumped immediately to behind me. Although the talking mouth of the person was clearly visible in front, the auditory object remained behind. The dynamic cues thus overrode the visual cues in this case.

One of the reasons why headphone reproduction of audio is, in most cases, internalized is the lack of dynamic cues. In such cases, the binaural cues in audio reproduction do not change at all when the listener moves his or her head. The only situation where cues remain constant when the head is moved is when the source is inside the head, and the localization mechanisms take this as strong evidence that the sources are indeed inside the skull. Note that localizing sounds inside one's own head is not at all unusual. Typical inside-the-head-localized sounds are, for example, one's own voice and eating and breathing sounds.

The dynamic cues thus seem to provide strong evidence as to whether the source is in front, behind, up, down, or inside the head. This effect is of great interest in headphone-based binaural reproduction of sound, where the position of the head is updated in the reproduction system. As will be discussed in Section 14.6.2, the correct reproduction of dynamic ITD and ILD cues mitigates directional errors perceived due to erroneous monaural cues. This underlines the effectiveness of dynamic cues.

12.3.5 Interaction Between Spatial Hearing and Vision

Hearing is only one of the sensory mechanisms for communication and receiving information from the environment. Spatial perception and sound-event localization may be influenced by non-acoustic cues as well, particularly by visual cues. In *multimodal perception*, the peripheral sensory mechanisms work relatively independently, but these partial percepts are fused together into a coherent internal representation, unless conflicting cues leave them apart or the dominating sensory organ overwhelms the weaker evidence.

A well-known example where vision modifies the auditory perception is ventriloquism (Alais and Burr, 2004), where listeners perceive the sound to be coming from a direction other than the one it actually comes from, if the visual movements associated with the sound generation are synchronized with the sound. The synchronized movements then 'capture' the spatial perception of sound. The effect is most powerful if the separation between the sound and visual sources is less than about 30°. However, the auditory cue may still be more salient than the visual cue in some cases. If the visual image is severely blurred, the perceived direction matches that of the sound source, and if both of them are blurred, an average direction is perceived.

12.4 Localization Accuracy

The previous sections described the cues available to human listeners to localize a sound source. This section discusses the accuracy of humans in terms of localization. However, the discussion is limited to listening to single real sources in a free field. Localization accuracy can be much worse in spaces with reflections and/or reverberation, as will be discussed in subsequent sections.

The signal content also has a strong effect on localization accuracy and localizability of sound sources. Accuracy is best with broadband sounds with strong temporal fluctuations, such as transients, speech, and noise in general. On the other hand, narrowband sounds, such as mosquito or certain bird sounds, can be very hard to localize, especially in rooms. This can be understood from the perspective of evolution. The correct localization of predators or prey is vitally important for survival, and the sounds generated by movements are typically broadband, impulse-like sounds.

12.4.1 Localization in the Horizontal Plane

In the best case, with broadband sounds in a free field, our directional hearing locates sound sources accurately. Figure 12.15 shows the results from an experiment where the perceived direction of a white noise sound source was estimated by subjects in four main directions (front, left, right, back). The sources on the left and right sides were placed approximately 10° towards the frontal direction. Localization blur, or the variance in the perceived angle, was lowest at the front, somewhat bigger at the back, and largest in the side directions.

One method to evaluate the localization resolution is to measure the JND of direction for different frequencies of tones. Figure 12.16 plots the results of such an experiment using sine-wave test signals (Stevens and Newman, 1936). The best resolution is about 1° for tones below 1 kHz coming from the front of the subject, and it deteriorates towards the sides. At around 1.7 kHz the resolution is the worst, especially for sounds from the sides, so much so that the subjects cannot make any reliable estimate at all. This is the frequency where the ILD does not depend monotonically on direction and the head size is about one wavelength, leading to maximal confusion in the ITD and ILD decoding. At higher frequencies, the resolution improves, but is never as good as for frequencies below 1 kHz. Note that the JND values do not reflect the localization accuracy, but only how much a source has to be moved for the change to be noticed.

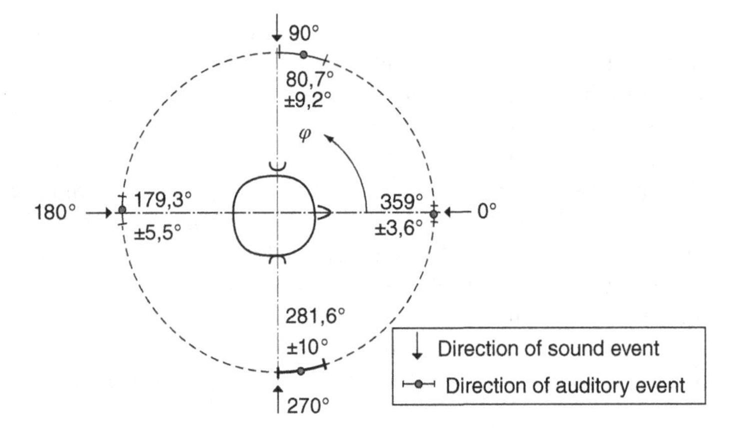

Figure 12.15 The perceived azimuth angle in the horizontal plane in free-field conditions for a white noise sound source in four primary directions (0°, 90°, 180°, and 270°). The average perceived angle and the localization blur are marked by a thick line. Adapted from Blauert (1996).

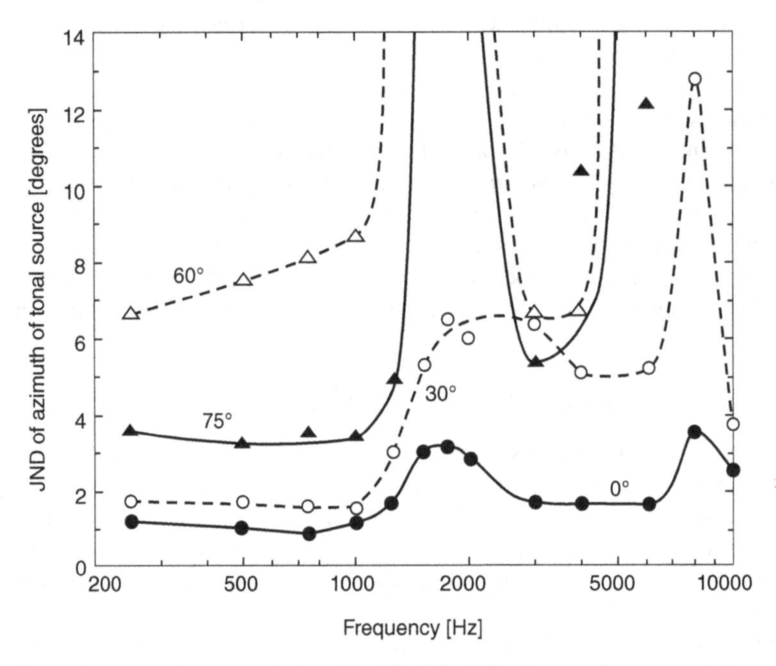

Figure 12.16 JND in the azimuth angle ($\varphi = 0°, 30°, 60°, 75°$) of a sound source emitting a tone in the horizontal plane as a function of frequency. Adapted from Mills (1958).

12.4.2 Localization in the Median Plane

The accuracy of perceived elevation of a sound event in the median plane is worse than the corresponding accuracy of perception of azimuth directions, especially with an immobilized head. Experimental results are shown in Figure 12.17 for a set of elevations using speech as the test signal. There is a noticeable bias to the frontal direction, and the localization blur increases for sources above and behind the subject. The lower accuracy compared to azimuth perception

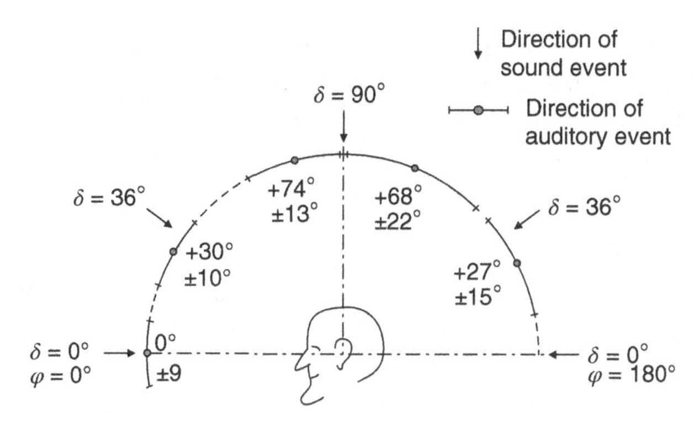

Figure 12.17 The perceived elevation and localization blur in the median plane for five physical elevation angles ($\delta = 0°$, 36°, 90° frontally, and 0°, 36° from the back) with speech as the test signal. Adapted from Damaske and Wagener (1969).

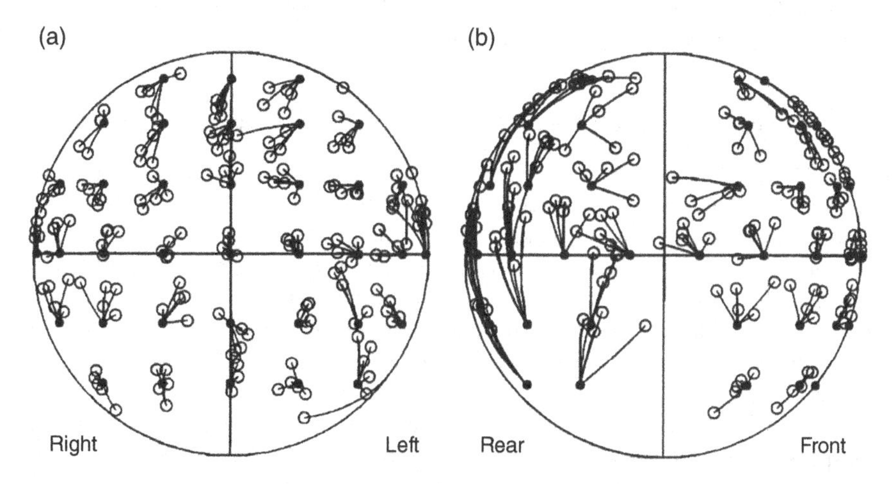

Figure 12.18 Localization results of 250-ms broadband sound sources in a low-echoic room with a spacing of 20° in azimuth and elevation. The filled dots show the positions of the sound sources and circles the responses of the same subject to five presentations of sound. The task of the listeners was to point towards the source with their nose. (a) Projection of the results in front of the listener. (b) Projection of the results to the front and rear of the listener. Adapted from Middlebrooks (1992). Reprinted with permission from The Acoustical Society of America.

is due to the fact that the ITD and the ILD cues cannot be utilized in the median plane, and spectral cues are the main source of information for the immobilized listener. If subjects are allowed to rotate or tilt their head, the change in ITD and ILD cues provides more information on the direction of the source.

12.4.3 3D Localization

A number of studies have been conducted to measure the accuracy of directional hearing when any direction in three dimensions is allowed. Figure 12.18 shows a result from a test where 250-ms noise bursts from different directions were presented to subjects, and their task was to point with their nose to the direction of the sound source (Middlebrooks, 1992). The results in Figure 12.18a show quite consistent accuracy within a few degrees for frontal directions, especially near the horizontal plane. Slightly larger errors occur when the source departs from the horizontal plane. Figure 12.18b shows the performance in the rear directions. It is clear that the subjects overestimate the elevation of the sources at the back, a result that is in agreement with the median plane localization discussed in the previous section.

The results show that the accuracy of perception of elevation degrades in directions outside the field of vision. This is in line with the fact that subjects continuously adapt to their spectral cues, as discussed in Section 12.3.4. The result also suggests that adaptation can occur only if the sources are visible. Using vision as the reference for adaptation of directional hearing is a viable option, as vision is the most accurate method to measure the direction of a sound source for adaptation. Furthermore, the auditory and visual pathways meet at a relatively low level in the brain, as discussed in Section 7.7.1, which could facilitate such adaptation.

The results on perceived direction and localization blur vary in different experiments with the test signals used, the test subjects and their familiarity with the task, the method of registration of the perceived direction, and so on. Thus, the values given in Figures 12.15 to 12.18 characterize only approximately the behaviour of directional hearing.

12.4.4 Perception of the Distribution of a Spatially Extended Source

White water, or waves hitting the seashore, forms a wide sound source from the perspective of an observer. A listener, indeed, perceives such sound to be wide. The ability to perceive the spatial distribution of a spatially extended sound source is not very accurate, especially if the type of sound signal is not optimal. Humans are at their best in this task when the sound from different sources is noise-like or impulsive, or if different parts of the source emit different frequency content. On the other hand, if all the parts of the source emit sinusoids with equal frequency, the subject perceives a very narrow auditory event that does not depend on the actual width of the source.

Figure 12.19 shows some results from a listening experiment where different subsets of 13 loudspeakers in the horizontal plane in an anechoic chamber were used to emit mutually incoherent pink noise at equal levels to the subject. The task of the subject was to indicate which loudspeakers emitted the sound. The loudspeakers which actually emitted the sounds are marked on the figure with black squares, and the grey bars show the percentage of 'sound on' indicated for each loudspeaker by the ten subjects for two repetitions of the task. The three topmost panels show that the subjects indicated the spatial distribution correctly only in the case when the number of sources was one, two, or three (Santala and Pulkki, 2011). For a wide and dense constellation of sources, as in the fourth panel, the edges of the distribution were perceived almost correctly, although a bit biased towards the centre. However, the perceived distribution in the central area of the source did not match at all with the actual distribution of the sound source. The two panels at the bottom show spatially complex scenarios, where the subjects clearly had no clue which loudspeakers were actually on.

The accuracy of the hearing system is thus clearly not at its best when analysing the spatial distribution of sources. Although the directional separation between loudspeakers was 15° or more, the listeners failed to report the distribution accurately. The task of perceiving the distribution in such a case could be thought to be a simple one, since the accuracy of direction perception for a single source is of the order of a few degrees at best. However, the distribution is analysed from signals from the sources summed in the ear canals, and naturally the presence of multiple temporally and spatially overlapping source signals arriving from different directions makes the analysis task difficult.

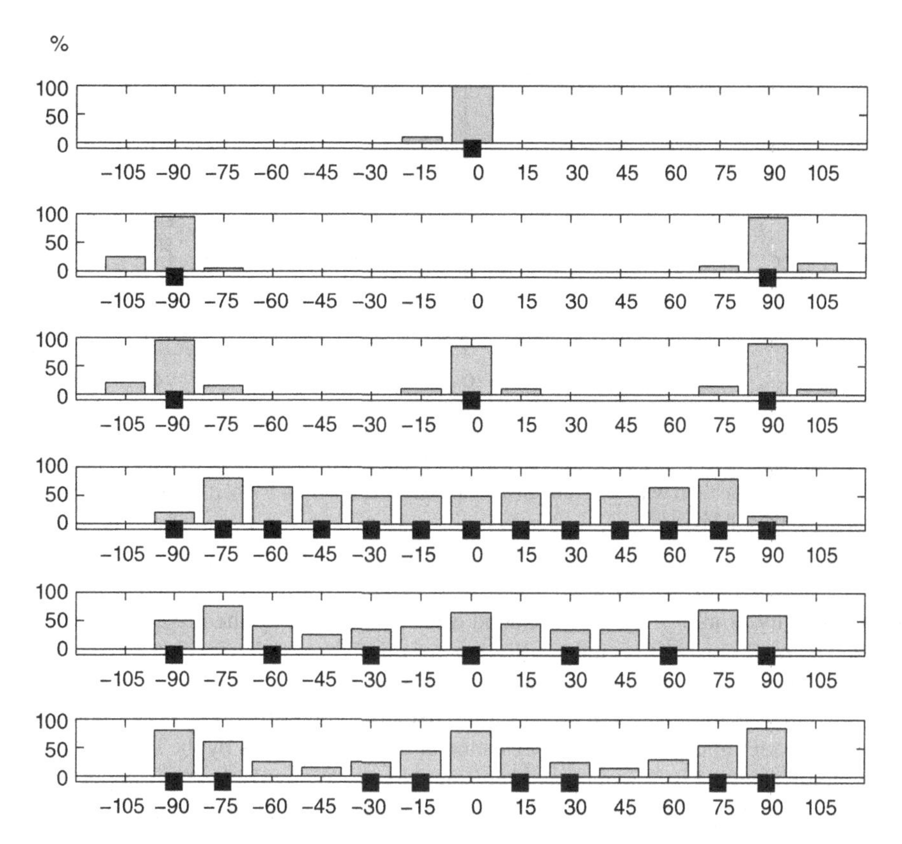

Figure 12.19 The results of a subjective test with 13 loudspeakers set up horizontally in front of the listener in an anechoic chamber. The loudspeakers, marked with black squares, emitted pink noise, and the task of the listeners was to indicate which loudspeakers were actually emitting sound. The grey bars denote the relative rate at which a specific loudspeaker was indicated to emit sound. The azimuth directions of the loudspeakers are represented on the abscissa. Courtesy of Olli Santala.

12.5 Directional Hearing in Enclosed Spaces

Spatial hearing is at its best in situations with point-like broadband sources in a free field, as shown in the previous section. Also of interest is the spatial resolution when strong reflections or reverberation exist in the environment. Our hearing has adapted to such cases with remarkable resilience. The directions of sound sources are perceived correctly in many challenging conditions. The ability of humans to perceive the geometry of an enclosed space, however, is quite limited, but on the other hand, the listener can compensate, at least partly, for the colouration caused by the room response in the detection of timbre.

12.5.1 Precedence Effect

The precedence effect is an assisting mechanism of spatial hearing (Litovsky *et al.*, 1999). It suppresses the effect of early reflections in source direction perception. This helps to localize sound sources in reverberant rooms. The reflections reaching the listener in a reverberant room are added to the direct sound in the ear canals, which changes the binaural cues considerably. Therefore, the only reliable cues are those generated by the direct sound without the presence of reflected sound in the ear canals. In enclosed spaces, direct sound is dominant in the ear

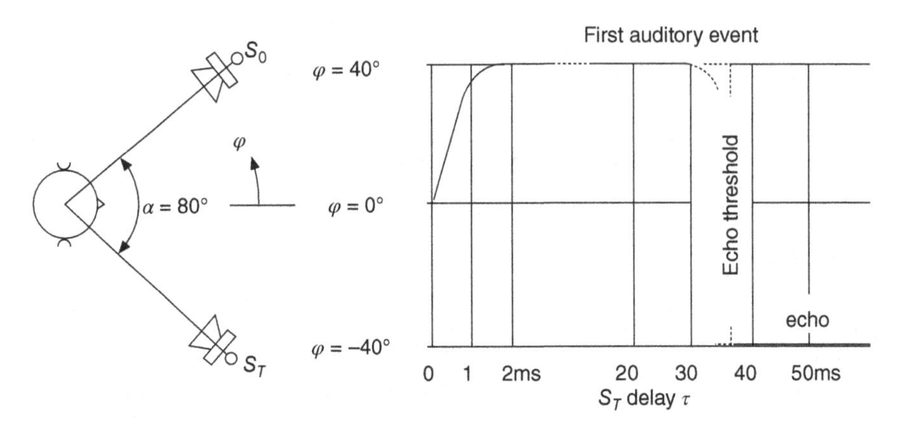

Figure 12.20 The perceived direction of an auditory event as a function of the delay between two impulsive sound events generated with different loudspeakers. Adapted from Blauert (1996), and reprinted with permission from MIT Press.

canals only shortly after the onset of sound after silence, before the reflections arrive at the listener. The precedence effect takes advantage of this.

The typical research scenario is shown in Figure 12.20, where two loudspeakers are placed in the azimuth directions of ±40° in anechoic listening conditions. The *lead* sound event S_0, typically with an impulsive sound, is presented at time instant 0 ms by the loudspeaker at 40°, and an identical signal is presented as a *lag* sound event, S_T, by the loudspeaker at −40°. The delay τ between the sound events is varied. When τ is zero, the listener perceives a single sound event in front, and when τ is increased to about 1 ms, the single spatial auditory event migrates towards the lead sound event. The spatial perception remains the same until the onset of the echo threshold, which occurs for values of τ of about 30–40 ms, depending on the signal. Only after this threshold is crossed does the lag sound event create an additive, spatially separated auditory event.

If a level difference is introduced between the lead and lag sound events, different thresholds are found in the precedence effect. As shown in Figure 12.21, the louder the lag sound is made compared to the lead, the more probable it is that it will be audible. The thresholds also depend on the delay and, in principle, the larger the delay is, the more prominent the lag. The lowest threshold is the detection threshold of the lag. When the level of the lag sound exceeds this threshold, its presence is noticed as a difference in timbre; it does not produce a spatially separated auditory event.

Evidently, our hearing responds strongly to onsets and transients, and the binaural cues decoded from the short response with a length of only about 1 ms have, correspondingly, a strong emphasis. After a strong response from most neurons, the refractory period makes the directional hearing mechanisms non-functional. Thus, the direction of the delayed sound does not affect the perception.

12.5.2 Adaptation to the Room Effect in Localization

As already mentioned, the localization of sources is based on signal analysis in the ear canals, and reflections and reverberation largely ruin the monaural and interaural cues. The precedence

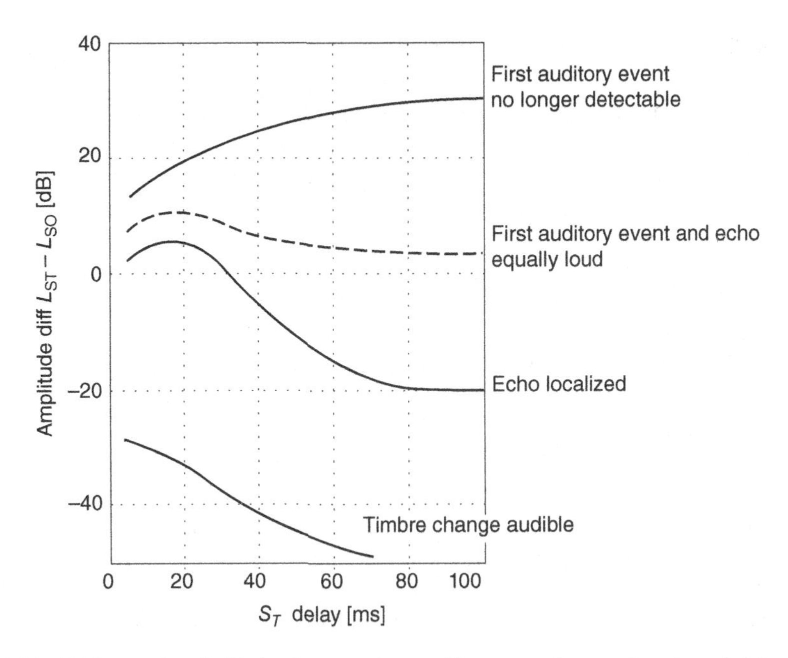

Figure 12.21 Different thresholds in the precedence effect scenario as a function of delay and level difference between sound events at time instants 0 ms and T. Adapted from Blauert (1996).

effect to be shows that transient information is judged most relevant in the detection of the left or right direction. In addition to the precedence effect, there are also other mechanisms aiding the localization tasks that are not known, but their existence has been shown by the fact that the accuracy of subjects' performance in localization tasks in rooms constantly improves with the number of trials, even after repeating several hundreds of trials (Shinn-Cunningham, 2000; Zahorik *et al.*, 2005). The subject gathers some data on the spatial cues generated by the room and somehow adapts to them, perhaps by using templates.

12.6 Binaural Advantages in Timbre Perception

A relevant question is how much the binaural hearing aids in decoding the sound spectrum emanated by multiple spatially separated sources, also in reverberant spaces. In vision, the spatial separation of sources definitely helps in the process, because when visual stripes are separated by more than about 0.1°, they are discriminated. The working principle of the eye and the ear, however, differ greatly. The eye has a lens and a multitude of receptors primarily sensitive to the different directions from which light comes, while in contrast the ear has only one cochlea. Spatial hearing is conducted as an analysis of the ear canal signals, and no primary sensitivity to direction exists in the cochlea. A lesser advantage is expected with hearing, but, as will be shown below, a significant benefit is still obtained.

12.6.1 Binaural Detection and Unmasking

How well humans can separate the signals emanated from spatially separated sources is an interesting subject. Since we have two ears, we should have some capability to listen selectively to different directions. Such a release of masking due to binaural hearing is called *binaural unmasking*.

Interestingly, humans are capable of detecting the content of sound coming from different directions in the presence of distracting sources. The speech signal has been widely used as a the test signal in a the study of this capability, as it presents a natural signal to which human hearing has adapted. In the studies, a speech signal S with masking noise(s) N is presented to the listener. The level of the signal is measured for both binaural listening and monaural reference listening so as to produce the same intelligibility in both cases, and the difference in level between the monaural and binaural conditions is taken as the *binaural intelligibility level difference* (BILD) (Blauert, 1996).

Figure 12.22a shows the BILD in the case when a broadband, speech-like masker noise is presented via six loudspeakers around the subject. The masking thresholds are measured for different azimuth directions of the signal source. When the measured thresholds are subtracted from the corresponding threshold of the signal in the direction of 180°, the plotted level differences are obtained. It can be seen that with binaural listening the advantage is of the order 5–7 dB compared to the reference case with the loudspeaker behind. When the same test is conducted with single-ear listening, the monaural intelligibility level difference (MILD) plot shown in the figure is obtained. It is remarkable that the difference between the MILD and the BILD is small when the sound source is ipsilateral to the unblocked ear. This leads to the

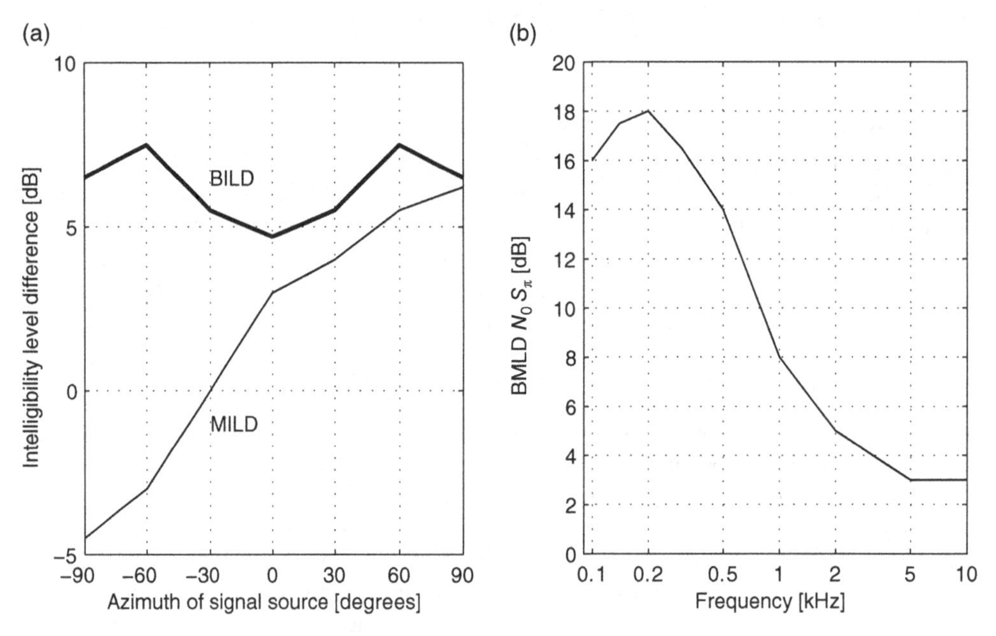

Figure 12.22 (a) Static broadband noise is presented to a subject through six surrounding loudspeakers. The level resulting in a certain degree of intelligibility has been measured using binaural and monaural listening. The direction of the signal source is shown on the abscissa and the BILD and the MILD on the ordinate. The level differences are computed in the reference case using a signal in the direction of 180° with monaural listening. (b) The binaural masking level difference with a broadband diotic masker and an antiphasic sinusoidal signal depending on frequency. Figures adapted from Blauert (1996) and reprinted with permission from MIT Press.

assumption that in such cases the signal detection relies on the 'better ear'; the signal is simply decoded from the ear canal signal that has a higher SNR. There is evidence, however, that subjects utilize instantaneous ILD to further improve the intelligibility from 'better ear' listening (Bronkhorst, 2000) by a few decibels.

A related phenomenon is the *cocktail party effect* already mentioned in Section 11.7.1, where the subjects have to follow a speech source among a number of distracting speech sources (Bronkhorst, 2000). In multi-talker conditions, the advantage provided by binaural hearing has been found to be 0–8 dB, which is in agreement with the results of BILD tests.

Certain synthetic signals reveal the significantly greater advantages of binaural listening. The binaural masking level difference (BMLD) has been researched extensively with headphone listening. The most dramatic effect is obtained when the subject is presented with a broadband diotic noise (N_0) and binaurally antiphasic sinusoidal signal S_π. The reference case is $S_m N_m$, where both the signal and noise are presented monotically. The BMLD is plotted as a function of frequency in Figure 12.22b, where it can be seen that when the frequency of the signal is about 200–300 Hz, the binaural advantage in signal detection is of the order of 18 dB (Blauert, 1996). Although such a big advantage is not obtained with distant sound sources, the result is interesting in the study of hearing.

Our relatively low capability to listen selectively to different directions may reveal something about the role of hearing in our evolution. Vision is used to precisely locate an object, and when looking in one direction, the perception of visual details of objects in other directions, is greatly diminished, and objects outside the field of vision are not perceived at all. We can assume that evolution has developed hearing as an omnidirectional alarm system with high sensitivity over a large frequency region to complement directionally selective vision. We can only guess at the capabilities of the auditory system if our hearing had developed towards better directional sensitivity. We would probably have more than two ears with poorer frequency resolution, or perhaps acoustic lenses coupled with more of ears.

12.6.2 Binaural Decolouration

Listeners are also able to adapt to spectral colouring caused by the room using binaural listening, which is called *binaural decolouration*. The goal of the mechanisms is interpreted to be the estimation of sound signals emanated by sound sources in an enclosed space. They show a remarkable ability to estimate the sound signal emanated by the source by ignoring the spectral contribution of the room response present in the ear canal signals (Brüggen, 2001). The same effect has been discussed in the context of sound reproduction (Toole, 2006). The decolouration effects are both monaural and binaural. Bilsen (1977) states that some pitch phenomena can be explained by assuming that listeners somehow up the spectra of the left and right ears to form a *central spectrum*. Such a summation would make the dips in the room responses less audible, since the dips typically occur at different positions in frequency, which could explain, at least partly, the decolouration effects.

12.7 Perception of Source Distance

The perception of distance is also an important part of the spatial hearing mechanism. It is a multi-faceted mechanism that is not perfectly understood yet (Zahorik *et al.*, 2005). Evidently, the auditory system uses different cues along with knowledge of the acoustic surroundings. Both monaural and binaural cues are used in the task.

12.7.1 Cues for Distance Perception

There are at least four cues used to extract the auditory distance:

- *Loudness*: The louder the auditory event is, the closer it is. Better distance estimates are extracted if an internal reference of the sound is used for comparison. In other words, familiar sounds aid the perception of distance.
- *Effect of room reflections and reverberation*: The more the room affects the sound, the farther away the source must be.
- *Spectral content*: The fewer high frequencies are in the auditory event, the farther away the source must be.
- *Binaural cues*: When the source is closer than about 1 m to the subject, the ILD between the ear canals grows larger, which is used as a distance cue.

The first of the mentioned distance cues, loudness, is naturally the most related to the SPL generated by the source at the listening position. This cue has been studied in an anechoic chamber by placing a real speaker at different distances and asking five subjects to estimate the distance. In this environment, the loudness with which the speaker talks is the only distance cue available to the subject. The perceived distance asymptotically approaches a finite distance less than 10 m, which can be called the *acoustic horizon*, as the physical distance increases, as seen in Figure 12.23.

The second distance cue in the list above is the *effect of room reflections and reverberation*. In everyday acoustic environments, the signals in the ear canals of a listener consist of direct sound and also a considerable amount of reflections from walls or obstacles and reverberation due to repeated reflections. In such conditions, can take advantage of the room cues and is typically more accurate than in free-field conditions (Nielsen, 1993; Zahorik *et al.*, 2005). Research on psychoacoustics often mentions the direct-to-reverberant ratio as a cue for distance perception. The more there is reverberant sound energy compared to direct sound energy, the farther away the listener perceives the source to be.

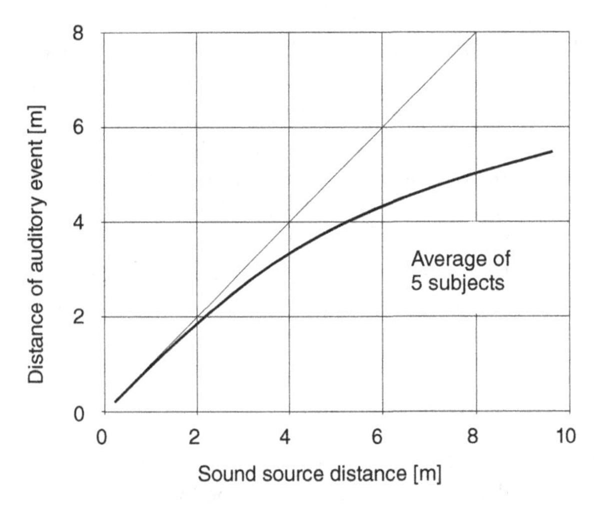

Figure 12.23 The dependency of perceived distance of a real speaker on the physical distance in free-field conditions. Adapted from von Békésy G 1949.

Added to this, audio engineers utilize another method to control distance perception using reverberators, wherein the auditory events caused by at least some types of sounds (typically voice) can be moved closer or farther away by changing the pre-delay parameter. Reverberators are audio effects, which, in principle, convolve a monophonic sound signal with the room response. The pre-delay parameter controls the temporal gap between the direct sound and the rest of the response. Shortening the pre-delay moves the perceived source farther away and making it longer brings it closer. Physically this makes sense, since a long gap would correspond to a situation where the direct path is much shorter than the reflected path, which can only happen if the source is close. This is different from the direct-to-reverberant ratio cue, since the ratio does not change. A very similar effect has been found in virtual reality simulations. When the first reflections are moved temporally in an impulse response without changing the properties of the late reverberation, the perceived distance changes systematically (Pellegrini, 2002).

The third distance cue mentioned in the list is the *spectral content* of sound. It changes when the sound travels long distances in the atmosphere. For example, the sound of lightning sounds like a 'crack' when the bolt is relatively close, and when heard from a distance of several kilometres it sounds more like 'rumbling'. This change is simply due to the fact that the air absorbs high frequencies more than low frequencies. The effect is quite mild, about 3 dB/100 m at 4 kHz. Thus, the listener may use this cue to detect the distances of sources that are far away.

The fourth distance cue in the list is binaural cues, which are essentially directional cues. However, if a source is brought near to a listener from afar in a constant direction, the ITD and ILD cues remain practically the same when the distance r is more than about 1 m. At smaller distances, the ILD cue depends significantly on the distance as well. This is explained by the distance attenuation of sound, which is proportional to the inverse of distance, $1/r$, as defined in Equation (2.21). At larger distances, the $1/r$-law attenuates sound by practically the same amount for both ears, since the distance between ears is negligible when compared to the distance of the source. With shorter distances, the sound entering the farther ear is attenuated more than the sound entering the closer ear due to the distance difference (Duda and Martens, 1998; Shinn-Cunningham *et al.*, 2000). The ITD values also change, since the sound has to bend around the head causing extra propagation delay, but they do not increase as drastically as the ILD values. Note that this effect is not dependent on frequency, and large ILD values are thus generated at low frequencies too. The exaggerated ILD cue compared to the ITD cue thus suggests a source near to the listener.

12.7.2 Accuracy of Distance Perception

A relevant question, then, is: how accurate is the distance perception mechanism? Zahorik *et al.* (2005) analysed 21 studies of perception of distance in different scenarios, and unfortunately found that very different results are obtained with different signals and in different listening conditions. Zahorik *et al.* (2005) concluded that the psychophysical function between the actual distance r and the perceived distance r' fits relatively well with the compressive power function

$$r' = kr^a, \tag{12.3}$$

where k varied between about 0.5 and 2, and a was most typically between 0.3 and 0.8 in the different research results analysed. In practice, the distances are underestimated more or less severely when the source is farther than about 3 m from the listener, and at distances less than 1 m, the distance of the auditory event overestimates the distance of the sound source.

Zahorik *et al.* (2005) also summarized the directional blur in different studies. In some cases, the blur was only 5–25% of the source distance, while in some other studies the error range was up to 60% to the distance to the source. Higher errors are obtained especially with distant sources.

The hearing mechanisms which are used in distance perception are not known in general. For example, it is not known precisely how the perceived distance is extracted from the effect of the room response in the ear canal signals. A research topic for auditory modelling would thus be to form signal-driven auditory models that measure the cues for distance perception and then combine them to form the final perception of distance.

Summary

This chapter introduced spatial hearing and related concepts. When a sound wave arrives at the listener, the head, the torso, and the external ear affect the sound signals entering the ear canals. These effects are included in head-related transfer functions measured from the ear canals. Based on this information, the auditory system analyses the signals and derives localization cues. The cues, such as interaural differences and the monaural spectrum, are used in associating direction, distance, and other spatial attributes to auditory events and to the internal representation of the acoustic environment. This enables the localization of sound sources with remarkable accuracy even in reverberant spaces.

Listeners are able, at least partially, to compenstate for the frequency response of the listening room; that is, our hearing tries to estimate the signal emitted by the sources without the effect of the room. Humans also have the ability to listen selectively to a certain direction in binaural listening. In multi-source situations, the sources are audible with a 0–8 dB lower level compared to monaural listening.

Further Reading

The book '*Spatial Hearing*' by Blauert (1996) is a broad overview of the field of spatial hearing. The reader might also find the books by Begault (1994), Gilkey and Anderson (1997), and Xie (2013) interesting, especially for HRTF technologies and for applications with virtual displays. Neural processing in humans as well as in some animals is described by Yost and Gourevitch (1987).

References

Alais, D. and Burr, D. (2004) The ventriloquist effect results from near-optimal bimodal integration. *Current Biol.*, **14**(3), 257–262.

Algazi, V.R., Duda, R.O., Thompson, D.M., and Avendano, C. (2001) The CIPIC HRTF database *Applications of Signal Processing to Audio and Acoustics, 2001 IEEE Workshop*, pp. 99–102 IEEE.

Begault, D.R. (1994) *3-D Sound for Virtual Reality and Multimedia*. AP professional.

Bilsen, F.A. (1977) Pitch of noise signals: Evidence for a "central spectrum". *J. Acoust. Soc. Am.*, **61**(1), 150–161.

Blauert, J. (1997) *Spatial Hearing – Psychophysics of Human Sound Localization*. MIT Press.

Blodgett, H.G., Wilbanks, W.A., and Jeffress, L.A. (1956) Effect of large interaural time differences upon the judgements of sidedness. *J. Acoust. Soc. Am.*, **28**(4), 639–643.

Borss, C. and Martin, R. (2009) An improved parametric model for perception-based design of virtual acoustics. *35th Int. Audio Eng. Soc. Conf.: Audio for Games*. AES

Bregman, A. (1990) *Auditory Scene Analysis*. MIT Press.

Bronkhorst, A.W. (2000) The cocktail party phenomenon: A review of research on speech intelligibility in multiple-talker conditions. *Acta Acustica United with Acustica*, **86**(1), 117–128.
Brüggen, M. (2001) Coloration and binaural decoloration in natural environments. *Acta Acustica United with Acustica*, **87**(3), 400–406.
Culling, J.F., Colburn, H.S., and Spurchise, M. (2001) Interaural correlation sensitivity. *J. Acoust. Soc. Am.*, **110**(2), 1020–1029.
Damaske, P. and Wagener, B. (1969) Directional hearing tests by the aid of a dummy head. *Acta Acustica United with Acustica*, **21**(1), 30–35.
Duda, R.O. and Martens, W.L. (1998) Range dependence of the response of a spherical head model. *J. Acoust. Soc. Am.* **104**, 3048–3058.
Faller, C. and Merimaa, J. (2004) Source localization in complex listening situations: Selection of binaural cues based on interaural coherence. *J. Acoust. Soc. Am.*, **116**(5), 3075–3089.
Gilkey, R.H. and Anderson, T.R. (eds)(1997) *Binaural and Spatial Hearing in Real and Virtual Environments.* Lawrence Erlbaum Associates.
Gómez Bolaños, J. and Pulkki, V. (2012) HRIR database with measured actual source direction data. *Audio Eng. Soc. Convention 133* AES.
Goupell, M.J. and Hartmann, W.M. (2007) Interaural fluctuations and the detection of interaural incoherence. III. Narrowband experiments and binaural models. *J. Acoust. Soc. Am.*, **122**, 1029–1045.
Grantham, D.W. (1995) Spatial hearing and related phenomena. *Hearing*, **6**, 297–346.
Hafter, E. and De Maio, J. (1975) Difference thresholds for interaural delay. *J. Acoust. Soc. Am.*, **57**(1), 181–187.
Hofman, P.M., Van Riswick, J.G., and Van Opstal, A.J. (1998) Relearning sound localization with new ears. *Nature Neurosci.*, **1**(5), 417–421.
Litovsky, R.Y., Colburn, H.S., Yost, W.A., and Guzman, S.J. (1999) The precedence effect. *J. Acoust. Soc. Am.*, **106**, 1633–1654.
Middlebrooks, J.C. (1992) Narrow-band sound localization related to external ear acoustics. *J. Acoust. Soc. Am.*, **92**, 2607–2624.
Mills, A. (1958) On the minimum audible angle. *J. Acoust. Soc. Am.*, **30**(4), 237–246.
Møller, H. (1992) Fundamentals of binaural technology. *Appl. Acoust.*, **36**(3), 171–218.
Nielsen, S.H. (1993) Auditory distance perception in different rooms. *J. Audio Eng. Soc.*, **41**(10), 755–770.
Pellegrini, R. (2002) Perception-based design of virtual rooms for sound reproduction *22nd Int. Audio Eng. Soc. Conf.: Virtual, Synthetic, and Entertainment Audio*. AES
Santala, O. and Pulkki, V. (2011) Directional perception of distributed sound sources. *J. Acoust. Soc. Am.*, **129**, 1522.
Sayers, B.M. (1964) Acoustic-image lateralization judgments with binaural tones. *J. Acoust. Soc. Am.*, **36**(5), 923–926.
Shinn-Cunningham, B. (2000) Learning reverberation: Considerations for spatial auditory displays. *Proc. Int. Conf. on Auditory Display*, pp. 126–134 ICAD.
Shinn-Cunningham, B.G., Santarelli, S., and Kopco, N. (2000) Tori of confusion: Binaural localization cues for sources within reach of a listener. *J. Acoust. Soc. Am.*, **107**(3), 1627–1636.
Stevens, S.S. and Newman, E.B. (1936) The localization of actual sources of sound. *Ame. J. Psychol.*, **48**(2), 297–306.
Toole, F.E. (2006) Loudspeakers and rooms for sound reproduction- A scientific review. *J. Audio Eng. Soc.*, **54**(6), 451–476.
Toole, F. and Sayers, B.M. (1965) Lateralization judgments and the nature of binaural acoustic images. *J. Acoust. Soc. Am.*, **37**(2), 319–324.
von Békésy, G. (1949) The moon illusion and similar auditory phenomena. *Ame. J. Psychol.*, **62**(4), 540–552.
Weiping, T., Ruimin, H., Heng, W., and Wenqin, C. (2010) Measurement and analysis of just noticeable difference of interaural level difference cue. *Int. Conf. Multimedia Technology*, pp. 1–3 IEEE.
Xie, B. (2013) *Head-Related Transfer Function and Virtual Auditory Display*, volume 2. J. Ross Publishing.
Yost, W.Y. and Gourevitch, G. (eds)(1987) *Directional Hearing.* Springer.
Zahorik, P., Brungart, D.S., and Bronkhorst, A.W. (2005) Auditory distance perception in humans: A summary of past and present research. *Acta Acustica United with Acustica*, **91**(3), 409–420.

13

Auditory Modelling

Previous chapters addressed hearing and its capabilities, mainly from an experimental point of view, and a few phenomena were formulated mathematically. Such formulas may be thought to represent a mathematical model of the corresponding phenomenon. However, it is unlikely that a holistic mathematical model or a theory about the complete auditory system can ever be derived due to the enormous complexity of the system. Having said that, simplified mathematical theories are essential for determining causalities and for predicting the perception evoked by a given stimulus, which provides the evident need for experimental analysis and modelling of hearing.

Due to the complexity of the auditory system, computational processing of digitized signals has proven to be the best method to model the functionality of the system. Moreover, computational simulations can be used to design experiments addressing a specific part of the auditory system, which can potentially result in new hypotheses about the physiological functionality. Typically, these computational models are employed to study information processing in the auditory pathway and different input–output relationships.

An even stronger motivation for modelling hearing computationally originates from the engineering point of view, wherein a functional model enables emulating the functionality of hearing in numerous practical applications, especially if the model runs in real time. Such applications include, for example, speech recognition, sound reproduction, spatial audio techniques, hearing aids, and cochlear implants. Mimicking brain functions in the computational domain is also very educating and an inspiring topic: the human brain is a good engineering solution, and reverse engineering it is a good exercise in signal processing.

The history of computational auditory models is relatively short, the first serious attempts being made in the 1960s and the 1970s (Chistovitch, 1974; Dolmazon *et al.*, 1976; Weiss, 1966). However, the number of researchers developing and applying auditory models has increased rapidly since the 1980s, making today's, field rich with a plethora of publications available. An overview of the current status is given by Meddis (2010).

This chapter describes several computational auditory models and their applications. The focus is first placed on the simpler models, moving gradually towards more complex ones.

Communication Acoustics: An Introduction to Speech, Audio, and Psychoacoustics, First Edition.
Ville Pulkki and Matti Karjalainen.

Here, the term *auditory model* is used as a general concept, while the terms *psychoacoustic model* and *perceptual model* are used to refer to models designed to explain results of psychoacoustic experiments without paying specific attention to the physiology.

This chapter briefly overviews existing auditory models classified as follows:

- Simple psychoacoustic models;
- Filter bank models;
- Cochlear models;
- Hair-cell models;
- Models for cognitive processing;
- Models of binaural interaction.

13.1 Simple Psychoacoustic Modelling with DFT

As discussed in previous chapters, the fundamental psychoacoustic theory of hearing describes the peripheral hearing system as a kind of spectral analyser that extracts several perceptual aspects like specific loudness and overall loudness, pitch, duration, sharpness, and roughness, to mention a few, from the ear canal input. Since each of these aspects can be described in a quantitative manner depending on the properties of the input signal, computational models can be designed to extract metrics related to these aspects and, consequently, to emulate the functionality of hearing at least to some extent.

13.1.1 Computation of the Auditory Spectrum through DFT

The most common approach in simple psychoacoustic models is based on the processing involved in loudness perception, as discussed in Section 10.2.5. Here, a discrete Fourier transform (DFT) -based approach is presented, which, instead of actually estimating the loudness, derives an 'auditory spectrum' describing the level of cochlear excitation in dB as a function of frequency on the ERB scale. Such a spectrum can be extracted with the Matlab script listed below.

```
fs=48000; % frequency of sampling
sig=(rand(1,fs/2)-0.5)*10; % 500 ms of white noise
winlen=round(fs/40); % 25 ms time window
[blp,alp]=butter(2,(500 / (fs/2)), 'high'); % high-pass filter
zE=[1:41];                        % utilized ERB channel numbers
fE=228.7*(10.^(zE/ 21.3)-1);  % corresponding frequencies
% gain to implement hump around 4 kHz (ERB 25 +- 7)
hump_coeffs=1+(max(0,7-abs(25-zE))/7)*6;
% approximated spreading function of excitation in ERB scale
spreadfunct=10.^([-80 -60 -40 -20 0 -8 -16 -24 -32 -40 -48 -56
                                                      -64]/10);
sig=filter(blp,alp,sig);%high-pass to simulate LF sensitivity
                                                          loss
a=1; % time position counter

for i=1:winlen/2:(length(sig)-winlen) % loop through the
                                                   signal
```

```
    % window the signal, and take FFT
    SIG=fft(hamming(winlen)' .* sig(i:(i+winlen-1)));
    POWSPECT=SIG.*conj(SIG);      % compute power spectrum
    % scaling linear frequency to ERB
    lowlimit=1; i=1;
    for z=zE(2:end)
        highlimit=round(fE(z)/fs*winlen); % upper FFT-bin for
                                                   ERB channel
        % sum the power inside ERB channel
        excitation(i)=sum(POWSPECT(lowlimit:highlimit))*hump_
                                                    coeffs(i);
        lowlimit=highlimit+1; i=i+1; % update counters and
                                                   lowlimit
    end
    % implement excitation spreading by convolution with
                                                spreading
    % function and store the excitation patter
    excitpattern(a,:)=conv(excitation,spreadfunct);
    a=a+1; %counter update
end
% computation of auditory spectrum
audspec=10*log10(mean(excitpattern,1)); % avg over time
hearthr=(zE-24).^2/15;  % crude approx. hearing threshold
figure(1); clf;  axes('Position',[0.1 0.1 0.5 0.3])
plot(zE(2:end)-0.5, max(hearthr(2:end), audspec(5:(end-8))),
                                                        '-')
hold on;  plot(zE(1:end), hearthr,'--'); set(gca,'XTick',
                                                [2:4:40]);
xlabel('ERB scale'); ylabel('Auditory spectrum [dB]')
```

The computation of an auditory spectrum begins with the power spectrum computation. The input signal is first divided into, say, 25-ms-long time frames that are then multiplied by a suitable window function, like a Hamming window. Thereafter, a short-term power spectrum is computed for each time frame using the DFT, which is implemented using the fast Fourier transform (FFT). This processing does not reflect the frequency-dependent sensitivity of hearing. In practice, the sensitivity can be emulated at any stage of the computation, but conceptually, it should be emulated at the beginning by filtering the input signal. The Matlab script yields a coarse approximation of the inverse of the equal loudness contour at 60 dB SPL (Figure 9.2) by high-pass filtering the signal at the beginning and then multiplying the short-term power spectra by frequency-dependent weights. As a consequence, both the poor sensitivity of hearing to low frequencies and the increased sensitivity around 4 kHz, resulting from the ear canal resonance, are emulated roughly.

After this, the power spectra on the Hz-scale are converted to the Bark or ERB scale. Such *frequency warping* can be implemented in various ways. For an example, see the explanation of the computation of the mel frequency cepstral coefficients below. The Matlab script above implements the frequency warping by simply summing the power spectrum values within each ERB band.

The next processing step emulates how the excitation evoked by a stimulus spreads to other frequency bands. The simplest emulation approach comprises a convolution with a spreading function, as in Equation (10.6), and such an approach is also exploited in the code above. The function approximates the shape of the simultaneous frequency masking curve at 60 dB SPL. In reality, the shape of such a curve is level dependent, which can be accounted for by selecting the spreading function in a level-dependent manner. This is neglected here, since it would make the structure of the auditory model far more complex. The excitation patterns following the convolution can then be scaled into short-term specific loudness spectra following Equation (10.7). However, the presented model omits this scaling and consequently the resulting auditory spectrum describes the level of the excitation pattern in dB in different frequency bands. In other words, the output provides an estimate of the sound spectrum that is available to hearing mechanisms.

The linear power spectra and auditory spectra for a pure tone, white noise, and two speech signals are shown in Figure 13.1. Overall, the auditory spectrum is seen to differ from the power spectrum due to the use of the ERB scale, the asymmetric spreading of the excitation, and the emulation of the hearing threshold. The pure tone case demonstrates how the excitation spreads to adjacent frequency bands. The white noise case, in turn, visualizes the combined effect of the frequency warping and the emulation of the frequency-dependent sensitivity of hearing on the spectrum. That is, the level of the auditory spectrum increases with frequency, and the boost around 4 kHz resulting from the ear canal resonance is visible as well. In addition, the auditory spectrum differs from the physical one in terms of frequency resolution. The auditory spectra of the speech signals illustrate how the harmonic fine structure in voiced phonemes is visible in the physical spectra but is smoothed in the auditory spectra due to frequency warping. Additionally, the warping increases the visibility of low frequencies in the spectra. However, the formants in the speech signals are clearly also present in the auditory spectra.

Applications of DFT-based auditory models

Several variations of the model above have been designed for different purposes. For instance, the two variations described below have been applied in feature extraction in speech recognition algorithms.

- *Mel frequency cepstral coefficients* (MFCCs) (Davis and Mermelstein, 1980) are commonly used to characterize speech sounds. Figure 13.2 shows how these features are extracted from an input signal that is first processed with a high-pass filter to differentiate the speech waveform. Subsequently, a power spectrum is computed for each windowed time frame. The power spectra are then warped onto the mel scale with a filter bank consisting of a set of M (here 20) triangular-shaped band-pass filters. Thereafter, the logarithm of the filter bank outputs gives the coefficients X_k. Finally, M MFCCs are derived with the discrete cosine transform:

$$c_n = \sum_{k=1}^{20} X_k \cos\left[n\left(k-\frac{1}{2}\right)\frac{\pi}{20}\right], \qquad \text{for } n = 1, 2, \ldots, M. \tag{13.1}$$

MFCCs have been found to characterize speech sounds efficiently, hence their frequent use in speech recognition algorithms, especially those employing statistical models for the actual identification.

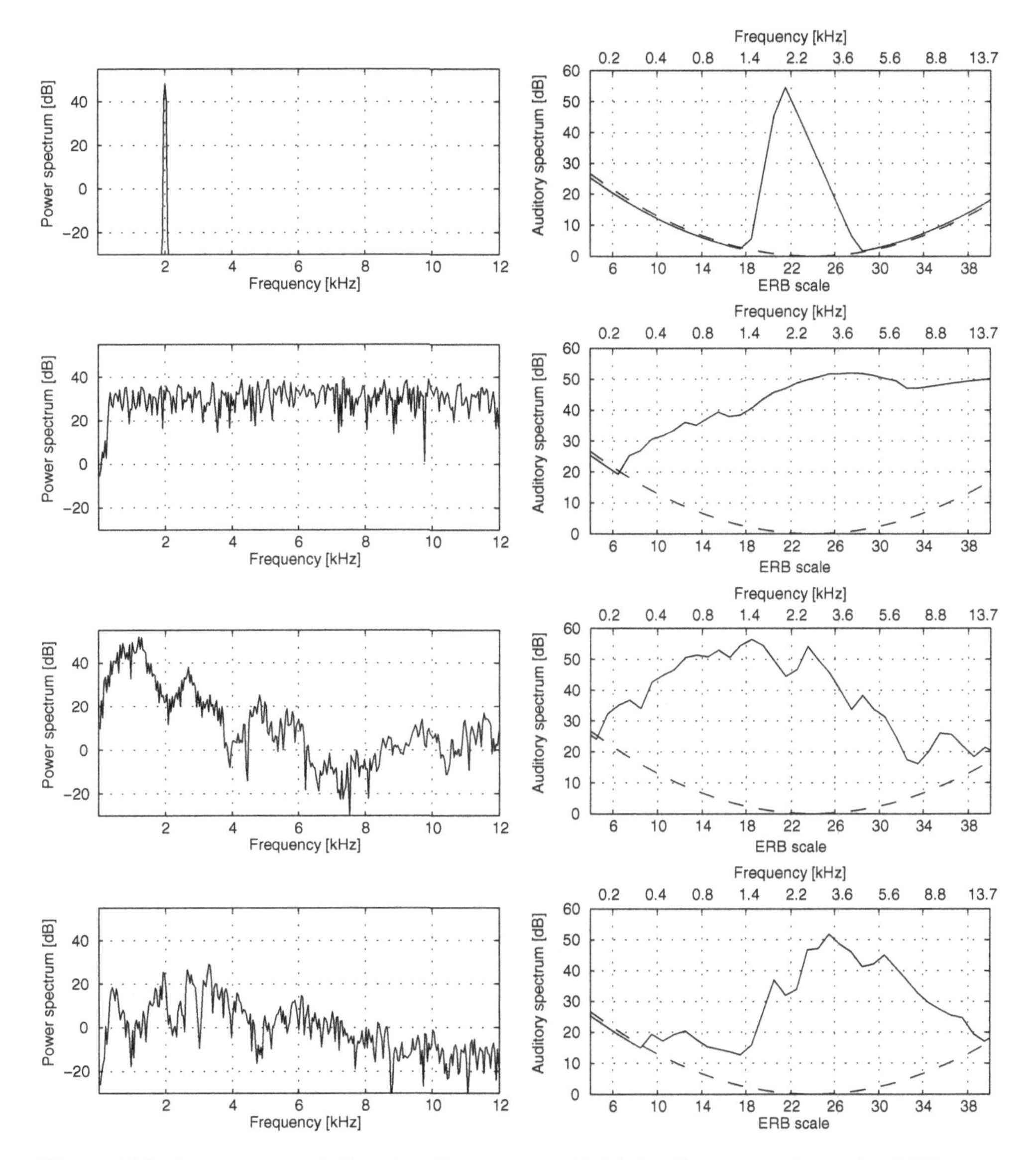

Figure 13.1 Power spectra (left) and auditory spectra (right) for (from top to bottom) a 2-kHz pure tone, white noise, the vowel /a/ and the fricative /s/. The auditory spectra were computed with the Matlab script presented in this section, which provides an implementation of a DFT-based auditory model. The auditory spectra have been computed as averages over several subsequent frames.

- *Perceptual linear prediction* (PLP) (Hermansky, 1990). Conceptually, the PLP coefficients and MFCCs are extracted in a similar manner. In PLP, however, the coefficients are extracted from a specific loudness spectrum. Moreover, the specific loudness spectra are transformed back into autocorrelation functions using the inverse Fourier transform, after which the traditional autocorrelation-based LP algorithm is used to extract the coefficients. The resulting coefficients are able to describe the spectral features of speech signals in a compact and rather speaker-independent manner.

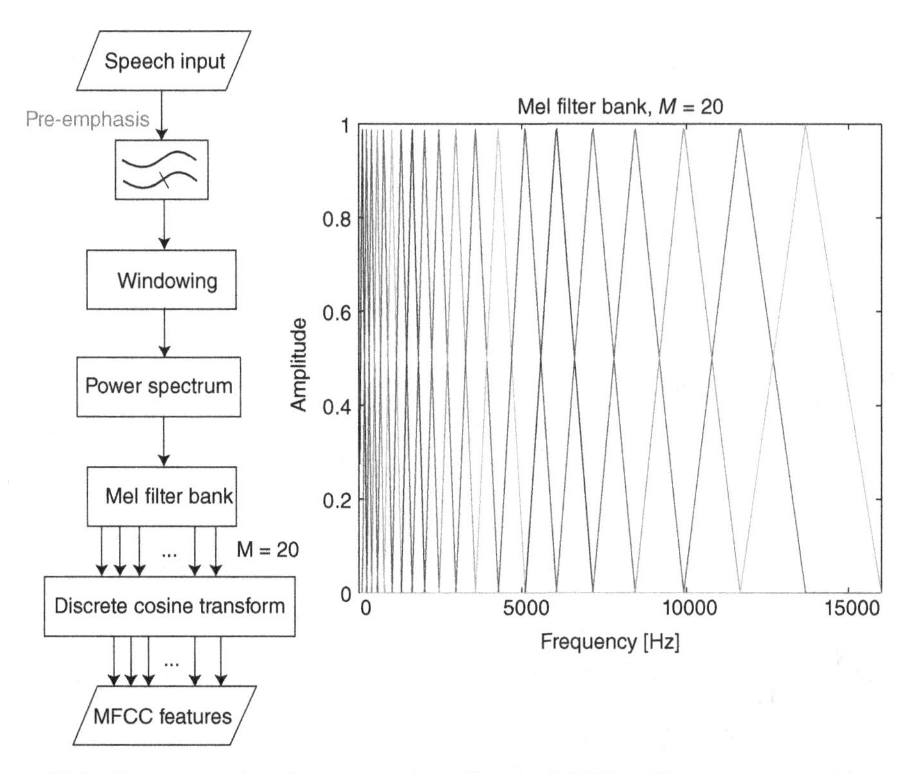

Figure 13.2 The computation of mel cepstral coefficients (MFCCs). The components of the power spectrum are combined with triangular weight functions to give spectral components in the mel frequency scale. The components are further transformed using a cosine transform (Equation (13.1)), which gives out the cepstral coefficients. Courtesy of Marko Takanen.

Similar computations are also performed in many *audio codecs* that map windowed frames of a signal into the frequency domain following the frequency resolution of human hearing. These codecs then perform further quantization or other operations in each time–frequency bin. Such techniques will be described in more detail in Chapter 15.

The technically straightforward and efficient computation of the auditory spectrum described above cannot be used to describe the functionality of hearing accurately, in part due to the following reasons:

- *Temporal resolution*. The length of the time frame and the type of the window function define the temporal resolution of the power spectrum derived with the DFT. However, the time–frequency resolution of hearing does not utilize a time frame of fixed length. The temporal resolution is about 1–2 ms at high frequencies and larger at low frequencies. Such a variation cannot be emulated with the above-mentioned procedure. The filter bank models described below can account for the time–frequency resolution of hearing more accurately.
- *Temporal dynamics*. The DFT-based model presented also ignores a few temporal effects in our hearing resolution. Such effects include, among others, temporal integration (Figure 10.16) and post-masking (Figures 9.10, 9.11, and 9.12). In principle, temporal integration and post-masking can be emulated by processing the time-framed signals, but not very accurately.

- *Level dependency*. If only a simple spreading function is used to simulate the spreading of the excitation to other frequencies, the level dependency cannot be emulated. However, as mentioned above, this can be fixed when a suitable level-dependent function is used.

13.2 Filter Bank Models

The fundamental problem in the above-mentioned simple psychoacoustic models is their inability to emulate the temporal resolution and dynamics of the auditory system. As mentioned previously, this problem originates from the use of fixed-length time frames in the DFT computation, the length of which defines the temporal resolution of such a model. The time–frequency resolution of hearing can be emulated more accurately with filter bank models that process the signals with a set of band-pass filters in the time domain.

Figure 13.3 illustrates how an auditory spectrum can be derived with a model employing a filter bank to emulate the time–frequency resolution. Both Bark and ERB resolutions can be emulated by selecting the filters appropriately. Such an auditory model can also emulate the spreading of the excitation and temporal masking effects.

13.2.1 Modelling the Outer and Middle Ear

The transfer functions of the external and middle ear must be emulated with appropriate filters before processing the signal with a filter bank model. Various approaches can be exploited for this purpose depending on the requirements of the application. The best accuracy in emulating the external ear is achieved using measured HRTFs of an individual subject or of a dummy head. Typically, the middle ear transfer function is considered a band-pass filter with, say, a 6-dB/octave decreasing frequency response at frequencies below 800 Hz as well as above 1.5 kHz, as shown in Figure 7.5.

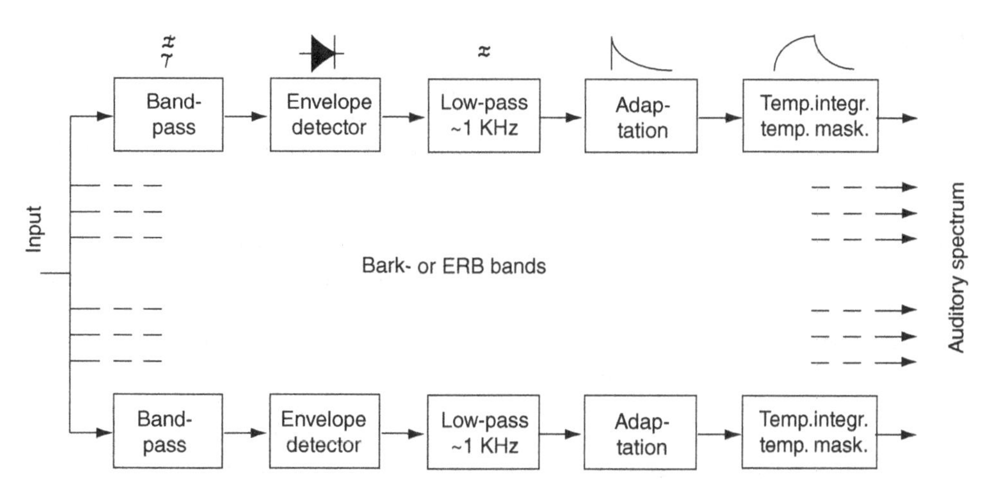

Figure 13.3 An auditory model implemented with a filter bank of multiple band-pass filters. The band-pass filters are followed by signal envelope detection using half-wave rectification and low-pass filtering corresponding to the monaural time resolution. The filters for short-term adaptation and temporal integration for temporal masking then simulate the low-level temporal effects in hearing. Section 13.6 gives two simple Matlab implementations of such auditory models.

13.2.2 Gammatone Filter Bank and Auditory Nerve Responses

The *gammatone filter bank* (Patterson, 1994) is the most commonly used method to emulate the frequency resolution of hearing. The physiological basis for such filters originates from the so-called *reverse correlation technique* measurements (De Boer, 1969) that yield an estimate of the impulse response of the auditory nerve fibre. Since this estimate resembles the shape of a pure tone that has been modulated with a gamma function, the corresponding filter is known as a gammatone filter. In addition, the frequency response of a gammatone filter is very similar to the human auditory filter as estimated by psychoacoustic notched-noise measurements (Glasberg and Moore, 1990). The popularity of the gammatone filter bank is also influenced by its computational efficiency and the relatively simple design. The impulse response of a gammatone filter is given by

$$g(t) = a\,t^{n-1}e^{-2\pi\, b(f_c)\, t}\cos(2\pi f_c t + \phi), \tag{13.2}$$

where a is the peak value of the response; t^{n-1}, which specifies the onset time of the response, together with the exponential term characterizes the bandwidth and decay of the response, f_c is the *characteristic frequency* of the filter, and ϕ is the initial phase of the response. As an example, the impulse response of a gammatone filter and the corresponding magnitude response as well as the magnitude responses of a 32-band gammatone filter bank (100 Hz $\leq f_c \leq$ 10 kHz) are shown in Figure 13.4. Typically, auditory models utilize about 42 bands in the gammatone filter banks, covering a frequency range from about 30 Hz to 18 kHz.

Although gammatone filters provide a good approximation of the human auditory filters, they suffer from a few shortcomings. They cannot emulate the level-dependent characteristics of the auditory filters. In addition, the impulse response of a gammatone filter has a relatively slow onset, which brings on problems when modelling phenomena involving temporally short sounds, such as the precedence effect.

13.2.3 Level-Dependent Filter Banks

As noted previously, the response of the cochlea shows level-dependent asymmetry in the form of compressive input or output functionalities. Various modelling approaches have been taken to form a filter bank that is able to emulate the suppressive and compressive characteristics of the auditory filters (see, for example, Carney 1993; Irino and Patterson 1997; Meddis *et al.* 2001; and Patterson *et al.* 2003).

An example of these approaches is the dual resonance non-linear (DRNL) filter bank (Meddis *et al.*, 2001). As shown in Figure 13.5, each band of a DRNL filter bank consists of two parallel processing paths, one of which employs a broadly tuned band-pass filter with a linear input–output relation. Additionally, a narrowly-tuned band-pass filter is used in the other processing path so that the gain of the filter compresses the output at higher levels. Both band-pass filters have an asymmetric frequency response, which is achieved by low-pass filtering the outputs of the gammatone filters. A weighted sum is then computed from the outputs of the two processing paths to acquire the DRNL filter bank output for a given frequency band. Moreover, the weights are set so that the outputs of the non-linear and the linear processing paths dominate the filter bank output at low and high levels, respectively. This implements the level dependence of the output.

Unfortunately, the response to a single impulse obtained by adding the outputs of the two processing paths is a double impulse. As a consequence, the output does not retain the temporal fine structure of the input, which may be problematic in some cases.

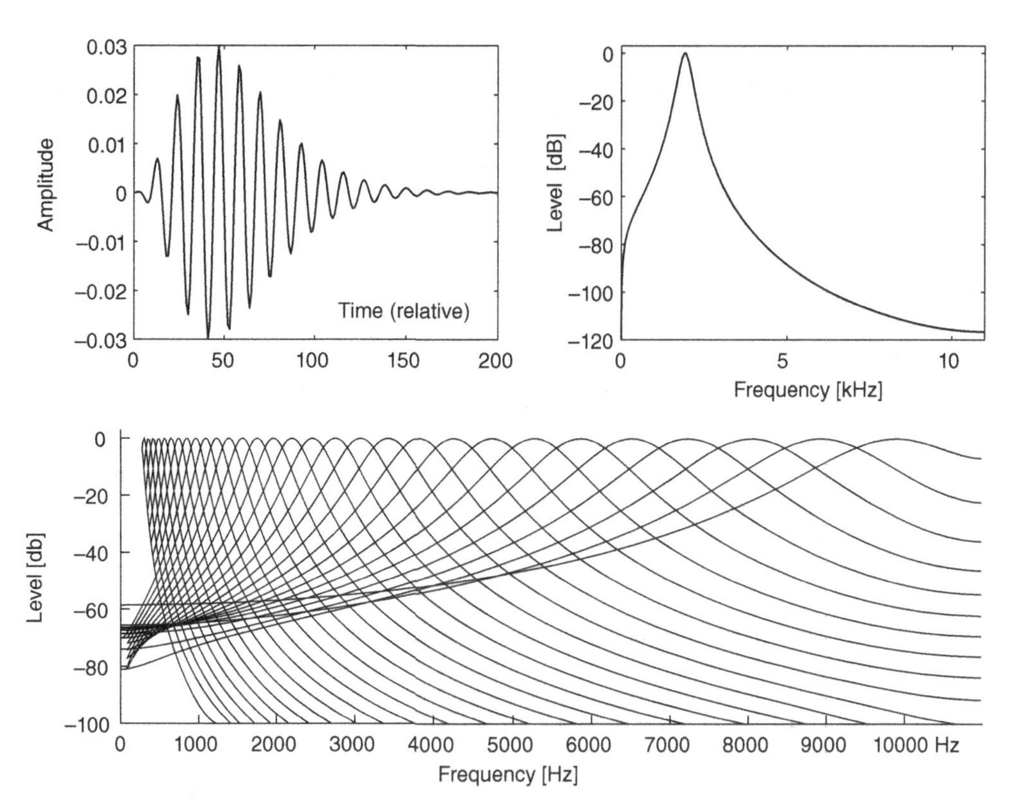

Figure 13.4 The characteristics of a gammatone filter bank: a) the impulse response of an individual filter, b) the corresponding magnitude response, and c) the magnitude responses of the filter bank on a linear frequency scale.

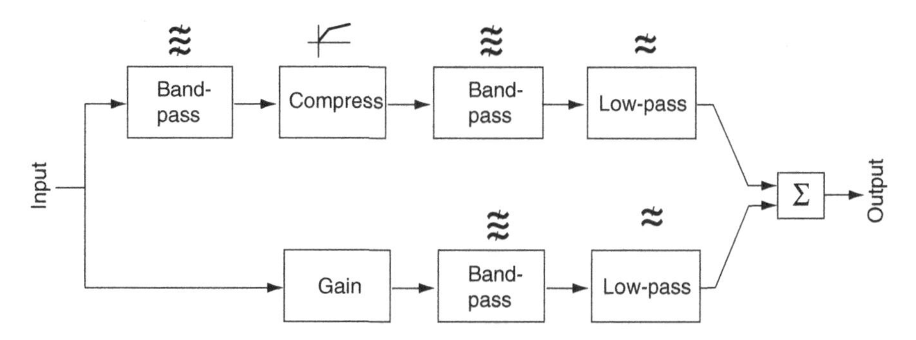

Figure 13.5 A dual-resonance filter, where the lower path employs a broader band-pass filter and the upper path a narrower band-pass filtering with compression. Such filters are used in DRNL filter banks to simulate cochlear processing since they emulate the level-dependent output of the cochlea better than a gammatone filter bank.

An alternative model that simulates the asymmetrical non-linear behaviour of the cochlea is the one by Zhang *et al.* (2001). Their model consists of a filter bank composed of filters that are time-varying, narrowly tuned, linear band-pass filters. Each of these is controlled by a non-linear, broadly tuned control filter. In particular, the output of the control filter sets the instantaneous gain and bandwidth of the corresponding filter, allowing the reproduction of cochlear non-linearities and phenomena like two-tone suppression. Furthermore, a variant of this model has been implemented by Zilany and Bruce (2006), in which an additional linear filter bank is used to emulate a second pathway of excitation of the inner hair cells. This model also allows the reproduction of the large phase changes in the inner-hair-cell responses at high SPLs. Moreover, it can be used to simulate the functionality of an impaired peripheral auditory system.

13.2.4 Envelope Detection and Temporal Dynamics

The inner hair cells and the auditory nerve fibres transform the mechanical vibrations of the basilar membrane into neural impulses. As noted earlier, the dependency of the rate of impulses on the basilar membrane displacement can be characterized with half-wave rectification (see Figure 7.18). The synchrony between the excitation and the firing rate is lost approximately at frequencies above 1 kHz. This can be modelled with a process that involves temporal integration with a certain temporal window. The 1-kHz limit corresponds to a time constant of 150 μs. Hence, the filter bank model shown in Figure 13.3 typically emulates the neural transduction by processing the half-wave rectified filter-bank outputs with first or second-order low-pass filters.

In the filter bank model of Figure 13.3, the next processing block emulates adaptation with a kind of high-pass filtering that strongly emphasizes the onset of a stationary stimulus. The various adaptation models, described, for instance, by Dau *et al.* (1996), Lyon (1982), and Seneff (1988), can be considered to be based on the idea of *automatic gain control* (AGC) that slowly reduces the amplification as the level of the input increases. Figure 13.6 illustrates an example of how the adaptation can be emulated with a series of feedback loops utilizing different time constants (Dau *et al.*, 1996). Specifically, the divisor elements control the

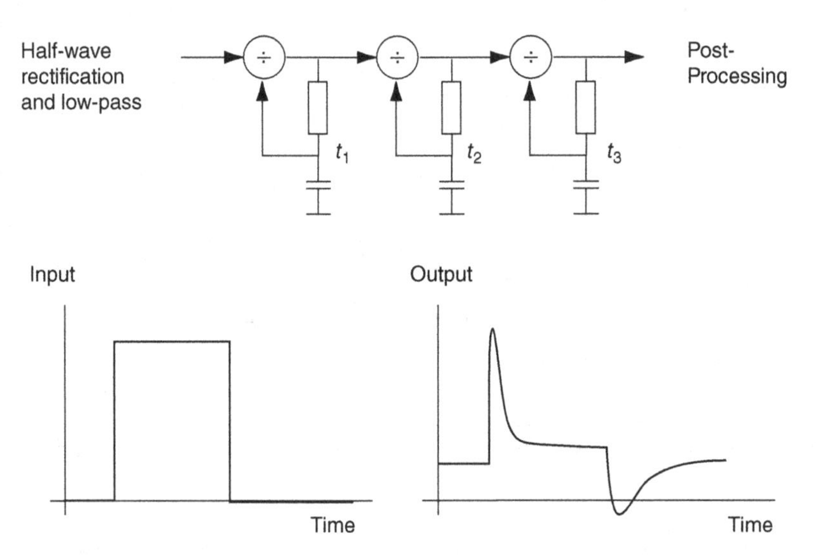

Figure 13.6 A model for adaptation, where the feedback loops act as automatic gain controls via the division operation. The time constants are typically selected to be between 5 ms and 500 ms.

amplification or attenuation by dividing the passing signal by the one coming from the corresponding feedback loop. For a continuous signal, the processing through a series of loops results in a nearly logarithmic output level, while the onsets are emphasized in a similar manner to Figure 3.13.

The last processing block in the model shown in Figure 13.3 utilizes larger time constants to emulate temporal integration and post-masking phenomena. Moreover, the energy of the input signal is low-pass filtered using time constants of 100–200 ms to simulate temporal integration, and post-masking is emulated using the same time constant in a non-linear filter that effectively prolongs the recovery time of the processing block. This block is needed, for example, when modelling dynamic loudness perception. Furthermore, a signal representing the specific loudness as a function of time can be obtained by suitably compressing the output. Optionally, the filter bank model can be designed to have parallel processing paths for temporal integration and adaptation, both of which receive the low-pass filtered envelope signals as input.

It should noted that Figure 13.3 shows only an overview of the functional elements, while an actual filter bank model implementation requires detailed design of compatible elements. In addition, some of the elements may be excluded from certain applications. For instance, not all the short time constants are necessary to evaluate loudness, whereas pitch analysis does not benefit from the use of large time constants.

Furthermore, the implementation is not restricted to splitting the processing in the aforementioned manner. For instance, adaptation, temporal integration, and post-masking can all be simulated effectively in a single element (Karjalainen, 1996), as shown in Figure 13.7. The element consists of an envelope detection unit (half-wave rectification and a low-pass filter), a multiplier controlling the amplification of the signal, two parallel low-pass filters utilizing different time constants, and a logarithmic feedback loop connecting the summed outputs of the low-pass filters to the multiplier. The primary output signal is the temporary loudness level (in phons or dBs) that can be transformed into specific loudness following Equation (10.2).

Figure 13.8 illustrates the two outputs of the above-described model (Figure 13.7) for a pure tone signal with a square wave envelope. The auditory nerve response (firing rate) is shown in Figure 13.8b, where the emphasis of the onset and the subsequent adaptation are clearly visible. Figure 13.8c shows the loudness-level output reflecting the temporal integration and post-masking effects. Interestingly, the two output signals that result from the same feedback process can be seen as complementary signals.

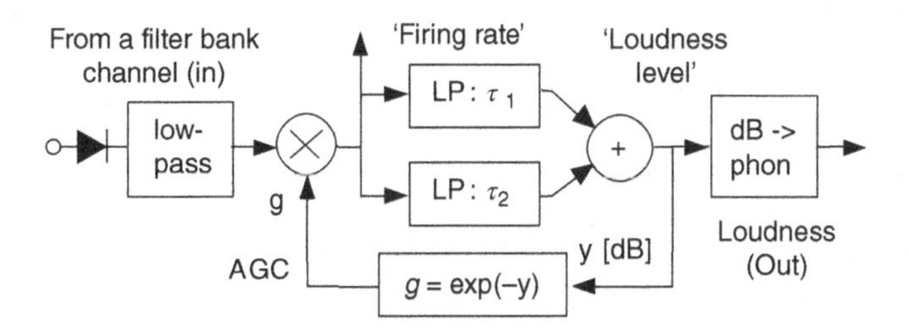

Figure 13.7 A model for adaptation and loudness for filter-bank-based auditory models.

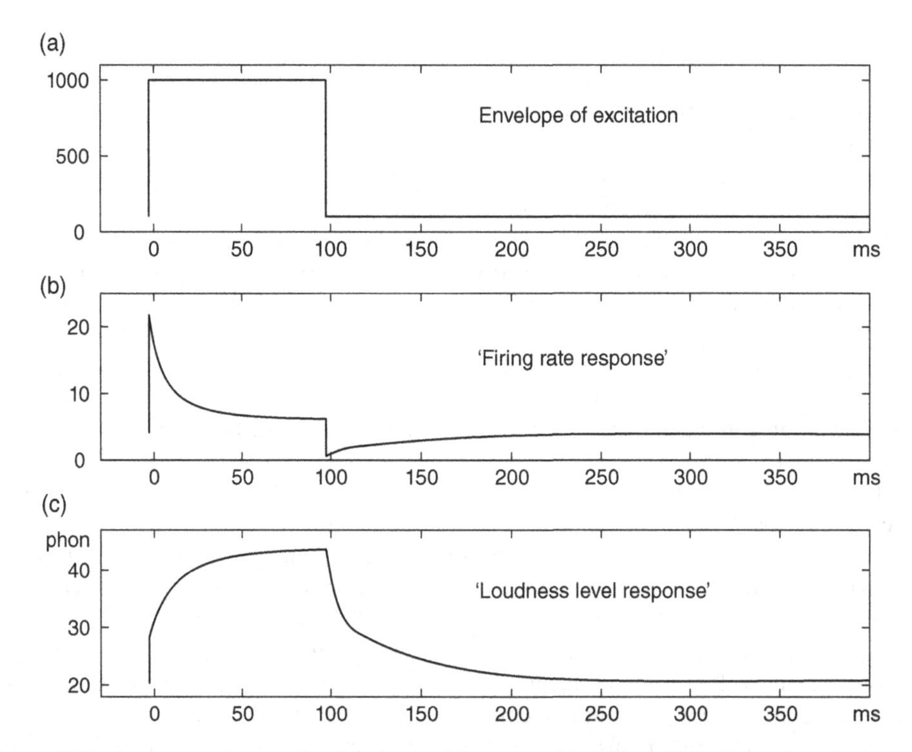

Figure 13.8 Responses computed using the model presented in Figure 13.7: (a) tonal excitation with stepped envelope; (b) fast response (onset response), and (c) slow response (loudness response).

13.3 Cochlear Models

So far, this chapter has presented auditory models that explain the functionalities of hearing, giving less emphasis to the physiological details. In some cases, more accurate modelling of the physiological characteristics is necessary for detailed investigations of the auditory system. This demand has been addressed in several models, many of them aiming to simulate the movement of the basilar membrane inside the cochlea.

13.3.1 Basilar Membrane Models

We saw earlier that the vibration of the stapes that is attached to the oval window generates pressure waves in the fluid inside the cochlea. Since these waves resonate at frequency-specific positions along the basilar membrane, accurate modelling of this phenomenon requires a model consisting of spatially distributed elements. Typically, one-dimensional (1D) travelling wave models are used, but 2D and 3D models may also be used at the expense of increased computational complexity. One option is, for instance, to use a *finite element method* (FEM) to simulate the phenomenon in the frequency domain, but this provides only a rough approximation, assuming the underlying system to be linear and time-invariant. Alternatively, a non-linear time-domain solution may be obtained with a *finite difference method*. However, most of the basilar membrane models see the membrane as a transmission line that can be modelled with electrical equivalent circuits.

Specifically, a transmission-line model represents the basilar membrane as a cascade of coupled mass–spring–damper systems. In the equivalent circuit illustrated in Figure 13.9

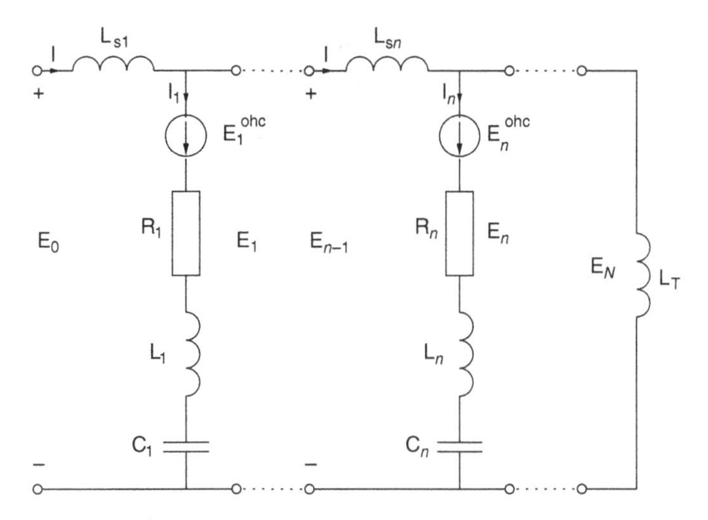

Figure 13.9 Equivalent circuit of the basilar membrane as a transmission line.

(Strube, 1985), the mass and damping characteristics along the basilar membrane are represented by inductors and resistors, respectively, whereas a capacitor is used to represent the energy storage capabilities, such as those of a spring. Using this approach, the vibrations in the membrane are simulated as longitudinal-wave propagation that resonates at a frequency-specific point, abating quickly thereafter. In a digital simulation, the analogue circuit is discretized using so-called *wave-digital filters*, or alternatively numerical methods are used to solve the system of ordinary differential equations that describe the model (Diependaal *et al.*, 1987; Elliott *et al.*, 2007).

Even though the circuit shown in Figure 13.9 is linear and time-invariant, the active role of the cochlear amplifier may be simulated by including negative damping elements (Zweig, 1991) and other non-linear and level-dependent elements (Shera, 2001). For instance, the detectability of amplitude modulation (Figure 10.14) cannot be modelled without accounting for the level dependency of the tuning curves.

13.3.2 Hair-Cell Models

As seen in Section 7.4.1, the bending of inner hair cell stereocilia due to cochlear vibrations modulates the potential difference across the membrane of the cell. This variation in the potentials drives non-deterministically the firing rates of the auditory nerve fibres synaptically connected to the hair cell. The functionality of an auditory fibre has a stochastic nature; one cannot accurately predict when the fibre fires. Therefore, the signal from a single auditory fibre contains somewhat noisy data, and the outputs of large numbers of fibres need to be combined to form the pure and clean sensation that a normal functional auditory system produces. In simple functional models, the combined functionality of inner hair cells and nerve fibres can be emulated deterministically with half-wave rectification and low-pass filtering. However, the stochastic nature of the nerve-fibre firing may be simulated more accurately with a probabilistic model of the inner hair cell and auditory nerve complex. Figure 13.10 depicts the working principle of the model by Meddis (1988), which is the most famous of these models.

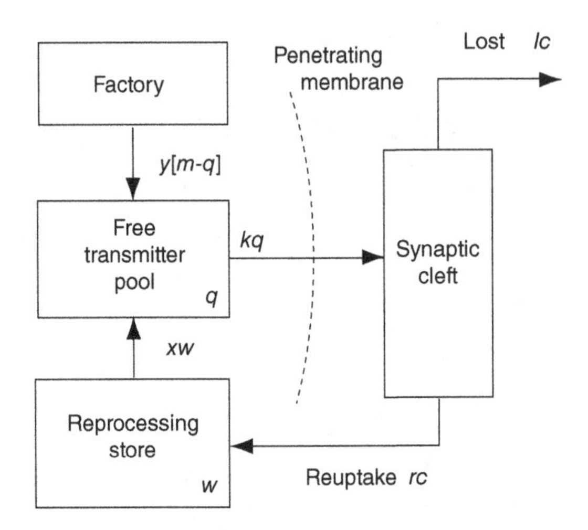

Figure 13.10 Flowchart of the inner-hair-cell model adapted from Meddis (1988).

The left-most blocks in the figure represent an inner hair cell that is synaptically connected to the auditory nerve fibre. The model assumes that the probability of a neural impulse $p(t)$ is linearly dependent on the amount of transmittal material $c(t)$ in the *synaptic cleft* between the hair cell and the nerve fibre:

$$p(t) = h\, c(t)\, \mathrm{d}t, \tag{13.3}$$

where h is a parameter of the model and $\mathrm{d}t$ corresponds to the computational sampling rate. Furthermore, the release of the transmittal material is designed to depend on the *permeability* of the hair-cell membrane $k(t)$, which is modulated by the amplitude of the excitatory stimulus $x(t)$:

$$k(t) = g\, \mathrm{d}t\, \frac{x(t) + A}{x(t) + A + B}, \quad \text{when } x(t) - A > 0, \text{ and} \tag{13.4a}$$

$$k(t) = 0, \quad \text{when } x(t) - A < 0. \tag{13.4b}$$

Here, g, A, and B are parameters of the model. Hence, the amount of released transmittal material at a time instant t corresponds to $k(t)\, q(t)\, \mathrm{d}t$, where $q(t)$ denotes the amount of transmittal material in the transmitter pool next to the membrane. The majority of the transmittal material returns from the cleft to the hair cell at the rate $r\, c(t)$, while some of the material is lost in the cleft and from the system with the speed $l\, c(t)$. This loss introduces an adaptation to the firing rate. Moreover, the returned transmittal material first spends some time in the reprocessing store before entering the transmitter pool again. The speed of transmittal material entering the pool corresponds to $x\, w(t)$, where $w(t)$ denotes the amount of transmittal material in the store. Additionally, the hair cell contains a factory producing the transmittal material at the speed $y\{m - q(t)\}$, where $m = 1$.

In practice, the functionality of the model can be characterized with three differential equations:

$$dq/dt = y\{m - q(t)\} + x\,w(t) - k(t)\,q(t) \tag{13.5}$$
$$dc/dt = k(t)\,q(t) + l\,c(t) - r\,c(t) \tag{13.6}$$
$$dw/dt = r\,c(t) - x\,w(t). \tag{13.7}$$

The model has been shown to yield accurate simulations of physiological responses.

13.4 Modelling of Higher-Level Systemic Properties

The above-mentioned auditory models are related to a relatively low level of neural processing. It is useful and necessary to simulate the functionality of hearing at higher levels to understand the functionality of the auditory system in detail. The functional models may either focus on a specific phenomenon or aim to describe the bigger picture of information processing. Unfortunately, there is a lack of precise knowledge and experimental data regarding phenomena requiring higher-level cognitive processing, and the models are based on high-level assumptions of neurophysiology and subsequent testing of the models against psychoacoustic data.

The following parts of this section describe a few functional models for higher-level processing. Some of them also have a limited physiological basis, but, in general, they are hypothetical models.

13.4.1 Analysis of Pitch and Periodicity

The existence of the two alternative theories for pitch perception is also reflected in the auditory models, that are based on either spectral (place theory) or periodicity (temporal theory) analysis (Plack *et al.*, 2005). Spectral analysis of pitch assumes that the auditory system can extract frequency information with a high resolution which is then analysed at the neural level by a central processor. Alternatively, the time-domain models bolster the idea that several pitch perception phenomena can be explained with simple, low-level time-domain processing based on periodicity. The models are generally based on the idea that the auditory system extracts pitch with neural processing resembling the computation of an autocorrelation function (Licklider, 1951, 1959; Meddis and Hewitt, 1991, 1992; Meddis and O'Mard, 1997).

Figure 13.11 shows the general concept for autocorrelation-based pitch analysis. The signals originating from the filter bank are first processed by a hair-cell model consisting of a half-wave rectifier and a subsequent low-pass filter. A separate autocorrelation function (ACF) is then computed for each signal in order to detect the periodicities in the different sub-bands. The ACF presentation is often called a *correlogram*. Thereafter, the separate ACFs are summed to obtain the *summary autocorrelation function*, characterizing the periodicities in the original stimulus.

The Matlab code in Section 13.6.1 produces the outputs plotted in Figure 13.12, demonstrating the effects of the different processing steps. Figure 13.12a shows the waveform of the vowel signal, and the outputs of the filter bank are shown in Figure 13.12b. The latter plot is often referred to as a *cochleogram*. In addition, the signals following the hair-cell processing are shown in Figure 13.12c, while the correlogram and the summary autocorrelation function are illustrated in Figure 13.12d. The last of the graphs shows a clear peak at 9 ms that corresponds to the fundamental frequency of 110 Hz of the vowel input.

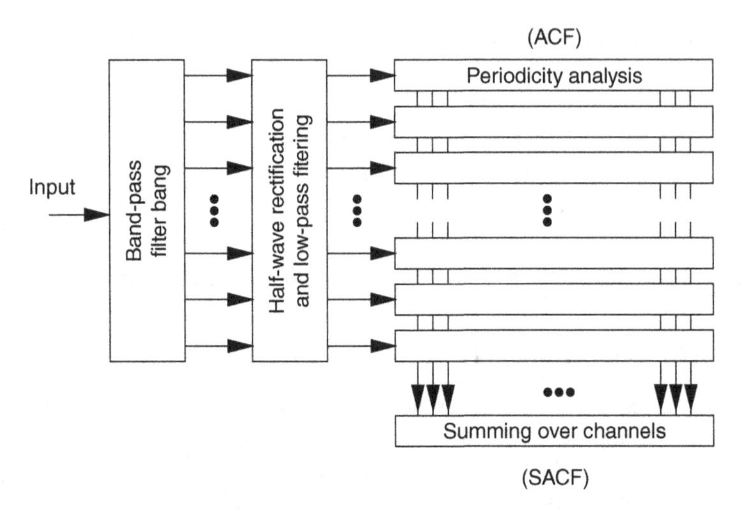

Figure 13.11 The general concept of pitch analysis with an autocorrelation-based auditory model. An autocorrelation function (ACF) is computed in the periodicity analysis, and the results are summed to form the summary autocorrelation function (SACF).

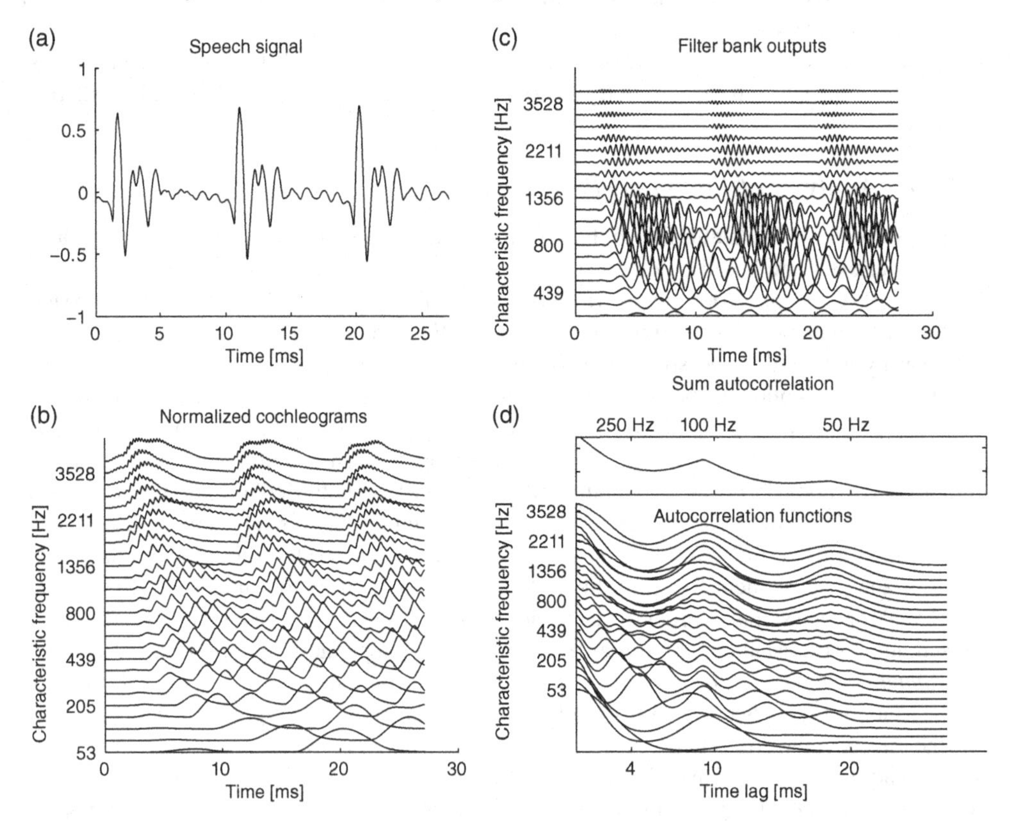

Figure 13.12 Auditory pitch analysis for a 27-ms segment of the vowel /a/: (a) the time-domain signal, (b) the cochleogram, (c) the hair-cell model output with the highest peaks normalized to unity, and (d) the normalized autocorrelation functions and the summary autocorrelation function.

Autocorrelation-based analysis has proven to be able to explain several phenomena of pitch perception (Plack *et al.*, 2005). In addition, such models can be applied to segregate different sound sources, particularly to segregate concurrent vowel sounds from each other when the sounds differ in terms of the fundamental frequency (Meddis and Hewitt, 1992).

13.4.2 Modelling of Loudness Perception

Section 10.2 introduced a simple loudness model that accurately estimates the loudness perception evoked by many relatively simple signals. Such models have been found to be unable to derive accurate loudness estimates for spectrally complex and time-variant signals. This shortcoming has motivated the design of more advanced loudness models (Florentine *et al.*, 2005), of which two approaches are briefly touched upon next.

The first one (Zwicker, 1977) processes the signal in the time domain and produces a continuous signal representing the loudness as a function of time. This approach opens up the possibility of predicting the loudness perception evoked by a time-variant signal based on the peak values in the model output. The second model (Glasberg and Moore, 2002) divides the signal into overlapping time frames and extracts short-term loudness values from each time frame. The model also emulates the temporal integration of loudness between adjacent time frames, and consequently, the model provides accurate estimates for time-variant signals as well, although this estimate is derived by simply averaging across the short-term loudness values. Despite the improved accuracy in predicting perceived loudness of complex signals, the different models cannot yet fully explain loudness perception. The objective audio and speech quality methods discussed in Sections 17.5.2 and 17.8.1 can also be seen as models that estimate the specific loudness depending on time, and the interested reader might find the references in those sections worth exploring.

13.5 Models of Spatial Hearing

As discussed in Chapter 12, human spatial hearing capabilities are based on the binaural and monaural analysis of the ear canal signals. Spatial hearing is able to localize sources with good accuracy, although the reflections and reverberations of the room may corrupt the directional cues in the ear canal signals. This remarkable ability has, for decades, inspired researchers to model spatial hearing. A plethora of binaural and monaural models of spatial hearing have been proposed (Blauert, 1996, 2013; Colburn, 1996; Stern and Trahiotis, 1995), and some of them are discussed next.

13.5.1 Delay-Network-Based Models of Binaural Hearing

The majority of the binaural processing algorithms are based on the coincidence detection model proposed by Jeffress (1948). The model suggests that certain neurons in the brain are narrowly tuned to specific ITDs between the ear canal signals. As illustrated in Figure 13.13, the model consists of an array of coincidence-detector neurons receiving excitatory signals from both ears, and delay lines are used to represent axons connecting the neuron to the cochlear nuclei of the left and right ears. The highest activity is then received from the coincidence-detector neuron where the propagation delay in the inputs effectively cancels out the ITD between the left and right ear inputs. This probably also facilitates channels sensitive to specific ITDs, as suggested by (Fastl and Zwicker, 2007).

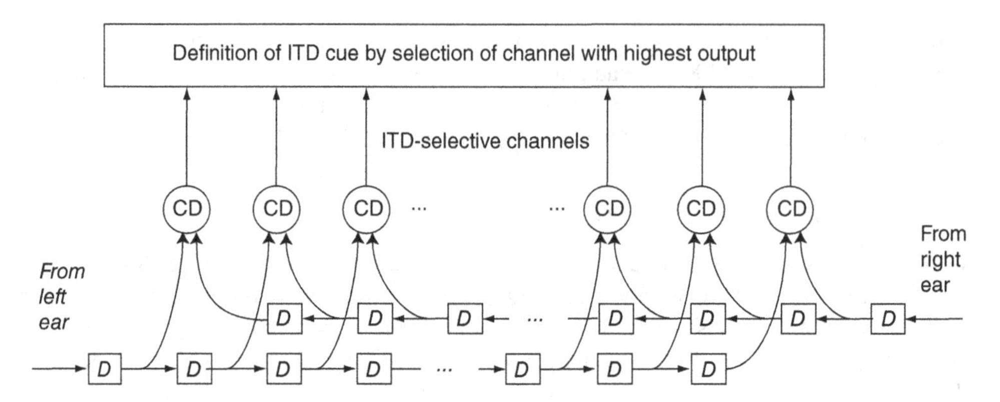

Figure 13.13 A schematic illustration of the coincidence detection model proposed by Jeffress (1948), where delay lines represent axons connecting the ear canal inputs to the coincidence-detector neuron (CD). Here, *D* is the unit delay. The outputs of the CD-neurons are then thought to be compared with each other, and the highest outputs are thought to define the ITD cue(s).

Such processing can be elegantly emulated by computing the normalized interaural cross-correlation (IACC) (Sayers and Cherry, 1957)

$$\gamma(t,\tau) = \frac{\int_t^{t+\Delta t} x_l(T-\tau/2)x_r(T+\tau/2)\,\mathrm{d}T}{\sqrt{\int_t^{t+\Delta t} x_l^2(T)\,\mathrm{d}T + \int_t^{t+\Delta t} x_r^2(T)\,\mathrm{d}T}}, \tag{13.8}$$

where t is time, τ is the interaural delay, Δt denotes the length of the integration window, and x_l and x_r are the signals from the left and right ears, respectively. An estimate of the ITD is then obtained as the interaural delay corresponding to the maximum of the IACC function. The output of the IACC computation may also be used to visualize the auditory scene as a cross-correlogram-type binaural activity map (Shackleton *et al.*, 1992). The Matlab script listed in Section 13.6.2 demonstrates how such a map can be extracted for a binaural input signal.

We will next discuss the output of the IACC computation, which is shown in Figure 13.14. Figure 13.12c shows the signals used as input in the computation, to make this discussion more comprehensible. The normalized cross-correlation function is plotted for each frequency channel. The plotting shows that the functions are tuned to a certain time lag, which corresponds to the ITD between the ear canal signals. In this case, the source was in the direction 30° azimuth, and the maximum value of the IACC depends relatively strongly on the frequency content in the interval between 0.2 ms and 0.5 ms. The value from the theoretical broadband curve shown in Figure 12.9 matches with the IACC-based estimate at 700 Hz. The time lags corresponding to the maxima of the IACC functions are plotted in the lower panel in Figure 13.14. This is the ITD function often seen in the literature, which is thought to represent the ITD cue accessible to the higher levels in processing (Blauert, 2013).

The IACC function is normalized with the power of signals, which means that it attains the value one only when the ear canal signals differ by the amount of the ITD and the ILD in the natural range, otherwise it gets a lower, non-negative value. The maximum value can thus be used to estimate the *interaural coherence* (Faller and Merimaa, 2004).

The IACC humps are broader at low frequencies than at high frequencies. This is because a constant change in time lag corresponds to a smaller change in phase at low frequencies, and

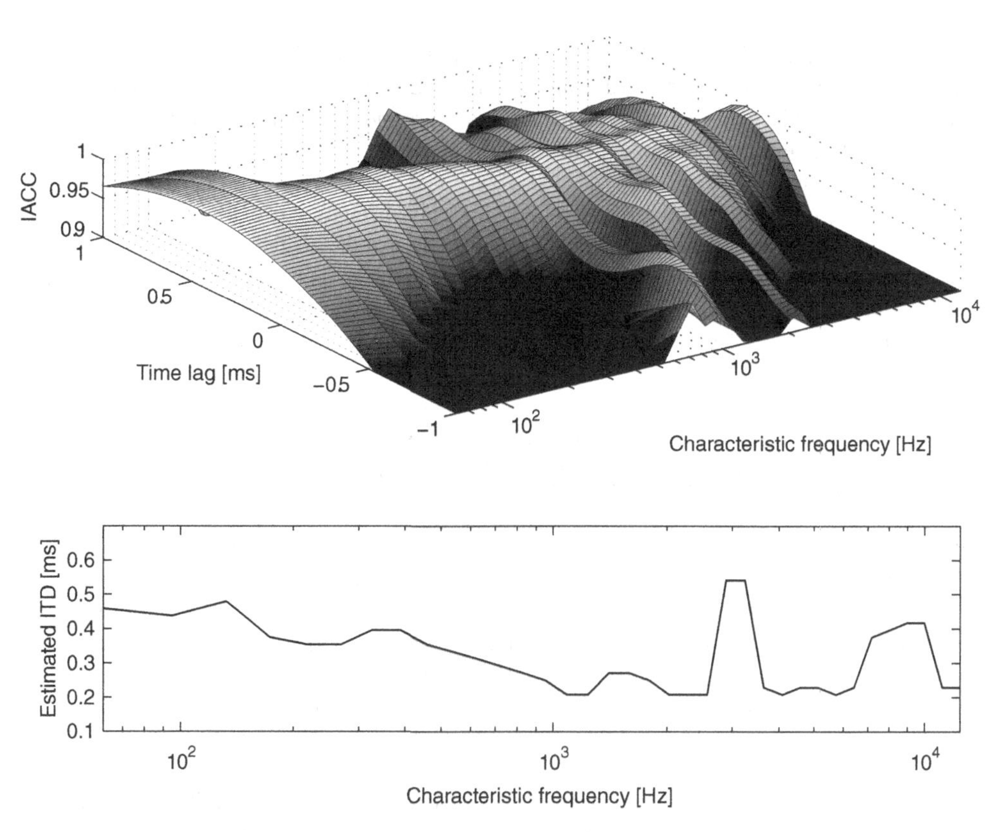

Figure 13.14 A cross-correlogram-type binaural activity map for a scenario where a single speech source at the azimuthal angle of 30° is simulated with HRTFs measured from a real subject. The bottom-most graph shows the estimated ITD in different frequency bands.

a larger change at higher frequencies. Near 1 kHz, sidelobes are seen in the functions, which occur because the period of the centre frequency of auditory bands is shorter than 1 ms, and the IACC analysis also finds high correlation in the signals when the time lags equal the delay or the phase advances by 2π. The effect of frequency on temporal spacing of peaks is clearly shown in Figure 13.12c.

At higher frequencies, the IACC no longer shows side bands, since the temporal details of the signals are mostly lost, which is also evident in Figure 13.12c. Instead, only the temporal envelope is preserved, and since, in this case, the signal was a short excerpt of the vowel /a/, a rather strong temporal structure remains. The cross-correlation between the left and right signals then results in only a single hump in the IACC functions. Such a clear unitary hump is not always found in the simulation results. If the input had been, for example, a high-frequency sinusoidal tone, the high-frequency IACC functions would be completely flat as a function of the time lag.

A number of extensions have been proposed to the coincidence detection model, such as the one presented by Breebaart *et al.* (2001). In their model, the delay lines are connected to a chain of attenuators and each coincidence detector of the original model (see Figure 13.13) is replaced by two excitation–inhibition cells, one receiving the excitation from the left ear and inhibition from the right ear, and the other with opposite connections. Effectively, they extend the coincidence detection model to account also for ILD sensitivity. For a binaural input signal, the model outputs an activity map with local minima around the positions corresponding to the ITD and ILD values, and the depths of the troughs depend on the interaural coherence between the ear canal signals.

Such correlation- or coherence-based models have also been used to explain the sensitivity of humans to the coherence between ear canal signals (Bernstein and Trahiotis, 1996). Additionally, in real rooms, when the coherence between ear canal signals varies temporally depending on the source signal and on the room response, it has been suggested that listeners utilize directional cues only when the interaural coherence value is larger than a threshold value (Faller and Merimaa, 2004).

13.5.2 Equalization-Cancellation and ILD Models

The above-mentioned delay-line models analyse the coherence of the ear canal signals to extract directional information. Alternative modelling concepts have also been exploited, some of them being based on subtracting the signals from each other. The equalization-cancellation model (Durlach, 1963) was designed to account for binaural signal detection in the presence of masking noise, and no attempts were made to emulate processing in the auditory pathway. In this model, the left and right inputs are first filtered with a set of band-pass filters so that the narrowband target can be more easily separated from the masker. Thereafter, the masker signal components are equalized in the two ears by adjusting the ITD and ILD values, and the ear canal signals are subtracted from each other, which ideally eliminates the masker from the signal.

A straightforward method to estimate the ILD cue accessible to the auditory system comprises the computation of the level difference between the ear canal signals (Blauert, 1996). Typically, separate ILD estimates are derived for each auditory channel and for each time frame of the signal to maximize accuracy. First, the signal levels in each auditory channel are measured within each time frame, after which the values obtained for the left and right ear signals are compared to each other to derive estimates of the ILD in each time frame and for each auditory channel. Thus, such an ILD estimation can be interpreted as an equalization cancellation model without the equalization phase.

The selection of the length of the time frame opens up possibilities to broaden the model performance for different applications. The output also becomes sensitive to interaural coherence when time frames as short as 5–10 ms are used in the ILD estimation. The ILD values fluctuate randomly in a diffuse sound field (Goupell and Hartmann, 2007; Pulkki and Hirvonen, 2009) and therefore also provide a potential cue for humans to sense spatial attributes related to reverberation.

13.5.3 Count-Comparison Models

Another group of binaural hearing models based on the count-comparison principle (van Bergeijk, 1962; von Békésy, 1930; 1960), which proposes that the nuclei in the two hemispheres encode the spatial direction of sound at the rate of their output (see Figure 13.15), and the spatial location is then indicated by the relative activation rates of the nuclei in the two hemispheres (Stecker *et al.*, 2005). Actually, the above-mentioned ILD models already follow the count-comparison principle, and now it is argued that the ITD is extracted similarly to the ILD. It has also been shown that the ‘comparison’ phase is not needed if the outputs of the

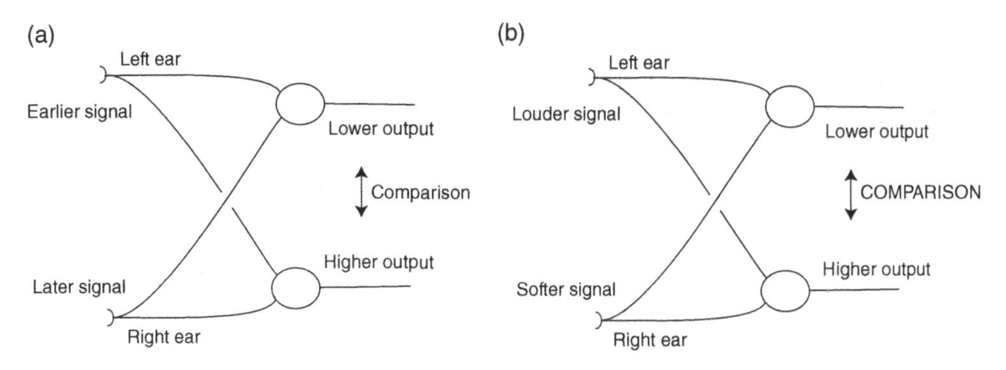

Figure 13.15 The count comparison principle for (a) ITD extraction (b) ILD extraction.

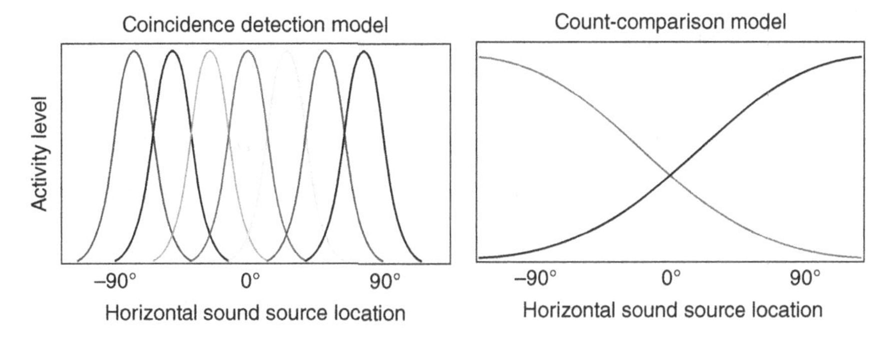

Figure 13.16 The schematic representation of the effect of a horizontal sound source location on the activity levels of different receptive fields in the brain as determined by the coincidence detection and count-comparison models.

nuclei models are self-normalized using their input signals (Pulkki and Hirvonen, 2009). Such an approach yields left- or right-coordinate outputs that depend only on the direction of the sound event and not on the sound pressure level or any other attribute.

One of the principal differences between the count-comparison and coincidence detection models lies in the nature of the output. A count-comparison model outputs a left or right coordinate, whereas each coincidence-detector neuron in a Jeffress-type model provides its own output. These neurons may be thought to be most sensitive to specific left or right directions. In other words, Jeffress-type models assume that receptive fields in the brain are narrowly tuned to specific horizontal directions, while the count-comparison principle bolsters the idea of two wide receptive fields spanning an entire hemifield, as illustrated in Figure 13.16. The left or right coordinate that a count-comparison model outputs cannot be used as such to visualize the surrounding auditory scene as an elegant binaural activity map, as in Figure 13.14. This issue was addressed in a study by Takanen *et al.* (2014), where the binaural cues in the ear canal signals were extracted following the count-comparison principle, and the resulting directional cues were utilized to steer the spectral content of the signals to specific locations on a topographically organized binaural activity map. Thus, the auditory scene is visualized similarly to the cross-correlogram of the Jeffress-type model output.

13.5.4 Models of Localization in the Median Plane

As seen in Section 12.3.4, the localization of a sound source in the median plane relies on spectral cues. Generically, a machine-learning approach can be used to simulate human performance in localizing the elevation of a sound source. In particular, an accurate prediction can be achieved using artificial neural networks trained with large data sets from a single subject (Jin *et al.*, 2000). Recently, a functional model that predicts listeners' performance in such localization tasks has been proposed by Baumgartner *et al.* (2013). This model computes the probability of perceiving a sound coming from a certain elevation angle by comparing the internal representation of the sound as processed by the peripheral auditory system with an internal representation of the HRTF.

13.6 Matlab Examples

This section shows two examples of filter-bank-based auditory models. The models rely on the implementation of a gammatone filter bank in the auditory modelling toolbox (AMT) that can be downloaded from `amtoolbox.sourceforge.net`.

13.6.1 Filter-Bank Model with Autocorrelation-Based Pitch Analysis

This example is explained in Section 13.4, and the plots of its outputs are shown in Figure 13.12.

```
%code for autocorrelation-based pitch analysis in auditory
                                                      models
clc;close all;clear;

%some parameters for the auditory model
fLow = 50;     % the lowest characteristic frequency of the
                                                 filter bank
fHigh = 4000; % and the highest
fCut = 1000;  % cut-off frequency of the low-pass filter

%load a speech file, the example contains 27 ms of /a/ vowel
[sample,fs] = wavread('kaksi.wav',[5900 7200]);

sampleLen = length(sample);
%create of a gammatone filter bank using a command from the
                                                    auditory
%modelling toolbox (http://amtoolbox.sourceforge.net)
[b,a] = gammatone(erbspacebw(fLow,fHigh),fs,'complex');

%processing the signal through the filter bank
filterOut = real(ufilterbankz(b,a,sample));

%emulation of the neural transduction with half-wave
                                       rectification and
```

```
%low-pass filtering of the filter bank output
rectified = filterOut.*(filterOut>0);
%a first-order IIR filter is used as the low-pass filter
beta = exp(-fCut/fs);
outSig = filter(1-beta,[1 -beta],rectified);

for freqInd=1:size(outSig,2) % autocorrelation for each band
    auCorr(:,freqInd)= xcorr(outSig(:,freqInd),'coeff')';
end

%plotting of the figures
auCorr=auCorr((size(auCorr,1)+1)/2:end,:);
lags=[[0:(size(auCorr,1)-1)]/fs*1000];
fcs = erbspacebw(fLow,fHigh);
h=figure;
%plot the input signal
g(1) = subplot('position',[0.13 0.6438 0.3326 0.3012]);hold on;
plot((1:sampleLen)./fs*1000,sample,'k');
xlabel('Time [ms]');title('a) speech signal');
set(gca,'xlim',[0 sampleLen/fs*1000]);

%plot the filter bank outputs
g(2) = subplot('position',[0.5803 0.6438 0.3326 0.3012]);hold
                                                            on;
for freqInd=size(outSig,2):-1:1
    plot((1:sampleLen)./fs*1000,(freqInd-1)/30+filterOut
                                            (:,freqInd),'k');
end
set(gca,'YTick',(0:4:(size(outSig,2)-1))/30);
set(gca,'YTickLabel',round(fcs(1:4:end)));
axis([0 30 0.2 0.9])
xlabel('Time [ms]');ylabel('Characteristic frequency [Hz]');
title('b) filter bank outputs');

%plot the cochleograms
g(3) = subplot('position',[0.13 0.115 0.333 0.3812]);  hold on;
for freqInd=size(outSig,2):-1:1
    plot((1:sampleLen)./fs*1000,(freqInd-1)/2+outSig
            (:,freqInd)/max(outSig(:,freqInd)),'k');
end
set(gca,'YTick',(0:4:(size(outSig,2)-1))/2);
set(gca,'YTickLabel',round(fcs(1:4:end)));
xlabel('Time [ms]');ylabel('Characteristic frequency [Hz]');
title('c) normalized cochleograms');
axis([0 30 0 13.5])

%plot the normalized autocorrelation functions
```

```
g(4) = subplot('position',[0.5803 0.1150 0.3826 0.3012]);hold
                                                              on;
for freqInd=size(outSig,2):-1:1
    plot(lags,(freqInd-1)*0.12+auCorr(:,freqInd),'k');
end

axis([0 30 0 4])
set(gca,'YTick',(0:4:(size(outSig,2)-1))*0.12+1);
set(gca,'YTickLabel',round(fcs(1:4:end)));
set(gca,'xTick',floor(1./[250 100 50]*1000));
ylabel('Characteristic frequency [Hz]');
xlabel('Time lag [ms]');
text(15,3.8,'d) autocorrelation functions',
          'HorizontalAlignment','center');

%plot the sum autocorrelation
g(5) = subplot('position',[0.5803 0.427 0.3826 0.07]);
plot(lags,sum(auCorr'),'k');
set(gca,'xaxisLocation','top')
set(gca,'YTickLabel',[]);
axis([0 30 0 25])
set(gca,'xTick',floor(1./[250 100 50]*1000));
set(gca,'xTickLabel',{'250 Hz','100 Hz','50 Hz'});
title('Sum autocorrelation');
```

13.6.2 Binaural Filter-Bank Model with Cross-Correlation-Based ITD Analysis

This example is explained in Section 13.5.1, and the plots of its outputs are shown in Figure 13.14.

```
%function for IACC-based binaural activity map computation

clear;clc;close all; fs=48000;
% the data files can be obtained from the web siteof this book
load hrir30.mat % a HRTF

%some parameters for auditory model
fLow = 50;%the lowest characteristic frequency of the filter
                                                          bank
fHigh = 13000;%and the highest
fCut = 1000; % cut-off frequency of the low-pass filter
maxLag= floor(0.001*fs); % IACC-values are computed within
                                                    -1...1 ms

%a 29-ms-long sample of speech (vowel /a/) is used as the
                                              source material
```

```
[stim] = wavread('kaksi.wav',[5000 6400]);

% convolve stimulus with HRIRs of left and right ear
insig = [conv(stim,hrir30(:,1)) conv(stim,hrir30(:,2))];

%create of a gammatone filter bank using a command from the
                                                  auditory
%modelling toolbox (http://amtoolbox.sourceforge.net)

cfs = erbspacebw(fLow,fHigh);%characteristic frequencies of
                                             the filter bank
[b,a] = gammatone(cfs,fs,'complex');

%processing the signal through the filter bank
filterOut.left = 2*real(ufilterbankz(b,a,insig(:,1)));
filterOut.right = 2*real(ufilterbankz(b,a,insig(:,2)));

%emulation of the neural transduction with half-wave
                                        rectification and
%low-pass filtering of the filter bank output
rectified.left = filterOut.left.*(filterOut.left>0);
rectified.right = filterOut.right.*(filterOut.right>0);

%a first-order IIR filter is used as the low-pass filter
beta = exp(-fCut/fs);
outSig.left = filter(1-beta,[1 -beta],rectified.left);
outSig.right = filter(1-beta,[1 -beta],rectified.right);

%compute interaural cross-correlation at each frequency band
iaccFuncts = zeros(2*maxLag+1,length(cfs));
lagValues = (-maxLag:maxLag)./fs;
for freqInd=1:length(cfs)
    iaccFuncts(:,freqInd) = xcorr(outSig.left(:,freqInd),...
        outSig.right(:,freqInd),maxLag,'coeff');
end

%compute the ITD estimate at different frequency bands based
                                                on the maxima
%of the IACC functions
[temp,lag] = max(iaccFuncts);
itdEst = lagValues(lag);

%plotting of figures
h=figure;
%for visualization purposes, only the IACC-values above 0.9
                                                  are plotted
iaccFuncts(iaccFuncts<=0.9)=0.9;
```

```
g(1) = subplot('Position',[0.1300 0.4838 0.7750 0.4412]);
surf(round(cfs),lagValues*1000,iaccFuncts);
set(gca,'XScale', 'log','xTick',[100 1000 10000]);
xlabel('Characteristic frequency [Hz]');ylabel('Time lag
                                                    [ms]');
zlabel('IACC');
xlim(round([min(cfs) max(cfs)]));
view(-35,70);colormap(colormap('gray'));
g(2) = subplot('Position',[0.1300 0.1100 0.7750 0.2412]);
semilogx(round(cfs),itdEst*1000,'k');ylabel('Estimated ITD
                                                      [ms]');
xlabel('Characteristic frequency [Hz]');
axis([round([min(cfs) max(cfs)]) 0.1 0.7])
```

Summary

This chapter reviewed different computational models of the auditory system, covering a wide range of modelling principles that aim to emulate the processing occurring within different regions of the auditory pathway in varying detail. The models based on windowing the ear canal signals and performing DFT-based processing can be used to explain the basic properties of auditory frequency resolution. Although DFT-based modelling has some drawbacks in temporal accuracy, it is interesting in the scope of this book, as many perceptual audio-coding methods are based on similar processing. The auditory models that model the cochlea using filter banks are more precise, and the best accuracy of peripheral modelling is obtained with transmission-line models. Unfortunately, the computational complexity increases drastically when the accuracy of modelling is increased. The models for binaural interaction are based either on sound being encoded into multiple direction-dependent channels, or on directional cue computation for each time–frequency position. The models for pitch and loudness perception succeed in some simple scenarios, however, none of the models can explain perception accurately in all cases.

Further Reading

The main directions of research are covered in the books by Blauert (2013) and Meddis (2010). Many models are also available publicly as part of computational toolboxes, like Majdak and Søndergaard (2013), and Slaney (1998). The auditory models have also found applications in speech and audio techniques, and the remaining chapters of this book review their use in many applications.

References

Baumgartner, R., Majdak, P., and Laback, B. (2013) Assessment of sagittal-plane sound localization performance in spatial-audio applications. In Blauert J. (ed.) *The Technology of Binaural Listening*. Springer, pp. 93–119.

Bernstein, L.R. and Trahiotis, C. (1996) The normalized correlation: Accounting for binaural detection across center frequency. *J. Acoust. Soc. Am.*, **100**(6), 3774–3784.

Blauert, J. (1997) *Spatial Hearing – Psychophysics of Human Sound Localization*. MIT Press.

Blauert, J. (2013) *The Technology of Binaural Listening*. Springer.

Breebaart, J., van de Par, S., and Kohlrausch, A. (2001) Binaural processing model based on contralateral inhibition. I. Model structure. *J. Acoust. Soc. Am.*, **110**(2), 1074–1088.

Carney, L. (1993) A model for the responses of low-frequency auditory-nerve fibers in cat. *J. Acoust. Soc. Am.*, **93**(1), 401–417.

Chistovich, I.A., Granstrem, M.P., Kozhevnikov, V.A., Lesogor, L.W., Shupljakov, V.S., Taljasin, P.A., and Tjulkov, W.A. (1974) A functional model of signal processing in the peripheral auditory system. *Acustica*, **31**(6), 349–354.

Colburn, H.S. (1996) Computational models of binaural processing. In Hawkins, H.L., McMullen, T.A., Popper, A.N., and Fay, R.R. (eds) *Auditory Computation*. Springer, pp. 332–400.

Dau, T., Püschel, D., and Kohlraush, A. (1996) A quantitative model of the "effective" signal processing in the auditory system. I. model structure, II. simulations and measurements. *J. Acoust. Soc. Am.*, **99**, 3615–3631.

Davis, S. and Mermelstein, P. (1980) Comparison of parametric representations for monosyllabic word recognition in continuously spoken sentences. *IEEE Trans. ASSP*, **ASSP-28**, 357–366.

De Boer, E. (1969) Encoding of frequency information in the discharge pattern of auditory nerve fibers. *Int. J. Audiol.*, **8**(4), 547–556.

Diependaal, R.J., Duifhuis, H., Hoogstraten, H., and Viergever, M.A. (1987) Numerical methods for solving one-dimensional cochlear models in the time domain. *J. Acoust. Soc. Am.*, **82**(5), 1655–1666.

Dolmazon, J.M., Bastet, L., and Shupljakov, V.S. (1976) A functional model of the peripheral auditory system in speech processing. *Proc. of IEEE ICASSP'77*, pp. 261–264.

Durlach, N.I. (1963) Equalization and cancellation theory of binaural masking-level differences. *J. Acoust. Soc. Am.*, **35**(8), 1206–1218.

Elliott, S.J., Ku, E.M., and Lineton, B. (2007) A state space model for cochlear mechanics. *J. Acoust. Soc. Am.*, **122**(5), 2759–2771.

Faller, C. and Merimaa, J. (2004) Source localization in complex listening situations: Selection of binaural cues based on interaural coherence. *J. Acoust. Soc. Am.*, **116**(5), 3075–3089.

Fastl, H. and Zwicker, E. (2007) *Psychoacoustics – Facts and Models*. Springer.

Florentine, M., Popper, A., and Fay, R.R. (eds) (2005) *Loudness*, volume 37. Springer.

Glasberg, B.R. and Moore, B.C.J. (1990) Derivation of auditory filter shapes from notched-noise data. *Hear. Res.*, **47**(1–2), 103–138.

Glasberg, B.R. and Moore, B.C.J. (2002) A model of loudness applicable to time-varying sounds. *J. Audio Eng. Soc.*, **50**(5), 331–342.

Goupell, M.J. and Hartmann, W.M. (2007) Interaural fluctuations and the detection of interaural incoherence. III. Narrowband experiments and binaural models. *J. Acoust. Soc. Am.*, **122**, 1029–1045.

Hermansky, H. (1990) Perceptual linear predictive (PLP) analysis of speech. *J. Acoust. Soc. Am.*, **87**(4), 1738–1752.

Irino, T. and Patterson, R. (1997) A time-domain, level-dependent auditory filter: The gammachirp. *J. Audio Eng. Soc. Am.*, **101**(1), 412–419.

Jeffress, L.A. (1948) A place theory of sound localization. *J. Comp. Physiol. Psychol.*, **41**(1), 35–39.

Jin, C., Schenkel, M., and Carlile, S. (2000) Neural system identification model of human sound localization. *J. Acoust. Soc. Am.*, **108**(3), 1215–1235.

Karjalainen, M. (1996) A binaural auditory model for sound quality measurements and spatial hearing studies. *Proc. of IEEE ICASSP'96*, pp. 985–988.

Licklider, J. (1951) A duplex theory of pitch perception. *Experientia*, **7**, 128–133.

Licklider, J.C.R. (1959) *Three Auditory Theories*. McGraw-Hill, pp. 41–144.

Lopez-Poveda, E., Fay, R.R., and Popper, A.N. (2010) *Computational Models of the Auditory System*, volume 35. Springer.

Lyon, R.F. (1982) A computational model of filtering, detection and compression in the cochlea. *Proc. of IEEE ICASSP'82*, pp. 1282–1285.

Majdak, P. and Søndergaard, P. (2013) The auditory modeling toolbox `http://amtoolbox.sourceforge.net`.

Meddis, R. (1988) Simulation of auditory neural transduction: Further studies. *J. Acoust. Soc. Am.*, **83**, 1056–1063.

Meddis, R. and Hewitt, M.J. (1991) Virtual pitch and phase sensitivity of a computer model of the auditory periphery. I: Pitch identification. *J. Acoust. Soc. Am.*, **89**(6), 2866–2882.

Meddis, R. and Hewitt, M.J. (1992) Modelling the identification of concurrent vowels with different fundamental frequencies. *J. Acoust. Soc. Am.*, **91**(1), 233–245.

Meddis, R. and O'Mard, L. (1997) A unitary model of pitch perception. *J. Acoust. Soc. Am.*, **102**(3), 1811–1820.

Meddis, R., O'Mard, L.P., and Lopez-Poveda, E.A. (2001) A computational algorithm for computing nonlinear auditory frequency selectivity. *J. Acoust. Soc. Am.*, **109**(6), 2852–2861.

Patterson, R.D. (1994) The sound of a sinusoid: Spectral models. *J. Acoust. Soc. Am.*, **96**(3), 1409–1418.

Patterson, R.D., Unoki, M., and Irino, T. (2003) Extending the domain of center frequencies for the compressive gammachirp auditory filter. *J. Acoust. Soc. Am.*, **114**(3), 1529–1542.

Plack, C.J., Oxenham, A.J., Fay, R.R., and Popper, A.N. (2005) *Pitch: Neural Coding and Perception*, volume 24. Springer.

Pulkki, V. and Hirvonen, T. (2009) Functional count-comparison model for binaural decoding. *Acta Acustica United with Acustica*, **95**, 883–900.

Sayers, B.M. and Cherry, E.C. (1957) Mechanism of binaural fusion in the hearing of speech. *J. Acoust. Soc. Am.*, **29**(9), 973–987.

Seneff, S. (1990) A joint synchrony/mean-rate model of auditory speech processing. In Waibel A. and Lee K-F, Readings in Speech recognition. Morgan-Kaufmann. pp. 101–113.

Shackleton, T.M., Meddis, R., and Hewitt, M.J. (1992) Across frequency integration in a model of lateralization. *J. Acoust. Soc. Am.*, **91**(4), 2276–2279.

Shera, C.A. (2001) Intensity-invariance of fine time structure in basilar-membrane click responses: Implications for cochlear mechanics. *J. Acoust. Soc. Am.*, **110**(1), 332–348.

Slaney, M. (1998) Auditory toolbox `https://engineering.purdue.edu/~malcolm/interval/1998-010/`.

Stecker, G.C., Harrington, I.A., and Middlebrooks, J.C. (2005) Location coding by opponent neural populations in the auditory cortex. *PLoS Biol*, **3**(3), 520–528.

Stern, R.M. and Trahiotis, C. (1995) Models of binaural interaction. *Handbook of Perception and Cognition*, **6**, 347–386.

Strube, H.W. (1985) A computationally efficient basilar-membrane model. *Acustica*, **58**(4), 207–214.

Takanen, M., Santala, O., and Pulkki, V. (2014) Visualization of functional count-comparison-based binaural auditory model output. *Hear. Res.*, **309**, 147–163.

van Bergeijk, W.A. (1962) Variation on a theme of Békésy: A Model of Binaural Interaction. *J. Acoust. Soc. Am.*, **34**(8), 1431–1437.

von Békésy, G. (1930) Zur theorie des Hörens. Über das Richtungshören bei einer Zeitdifferenz oder Lautstärkeungleighheit der beiderseitigen Schalleinwirkungen. *Physik. Zeitschr.* pp. 824–835, 857–868.

von Békésy, G. (1960) *Experiments in Hearing*. McGraw-Hill and Acoustical Society of America.

Weiss, T.F. (1966) A model of the peripheral auditory system. *Kybernetik*, **3**(4), 153–175.

Zhang, X., Heinz, M.G., Bruce, I.C., and Carney, L.H. (2001) A phenomenological model for the responses of auditory-nerve fibers: I. Nonlinear tuning with compression and suppression. *J. Acoust. Soc. Am.*, **109**(2), 648–670.

Zilany, M.S. and Bruce, I.C. (2006) Modelling auditory-nerve responses for high sound pressure levels in the normal and impaired auditory periphery. *J. Acoust. Soc. Am.*, **120**(3), 1446–1466.

Zweig, G. (1991) Finding the impedance of the organ of Corti. *J. Acoust. Soc. Am.*, **89**(3), 1229–1254.

Zwicker, E. (1977) Procedure for calculating loudness of temporally variable sounds. *J. Acoust. Soc. Am.*, **62**(3), 675–682.

14

Sound Reproduction

Sound reproduction denotes the process of recording, processing, storing, and recreating sound, such as speech, music, or other sounds. When recording an acoustic scenario, one or more microphones are used to capture sound in single or multiple positions for a recording device. When recording electronical or digital sound sources, microphones are not necessarily required, since the recording devices can directly store the electrical or digital signals. The signals may be processed and stored, and finally made audible to a human listener with loudspeakers or headphones. Note that during this process many choices must be made in order to capture, process, and play back the sound. However, the unifying factor is the common endpoint of the chain, the human listener. This is a distinctive property of the field of communication acoustics compared to some other fields of acoustics. In communication acoustics, the sound is the desired signal, which brings some information or added value to the listener.

Historically, reproduction of sound has come a long way from the first phonograph built by Thomas Edison in 1877. A large variety of sound reproduction applications is currently in use, and they differ both in implementation and the purpose for which they were developed. We begin this chapter by discussing the needs and challenges faced in sound reproduction and continue to present various solutions and technologies for this purpose.

14.1 Need for Sound Reproduction

A wide variety of applications in which sound needs to be reproduced such as:

- *Public address* – the sound amplification systems that make speech or sound audible to a large audience indoors or outdoors.
- *Full-duplex speech communication* over technical channels such as in telephone or teleconference systems.
- Audio content production for the *music and cinema industries.*
- *Broadcasting* of sound in radio or of audiovisual content in TV.

Communication Acoustics: An Introduction to Speech, Audio, and Psychoacoustics, First Edition.
Ville Pulkki and Matti Karjalainen.

- *Computer games and virtual reality*, where sound is captured to be reproduced in association with virtual objects.
- *Accurate reproduction of sound*, where as authentic a replica of the perception of the original sound event as possible is the goal for technical or scientific purposes.
- *Enhancement of acoustics and active noise cancellation.* Sound is captured with a microphone and immediately reproduced after possible processing, resulting in a change in the properties of the sound field. For example, the reverberation properties of a room can be changed, or the noise level can be attenuated.
- In *aided hearing*, devices and processes are used to make sound more audible for the hearing-impaired subject, usually with the aim of making speech more intelligible.

The applications are very different, and the required technical specifications for acceptable functioning may also be very different. Starting from the most basic cases, in certain environmental monitoring systems where the goal is simply to be aware of some sound events, very large deviations in the magnitude and the phase spectrum of the reproduced sound may be allowed. When the requirements are more demanding, the acoustic attributes of the original sound event have to be transmitted with higher fidelity to the listener. For example, with minimum requirements in speech communication, speech is barely understandable without the delivery of prosodic or personal features. In general, the better the original sound signals are preserved in sound reproduction, the higher the quality of reproduction obtained.

In some special cases, full authenticity is sought in reproduction of sound. In communication acoustics, authenticity is determined with human perception as the reference; for authentic reproduction, the auditory perception of a sound scenario in the reproduced conditions should be identical to those in the original conditions, as illustrated in Figure 14.1. Some example cases where authentic reproduction is essential are the reproduction of natural sound scenarios for hearing aid testing and audiology, different training tasks, telepresence applications, and the evaluation of room acoustic parameters of different venues.

The terms *immersion* and *immersive* are often used in connection with sound reproduction, especially in the context of computer games (Björk and Holopainen, 2005), to mean that the

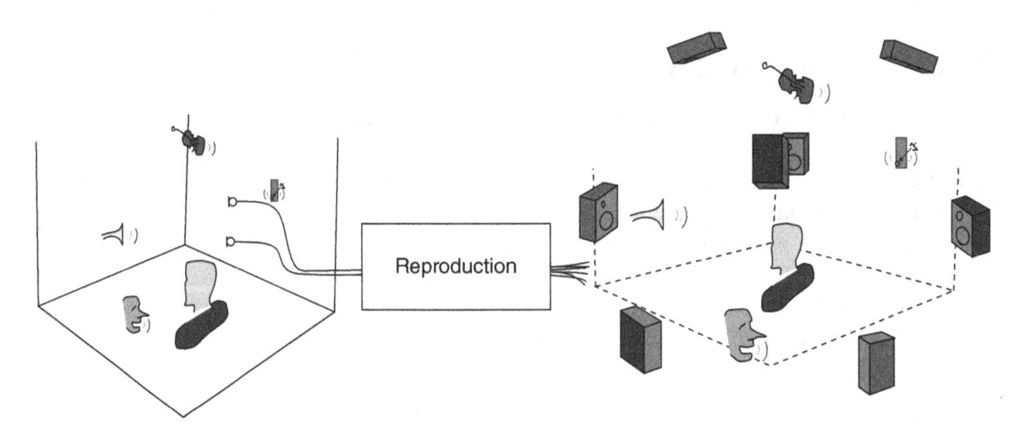

Figure 14.1 The sound reproduction set-up targeting faithful reproduction of an acoustic scenario. Listeners on the left and right ideally should perceive the auditory events identically, that is, the pitch, loudness, duration, timbre, and spatial characteristics of the auditory events should match.

sound reproduced is perceived as convincing. The subject feels that he or she is really 'in' the reproduced sound scenario, and the scenario is perceived as if the reproduced sources were real (Kyriakakis, 1998).

14.2 Audio Content Production

The term *audio content* refers here to sound signals produced that have meaning or value to a listener. For example, music recorded in the studio is audio content that can be delivered to a listener. The term *audio engineering* refers to the production of audio content. An *audio engineer* concerns himself or herself with the recording, manipulation, mixing, mastering, and reproduction of sound. Audio engineers creatively use technologies to produce sound for radio, television, film, public address, electronic publishing, and computer games.

Some important terms used in audio engineering that must be understood are:

- *Recording* is the process of capturing sound in a real acoustic scenario, or capturing the output signals of electrical or digital sources. Capturing of sound from an acoustic scenario can be conducted using one or more microphones, close to or far from the sound source, indoors or outdoors, and storing it in a recording device. Typically, each microphone signal is stored on one channel, or *track*, of the device. The capturing of digital and electrical signals is trivial with such devices, as they have electrical and/or digital inputs.
- *Mixing* is the process of adding different recorded tracks together after they are amplified, and possibly processed with systems that modify audio signals (audio effects). *Audio effects* are, for example, equalization, dynamic range control, panning, and reverberation. The following sections describe these effects in more detail. Mixing is done either on a *mixing console*, which is a dedicated device for this process, or using dedicated computer software. The outcome of mixing is a single- or multi-channel audio track that is meant to be played over a *loudspeaker set-up*, where each loudspeaker reproduces one channel.
- *Mastering* is the process of preparing and transferring the mixed audio track to its final form, ready to be copied to media, such as a CD, DVD, or the Internet, or for broadcasting. The sound is often processed further during mastering, where typically the sound signals are equalized and some dynamic range control is applied.
- *Live sound* is the on-line mixing and mastering of the audio signals by audio engineers during live concerts or live broadcasts for the public-address loudspeakers, for the signal stream for the purpose of broadcasting, or for both. The audio systems are typically set up earlier, and the desired settings for devices are found during a *sound check*.
- *Studio* – a facility that generally consists of at least two rooms: the studio(s) or *live room(s)*, where the music or other sound is generated, and the *control room(s)*, to where the sound signals from the microphones in the live rooms are routed, stored, and later mixed. A room may also be dedicated to the mastering process.
- *Audio format* defines the number of tracks, the loudspeaker set-up, and the encoding/decoding method used to record the audio content. In most cases, each track is meant to be listened to using a dedicated loudspeaker, although some exceptions exist.

Audio content production seldom strives to relay faithfully the listening experience of a real acoustic scenario, such as a musical concert. Even the recordings from live concerts are processed to produce a 'good-sounding' and plausible result, and typically an authentic reproduction of a live event is not the goal. By *plausible* we mean that the perceived features of the reproduced audio are believable and correspond well with listeners' expectations in the given context.

Audio engineering is a form of art, where an audible piece of art – music, speech, or other soundscape – is turned into audio tracks, ready for later listening. For authentic reproduction of audio content, the final listening experience created by the audio engineer in the mastering (or mixing) studio should be conveyed to listening consumers.

14.3 Listening Set-Ups

Sound can be made audible over different loudspeaker set-ups, ranging from a monophonic set-up to multi-channel systems and headphones. The best-known systems will be discussed below. The acoustic effect of the listening room also affects the perception of the reproduced sound. The loudspeaker set-up may also be accompanied by a visual display and/or vibroacoustic transducers, which add different cross-modal effects to the process.

14.3.1 Loudspeaker Set-ups

A *loudspeaker set-up*, or *loudspeaker layout*, defines the number of loudspeakers and their directions with respect to the *best listening position*. The term *listening area* refers to the area where the system is listened to. The distances from the best listening position to the loudspeakers and the responses of each loudspeaker are assumed to be identical. If they are not identical, some delaying and gaining may be applied to loudspeaker channels to compensate for the differences.

The most common loudspeaker set-up is the two-channel stereophonic set-up. Its use became widespread after the development of the single-groove 45°/45° two-channel record in the late 1950s. Two loudspeakers are positioned in front of the listener 60° apart, as illustrated in Figure 14.2a. The set-up enables the positioning of virtual sources between the loudspeakers, and it also makes the timbral quality of reproduction better when compared to monophonic reproduction, as discussed later in Section 14.4.1.

The motivation for using more than two loudspeakers in the reproduction is the potentially better spatial quality in a larger listening area. Different multi-channel loudspeaker set-ups have been specified in the history of multi-channel audio (Davis, 2003; Steinke, 1996; Torick, 1998). In the 1970s, the quadraphonic set-up was proposed, in which four loudspeakers are positioned evenly around the listener at azimuth angles ±45° and ±135°. This layout was never successful because of problems related to content delivery techniques at that time, and because the layout itself has too few loudspeakers to provide good spatial quality in all directions around the listener (Rumsey, 2001).

A sound reproduction system was developed for cinema whereby the front image stability of the standard stereophonic set-up was enhanced by an extra centre channel and two surround channels were added to create atmospheric effects and room perception. This surround sound system for cinemas was first used in 1976 (Davis, 2003), and the ITU made a recommendation for the layout in 1992 (BS.775-2, 2006). The late 1990s saw households also acquiring this 5.1 surround system, where the figure before the dot stands for the number of loudspeakers and the figure after the dot is the number of low-frequency channels. In the ITU recommendation, the three frontal loudspeakers are in the directions 0° and ±30°, and the two surround channels

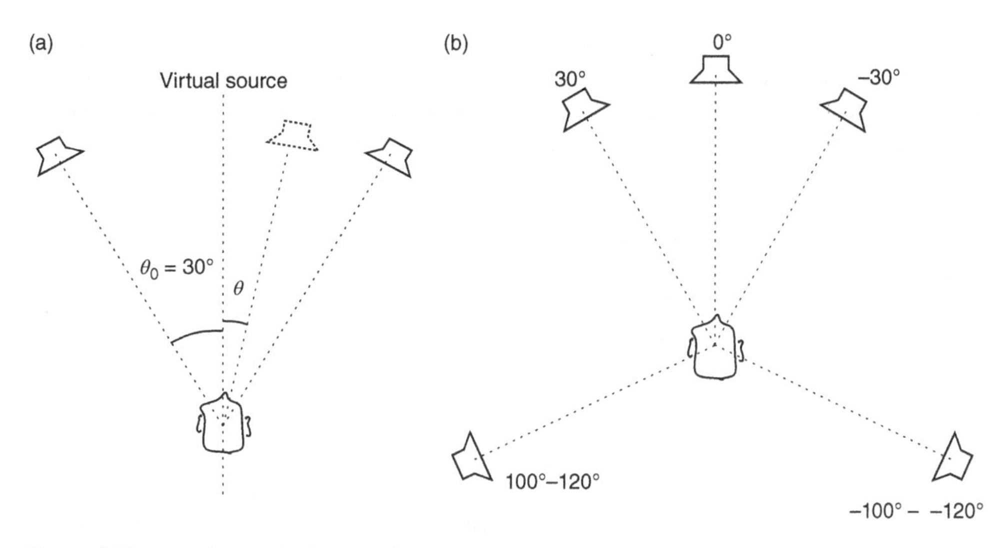

Figure 14.2 (a) The standard stereophonic listening configuration. (b) The 5-channel surround loudspeaker set-up based on the ITU recommendation BS775-2.

in directions $\pm 110 \pm 10°$, as shown in Figure 14.2b. The system has been criticized for not being able to deliver good directional quality anywhere else other than in the front (Rumsey, 2001). So other layouts with 6–12 loudspeakers have been proposed to enhance the directional quality in other directions as well.

A factor limiting the use of loudspeakers is their physical size. Optimally, the loudspeakers should be relatively large in order to be able to reproduce low frequencies. The smaller the loudspeaker, the higher is the lowest frequency it can reproduce. Unfortunately, larger size also means increased costs and more complicated installation. The idea of using a subwoofer(s) is to reproduce the low frequencies of the stereophonic or multi-channel audio track using one or many loudspeakers dedicated for reproducing frequencies typically under 80–200 Hz (Borenius, 1985; Welti, 2004). Broadband loudspeakers can then be designed to be relatively small without the need to reproduce low frequencies. The motivation for this design comes from psychoacoustics. Our directional hearing at low frequencies is rather poor, and thus not having broadband loudspeakers reproducing low frequencies is assumed not to severely impair the overall quality of reproduction.

All the loudspeaker layouts described above have loudspeakers only in the horizontal plane. However, there are systems, for use in theaters and virtual environments, in which loudspeakers are placed above and/or below the listener too, thus enhancing the perceived realism, especially in situations where the 3D position of a virtual source is important (Silzle *et al.*, 2011). Typical examples of such situations are virtual sources for flying vehicles or the sound of raindrops on a roof. Such 3D set-ups have been proposed for use in domestic listening too, and are currently being standardized (ISO/IEC 23008-1, 2014). For example, Japan Broadcasting Corporation has proposed a 22.2 loudspeaker set-up (Hamasaki *et al.*, 2005), which has 22 loudspeakers in planes at three heights and two subwoofers, or a common set-up has four elevated loudspeakers added to the 5.1 set-up. Note that the positioning of loudspeakers in domestic or other real listening set-ups may be very different from the theoretical specifications of the layouts. For example, the loudspeakers of a stereophonic set-up may, in practice, be in any direction and at arbitrary distances from the listener. In addition, in cars the loudspeaker set-ups are different for different car models, and none of the passengers are situated equidistantly from the loudspeakers. The audio content thus has to be created in such a way that such deviations from the theoretical set-ups do not affect the listening experience too much.

Some audio formats meant primarily for cinema also allow the use of different multi-channel systems to reproduce the audio content. The audio is sent with some metadata, which then defines how the sound is rendered to the loudspeakers. The corresponding methods are called *object-based audio* techniques. Such systems, like those covered by Robinson *et al.* (2012) and Lemieux *et al.* (2013), are called *loudspeaker-set-up agnostics*. The first movies with sound tracks in such formats were released in 2012.

14.3.2 Listening Room Acoustics

The acoustics of listening rooms vary significantly. Car audio systems and headphones have very short or non-existent reverberation. Small rooms and big rooms produce very different room responses. These differences make the goal of providing an identical listening experience in every listening environment appear impossible to achieve. However, despite the great variations in listening room acoustics, a certain piece of audio 'sounds' very similar in many different listening conditions. Listeners are typically able to identify the sound sources of a musical piece, recognize, say, the vocalist, the lyrics, and the music itself correctly, although the signals in their ear canals may have very different spectral content in different rooms (Toole, 2012). The ability of human hearing to adapt to diverse listening room acoustics is remarkable, as discussed briefly in Sections 11.5.1 and 12.6.

As a matter of fact, mixing and mastering studios around the world have different acoustics and use different audio devices. In principle, the audio content should produce a similar perception of sound to how the audio engineer perceived it in the final stage of the audio content production, as reasoned at the end of Section 14.1. But since each and every studio has a different listening set-up, there is no single answer to the question of what kind of acoustic properties domestic listening rooms should have ideally. However, the fact that humans are able, at least partly, to eliminate the effect of room acoustics in timbre perception mitigates the issue.

The acoustics of the listening room may have larger implications on certain perceptual attributes. For example, the perception of spatial characteristics of sound is largely affected by the listening room. The perceived directions of sources may be smeared if the room has strong early reflections, and the perception of reverberation in recorded sound is generally impossible if the listening room has stronger reverberation in itself. In addition, certain non-linear distortions, such as pre-delays and smearing of transients that may occur in lossy perceptual coding, may be more audible in less reverberant rooms. Our current knowledge of the effect of the listening room and loudspeaker characteristics on listening experience is summarized by Toole (2012) and Bech and Zacharov (2006).

A number of specifications for listening room acoustics and loudspeaker set-ups therein have been proposed. The main purpose for such specifications is to make the results of psychoacoustic listening tests conducted in rooms comparable between different academic and industrial sites. The specifications are relatively detailed. For example, ITU-R BS.1116-1 (1997) gives the details for the geometry of the room, the reverberation time, the amount of allowed background noise, the loudspeaker set-up geometry, listening positions, and audio system performance. Most of the values are given with tolerance rates, like how much the reverberation time can deviate from the optimal value depending on frequency. The resulting rooms resemble living rooms with relatively short reverberation time T_{60} (see Section 2.4.2 for a definition of T_{60}). T_{60} is also required to be almost constant over a large frequency range.

14.3.3 Audiovisual Systems

Sound is often reproduced in the presence of a moving picture, as in TV or cinema. Historically, starting from the 1890s, the first three decades of the cinema industry were the era of 'silent movies'. The technological advances in sound reproduction in the late 1920s made it possible to synchronize monophonic sound with the moving picture, soon replacing the silent movies with 'talking movies'. Cinema sound reproduction has progressed through two-channel stereo in the 1950s to the delivery of multiple discrete loudspeaker playback channels in the 1990s, and in the 2010s to loudspeaker-set-up-agnostic object-based audio formats. New developments of audiovisual systems have often been taken into use first in the cinema, and later they have been introduced for domestic use too.

The effect of the degradation of different features of sound on the quality of audio and video has been studied a lot, and an overview of these studies is given by Kohlrausch and van de Par (2005). The studies show that the effect of degradation on the quality of either audio or video depends on the content, implying that, in some cases, degradation in audio is more easily noticed than in video and vice versa (Hollier *et al.*, 1999). It is quite natural that degradation is more audible in the modality on which the subject is focusing (Rimell and Owen, 2000). Cross-modal effects also exist. Joly *et al.* (2001) showed that better audio quality can make video degradation less annoying, but good video quality was not found to improve the perceived audio quality.

A common problem found in audiovisual reproduction is the lack of synchronicity between audio and video. The lack of synchronicity is often caused because the audio and video signal routes produce different delays. The detection of asynchronicity is not symmetric across the modalities. Audio that is presented too early is noticed when the time shift is shorter than when it is presented after the corresponding video. Different studies propose different numbers for the threshold of detection of the shift: an audio lead of 25–75 ms and lag of 40–90 ms are said to be detected (Levitin *et al.*, 2000). This can be understood from the observation that in nature, for larger distances between the subject and the source, sound arrives at the ears considerably later than light at eyes from the source. Thus, humans are quite accustomed to sound stimuli lagging visual stimuli. The ITU has published a recommendation concerning synchronization thresholds for audio and video components in television signals (ITU-T, 1990). The maximum tolerated lead of audio in the recommendation is 20 ms, and correspondingly the maximum tolerated lag is 40 ms.

The presentation of both audio and video to the subject creates many types of interactions between modalities. For example, the sound of a red car is perceived to be louder than a blue or green one, with a difference corresponding to 1–3 dB of SPL (Menzel *et al.*, 2008). The audio track also influences eye movements and the direction of gaze, and sound has been shown to strengthen the salience of corresponding visual events (Coutrot *et al.*, 2012).

The perception of audio and video alone often differ in their spatial characteristics. Auditory objects may be localized to a position different from their visual counterparts, and the space perceived visually may be larger or smaller than that perceived by hearing. When the audio and video are reproduced simultaneously, the perceptual mechanisms try to resolve such conflicts. Ventriloquism exploits how our brain resolves these conflicts, as already discussed in Section 12.3.5 on page 234, where it was shown that correlated sound and video objects are perceived to be in the most probable direction, based on the cues available.

14.3.4 Auditory-Tactile Systems

Sound can also be perceived through the sense of touch. The human skin has a large number of mechanoreceptors, which are sensitive to vibrations. The vibration frequencies perceived by humans range from a few hertz to a few hundred hertz. The vibrations perceived by humans exist in diverse situations, from inside vehicles to live concerts. Some musical instruments also make the structures in the listening room vibrate. The airborne sound and structural vibrations are then transferred to the listener and may produce tactile perceptions. Examples include PA systems in loud rock concerts, timpani, and church organs. However, listeners may not consciously notice the vibrations, as they are typically in synchronicity with the music or other sounds, so that different percepts fuse, causing a holistic perception of events.

The degree of accuracy in the perception of the frequency spectrum of vibrations is nothing near that of the auditory system. However, humans are able to perceive the presence of vibration, and its frequency is analysed with very low selectivity. The JND of the level of stimulus is thought to be of the order of 1 dB, which is similar to that for auditory perception. In frequency-matching tasks, subjects identify relatively accurately or misjudge to be an octave lower a tactile sinusoidal stimulation between 60 and 180 Hz (Altinsoy and Merchel, 2010).

The interaction between presented tactile and auditory stimuli is of interest in this book. The loudness of a sound presented via headphones is perceived to be higher in the presence of a simultaneous tactile vibration produced with a whole-body shaker. The change in loudness corresponds to a change in level of about 1 dB (Merchel *et al.*, 2009). In bass reproduction, the preferred level of acoustic sound is lower if a shaker is also used to reproduce low frequencies (Simon *et al.*, 2009). Interestingly, this auditory–tactile loudness interaction does not depend significantly on the power of the vibration (Merchel *et al.*, 2009).

The JND of detection of haptic–audio asynchronicity has been found to be 24 ms with impulsive stimuli (Adelstein *et al.*, 2003), which is a value similar to that obtained with tests on the detection of audio–visual asynchronicity. However, in many cases, the auditory–visual asynchronicity threshold values are higher than auditory–tactile asynchronicity threshold values. Therefore, auditory–tactile asynchronicity may be even more critical than auditory–visual asynchronicity. In particular, musicians have lower auditory–tactile asynchronicity thresholds, about 10 ms, than the general population, possibly because of their training.

There are some interesting applications designed for auditory-tactile systems. The perceived quality of music reproduction has been found to be higher when vibrations are reproduced to the listener through a chair (Merchel and Altinsoy, 2013). A haptic device may also be used to provide the user with feedback. For example, mixing desks typically have a large number of input channels, each having a fader to control the level of the associated instrument. If the sound of the instrument is used to vibrate the faders, the audio engineer can recognize the instrument in each channel simply by touching the fader. This enables heads-up mixing, or mixing in low-light conditions (Merchel *et al.*, 2010).

14.4 Recording Techniques

The term *recording technique* is used here rather loosely to refer to the number of microphones used and how they are positioned in relation to each other, to the sound source(s), and to the recording room. A technique may also specify how the recorded signals are processed and how they are routed to the loudspeaker(s). The most common recording techniques are described below. The largest differences between them are in how they reproduce the spatial characteristics of sound. The perception of a reproduced spatial sound scene depends significantly on how the microphones are positioned and how their signals are processed and routed to loudspeakers.

14.4.1 Monophonic Techniques

The simplest recording technique is monophonic recording, where one or more microphones are used near to or far from the sources. The single recorded track is reproduced over a loudspeaker, or the recorded signals are mixed together and reproduced via a single loudspeaker. The reproduction is called monophonic even if this single signal is applied to a larger number of loudspeakers, such as in public address or general paging systems. The non-spatial characteristics of the sound can be captured using the monophonic recording technique, which is often good enough for many applications, such as speech communication.

A major disadvantage of monophonic reproduction is that the colouration caused by the recording room is exaggerated in the listening when compared to binaural listening of the recording venue. The reason for this is that the reproduced single sound signal gets filtered by the room, manifesting as a complex structure in the frequency response, as illustrated in Figure 2.23 on page 39. The sound reaching the two ears of a listener through the room has *different* magnitude and phase spectra, which results in binaural decolouration, as discussed in Section 12.6.2. The spectrum emanated by the single loudspeaker during monophonic reproduction is already coloured by the recording room response, but the binaural decolouration mechanisms can only try to compensate for the *listening room* acoustics, not the *recording room* acoustics. This emphasizes the acoustic effect of the recording room. When the recording is made with at least two microphones and reproduced over at least two loudspeakers, the recording room effect is different in the different loudspeaker signals, which enables at least partial decolouration of the recording room effect.

However, monophonic sound reproduction is evidently the most used sound reproduction method, since it is used in telephony. When the microphone is at a distance of a few centimetres from the source, the mouth of the person speaking, the acoustics of the recording room do not have much of an effect on the result, and the timbral quality of the reproduction is good. The disadvantage discussed in the previous paragraph is thus valid only when the captured effect of the room is significant, or, in technical terms, the level of the direct sound is comparable to or lower than that of the reverberant field.

14.4.2 Spot Microphone Technique

The *spot microphone technique* (also called close miking) records a number of channels, and the final audio content is mixed by an audio engineer. The technique is commonly used to capture a concert or studio session where many instruments are played. A microphone, called a *spot microphone*, is placed near each source to be captured (Rumsey, 2001). The microphone signals are often required to be as independent as possible, each signal containing sound from one source only.

The positioning of the microphones is critical, since the frequency-dependent directional patterns of many instruments have a major effect on the magnitude spectrum of the sound radiated in different directions (Pätynen and Lokki, 2010). Such recording is often complemented with distant microphones, which are used to record the sound from the sources through the response of the room. Such microphones are often called *ambience* or *ambient* microphones. The spot microphone signals together with the ambience microphone signals are then used to create the final audio content through the mixing process.

14.4.3 Coincident Microphone Techniques for Two-Channel Stereophony

Coincident microphone techniques for two-channel stereophonic systems are often referred to as *XY techniques*. Two directional microphones are placed with their diaphragms as close to each other as possible, but facing different directions. Typically, two cardioid or hypercardioid microphones are used, and the angle between the directions of the microphones is −60° − 120°, although angles up to 180° may also be used (Streicher and Dooley, 1985). Each of the recorded signals is applied to the corresponding loudspeaker of the stereophonic set-up. The recorded sources perceptually appear relatively point-like in the reproduced sound (Lipshitz, 1986) when listened to in the best listening position. When the listener moves away from the best listening position by a few tens of centimetres, the perceived sources migrate to the nearest loudspeaker.

However, these techniques have been characterized to lack a 'sense of space' (Streicher and Dooley, 1985). One reason for the weak perception of space is that the microphones are directional and are directed typically towards the instruments. The effect of the room, in turn, arrives evenly from all directions, and thus the directional pattern effectively reduces the level of reverberation in the recording.

The first XY techniques were presented by Blumlein (1931) in the 1930s. A particular XY technique is still known as the *Blumlein pair*, where the positive lobes of the two dipole microphones are in the directions 45° and −45° azimuth. The sound source(s) to be recorded is situated between the directions of the positive lobes. The technique reproduces the sound energy equally in all directions and does not suffer from attenuated capture of the reverberant field. However, the sound arriving from the sides produces a 180° phase reversal between the microphone signals. The sound is thus reproduced in loudspeakers out of phase, which may produce colouration and directional artefacts.

A microphone technique widely known as *MS stereo* was also presented by Blumlein (1931), where a directional microphone (M) is directed towards the sound source(s) and a figure-of-eight microphone (S) to the side. Then, a weighted sum of the microphone signals is made for each loudspeaker, thus creating two virtual microphone signals. Such a computation is also called *matrixing*. The directional patterns of these virtual microphone signals can be adjusted during mixing by changing the weights in the summing, as explained in more detail in the context of Ambisonics in Section 14.4.6.

14.4.4 Spaced Microphone Techniques for Two-Channel Stereophony

Spaced microphone techniques differ from coincident techniques in that the microphones are typically placed 20 cm to a few metres apart. These techniques are also known as *AB techniques*. Often, two microphones with omnidirectional patterns are used, although directional microphones may also be utilized. The signals from each source arrive at different times at the two microphones, and thus the time difference between the microphone signals depends on the direction of arrival. Also, differences in amplitude might exist if the microphones are far away from each other.

If the recorded signals are impulse-like, virtual sources may be perceived to be point-like and arriving from a direction that depends on the time delays between the microphone signals.

For tonal signals, such as harmonic complexes, the virtual sources may be localized inconsistently; the localization varies with frequency, and a virtual source may be perceived to be wide. However, sound recorded with spaced microphone techniques is often considered to be more 'ambient', 'airy', or 'warm' than sound recorded with coincident microphone techniques (Lipshitz, 1986). A reason for the more pronounced room response in AB techniques than in XY techniques is the omnidirectional response, which does not attenuate the diffuse field. Another effect that emphasizes the room response is that, due to the considerable distance between the microphones, the captured signals are incoherent within a 1-ms window for most directions of arrival. When such incoherent signals are applied to loudspeakers, the interaural cues of the listener suggest directions exceeding the directions between the loudspeakers in the stereophonic set-up, which may lead to higher detectability of the reverberant sound (Pulkki, 2002).

There also exist methods, such as the *ORTF* technique, that can be classified as falling this technique between the coincident and spaced microphone techniques. In it, two cardioid microphones are positioned 17 cm apart with an angle of 110°. The captured signals may thus differ both in time and in amplitude. At low frequencies, the system corresponds to the XY cardioid technique, since the distance between the microphones is small compared to the wavelength. At high frequencies, the microphone signals have considerable phase differences. The perceptual attributes of the reproduced virtual sources and reverberation are reported to be somewhere between the attributes of the coincident and the spaced techniques (Lipshitz, 1986).

14.4.5 Spaced Microphone Techniques for Multi-Channel Loudspeaker Systems

When recording a live acoustic event for reproduction over a multi-channel loudspeaker set-up, typically a number of microphones are positioned in a spaced configuration at the recording venue. The number of microphones is at least the same as the number of loudspeakers, and they are separated by distances ranging from 10 cm to several metres. Often, the microphones face approximately towards the corresponding loudspeaker directions (Rumsey, 2001). Microphones with different directional patterns are used in the set-ups. Some proposed arrangements that achieve different properties in reproduction of sound exist, such as the layout shown in Figure 14.3a. Different microphone arrangements can also be used near and far from the sources. Such a layout is shown in Figure 14.3b, where several microphones are located close to the sources and a Hamasaki square of four figure-of-eight microphones is located far from the sources, well outside the critical distance. The recorded signals are then mixed and mastered into the final multi-channel audio format.

There is no single microphone technique that is the best solution in all situations. In practice, audio engineers decide microphone positions separately for every recording venue and for every ensemble to be recorded, tuning the set-up using monitoring loudspeakers in a separate listening room. The resulting surround audio content is, at best, plausible and enjoyable, some kind of spaced microphone set-up being the most often used technique for surround sound recordings.

14.4.6 Coincident Recording for Multi-Channel Set-up with Ambisonics

The *Ambisonics* reproduction technique (Gerzon, 1973) provides a theoretical framework for coincident recording techniques for 2D and 3D multi-channel loudspeaker set-ups. In theory, Ambisonics is a compact-format, loudspeaker-set-up-agnostic, efficient, and comprehensive

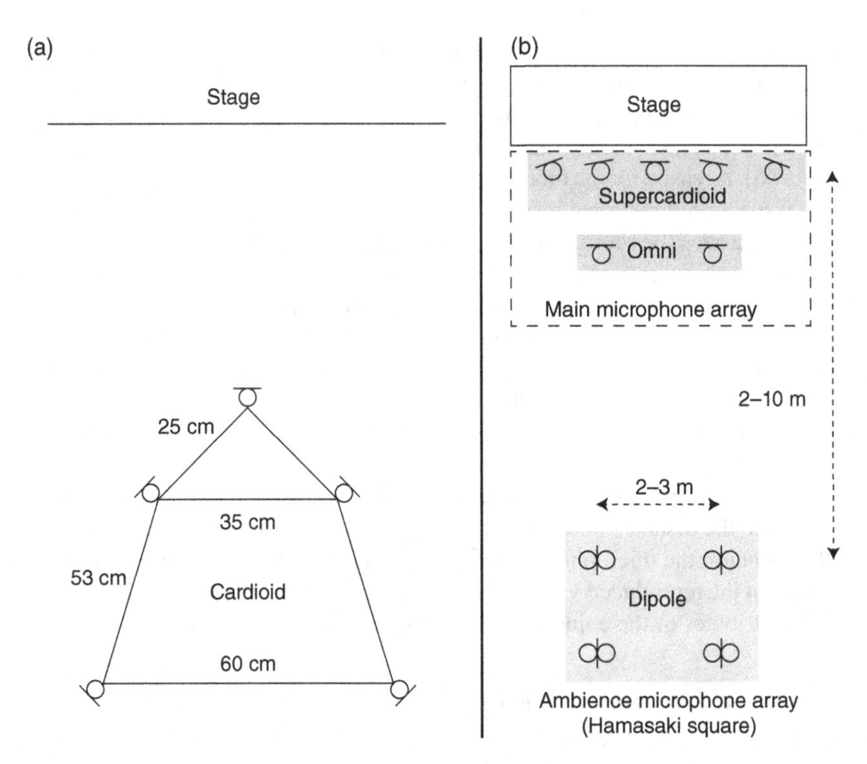

Figure 14.3 Two microphone configurations for recording for 5.1 surround reproduction. (a) The recording of five signals for five loudspeakers using the 'optimal cardioid array', as described by Rumsey (2001). (b) The main microphone array near the sources and an ambience array far from the sources used to record more than five signals, which are then mixed into loudspeaker signals (Hamasaki and Hiyama, 2003).

method for capture, storage, and reproduction of spatial sound. Unfortunately, it has some unsolved technical drawbacks in recording of real acoustic scenarios, which, however, can be at least partly mitigated using non-linear techniques.

All coincident multi-channel microphone techniques produce signals that can be transformed into *B-format* signals. B-format signals have directional patterns that have *spherical harmonics*. The zeroth-order harmonic is the omnidirectional pattern, the three first-order patterns are those of a dipole facing in each of the three directions of the axes of the Cartesian coordinate system, and second order patterns are more complicated, having the form of a quadrupole or a dipole-with-ring form, as shown in Figure 14.4. The directional patterns of signals are additive; when two signals are added, the directional pattern of the combined signal follows the combined directional pattern. For example, combining the omnidirectional and dipole patterns with equal gains produces a signal with a cardioid pattern. When higher-order components are recorded, more complex directional patterns are formed for the combined signals. The spherical harmonics can thus be seen as basis functions for the design of arbitrary patterns.

The most common microphone device for Ambisonics is the first-order, four-capsule *B-format* microphone, producing signals with directional patterns having spherical harmonics up to the first order. The omnidirectional signal is denoted $w(t)$, and the three dipole signals are denoted $x(t)$, $y(t)$, and $z(t)$. The $w(t)$ signal is usually scaled down by a factor of $\sqrt{2}$.

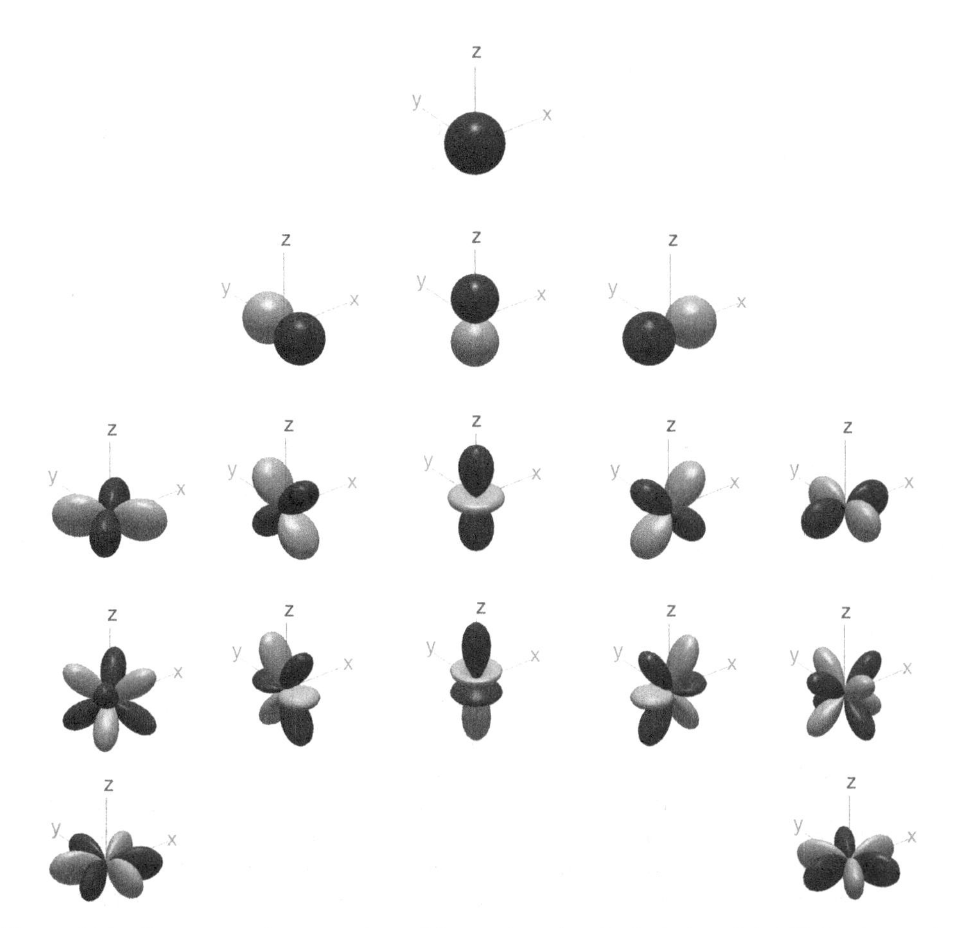

Figure 14.4 From top to bottom: spherical harmonics of order 0, 1, 2, and 3. Third-order sperical harmonics are shown in the two bottom rows. Courtesy of Archontis Politis.

Higher-order microphones with more capsules have also been developed and are commercially available. The higher-order components can then be derived in a specific frequency window. There are some higher-order microphones available which can extract harmonics up to about the fifth order in a certain frequency window. The number of good-quality microphone capsules in such devices is relatively high, something like 32. Outside the frequency window, the microphones suffer from low-frequency noise and deformation of directional patterns at high frequencies (Moreau *et al.*, 2006; Rafaely *et al.*, 2007).

First-order microphones have been available for decades, and we use as an example of how to compute signals for loudspeakers. The channels are matrixed that is, added together with different gains. Thus, each loudspeaker signal can be considered as a virtual microphone signal having first-order directional characteristics. This is expressed as

$$s(t) = (1-\kappa)w(t) + \frac{\kappa}{\sqrt{2}}[\cos(\theta)\cos(\phi)x(t) + \sin(\theta)\cos(\phi)y(t) + \sin(\phi)z(t)], \quad (14.1)$$

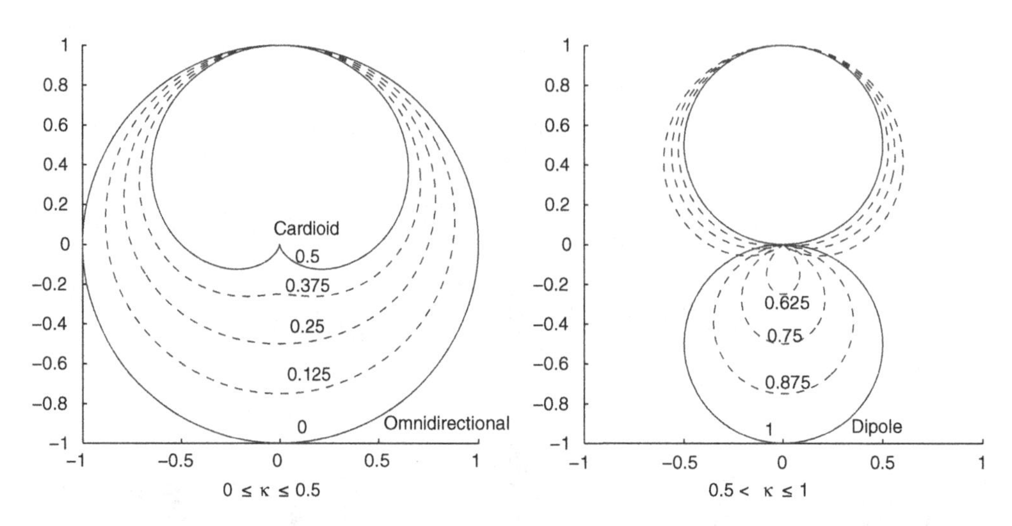

Figure 14.5 The directional patterns of virtual microphones that can be generated from first-order B-format recordings.

where $s(t)$ is the produced virtual microphone signal oriented at azimuth angle θ and elevation ϕ. The parameter $\kappa \in [0, 1]$ defines the directional characteristics of the virtual microphone from omnidirectional to cardioid and dipole, as shown in Figure 14.5.

In principle, first-order Ambisonics could be used for any loudspeaker set-up, but unfortunately it has a very limited range of use. The broad first-order directional patterns make the listening area very small, extending the size of the head of the listener only at frequencies below about 700 Hz (Solvang, 2008). At higher frequencies, the high coherence between the loudspeaker signals leads to undesired effects, such as colouration and loss of spaciousness. The number of loudspeakers in a horizontal set-up should not exceed $2N + 1$, where N is the order of the B-format microphone, to avoid too high a coherence between the loudspeaker signals. Thus, first-order microphones can be used only with three-loudspeaker set-ups, which is far too few to produce the perception of virtual sources between the loudspeakers. This calls for the use of microphone set-ups able to capture signals with higher-order spherical harmonic directional patterns. With higher-order directional signal components, the directional patterns of the loudspeaker signals can be made narrower, which solves these issues.

14.4.7 Non-Linear Time–Frequency-domain Reproduction of Spatial Sound

The spatial resolution of hearing is limited within the auditory frequency bands (Blauert, 1996). In principle, all sound within one critical band can only be perceived as a single source with a broader or narrower extent. The limitations of spatial auditory perception raise the question of whether the spatial accuracy in reproduction of an acoustic wave field can be compromised without a decrease in perceptual quality. When some assumptions on the resolution of human spatial hearing are used to derive reproduction techniques, potentially an enhanced quality of reproduction is obtained.

The audio recording and reproduction technology called *directional audio coding* (DirAC) (Merimaa and Pulkki, 2004; Pulkki, 2007) assumes that the spatial resolution of the auditory system at any one time instant and in one critical band is limited to extracting one cue for direction and another for interaural coherence. It further assumes that if the direction and diffuseness of the sound field are measured and reproduced correctly with a suitable time resolution, a human listener will perceive the directional and coherence cues correctly.

An example of the implementation of DirAC is shown in Figure 14.6. In the analysis, the direction and diffuseness of the sound field are estimated using temporal energy analysis in the auditory frequency bands, as described below. The direction is expressed in azimuth and elevation angles, indicating the most important direction of arrival of sound energy. Diffuseness is a real number between zero and one, which indicates whether a sound field more closely resembles a plane wave or a diffuse field. Virtual microphones are then formed from B-format signals, which are divided into a *diffuse stream* and a *non-diffuse stream* using the diffuseness parameter. The non-diffuse stream is assumed to contain sound that originates primarily from one source, allowing it to be applied in a single direction. This is implemented here by computing amplitude-panning gain factors (see Section 14.5.3) and by gaining the virtual microphone signals with them. The sound is thus effectively reproduced only by loudspeakers near the analysed direction.

The diffuse stream, in turn, is assumed to contain sound originating from reverberation or from multiple concurrent sources from different directions, which should produce low interaural coherence. The virtual microphone signals in the diffuse stream divide the energy around the listener, which fits nicely the assumption of diffuse reverberation. Unfortunately, the virtual microphone signals have too high coherence between each other, which produces similar shortcomings to the first-order Ambisonics. A decorrelation process can be used to reduce the coherence between the channels, so that the phase spectrum of the signals is altered, ideally without affecting the magnitude spectrum, as explained in Section 15.2.8. The shortcoming of decorrelation, time-smearing of transients, is typically not audible, since diffuse sound does not typically contain strong transients.

This implementation of DirAC has been proven to produce better perceptual quality in loudspeaker listening than other available techniques using the same microphone input (Vilkamo *et al.*, 2009). An almost authentic reproduction is obtained if the acoustic scenario fits the DirAC signal model, where sound from one source dominates in one frequency band with potentially some diffuse sound arising from room reverberation. On the other hand, two broadband sources with temporally relatively smooth envelopes that arrive at the microphone from opposite directions cause some quality degradation (Laitinen, 2014). The degradation may be audible as smearing of transients or a perception of added room effect. A number of techniques to avoid such degradation with first-order B-format input have been proposed (Laitinen, 2014; Vilkamo and Pulkki, 2013).

Similar assumptions have been used in the development of other time–frequency-domain methods for spatial sound reproduction (Alexandridis *et al.*, 2013; Berge and Barrett, 2010; Cobos *et al.*, 2010; Tournery *et al.*, 2010). DirAC has also been developed for higher-order B-format microphone input by dividing the sound field into sectors and performing the analysis and synthesis individually for each sector, which makes the acoustic conditions less challenging for the analysis–synthesis methods (Politis *et al.*, n.d.).

A basic method of analysing the sound field to compute the DirAC metadata is described briefly. Let us assume that frequency-domain signals for pressure P and 3D particle velocity

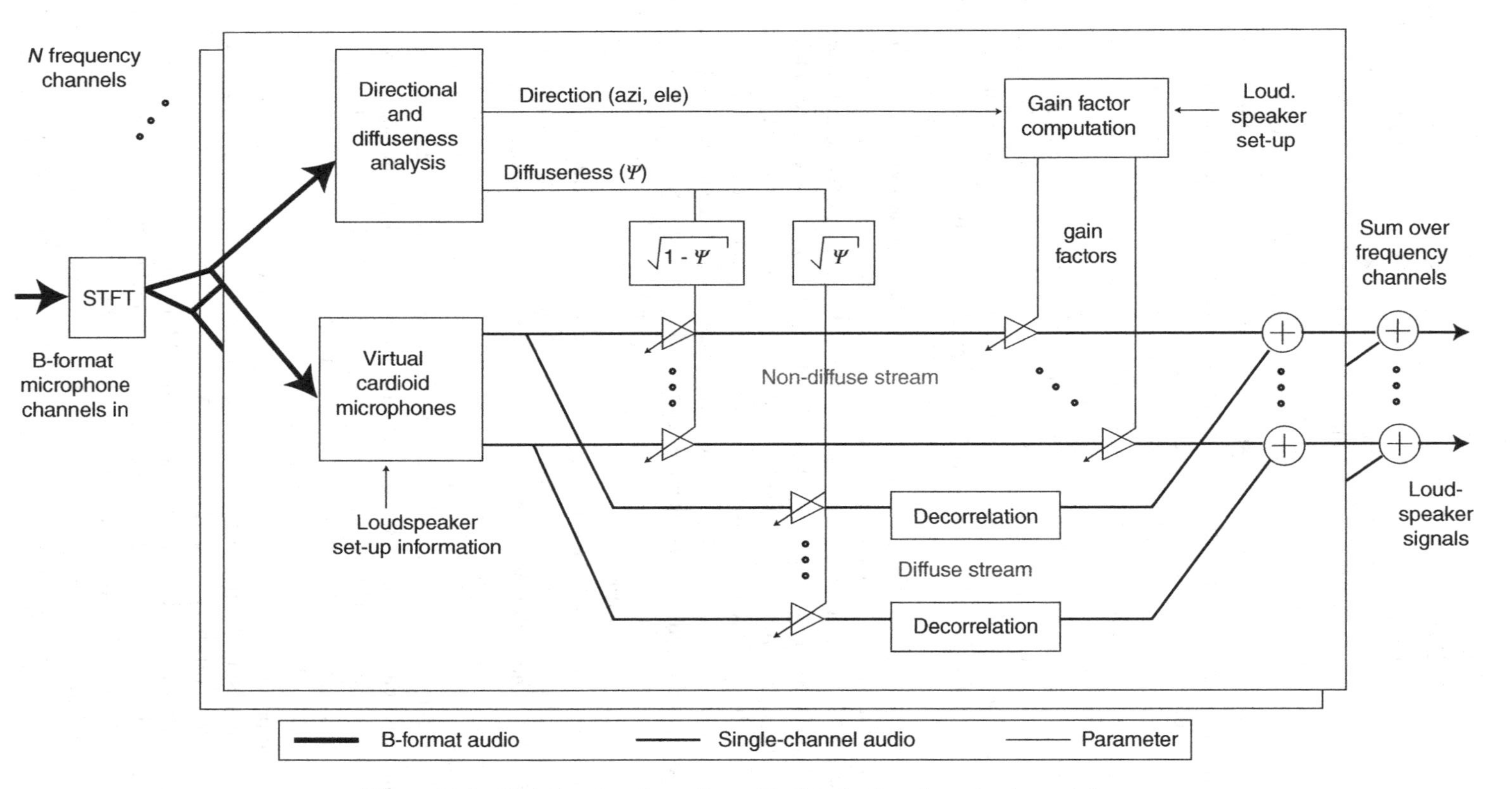

Figure 14.6 Directional audio coding with virtual microphone implementation.

U are available. They can be estimated from B-format signals, but the mathematical derivation of the estimates is beyond the scope of this book. The energy E of the sound field can be computed as

$$E = \frac{\rho_0}{4}||\mathbf{U}||^2 + \frac{1}{4\rho_0 c^2}|P|^2, \tag{14.2}$$

where ρ_0 is the mean density of air and c is the speed of sound.

The *intensity vector* **I** expresses the net flow of sound energy as a 3D vector, and can be computed as

$$\mathbf{I} = P^*\mathbf{U}, \tag{14.3}$$

where $(\cdot)^*$ denotes complex conjugation. The direction of arrival of sound is defined to be the opposite of that of the intensity vector in each frequency band. The direction is denoted as corresponding angular azimuth and elevation values in the transmitted metadata. The diffuseness of the sound field is computed as

$$\psi = 1 - \frac{|\mathrm{E}\{\mathbf{I}\}|}{\mathrm{cE}\{E\}}, \tag{14.4}$$

where E is the expectation operator. Typically, the expectation operator is implemented as an integration in time. This process is also called 'smoothing'. The analysis is repeated as often as necessary for the application, typically with an update frequency of 100–1000 Hz.

14.5 Virtual Source Positioning

Virtual source positioning is a method to control perceived localization using the appropriate reproduction of a monophonic signal. A *virtual source* is an auditory object perceived at a location that does not have any real sources. Virtual source positioning aims to control only the perceived direction of virtual sources, although sometimes the distance and the spatial width of sources may also be controlled.

14.5.1 Amplitude Panning

Amplitude panning means feeding a sound signal $x(t)$ to loudspeakers with different amplitudes, mathematically expressed as

$$x_i(t) = g_i x(t), \quad i = 1, \ldots, N, \tag{14.5}$$

where $x_i(t)$ is the signal fed to loudspeaker i, g_i is the gain of the corresponding channel, N is the number of loudspeakers, and t is time. The listener perceives a virtual source in a direction that depends on the gains. A *panning law* estimates the perceived direction θ from the gains of the loudspeakers. The estimated direction is called the *panning direction* or the *panning angle*. Amplitude panning is conceptually equivalent to the coincident microphone techniques discussed in Section 14.4.3, since the captured signals in these techniques differ, in principle, only in amplitude.

14.5.2 Amplitude Panning in a Stereophonic Set-up

When a virtual source is generated using amplitude panning for the two-channel stereophonic listening set-up, the same sound is applied to the loudspeakers with potentially different amplitudes. The sound arrives from both loudspeakers at both ears, the ipsilateral sound arriving a little earlier than the contralateral, a phenomenon called *cross talk*. A difference of about 0.5 ms in the arrival time at one ear effectively results in a weighted average of the phase of the signals. The *level* differences between the loudspeakers are thus turned, a little surprisingly, into *phase* differences between the ears (Bauer, 1961). This effect is valid only at low frequencies, below about 1 kHz. At high frequencies, above about 2 kHz, the level differences remain as level differences due to the lack of cross talk caused by the shadowing of the head.

The tangent law by Bennett *et al.* (1985) estimates the perceived direction of a virtual source, and it is expressed as

$$\frac{\tan\theta}{\tan\theta_0} = \frac{g_1 - g_2}{g_1 + g_2}, \tag{14.6}$$

which has been found to estimate the direction best in conditions tests in anechoic conditions. Other panning laws also exist and are reviewed by Pulkki (2001b).

The panning laws only determine the ratio between the gains. To prevent an undesirable change in loudness of the virtual source, depending on panning direction, the sum-of-squares of the gains should be normalized:

$$\sum_{n=1}^{N} g_n^p = 1, \tag{14.7}$$

where $p = 2$. This value of p has been found to be the best in real rooms with some reverberation. Depending on listening room acoustics, different normalization rules may be used (Laitinen *et al.*, 2014; Moore, 1990).

The presented analysis is valid only if the loudspeakers are equidistant from the listener and if their angular distance is no larger than about 60°. These criteria define the best listening area in terms of where the virtual sources are localized between the loudspeakers, which, in practice, is only a few tens of centimetres in the left–right direction. When the listener moves outside this area, the virtual source is localized towards the nearest loudspeaker, which emanates a considerable amount of sound. Such erroneous localization occurs due to the precedence effect.

In principle, amplitude panning methods create a comb-filter effect in the ear canal signal spectra, since the same sound arrives from both loudspeakers at each ear with a small time difference. This effect is clearly audible in an anechoic chamber, where it produces a notch in the spectrum between frequencies of 1 kHz and 2 kHz. Fortunately, this effect is not present when listening in a normal room, since the room reverberation mitigates the colouring effect largely (Pulkki, 2001a). The lack of prominent colouring and the relatively robust directional effect provided by amplitude panning are very probably the reasons why it is included in all mixing consoles as a 'panpot' control, making it the most widely used technique to position virtual sources.

14.5.3 Amplitude Panning in Horizontal Multi-Channel Loudspeaker Set-ups

In many cases more than two loudspeakers are placed around the listener. Pairwise amplitude panning (Chowning, 1971) is commonly used to position virtual sources with multi-channel set-ups by applying the sound signal only to the loudspeaker pair between which the virtual source lies. Vector base amplitude panning (VBAP) (Pulkki, 2001b) is a commonly used method to formulate pairwise panning. In 2D VBAP, a loudspeaker pair is specified with two vectors. The unit-length vectors $\mathbf{l}_m$ and $\mathbf{l}_n$ point from the listening position to the loudspeakers. The intended direction of the virtual source (panning direction), represented by the unit vector $\mathbf{p}$, is expressed as a linear weighted sum of the loudspeaker vectors

$$\mathbf{p} = g_m \mathbf{l}_m + g_n \mathbf{l}_n. \tag{14.8}$$

Here, g_m and g_n are the gain factors of the respective loudspeakers. The gain factors can be computed from

$$\mathbf{g} = \mathbf{p}^{\mathrm{T}} \mathbf{L}_{mn}^{-1}, \tag{14.9}$$

where $\mathbf{g} = [g_m \; g_n]^{\mathrm{T}}$ and $\mathbf{L}_{mn} = [\mathbf{l}_m \; \mathbf{l}_n]$. The calculated factors are used in amplitude panning as gains of the signals applied to the respective loudspeakers after appropriate normalization, say $||\mathbf{g}|| = 1$.

When a virtual source is panned between loudspeakers, the binaural cues are more or less unnatural, since the summing localization mechanism produces somewhat unnatural cues even in the stereophonic case. With loudspeaker pairs not symmetric about the median plane, the produced ITD and ILD cues suggest different directions at different frequencies and are also biased somewhat towards the median plane (Pulkki, 2001b). As a result, the perceived spatial width of the virtual source varies with the panning direction. Thus, the directions of the loudspeakers are perceived if a moving source is created, since the virtual sources are more point-like in their directions. Such an uneven directional width of the sources can be compensated for by blurring the virtual sources slightly in the directions of the loudspeakers. In practice, the sound is always applied to more than one loudspeaker (Pulkki, 2001b; Sadek and Kyriakakis, 2004).

14.5.4 3D Amplitude Panning

A three-dimensional loudspeaker set-up here means a set-up in which the loudspeakers are not all in the same plane. Thus, the set-up has some loudspeakers above and/or below the plane of the horizontal loudspeaker set-up. Triplet-wise panning can be used in such set-ups (Pulkki, 1997), wherein a sound signal is applied to at most three loudspeakers at a time, forming a source triangle from the listener's viewpoint. If more than three loudspeakers are available, the set-up is divided into triangles, one of which is used in the panning of a single virtual source at any one time, as shown in Figure 14.7.

Three-dimensional vector base amplitude panning (3D VBAP) is a method to position virtual sources using such set-ups (Pulkki, 1997). It is formulated similarly to the pairwise panning in the previous section. However, now the gain factors are $\mathbf{g} = [g_m \; g_n \; g_k]^{\mathrm{T}}$, the direction vectors are 3D Cartesian vectors, and the loudspeaker vector base is $\mathbf{L}_{mnk} = [\mathbf{l}_m \; \mathbf{l}_n \; \mathbf{l}_k]$ in

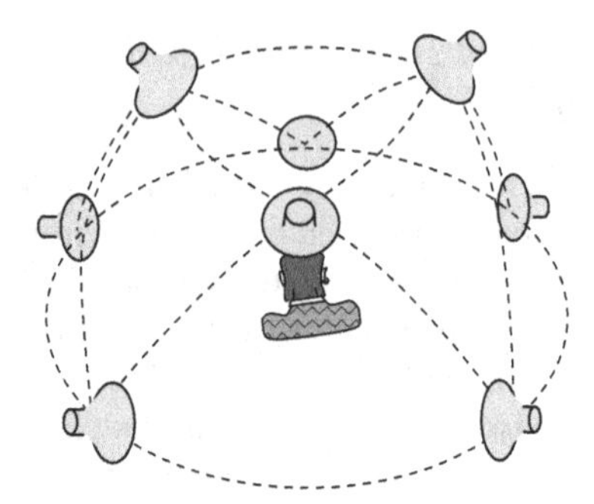

Figure 14.7 A 3D triangulated loudspeaker system for triplet-wise panning.

Equation (14.9). The equation estimates the perceived virtual source direction with relatively good accuracy. The confusion cone azimuth φ_{cc} is estimated to an accuracy of a few degrees in most cases. The perceived confusion cone elevation of a virtual source δ_{cc} is personal for each subject (Pulkki, 2001b), but it is typically perceived inside the loudspeaker triplet.

14.5.5 Virtual Source Positioning using Ambisonics

The Ambisonics microphone technique, discussed in Section 14.4.6, can also be simulated in the computational domain to position virtual sources for 2D or 3D loudspeaker set-ups (Furse, 2009; Malham and Myatt, 1995). The quality issues of Ambisonics as a recording technique, discussed in Section 14.4.6, can be avoided by choosing the order for B-format signals that matches the loudspeaker set-up. The B-format signals produced by a single virtual source can be simulated simply by using such directional patterns as shown in Figure 14.5 ideally, as virtual higher-order microphones do not suffer from self-noise or deformed directional patterns. When multiple concurrent virtual sources exist, their signals can be simply summed to the same B-format audio bus, which corresponds to ideal B-format recording of the corresponding real multi-source scenario. The B-format audio can then be manipulated efficiently. For example, the rotation of the spatial audio scene can be done with simple operations.

14.5.6 Wave Field Synthesis

An alternative goal of sound reproduction is to control the sound field within the loudspeaker set-up. This means that the pressure and velocity waves originating from the virtual sources are to be synthesized as plane waves or as spherical waves. The targeted sound field would then be well defined, and the reproduced sound field can also be simulated, or even measured. Unfortunately, the errors in the synthesis of the sound field between the targeted and reproduced fields do not say much about the perceptual significance of the deviation. Sometimes a small difference in the wave field causes a large perceptual deviation, and a large deviation in the wave field may be perceptually irrelevant in some cases. However, the advantage of this starting point is that the reproduction techniques can be derived mathematically.

Wave field synthesis (WFS) is a technique that requires a large number of carefully equalized loudspeakers (Ahrens, 2012; Berkhout *et al.*, 1993; Vries and Boone, 1999). It aims to reconstruct the entire sound field in a listening room. When a virtual source is reproduced,

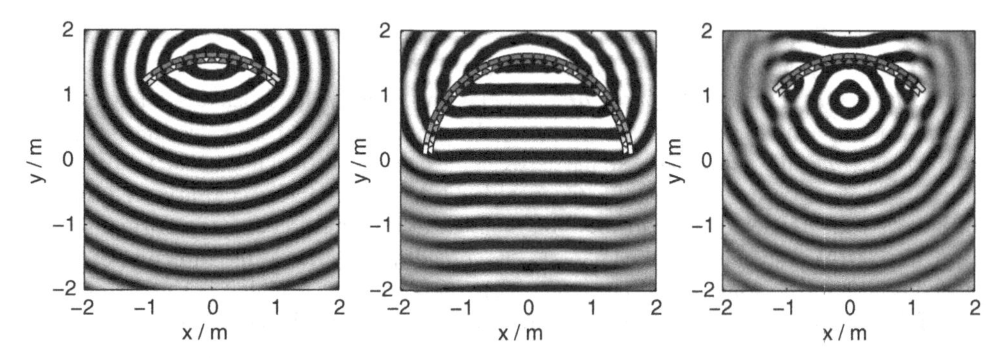

Figure 14.8 The wave field synthesis concept. A circular 56-loudspeaker set-up with radius of 1.5 m is used to reproduce a virtual source emitting a 1 kHz sinusoid 0.5 m behind the set-up (left), a plane wave (centre), and a virtual source 0.5 m inside the set-up (right). The colour of the loudspeaker indicates the level of radiation. Inactive loudspeakers are not shown at all. The illustrations were rendered using the sound field synthesis toolbox (Wierstorf and Spors, 2012).

the sound for each loudspeaker is delayed and amplified so that a desired circular or planar sound wave results as a superposition of sounds from each loudspeaker. The virtual source can be positioned far behind the loudspeakers or, in some cases, even in the space inside the loudspeaker array, as shown in Figure 14.8.

Usually, WFS is used to synthesize virtual sources by specifying the location of the source in the virtual world and using the loudspeaker set-up to generate the wave field. The use of actual recordings for WFS is very rare, probably because setting up the hundreds of carefully equalized microphones required is technically very demanding. This is the reason why WFS is discussed in this book in the section on virtual source positioning and not in the section on microphone techniques. In practice, the microphone techniques discussed in the Section 14.4.5 are used for recording, and the recorded signals are then reproduced with the appropriate system.

Theoretically, the WFS is superior as a technique, because the perceived position of the sound source is correct within a very large listening area. Unfortunately, to create the desired wave field in the total area inside the array requires that the loudspeakers are at a distance of at most half a wavelength from each other. Arrays for wave field synthesis have been built for room acoustics control and enhancement to be used in theatres and multi-purpose auditoria (Vries and Boone, 1999).

Another application of wave field synthesis techniques using a large number of loudspeakers is a method called *sound field control* (Francombe *et al.*, 2013; Nelson and Elliott, 1992). Sound field control is used for various applications. For example, different signals can be made audible in different zones of a large listening area in such a manner that only one of the signals is heard in one zone. The method may also be used to attenuate noise in a large area, for example, to reduce background noise for all passengers in the cabin of an aeroplane.

14.5.7 Time Delay Panning

When the signal to one loudspeaker in a stereophonic set-up is delayed by a constant amount, virtual sources with transient signals are perceived to migrate towards the loudspeaker that radiates the earlier sound signal (Blauert, 1996). The maximal effect is achieved asymptotically when the delay is approximately 1.0 ms or more.

Such processing converts the *phase or time* delays between the loudspeakers at low frequencies to a perception of *level* differences between the ears, and at high frequencies the perception remains as a *time* difference in the signal between the ears. This effect makes the virtual source direction depend on the signal itself (Cooper, 1987; Lipshitz, 1986). The produced binaural cues vary with frequency, and different cues suggest different directions for virtual sources (Pulkki, 2001b). The cue may thus generate a 'spread' perception of the direction of sound, which is desirable in some cases. The effect is also dependent on listening position. For example, if the sound signal is delayed by 1 ms in one loudspeaker, the listener can compensate for the delay by moving a bit towards the delayed loudspeaker. Time delay panning thus resembles spaced microphone techniques, since the spacing between the microphones causes time delays between microphone signals and since the resulting spatial image is similar.

A special case of phase difference in stereophonic reproduction is the use of antiphasic signals in the loudspeakers, where the same signal is applied to both loudspeakers but with the polarity inverted at one loudspeaker, producing a constant 180° phase difference between the signals at all frequencies. This phase difference changes the perceived sound colour, and also spreads the virtual sources. Depending on the listening position, low frequencies may be cancelled out. At higher frequencies this effect is milder, since when the wavelength becomes shorter, the listening area where the loudspeaker signals cancel each other shrinks, and beyond some frequency one of the ears will be outside this listening region where the signals cancel. This effect is also milder in rooms with longer reverberation. The directional perception of the antiphasic virtual source depends on the listening position. In the best listening position, the high frequencies are perceived at the centre and low frequencies in random directions. Outside the best position, the direction is either random or towards the closest loudspeaker. In the language of audio engineers, this effect in the sound is called 'phasy', or 'there is phase error in here'.

14.5.8 Synthesizing the Width of Virtual Sources

A loudspeaker layout with loudspeakers around the listener can also be used to control the width of a virtual source or even to produce an enveloping perception of the sound source. A simple demonstration can be made by playing back pink noise through all loudspeakers independent of each other (Blauert, 1996). The sound source is then perceived to surround the listener completely.

Time–frequency-domain spatial audio processing also provides a means to control the source width effectively. The input signal is divided into frequency channels that are then positioned in different directions around the listener (Pihlajamäki *et al.*, 2014).

14.6 Binaural Techniques

Binaural techniques are loosely defined to be methods that aim to reproduce accurately ear canal signals recorded in a real acoustic scenario with a real subject, or to reproduce ear canal signals which would occur in a virtual world. This is done by recording sound from ear canals or by utilizing measured or modelled head-related transfer functions (HRTFs) and acoustic modelling of listening spaces.

14.6.1 Listening to Binaural Recordings with Headphones

The basic *binaural recording* technique is to reproduce a recorded binaural sound track through headphones. The recording is made by inserting miniature microphones in the ear canals of a real human listener, or by using a manikin with microphones in the ears (Blauert, 1996; Wilska, 1938). Such a recording is reproduced by playing the recorded signals to the ears of the listener. In principle, this is a very simple technique and can provide effective results. A simple implementation is to replace the transducers of in-ear headphones with miniature microphones, use a portable audio recorder to record the sounds of the surroundings, and play back the sound with headphones. Without any further processing, a convincing spatial effect is already achieved, as the left–right directions of the sound sources and the reverberant sound field are reproduced naturally. If the person who did the recording is the listener, the effect can be particularly striking.

Unfortunately, there are also technical challenges with the technique. The sound may appear coloured, the perceived directions move from front to back, and everything may be localized inside the head. To partially avoid these problems, the recording and the reproduction should be carefully equalized, because headphone listening typically produces a different magnitude spectrum to the ear drum than natural listening. Careful equalization of headphone listening is, unfortunately, a complicated business, and it requires very careful measurements of the acoustic transmission of sound from the headphone to the ear drum (Xie, 2013).

A further challenge in binaural reproduction is that the auditory system also utilizes dynamic cues to localize sound sources, as discussed in Section 12.3.4 on page 232. When listening to a binaural recording with headphones, the movements of the listener do not change the binaural reproduction at all. This is one reason why headphone reproduction easily tends to be localized inside the head of the listener (Blauert, 1996).

Another issue is the problem of individuality. Every listener has a unique pinna and head size, and sound in similar conditions seems different in different individuals' ears. When listening to a binaural recording made by another individual, similar problems occur as with non-optimal equalization (Rumsey, 2011; Wenzel *et al.*, 1993).

14.6.2 HRTF Processing for Headphone Listening

A monophonic sound signal can be virtually positioned in any direction in headphone listening if the HRTFs for both ears are available for the desired virtual source direction (Møller *et al.*, 1995; Xie, 2013). A signal x_m, meant to be perceived to be arriving from a certain direction, is convolved with the HRIR pair $\{H_\mathrm{r}, H_\mathrm{l}\}$ measured with the source in the same direction, and the convolved signals

$$y_\mathrm{l} = H_\mathrm{l} \star x_m, \quad \text{and} \tag{14.10}$$

$$y_\mathrm{r} = H_\mathrm{r} \star x_m \tag{14.11}$$

are applied to the headphones, as shown in Figure 14.9a. Since the decay time of the HRIR is always less than a few milliseconds, 256 to 512 taps in the filters are sufficient at a sampling rate of 44.1 kHz. The method ideally reproduces the ear canal signals that would have been produced had the sound source existed in the desired direction. The response of the headphones is assumed to be equalized to be the ideal dirac impulse.

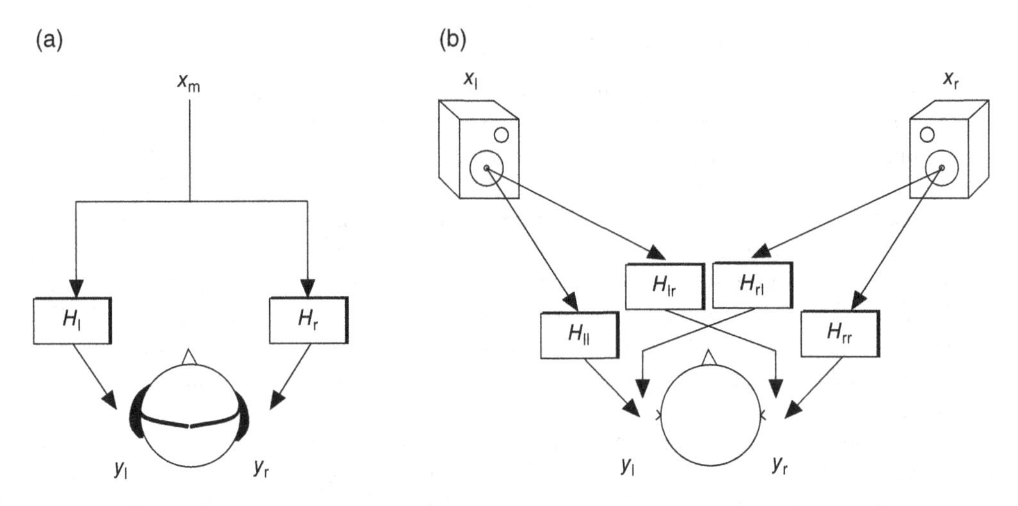

Figure 14.9 (a) Creating a virtual source with HRTF processing. (b) HRTFs in stereophonic listening.

The simple HRTF processing described above very probably produces the perception of an inside-head virtual source, which may also be perceived to be coloured if the headphones are not calibrated carefully. Using a head tracker, a far more realistic perception of external sound sources can be obtained. In *head tracking*, the direction of the listener's head is monitored about 10–100 times a second, and the HRTF filter is changed dynamically to keep the perceived direction of sound constant in the global coordinates (Begault *et al.*, 2001; Breebaart and Schuijers, 2008). In practice, the updating of the HRTF filter has to be done carefully so as not to produce audible artefacts.

The above-discussed technique simulates anechoic listening. It is also possible to simulate the binaural listening of a sound source in a real room. In this approach, the *binaural room impulse responses* (BRIRs) are measured from the ear canals of a subject in a room with a relatively distant source (Blauert, 1996; Møller, 1992). The response thus contains the contributions of the direct sound, reflections, and reverberation. When the response is convolved with a signal and reproduced over headphones, listeners would optimally perceive the sound as if they were in the room where the BRIR was measured. Additionally, the presence of the room response in the reproduction may cause externalization of the perceived auditory scene.

14.6.3 Virtual Listening of Loudspeakers with Headphones

An interesting application for HRTF technologies with headphones is listening to of existing multi-channel audio content. Here, each loudspeaker of a multi-channel loudspeaker layout is simulated using an HRTF set. For example, to listen to stereophonic signals $\{x_l, x_r\}$ over virtual loudspeakers in the directions of $\{-30°, 30°\}$, the signals are convolved with the HRTFs for each loudspeaker measured from the corresponding directions, as shown in Figure 14.9b (Blauert, 1996; Kirkeby, 2002). The convolved signals

$$y_l = H_{ll} \star x_l + H_{rl} \star x_r, \quad \text{and} \tag{14.12}$$

$$y_r = H_{rr} \star x_r + H_{lr} \star x_l \tag{14.13}$$

are applied to the headphones. The headphones are not shown in the figure.

The use of HRTFs measured in anechoic conditions is problematic in many cases. In most cases the listener is situated in a normal room or outdoors, and reproducing anechoic binaural sound to the listener may cause the virtual sources to be localized inside the head. The use of BRIRs, which optimally have a similar response as in the room where the listener is located is beneficial here. Correcting the effect of the HRTFs measured in anechoic conditions can also be done using measured room impulse responses, or by simulating the effect or room with a reverberator (Mackensen *et al.*, 1999).

14.6.4 Headphone Listening to Two-Channel Stereophonic Content

Headphone listening to stereophonic content without HRTF processing is significantly different from loudspeaker listening to the same stereophonic content. The cross talk present in loudspeaker listening is missing in headphone reproduction meaning that the sound from the left headphone only enters the left ear canal and likewise on the right side. Typically, audio engineers create the stereophonic audio content in studios with two-channel loudspeaker listening. A relevant question, then, is how does the spatial perception of the content change when listened to with headphones?

With amplitude-panned virtual sources the level difference between headphone channels is converted directly into an ILD, and the ITD remains zero. This is very different from loudspeaker listening, where the direction of the amplitude-panned sources is implied by ITD cues with the ILD at zero for low frequencies. Although this appears to be a potential source of large differences in spatial perception of the resulting virtual sources, the resulting spatial image is similar. The virtual sources are perceived in about the same order in the left–right direction as in loudspeaker listening. However, in headphone listening, the sources are perceived inside the listener's head due to erroneous monaural spectra and a lack of dynamic cues, as explained previously.

If the stereophonic content includes virtual sources that have been reproduced with time delays between the loudspeaker channels, as explained in Section 14.5.7, the result may be a vastly different spatial perception in headphone listening. For example, a 3-ms delay in the left loudspeaker may produce spread perception of the sound in loudspeaker listening, but in headphone listening the sound most probably is perceived to originate only from the right headphone.

14.6.5 Binaural Techniques with Cross-Talk-Cancelled Loudspeakers

Binaural recordings are meant to be played back in such a manner that the sound originating at the left ear should be applied to the left ear only, and correspondingly with the right ear, as depicted in Figure 14.9a. If such a recording is played back with a stereophonic set-up of loudspeakers, the sound from the left loudspeaker also enters the right ear, and vice versa, as shown in Figure 14.9b. This mixing of signals is called *cross talk*, which should be avoided in binaural playback.

In order to be able to listen to binaural recordings over two loudspeakers, methods to cancel cross talk have been proposed (Cooper and Bauck, 1989; Kirkeby *et al.*, 1998; Mouchtaris *et al.*, 2000). A system can be built to deliver binaurally recorded signals to the listener's ears using two closely spaced loudspeakers with crosstalk cancellation. The binaural signals are represented as a 2×1 vector $\mathbf{x}(n)$ and the produced ear canal signals also as a 2×1 vector $\mathbf{d}(n)$. The system can be described in the Z domain as

$$\mathbf{d}(z) = \mathbf{H}(z)\mathbf{G}(z)\mathbf{x}(z), \tag{14.14}$$

where $\mathbf{H}(z) = \begin{bmatrix} H_{\mathrm{ll}}(z) & H_{\mathrm{lr}}(z) \\ H_{\mathrm{rl}}(z) & H_{\mathrm{rr}}(z) \end{bmatrix}$ contains the electro-acoustic responses of the loudspeakers measured in the ear canals, as shown in the figure, and $\mathbf{G}(z) = \begin{bmatrix} G_{\mathrm{ll}}(z) & G_{\mathrm{lr}}(z) \\ G_{\mathrm{rl}}(z) & G_{\mathrm{rr}}(z) \end{bmatrix}$ contains the responses for performing inverse filtering to minimize the cross talk.

Ideally, $\mathbf{x}(z) = \mathbf{d}(z)$, which is obtained if $\mathbf{G}(z) = \mathbf{H}(z)^{-1}$. Unfortunately, direct inversion of the matrix is not feasible due to the non-idealities of the loudspeakers and the listening conditions. A regularized method to find an optimal $\mathbf{G}_{\mathrm{opt}}(z)$ has been proposed by Kirkeby *et al.* (1998):

$$\mathbf{G}_{\mathrm{opt}}(z) = \left[\mathbf{H}^{\mathrm{T}}(z^{-1})\mathbf{H}(z) + \beta\mathbf{I}\right]^{-1} \mathbf{H}^{\mathrm{T}}(z^{-1})z^{-m}, \tag{14.15}$$

where β is a positive scalar regularization factor and z^{-m} models the time delay due to the sound reproduction system. If β is selected to be very small, sharp peaks will result in the time-domain inverse filters, which may exceed the dynamic range of the loudspeakers. If β is larger, the impulse response of the inverse filter will have a longer duration in time, which is less demanding on the loudspeakers, but the price paid is that the inversion is less accurate (Kirkeby *et al.*, 1998).

In practice, this method performs best with loudspeakers near each other, since a larger loudspeaker base angle leads to colouration at lower frequencies. The listening area where the effect is audible is very small, because if the listener departs from the mid-line between the loudspeakers, a region about 1–2 cm wide, the effect is lost.

A nice feature of this technique is that the sound is typically externalized. This may be because head movements of the listener produce somewhat relevant cues, and because the sound is reproduced using far-field loudspeakers generating correct monaural directional spectral cues. However, although the sound is externalized, it is hard to reproduce the virtual sources in all directions with this technique. With a stereo dipole in the front, the reproduced sound scene is typically perceived only in the frontal hemisphere.

The technique is also sensitive to the reflections and reverberation of the listening room. It performs at its best only in spaces without prominent reflections. To obtain the best results, the HRTFs of the listener should be known, but very plausible results can be obtained with generic responses.

14.7 Digital Audio Effects

Digital audio effects are systems that modify audio signals fed at their inputs according to set control parameters and make the modified signal available at their outputs (Zölzer, 2011). The purpose of the processing is to modify perceptual characteristics of sound to meet artistic needs in audio engineering. The settings of the parameters of the effects are made by audio engineers, musicians, and even by the listeners of music or other audio. Some of the techniques already described in this book, such as equalization and virtual source positioning, can be classified as audio effects. Other examples of audio effects are:

- *Dynamic range control*, *dynamic processing*, or *automatic gain control*: The effect comprises a gain that is automatically controlled by the level of the input signal (Zölzer, 2008)

(see Figure 19.10 on page 412). A *dynamic range controller*, a.k.a. a *compressor*, attenuates sound the more the level increases beyond a certain threshold. A *limiter* is a similar effect, but limits the level of signal to a preset value. An *expander*, on the other hand, leaves large amplitudes untouched, but increasingly amplifies the signal the smaller the amplitude is. These techniques will be discussed later in this section, and also in the context of hearing aids in Section 19.5.2.
- *Pitch shifting*: The pitch of harmonic complexes can be shifted with different methods. A basic method is to re-sample the signal with time-scaling, which also scales the magnitude spectrum up or down. There are many methods to change only the pitch, leaving the spectral characteristics unchanged (Bristow-Johnson, 1995; Moulines and Laroche, 1995). This effect enables, for example, the correction of out-of-tune singing or the creation of different pitches from a sampled note of a musical instrument.
- *Chorus*, *flanger*, and *phaser*: These are effects where at least one copy of the original signal is modulated or changed in phase or pitch and the modified signal(s) is added to the original signal. This effectively cancels and enhances partials of the original signal and/or changes somewhat the spectral structure of the original signal (Orfanidis, 1995). For example, when the spectral structure of a single voice is smeared slightly in frequency, the processed voice may resemble the sound of a choir, which is the principle of the *chorus* effect.
- *Room effects*: These effects simulate the effect of a room on audio signals and are discussed in the next section.

The interested reader is referred to Zölzer (2011) for a more complete list of audio effects.

The dynamic range control methods have evoked a lively discussion among audio engineers – the *loudness war* (Vickers, 2011). A basic purpose of compression is to make loudness evoked by processed sound events roughly constant, and often also to maximize the loudness of the piece of audio content delivered. This processing is advantageous when audio content should be audible over loud background noise existing in the listening conditions. This is typically the case in, say, car audio. A downside of this processing is that the non-linear processing changes the spectrum of the input signal, typically generating harmonic distortion in the output. A problem arises when several compressors are in the signal path: the resulting attributes of perceived audio can differ immensely from those in the mastering studio and may lead to large changes in perceptual attributes of the reproduced sound. This situation can exist in broadcasting, because most broadcasters compress all broadcast audio regardless of whether the content has already been compressed or not; this is possibly followed by further compression by audio devices like those in a car, which also compress their output.

14.8 Reverberators

Generating the effect of reverberation in the produced sound is often desirable in audio content production. A digital audio effect device called a *reverberator* is able to *reverberate* the signal, causing a human listener to perceive the sound to be *reverberant*.

An intuitive method of creating such a room effect is to measure the impulse response of a real room and to compute the convolution of the signal and the impulse response. An alternative approach to obtaining the impulse response of a room is to simulate its acoustics using computational models of room acoustics (see Section 2.4.5). Unfortunately, convolution-based reverberation is not feasible in most cases due to the computational complexity of the filtering.

Instead of convolution, computationally less demanding DSP structures are often used to create the perception of reverberation. There are several methods of obtaining a room effect to make the processed sound reverberant, although the processing itself does not mimic the physical propagation of sound in rooms. These approaches are briefly summarized below.

14.8.1 Using Room Impulse Responses in Reverberators

If the impulse response of a target room is readily available, the most faithful reverberation method is to convolve the input signal with the impulse response. The result corresponds to the imaginary case where the input signal to be reverberated is applied to the loudspeaker in the measurement room and the listener listens to the signal recorded using the microphone used in the measurement. Note that the directional responses of the microphone and the sound source used in the measurement affect the result. For example, the reverberation effect will be considerably different if the directivity of the microphone is changed. An omnidirectional microphone captures the reverberation from all directions equally, while a directional microphone facing towards the source significantly suppresses sound arriving from the reverberant field. If these responses are used in convolving reverberators, the omnidirectional response will be perceived as much more reverberant than the directional response.

Direct convolution can be implemented by storing each sample of the impulse response as a coefficient of an FIR filter whose input is the signal recorded in a free field. Direct convolution easily becomes impractical if the length of the target response exceeds small fractions of a second, since it would translate into several hundreds of taps in the filter structure. A solution is to perform the convolution block by block in the frequency domain. Given the Fourier transforms of the impulse response and of a block of the input signal, the two can be multiplied point by point and the result transformed back to the time domain. As this kind of processing is performed on successive blocks of the input signal, the output signal is obtained by overlapping and adding the partial results (Oppenheim and Schafer, 1975). Thanks to the FFT computation of the discrete Fourier transform, this technique is significantly faster. A drawback is that, in order to be operated in real time, a block of N samples must be read and then processed while a second block is being read. Therefore, the input–output latency in samples is twice the size of a block, and this is not tolerable in practical real-time environments.

When a monophonic signal is convolved with a monophonic response, the room response is perceived to be in the direction of the loudspeaker. In reproduction with multiple loudspeakers, localizing the reverberation to the directions reproducible with the loudspeaker set-up is often desired. The microphone techniques discussed in Section 14.4 can be used for impulse response recording, and their pros and cons also hold in impulse-response-based reverberation. Some dedicated non-linear techniques for impulse response reproduction for multi-channel playback have also been proposed (Farina and Ugolotti, 1999; Merimaa and Pulkki, 2005; Tervo *et al.*, 2013), which overcome some flaws in the microphone techniques. In the non-linear techniques, measured impulse responses are processed into impulse responses for each loudspeaker in multi-channel layouts using instantaneous spatial analysis from the measured response. Although good results have already been obtained, the technical requirements for authentic spatial reproduction of impulse responses with such techniques are an on-going research topic.

The room effect, that is, the room impulse response, can be created by modelling how sound propagates and reflects from surfaces if the geometry of the room is known. This process is called room acoustics modelling, and several techniques are discussed in Section 2.4.5. When the impulse response is computed with such a model and used to reverberate sound, the process is often referred to as *auralization*, meaning making a room effect audible. Unfortunately, the impulse responses computed from models often do not create natural-sounding reverberation. The frequency range of the modelling may not be sufficient, and the late reverberant tail may sound unrealistically bright. In some cases, only the early reflections are modelled accurately, and the late response is generated using DSP structures (Savioja *et al.*, 1999).

14.8.2 DSP Structures for Reverberators

In the second half of the twentieth century, several engineers and acousticians tried to invent electronic devices capable of simulating the long-term effects of sound propagation in enclosures. The most important pioneering work in the field of *artificial reverberation* was that of Manfred Schroeder at the Bell Laboratories in the early 1960s (Schroeder, 1970). Schroeder introduced the recursive *comb filters* and the delay-based *all-pass filters* as computational structures suitable for the inexpensive simulation of complex patterns of echoes. In particular, the all-pass filter based on the recursive delay line has the form

$$y(n) = -g \cdot x(n) + x(n-m) + g \cdot y(n-m), \tag{14.16}$$

where m is the length of the delay in samples. The filter structure is depicted in Figure 14.10, where $A(z)$ is usually replaced by a delay line. The filter results in a dense impulse response and a flat frequency response. The structure rapidly became a standard component used in almost all the artificial reverberators designed until today (Moorer, 1979). It is usually assumed that all-pass filters do not introduce colouration in the input sound. However, this assumption is valid from a perceptual point of view only if the delay line is much shorter than the integration time of the ear, which is about 50–100 ms. If this is not the case, the time-domain effects become much more relevant, and the timbre of the incoming signal is significantly affected.

The shortcomings of such simple methods have been worked on in later generations of reverberators. There exists a vast number of different DSP structures proposed to create the effect of room reverberation (see Välimäki *et al.* 2012). The best reverberators are able to deliver natural-sounding room effects with significantly lower computational complexity than FIR-based reverberators. An advantage of such reverberators is the simpler control of the perceptual

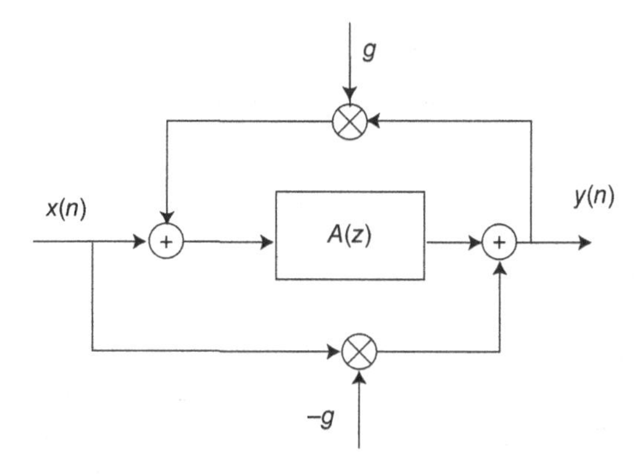

Figure 14.10 The all-pass filter structure.

attributes of the reverberation effect, which can be affected by changing the few parameters of the reverberator. In contrast, to change the parameters of an FIR-based reverberator requires the computation of the modified response and also changing the FIR filter coefficients on the fly, which is less flexible by default.

Summary

Many applications exist where sound is captured, processed, and reproduced. Perhaps the best-known application is audio content production, which serves the cinema, music, and gaming industries. The capturing of sound is performed with microphone techniques designed for different listening set-ups, which may be associated with visual or tactile displays. The microphone techniques have a major effect on the spatial characteristics of the reproduced sound, which can also be emulated using virtual source positioning techniques. Some shortcomings of traditional microphone techniques can be circumvented using non-linear time–frequency-domain reproduction techniques that have been developed by exploiting the knowledge of human spatial hearing resolution. The recorded sounds may also be processed using audio effects to modify the perceptual properties.

Further Reading and Available Toolboxes

The reproduction of sound is a wide topic with many directions in which the interested reader may continue his or her studies. Audio engineering and microphone techniques are covered by Katz and Katz (2007), Owsinski and O'Brien (2006), and Toole (2012). The time frequency-domain techniques for spatial sound reproduction are elaborated in Ahonen (2013), Laitinen (2014), and Vilkamo (2014). A concise summary of recent activities in binaural technologies is made Xie (2013). Digital audio effects can be studied further in Pirkle (2012), Reiss and McPherson (2014), and Zölzer (2011).

A number of software packages exist to perform spatial sound synthesis:

- *SoundScape Renderer* (TU Berlin/Universität Rostock). WFS, VBAP, Ambisonics, and binaural synthesis:
  `http://spatialaudio.net/ssr/.`
- *Panorama* (WaveArts/William Gardner). Binaural rendering with room modelling, and cross talk cancellation:
  `http://wavearts.com/products/plugins/panorama/`
- *Ambdec* (Fons Andriensen). Traditional, high-order, and optimized Ambisonics:
  `http://kokkinizita.linuxaudio.org/linuxaudio/index.html`
- *Blue Ripple Sound* (Richard Furse). Traditional and higher-order Ambisonics, HRTF rendering, and cross talk cancellation:
  `http://www.blueripplesound.com/product-listings/pro-audio`
- *Harpex* (Svein Berge). Time–frequency-domain parametric reproduction of B-format signals:
  `http://www.harpex.net/`
- *VBAP* (Aalto University). VBAP for 2D and 3D loudspeaker layouts:
  `http://www.acoustics.hut.fi/software/vbap/`
- *SPAT* (IRCAM). Panning, Ambisonics, room acoustics modelling, and sound diffusion:
  `http://www.fluxhome.com/products/plug_ins/ircam_spat`

- *Ambitools* (NTNU/Peter Svensson). Higher-order Ambisonics: `http://www.iet.ntnu.no/s̃vensson/software/index.html#AMBI`
- *sWonder* (Marije Baalman). WFS and binaural synthesis: `http://sourceforge.net/projects/swonder/`

References

Adelstein, B.D., Begault, D.R., Anderson, M.R., and Wenzel, E.M. (2003) Sensitivity to haptic–audio asynchrony *Proc. 5th Int. Conf. on Multimodal Interfaces*, pp. 73–76 ACM.

Ahonen, J. (2013) *Microphone front-ends for spatial sound analysis and synthesis with Directional Audio Coding*. PhD thesis, Aalto University.

Ahrens, J. (2012) *Analytic Methods of Sound Field Synthesis*. Springer.

Alexandridis, A., Griffin, A., and Mouchtaris, A. (2013) Capturing and reproducing spatial audio based on a circular microphone array. *J. Elec. Comp. Eng.* **2013**, 7.

Altinsoy, M.E. and Merchel, S. (2010) Cross-modal frequency matching: Sound and whole-body vibration. In Nordahl, R., Serafin, S., Fontana, F., and Brewster, S. (eds) *Haptic and Audio Interaction Design*. Springer, pp. 37–45.

Bauer, B.B. (1961) Phasor analysis of some stereophonic phenomena. *J. Acoust. Soc. Am.*, **33**, 1536–1539.

Bech, S. and Zacharov, N. (2006) *Perceptual Audio Evaluation – Theory, Method and Application*. John Wiley & Sons.

Begault, D.R., Wenzel, E.M., and Anderson, M.R. (2001) Direct comparison of the impact of head tracking, reverberation, and individualized head-related transfer functions on the spatial perception of a virtual speech source. *J. Audio Eng. Soc.*, **49**(10), 904–916.

Bennett, J.C., Barker, K., and Edeko, F.O. (1985) A new approach to the assessment of stereophonic sound system performance. *J. Audio Eng. Soc.*, **33**(5), 314–321.

Berge, S. and Barrett, N. (2010) A new method for B-format to binaural transcoding. *40th Int. Audio Eng. Soc. Conf.: Spatial Audio*. AES.

Berkhout, A., Vries, D., and Vogel, P. (1993) Acoustics control by wave field synthesis. *J. Acoust. Soc. Am.*, **93**(5), 2764–2778.

Björk, S. and Holopainen, J. (2005) *Patterns in Game Design*. Cengage Learning.

Blauert, J. (1996) *Spatial Hearing – Psychophysics of Human Sound Localization*. MIT Press.

Blumlein, A.D. (1931) U.K. Patent 394,325. Reprinted in 1986 *Stereophonic Techniques*, AES.

Borenius, J. (1985) Perceptibility of direction and time delay errors in subwoofer reproduction. *79th Audio Eng. Soc. Convention*. AES.

Breebaart, J. and Schuijers, E. (2008) Phantom materialization: A novel method to enhance stereo audio reproduction on headphones. *IEEE Trans. Audio, Speech, and Language Proc.*, **16**(8), 1503–1511.

Bristow-Johnson, R. (1995) A detailed analysis of a time-domain formant-corrected pitch-shifting algorithm. *J. Audio Eng. Soc.*, **43**(5), 340–352.

BS.775-2, I. (2006) Multichannel stereophonic sound system with and without accompanying picture. Recommendation, International Telecommunication Union, Geneva, Switzerland.

Chowning, J. (1971) The simulation of moving sound sources. *J. Audio Eng. Soc.*, **19**(1), 2–6.

Cobos, M., Lopez, J., and Spors, S. (2010) A sparsity-based approach to 3D binaural sound synthesis using time–frequency array processing. *EURASIP J. Adv. Sig. Proc.* **2010**, 1–13.

Cooper, D.H. (1987) Problems with shadowless stereo theory: Asymptotic spectral status. *J. Audio Eng. Soc.*, **35**(9), 629–642.

Cooper, D.H. and Bauck, J.L. (1989) Prospects for transaural recording. *J. Audio Eng. Soc.*, **37**(1/2), 3–39.

Coutrot, A., Guyader, N., Ionescu, G., and Caplier, A. (2012) Influence of soundtrack on eye movements during video exploration. *J. Eye Movement Res.*, **5**(4), 1–10.

Davis, M.F. (2003) History of spatial coding. *J. Audio Eng. Soc.*, **51**(6), 554–569.

Farina, A. and Ugolotti, E. (1999) Subjective comparison between stereo dipole and 3D ambisonic surround systems for automotive applications. *16th Int. Audio Eng. Soc. Conf: Spatial Sound Reproduction*. AES.

Francombe, J., Coleman, P., Olik, M., Baykaner, K., Jackson, P.J., Mason, R., Dewhirst, M., Bech, S., and Pederson, J.A. (2013) Perceptually optimized loudspeaker selection for the creation of personal sound zones. *52nd Int. Audio Eng. Soc. Conf.: Sound Field Control-Engineering and Perception*. AES.

Furse, R.W. (2009) Building an open AL implementation using ambisonics. *35th Int. Audio Eng. Soc. Conf.: Audio for Games* AES.
Gerzon, M.J. (1973) Periphony: With height sound reproduction. *J. Audio Eng. Soc.*, **21**(1), 2–10.
Hamasaki, K. and Hiyama, K. (2003) Reproducing spatial impression with multichannel audio. *24th Int. Conf. Audio Eng. Soc.: Multichannel Audio, The New Reality*. AES.
Hamasaki, K., Hiyama, K., and Okumura, R. (2005) The 22.2 multichannel sound system and its application. *Audio Eng. Soc. Convention 118* AES.
Hollier, M.P., Rimell, A.N., Hands, D.S., and Voelcker, R.M. (1999) Multi-modal perception. *BT Technol. J.*, **17**(1), 35–46.
ISO/IEC 23008-1 (2014) High efficiency coding and media delivery in heterogeneous environments. Standard.
ITU-R BS.1116-1 (1997) Methods for the subjective assessment of small impairments in audio systems including multichannel sound systems. Recommendation, International Telecommunication Union, Geneva, Switzerland.
ITU-T (1990) Tolerances for transmission time differences between the vision and sound components of a television signal. Recommendation J.100, International Telecommunication Union, Geneva, Switzerland.
Joly, A., Montard, N., and Buttin, M. (2001) Audio-visual quality and interactions between television audio and video *Sixth International Symposium on Signal Processing and its Applications 2001*, volume 2, pp. 438–441 IEEE.
Katz, B. and Katz, R.A. (2007) *Mastering Audio: The art and the science*. Taylor & Francis US.
Kirkeby, O. (2002) A balanced stereo widening network for headphones. *22nd Int. Audio Eng. Soc. Conf.: Virtual, Synthetic, and Entertainment Audio* AES.
Kirkeby, O., Nelson, P., and Hamada, H. (1998) Local sound field reproduction using two closely spaced loudspeakers. *J. Acoust. Soc. Am.*, **104**, 1973–1981.
Kohlrausch, A. and van de Par, S. (2005) Audio-visual interaction in the context of multi-media applications. In Blauert, J. (ed.) *Communication Acoustics*. Springer, pp. 109–138.
Kyriakakis, C. (1998) Fundamental and technological limitations of immersive audio systems. *Proc. of the IEEE*, **86**(5), 941–951.
Laitinen, M-V., (2014) Techniques for versatile spatial-audio reproduction in time-frequency domain. PhD thesis, Aalto University.
Laitinen, M-V., Vilkamo, J., Jussila, K., Politis, A., and Pulkki, V. (2014) Gain normalization in amplitude panning as a function of frequency and room reverberance. *55th Int. Audio Eng. Soc. Conf.: Spatial Audio*. AES.
Lemieux, P.A.S., Dressler, W., and Jot, J.M. (2013) Object-based audio system using vector base amplitude panning. US Patent App. 13/906,214.
Levitin, D.J., MacLean, K., Mathews, M., Chu, L., and Jensen, E. (2000) The perception of cross-modal simultaneity (or 'the Greenwich observatory problem' revisited). *AIP Conference Proceedings*, volume 517, pp. 323–329.
Lipshitz, S.P. (1986) Stereophonic microphone techniques... are the purists wrong? *J. Audio Eng. Soc.*, **34**(9), 716–744.
Mackensen, P., Felderhoff, U., Theile, G., Horbach, U., and Pellegrini, R.S. (1999) Binaural room scanning–A new tool for acoustic and psychoacoustic research. *J. Acoust. Soc. Am.*, **105**(2), 1343–1344.
Malham, D.G. and Myatt, A. (1995) 3-D sound spatialization using ambisonic techniques. *Comp. Music J.*, **19**(4), 58–70.
Menzel, D., Fastl, H., Graf, R., and Hellbrück, J. (2008) Influence of vehicle color on loudness judgments. *J. Acoust. Soc. Am.*, **123**, 2477–2479.
Merchel, S. and Altinsoy, M.E. (2013) Vibration in music perception. *Audio Eng. Soc. Convention 134* AES.
Merchel, S., Altinsoy, E., and Stamm, M. (2010) Tactile music instrument recognition for audio mixers. *Audio Eng. Soc. Convention 128* AES.
Merchel, S., Leppin, A., and Altinsoy, E. (2009) Hearing with your body: the influence of whole-body vibrations on loudness perception. *16th Int. Conf. Sound and Vibration*.
Merimaa, J. and Pulkki, V. (2004) Spatial impulse response rendering. *7th Intl. Conf. on Digital Audio Effects (DAFXÕ04)*.
Merimaa, J. and Pulkki, V. (2005) Spatial impulse response rendering I: Analysis and synthesis. *J Audio Eng. Soc.*, **53**(12), 1115–1127.
Møller, H. (1992) Fundamentals of binaural technology. *Appl. Acoust.* **36**(3), 171–218.
Møller, H., Sørensen, M.F., Hammershøi, D., and Jensen, C.B. (1995) Head-related transfer functions of human subjects. *J. Audio Eng. Soc.*, **43**(5), 300–321.
Moore, F.R. (1990) *Elements of Computer Music*. Prentice Hall.
Moorer, J.A. (1979) About this reverberation business. *Computer Music J.*, **3**(2), 13–28.
Moreau, S., Daniel, J., and Bertet, S. (2006) 3D sound field recording with higher order Ambisonics – objective measurements and validation of spherical microphone. *Audio Eng. Soc. Convention 120*. AES.

Mouchtaris, A., Reveliotis, P., and Kyriakakis, C. (2000) Inverse filter design for immersive audio rendering over loudspeakers. *IEEE Trans. Multimedia*, **2**(2), 77–87.

Moulines, E. and Laroche, J. (1995) Non-parametric techniques for pitch-scale and time-scale modification of speech. *Speech Commun.* **16**(2), 175–205.

Nelson, P. and Elliott, S. (1992) *Active Control of Sound*. Academic Press.

Oppenheim, A.V. and Schafer, R.W. (1975) *Digital Signal Processing*. Prentice-Hall.

Orfanidis, S.J. (1995) *Introduction to Signal Processing*. Prentice-Hall,

Owsinski, B. and O'Brien, M. (2006) *The Mixing Engineer's Handbook*. Thomson Course Technology.

Pätynen, J. and Lokki, T. (2010) Directivities of symphony orchestra instruments. *Acta Acustica United with Acustica*, **96**(1), 138–167.

Pihlajamäki, T., Santala, O., and Pulkki, V. (2014) Synthesis of spatially extended virtual sources with time–frequency decomposition of mono signals. *J. Audio Eng. Soc.* **62**(7/8), 467–484.

Pirkle, W. (2012) *Designing Audio Effect Plug-Ins in C++: With Digital Audio Signal Processing Theory*. Taylor & Francis.

Politis, A., Vilkamo, J., and Pulkki, V. n.d. Sector-based perceptual sound field reproduction in the spherical harmonic domain. Unpublished manuscript.

Pulkki, V. (1997) Virtual source positioning using vector base amplitude panning. *J. Audio Eng. Soc.*, **45**(6), 456–466.

Pulkki, V. (2001a) Coloration of amplitude-panned virtual sources. *Audio Eng. Soc. Convention 110*. AES.

Pulkki, V. (2001b) *Spatial Sound Generation and Perception by Amplitude Panning Techniques*. PhD thesis, Helsinki University of Technology, Laboratory of Acoustics and Audio Signal Processing.

Pulkki, V. (2002) Microphone techniques and directional quality of sound reproduction. *Audio Eng. Soc. Convention 112* AES.

Pulkki, V. (2007) Spatial sound reproduction with directional audio coding. *J. Audio Eng. Soc.*, **55**(6), 503–516.

Rafaely, B., Weiss, B., and Bachmat, E. (2007) Spatial aliasing in spherical microphone arrays. *IEEE Trans. Signal Proc.*, **55**(3), 1003–1010.

Reiss, J.D. and McPherson, A.P. (2014) *Audio Effects: Theory, Implementation and Application*. Taylor & Francis/CRC Press.

Rimell, A. and Owen, A. (2000) The effect of focused attention on audio-visual quality perception with applications in multi-modal codec design. *IEEE Int. Conf. Acoustics, Speech, and Signal Proc.*, volume 6, pp. 2377–2380 IEEE.

Robinson, C.Q., Mehta, S., and Tsingos, N. (2012) Scalable format and tools to extend the possibilities of cinema audio. *SMPTE Motion Imag. J.* **121**(8), 63–69.

Rumsey, F. (2001) *Spatial Audio*. Taylor & Francis.

Rumsey, F. (2006) *Sound and Recording: An introduction*. Taylor & Francis US.

Rumsey, F. (2011) Whose head is it anyway? Optimizing binaural audio. *J. Audio Eng. Soc.*, **59**(9), 672–675.

Sadek, R. and Kyriakakis, C. (2004) A novel multichannel panning method for standard and arbitrary loudspeaker configurations. *Audio Eng. Soc. 117th Convention*, AES.

Savioja, L., Huopaniemi, J., Lokki, T., and Väänänen, R. (1999) Creating interactive virtual acoustic environments. *J. Audio Eng. Soc.*, **47**(9), 675–705.

Schroeder, M. (1970) Digital simulation of sound transmission in reverberant spaces. *J. Acoust. Soc. Am.*, **47**(2), 424–431.

Silzle, A., George, S., Habets, E., and Bachmann, T. (2011) Investigation on the quality of 3D sound reproduction. *Proceedings of ICSA*, p. 334.

Simon, G., Olive, S., and Welti, T. (2009) The effect of whole-body vibration on preferred bass equalization in automotive audio systems. *Audio Eng. Soc. Convention 127* AES.

Solvang, A. (2008) Spectral impairment of two-dimensional higher order Ambisonics. *J. Audio Eng. Soc.*, **56**(4), 267–279.

Steinke, G. (1996) Surround sound – the new phase. An overview *Audio Eng. Soc. 100th Convention*, AES.

Streicher, R. and Dooley, W. (1985) Basic stereo microphone perspectives – a review. *J. Audio Eng. Soc.*, **33**(7/8), 548–556.

Tervo, S., Pätynen, J., Kuusinen, A., and Lokki, T. (2013) Spatial decomposition method for room impulse responses. *J. Audio Eng. Soc.*, **61**(1/2), 17–28.

Toole, F. (2012) *Sound Reproduction: the Acoustics and Psychoacoustics of Loudspeakers and Rooms*. Focal Press.

Torick, E. (1998) Highlights in the history of multichannel sound. *J. Audio Eng. Soc.*, **46**(1/2), 27–31.

Tournery, C., Faller, C., Kuech, F., and Herre, J. (2010) Converting stereo microphone signals directly to MPEG-surround. *Audio Eng. Soc. Convention 128* AES.

Välimäki, V., Parker, J.D., Savioja, L., Smith, J.O., and Abel, J.S. (2012) Fifty years of artificial reverberation. *IEEE Trans. Audio, Speech, and Language Proc.*, **20**(5), 1421–1448.

Vickers, E. (2011) The loudness war: Do louder, hypercompressed recordings sell better? *J. Audio Eng. Soc.*, **59**(5), 346–351.

Vilkamo, J. (2014) *Perceptually Motivated Time–Frequency Processing of Spatial Audio*. PhD thesis, Aalto University.

Vilkamo, J. and Pulkki, V. (2013) Minimization of decorrelator artifacts in directional audio coding by covariance domain rendering. *J. Audio Eng. Soc.*, **61**(9), 637–646.

Vilkamo, J., Lokki, T., and Pulkki, V. (2009) Directional audio coding: Virtual microphone-based synthesis and subjective evaluation. *J. Audio Eng. Soc.*, **57**(9), 709–724.

Vries, D. and Boone, M. (1999) Wave field synthesis and analysis using array technology. *IEEE Workshop on Applications of Signal Processing to Audio and Acoustics*, pp. 15–18, Mohonk Mountain House, New Paltz.

Welti, T. (2004) Subjective comparison of single channel versus two channel subwoofer reproduction. *117th Audio Eng. Soc. Convention*. AES.

Wenzel, E.M., Arruda, M., Kistler, D.J., and Wightman, F.L. (1993) Localization using nonindividualized head-related transfer functions. *J. Acoust. Soc. Am.*, **94**(1), 111–123.

Wierstorf, H. and Spors, S. (2012) Sound field synthesis toolbox. *Audio Engineering Society Convention 132* AES.

Wilska, A. (1938) *Untersuchungen über das richtungshoeren (Studies on Directional Hearing)*. PhD thesis, Helsinki University. English translation available: `http://www.acoustics.hut.fi/publications/Wilskathesis/`.

Xie, B. (2013) *Head-Related Transfer Function and Virtual Auditory Display*, volume 2. J. Ross Publishing.

Zölzer, U. (2008) *Digital Audio Signal Processing*. John Wiley & Sons.

Zölzer, U. (2011) *DAFX: Digital Audio Effects*. John Wiley & Sons.

15

Time–Frequency-domain Processing and Coding of Audio

A common trend in the field of audio is to process the audio signal in the time–frequency domain. In other words, the input audio signal is divided into a number of frequency bands which are processed separately and also depending on time. The aim in such processing is, for example, data compression, audio effects, or the enhancement of audio quality. The benefit of time–frequency processing in such tasks is that the structure of human hearing mechanisms is based on similar time–frequency analysis of the ear canal signals. Already many applications, such as the perceptual coding of audio, take advantage of the human hearing resolution in the time–frequency domain. An emerging field is multi-channel and spatial applications utilizing time–frequency processing.

15.1 Basic Techniques and Concepts for Time–Frequency Processing

The use of time–frequency transforms to visualize audio signals was already touched on in Section 3.2.6 on page 53. This chapter elaborates on the techniques and introduces some phenomena, concepts, and issues related to the processing of audio in the time–frequency domain. We will describe the time–frequency processing methods first using the concepts of frame-based analysis, and second using the concepts of downsampled filter banks.

15.1.1 Frame-Based Processing

Many time–frequency-domain audio techniques are implemented such that an input signal(s) is first divided into overlapping time frames, after which the frames are processed separately. We will first review the basic concepts and techniques in such approaches, which will then make it easier to understand the methods better.

Let us first denote the original signal as $x_{\mathrm{a}}(n_a)$, where n_a is a whole number running from zero to the number of samples R in the audio data in a file. The signal is divided into *frames*, which are short portions of the signal, typically with a length between 1 ms and 100 ms, as shown in Figure 15.1.

Communication Acoustics: An Introduction to Speech, Audio, and Psychoacoustics, First Edition.
Ville Pulkki and Matti Karjalainen.

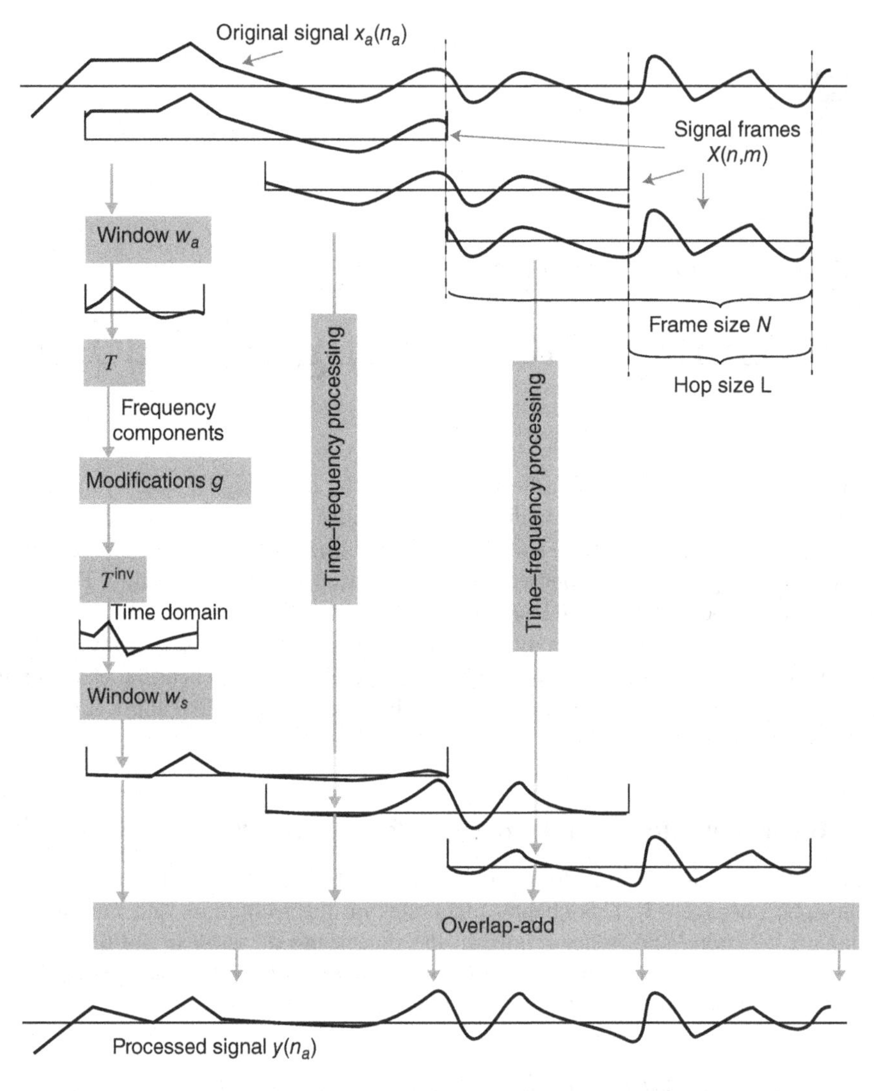

Figure 15.1 The principle of frame-based time–frequency-domain processing of audio. In this schematic example, the modifications of the spectral components implement a high-pass filter.

For each frame index m, a set of N samples is chosen from the input signal so that $x(n, m) = x_a(mL + n - 1)$ where $n \in [0, N - 1]$ is the time index within the frame, and L is the *hop size*, indicating by how much the frame is advanced in x_a. The difference between x_a and x is that x_a contains the whole data to be processed and x is a short portion of x_a that is currently being processed.

A frame of the signal, which may be windowed, is transformed into its frequency-domain representation $X(k, m)$ as

$$X(k, m) = \mathcal{T}(w_a(n)x(n, m)), \tag{15.1}$$

where $k = 0, \ldots, N-1$ is the frequency index, $w_a(n)$ is the analysis time window, and $\mathcal{T}$ is the transformation operation. $X(k, m)$ is a real- or complex-valued sample that represents the original signal in the time–frequency domain. $X(k, m)$ is often also called a *frequency bin*; however, note that the term in some contexts may also mean the spacing of the samples in frequency.

The frequency-domain representation $X(k, m)$ is transformed back into the time domain by

$$\hat{x}(n, m) = w_s(n)\mathcal{T}^{\text{inv}}(X(k, m)), \tag{15.2}$$

where $w_s(n)$ is the synthesis time window and $\mathcal{T}^{\text{inv}}$ is the inverse transformation operation. Assuming $L = N/2$, the design of $w_a(n)$ and $w_s(n)$ is typically such that it preserves the amplitudes across the overlapping frames:

$$w_a(n)w_s(n) + w_a(n+L)w_s(n+L) = 1, \qquad \text{for } n = 0, ..., (N/2-1), \tag{15.3}$$

which is known as the constant overlap-add (COLA) (Smith, 2011). An example fulfilling Equation (15.3) is to design both $w_a(n)$ and $w_s(n)$ as sequences which have entries that are the square roots of the corresponding entries of a Hann window sequence.

The final synthesized signal is computed as the sum of all signals

$$y(n_a) = \sum_m \hat{x}(n_a - mL, m), \tag{15.4}$$

where $\hat{x} = 0$ when $n_a - mL < 0$ or $n_a - mL \geq N$, and $0 \leq n_a < R$ runs through all samples in the original data. The summing operation thus takes subsequent frames of audio and adds them together, overlapping in time in a process called *overlap-add* (OLA), as also shown in Figure 15.1.

Audio coding applications aim to get the approximation $y(n_a) \approx x_a(n_a)$, striving to reconstruct the original signal exactly or to make any error due to quantization to have the least perceptual impact. In many other applications, such as those involving audio enhancement of any kind, the signal has to be modified somehow, and Equation (15.2) must thus be rewritten as

$$\hat{x}(n, m) = w_s(n)\mathcal{T}^{\text{inv}}(g(k, m)X(k, m)), \tag{15.5}$$

where $g(k, m)$ are frequency- and/or time-dependent real or complex gain values. The values of $g(k, m)$ affect the magnitudes and/or phases of corresponding frequency components, which, in principle, enables arbitrary filtering operations. However, depending on the applied transformation, time–frequency processing can be prone to aliasing artefacts, which must be avoided in the processing, as discussed below.

15.1.2 Downsampled Filter-Bank Processing

Various *downsampled filter banks* are used in time–frequency processing of audio. As briefly mentioned in Section 3.2.7, a filter bank is an array of band-pass filters which divide a broadband time-domain input signal into time-domain signals with narrowband frequency content. The process of downsampling in the context of filter banks is now discussed.

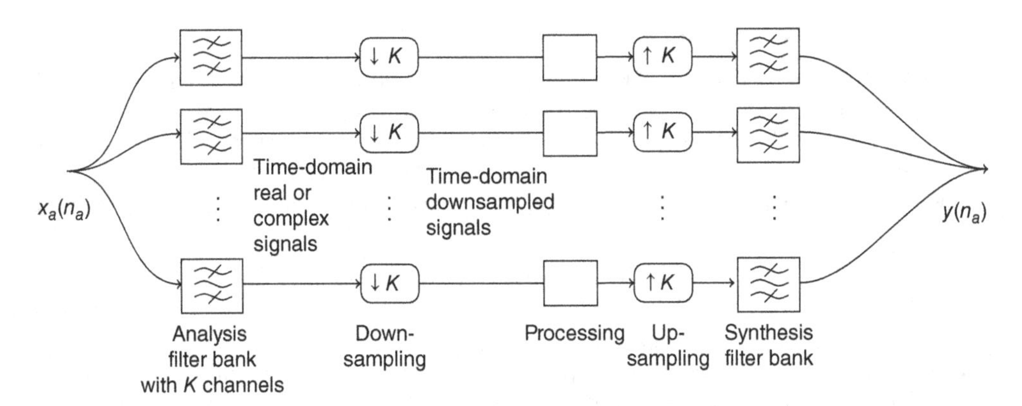

Figure 15.2 A schematic diagram of time–frequency processing of audio utilizing a downsampled filter bank.

A schematic diagram of a downsampled filter bank is shown in Figure 15.2. The input signal $x_a(n_a)$ is divided into K narrowband frequency bands using real- or complex-valued analysis filters, and the bands are downsampled by the same factor K. Downsampling is a procedure that is necessary for efficiency. Firstly, downsampling reduces the data rate for the processing or coding in the time–frequency domain. Without downsampling, dividing a signal into K frequency bands multiplies the number of data points per second by the factor K. Downsampling by a factor of K means that only every Kth sample is preserved while the others are discarded. This changes the sampling rate and causes all frequencies beyond the new Nyquist frequencies $\pm F_s/(2K)$ to fold back to between this interval.

Downsampling discards excessive data by an amount where the core signal information is still preserved. Secondly, the computational efficiency of the transform operation itself can be optimized when the output channels are downsampled. It is typical of symmetric downsampled structures, such as that in Figure 15.2, that they can be implemented with efficient DSP techniques. Similarly, the fast Fourier transform (FFT) is an efficient implementation of the discrete Fourier transform (DFT). Although various implementations can be realized, the working principles and properties of such filter banks can be considered in terms of this basic schematic diagram, as it is easy to comprehend.

The mechanism of how downsampling preserves the signal information with narrowband signals is discussed next. Discrete signals with the sampling rate F_s can only represent frequencies below the Nyquist frequency $f = F_s/2$, which is the same as the normalized angular frequency $\omega = \pi$ radians per sample. This property is observed in Equation (3.17b), where the tone sequences with frequencies beyond π radians per sample produce spectral coefficients that are equivalent to specific tone sequences with frequencies within the interval $\pm\pi$.

Upsampling is applied to re-map the frequencies to their original intervals. When upsampling by a factor K, a sequence of $K - 1$ zeros is appended, or concatenated, after every sample, in which case the repetitive nature of the spectrum is actualized as a set of duplicate frequency components in the new sampling interval. The spectral effects of downsampling and upsampling are illustrated in Figure 15.3.

The original spectral content is thus preserved, but the aliased components are spread all over the spectrum. In the design of time–frequency transform methods, this effect must be taken into account, and the aliased components must be removed either with a synthesis filter

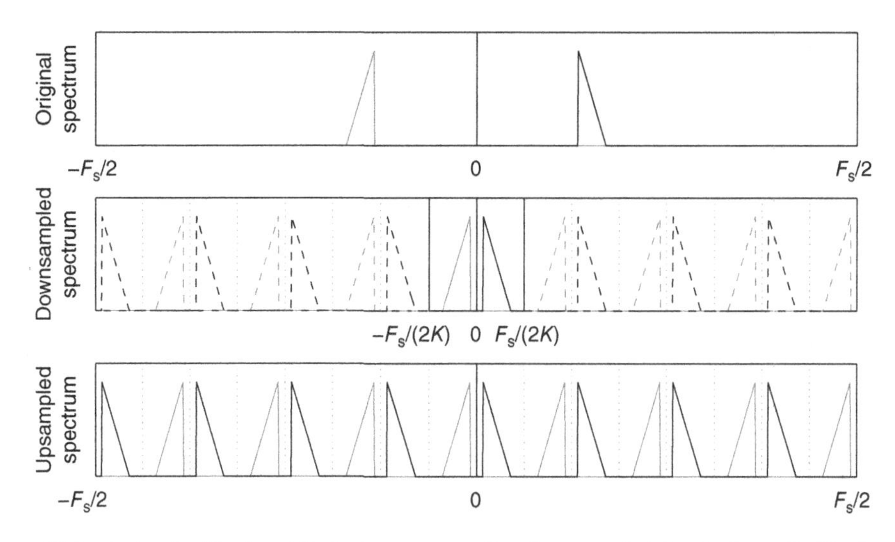

Figure 15.3 A signal with a band-limited spectrum is first downsampled (by the factor 8). The frequencies are folded to the limits set by the new sampling frequency. After upsampling by the same factor, the frequencies are replicated repetitively. They contain the original spectrum and a set of aliased spectra. Courtesy of Juha Vilkamo.

or through interference between frequency channels or time frames. In the downsampled filter bank processing shown in Figure 15.2, the aliased components are removed with a narrowband synthesis filter after upsampling, and the remaining channels are summed to obtain the original or modified signal. The modifications can again be implemented by some processing of the sub-band signals, such as by multiplication with real or complex values.

15.1.3 Modulation with Tone Sequences

The filter-bank processing discussed previously included specific filters for analysis and synthesis, but the design of the filters was not discussed at all. *Modulation* is often used as a tool in the design of the analysis and synthesis filter banks for time–frequency transforms, as will be shown in Section 15.2.4, and so the basics of modulation are reviewed in this section.

The modulation of a signal sequence $x(n)$ with a tone sequence refers to an operation in which each of its samples is multiplied by the corresponding sample from the tone sequence. In general, the signal sequence can be any signal, and in this context one may consider it to be the impulse response of an FIR filter. The real-valued modulation is

$$x_{\mathrm{modR}}(n) = x(n)\cos(\omega n) = x(n)\frac{e^{\mathrm{j}\omega n} + e^{-\mathrm{j}\omega n}}{2}, \tag{15.6}$$

and the complex-valued modulation is

$$x_{\mathrm{modC}}(n) = x(n)e^{\mathrm{j}\omega n}. \tag{15.7}$$

When modulated, the spectrum of the signal is centred with respect to the frequency of the modulator. A real modulator produces positive and negative frequencies, while a complex

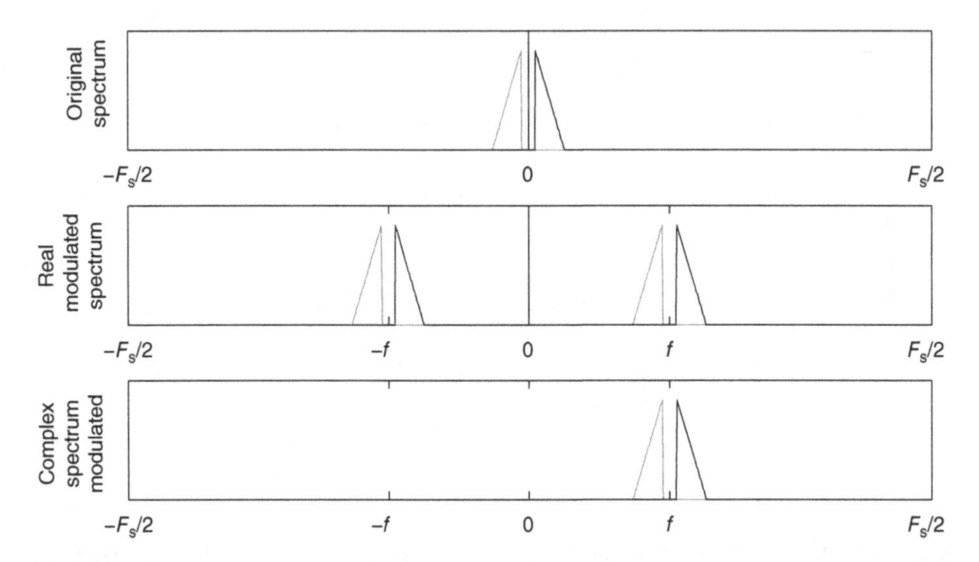

Figure 15.4 When a signal is modulated by a tone sequence with frequency $f = \omega F_S/(2\pi)$, its spectrum is centred with respect to that frequency. A real-valued modulator produces positive and negative frequencies, while a complex-valued modulator produces only positive frequencies. Courtesy of Juha Vilkamo.

modulator produces only one of the polarities, usually the positive frequencies, as shown in Figure 15.4. In time–frequency processing, modulation is applied to the impulse response of an FIR filter. The band-pass filters for time–frequency transforms can be designed by shifting the spectrum of a low-pass FIR filter using modulation to the desired centre frequencies.

Note that modulation as defined here is different from the amplitude modulation defined in Section 3.1.2. The modulation used in this chapter is real- or complex-valued sinusoid, in contrast to amplitude modulation, where the sinusoidal modulator is defined to be positive-valued – a sinusoid offset by a positive value.

15.1.4 Aliasing

Aliasing is a non-linear effect in digital signal processing by which signal components appear where none existed in the original signals. The effect is similar to non-linear distortion, but the term ‘aliasing’ is used to mean those typical distortions created by sampling. In time–frequency DSP, aliasing must be avoided, cancelled, or suppressed sufficiently to avoid perceived distortions in the sound.

Aliasing may occur in signals that are processed considerably in the frequency–domain. Aliasing components emerge especially if the applied time–frequency transform relies significantly on signal cancellation properties of the different samples in the time–frequency domain, which is a typical feature of non-redundant time–frequency representations. *Non-redundant transforms* are transforms that provide the lowest possible amount of data points per second without losing any signal information. A transform is *redundant* if some of the information it produces is repeated more than once, which may be necessary to avoid aliasing if the signal is processed in the frequency domain.

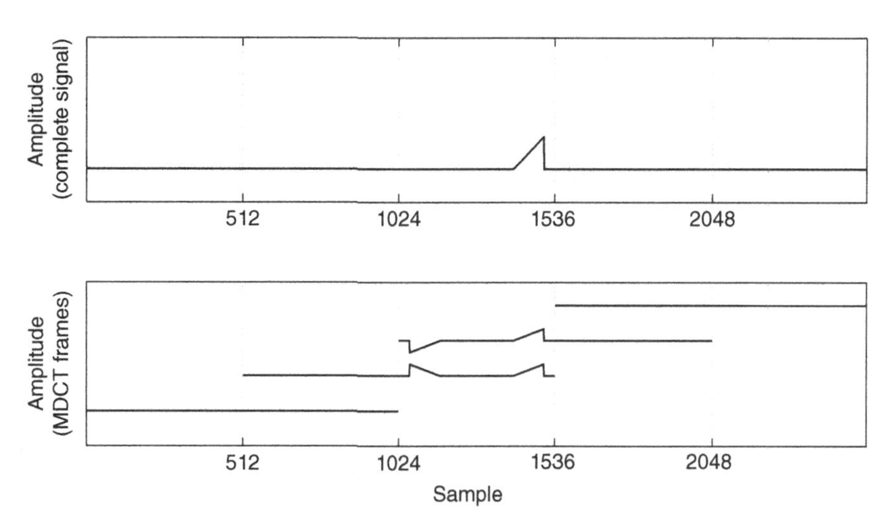

Figure 15.5 Illustration of a condition where time aliasing can occur during adaptive processing with frame-based time–frequency processing of audio. The upper graph is the original signal, and the lower one shows the time-domain counterparts of the modified discrete cosine transforms (MDCTs), described in Section 15.2.3, of frames of length 1024 samples each. The time-domain aliasing component is cancelled if the signals are unprocessed. In this illustration, the analysis and the synthesis windows of the MDCT are rectangular. Courtesy of Juha Vilkamo.

Consider Figure 15.5, where the upper graph shows the original signal and the lower one shows four subsequent time frames reproduced. Note that the second and third frames have a triangular component near sample 1024 that does not correspond to the original signal and is cancelled out when the frames are overlap-added. If the level of the second or third frame is modified in the processing of frequency-domain signals, the triangular component of the second frame produces a clearly audible aliasing component.

Similarly, a potential case of frequency aliasing is shown in Figure 15.6. A single frame of a static sinusoid is transformed into time-domain sub-bands using a filter bank. An added sinusoid of slightly higher frequency occurs in the two sub-bands with the highest energy, but the added sinusoids are exactly out of phase. If and when the sub-bands are added together during synthesis, the added sinusoids will cancel each other, removing the potential aliasing component. If, however, the gains g for the sub-bands differ, the cancellation is not perfect, and an aliasing component will remain. Note that these were examples of non-redundant transforms. Further on in this chapter, more robust transforms, with some redundancy, will also be discussed.

15.2 Time–Frequency Transforms

This section describes a few *time–frequency transforms* that are commonly used in the audio industry. The following concepts are applied to discuss the properties of filter banks:

- *Critical sampling* is a property whereby the combined sampling rate of the transformed frequency bands is the same as that of the original band. Critical sampling is exploited to minimize redundancy, which is a relevant and desirable property for audio coding. An

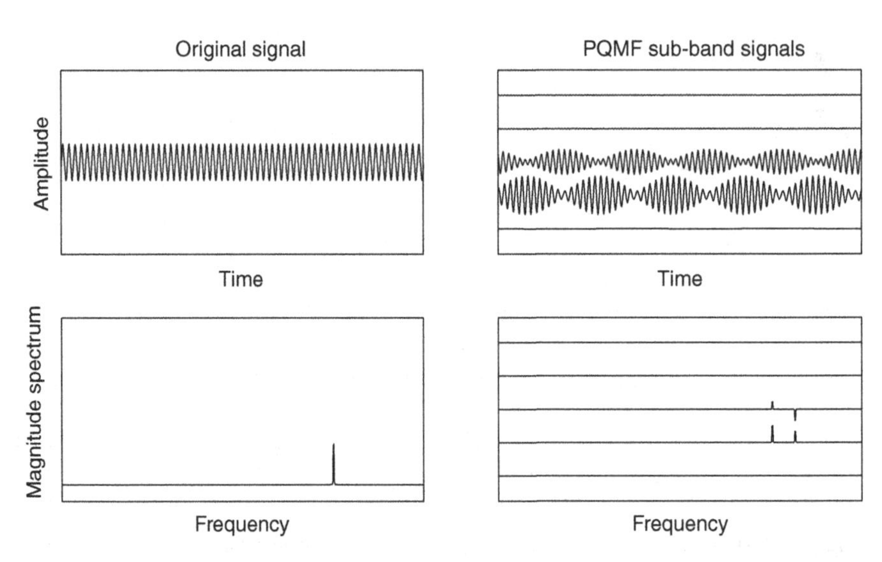

Figure 15.6 Illustration of a condition where frequency aliasing can occur during adaptive processing with filter-bank processing. The original tone and its spectrum are shown in the left column, and the time-domain counterparts of the pseudo-quadrature mirror filter (PQMF) bank sub-band signals, described in Section 15.2.4, and their spectra are shown in the right column. The right-hand peaks in the PQMF band spectra show the aliased frequency components, which are antiphasic with respect to each other in the sub-bands, visualized as polarity-up or polarity-down. The antiphasic peaks are cancelled if the sub-band signals are not processed. The complementary nature of the band signals is also visible in their waveforms. Courtesy of Juha Vilkamo.

example of a critically sampled transform is one that decomposes a real-valued signal using the sampling rate F_s into K real-valued bands, each with a sampling rate of F_s/K.

- *Oversampling* is a property whereby the combined sampling rate of the transformed bands is higher than that of the original signal. This property is typical of transforms intended for robust signal adjustments. Consider the previous example with the sampling rate of F_s and the decomposition into K bands, each with a sampling rate of F_s/K. However, if the produced frequency-band signals are complex-valued, the transformed signal is oversampled by a factor of two. An example of such a transform is the complex-modulated QMF described in Section 15.2.5.
- *Perfect reconstruction* means that the original signal waveform is retrieved exactly after the inverse transform, if the signals are not otherwise altered during the process.
- *Near-perfect reconstruction* implies that the original signal can be retrieved with a high degree of accuracy, but not exactly. The distortion produced by the transform itself is present but negligible in practice. The difference in properties of perfect and near-perfect reconstruction is mostly descriptive when it comes to audio.

15.2.1 Short-Time Fourier Transform (STFT)

The *short-time Fourier transform* (STFT) is an approach often applied in the field of audio. In STFT, an analysis window sequence $w_a(n)$ is applied to a signal sequence $x(n)$, after which the result is transformed into frequency bands using the DFT in Equation (3.17b). For a single frame of OLA processing, STFT analysis can be expressed as

$$X(k) = \sum_{n=0}^{N-1} w_{\mathrm{a}}(n)x(n)e^{-\mathrm{j}2\pi kn/N}, \tag{15.8}$$

and the inverse STFT, with the synthesis window sequence $w_{\mathrm{s}}(n)$, as

$$y(n) = \frac{1}{N} w_{\mathrm{s}}(n) \sum_{k=0}^{N-1} X(k)e^{\mathrm{j}2\pi kn/N}. \tag{15.9}$$

In practice, the STFT is computed using the fast Fourier transform (FFT). Due to the windowing and the overlapping frames, the representation is oversampled. STFT can be configured to be a perfect reconstruction transform by designing the windowing functions as in Equation (15.3).

Often, the STFT processes only the positive, DC, and Nyquist frequencies. Prior to the inverse FFT, the positive frequencies are complex-conjugate mirrored to the corresponding negative frequencies so as to obtain real-valued output signals, or, alternatively, an inverse FFT that assumes a real-valued output is applied.

An illustrative feature of the STFT representation for time–frequency adaptive processing is that each STFT sample, due to the definition of the DFT, corresponds to a circularly continuous tone (see Figure 15.7). If the forward transform involves windowing or zero padding prior to the FFT, the STFT components are simply phase-organized so that the frequency band signals cancel or amplify each other to form the envelope of the window and zero padding.

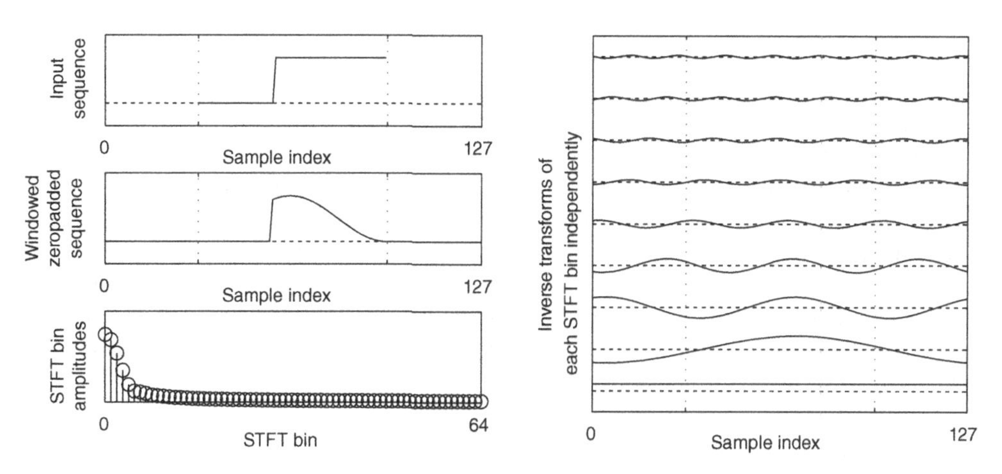

Figure 15.7 STFT analysis of a Hann-windowed and zero-padded step function. The STFT bins $X(k)$, $k \in [0, 1, \ldots, 63]$ correspond to circular tones through the entire window, but organized in phase and amplitude such that when summed, the original signal shape is obtained. Significant magnitude or phase processing removes the phase organization, which potentially spreads the waveform across the whole frame. In a typical implementation, such effects are suppressed by a synthesis window. Courtesy of Juha Vilkamo.

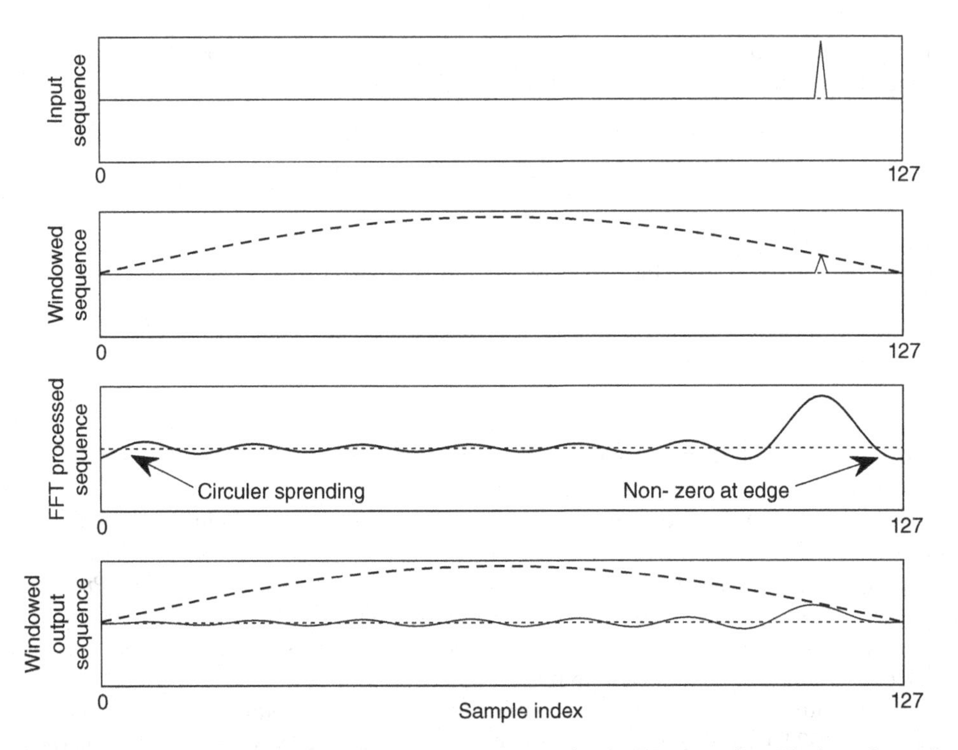

Figure 15.8 An example of circular effects that can occur during STFT processing. The input is a unity impulse. The signal is windowed, in this case using a square-root Hann window, and then processed using a zero-phase low-pass filter. The resulting signal is aliased in time. The synthesis filter suppresses the effects of the edge discontinuity. Courtesy of Juha Vilkamo.

Zero padding refers to the adding of a certain number of zeros before and/or after the windowed frame prior to computing the frequency transform.

When the STFT data are arbitrarily modified, the time-domain representation of the same data no longer has the original windowed shape. Furthermore, due to the circular representation, the values at the edges of the frame may deviate from zero, as shown in Figure 15.8. The non-zero edges then produce wideband transient artefacts in the output signal, and the conventional method for suppressing such effects is to apply a synthesis window, as shown in Figure 15.8. The synthesis window forces the time-domain signal values to zero at the start and at the end of the frame. The suppressed components of the signal are basically lost, which again may cause quality degradation.

15.2.2 Alias-Free STFT

A variant of STFT processing has been proposed, in which potential aliasing components can be avoided (Vickers, 2012). With a compromise in the form of an increased CPU load, it is possible to process the STFT bins so that the processing is equivalent to a time-domain convolution with an adaptive FIR filter. Such a linearized approach avoids the circular effects of the

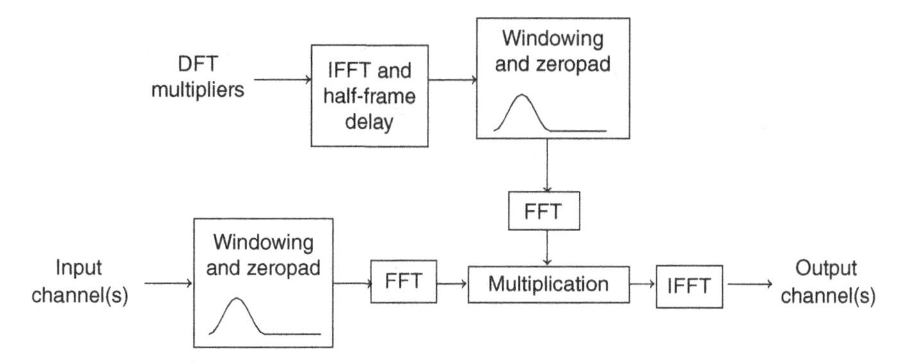

Figure 15.9 Alias-free STFT processing. The multipliers at the individual STFT lines are pre-processed: first, they are transformed into the time domain, and the response is shifted circularly by a half-frame delay, which centres the response to the middle of the window. The coefficients are then time-windowed so that abrupt window edges do not affect the spectrum. The response is appended with zero padding, and the result is transformed using FFT to obtain the processed STFT-domain multipliers. The combined length of the zero padding must be at least the combined length of the non-zero parts to obtain alias-free STFT processing. No synthesis window is necessary in this approach.

traditional STFT. The method involves extending the signal frame with zero padding and also pre-processing the complex processing multipliers (the STFT-domain magnitude and phase operators) in such a way that the non-zero part of their time-domain counterpart is limited in length. The steps for processing the multipliers involve applying the inverse FFT, windowing, zero padding, and FFT prior to applying them to the STFT signal frame, as shown in Figure 15.9. The combined length of the non-zero parts of the time-domain counterparts needs to be at most the same as the combined length of the zero-padding parts to avoid the circular convolution effects completely.

15.2.3 Modified Discrete Cosine Transform (MDCT)

The *modified discrete cosine transform* (MDCT) is a perfect reconstruction, real-valued, and critically sampled transform. The MDCT is widely applied in audio coding due to its non-redundancy and its property of representing narrowband signals with a relatively small number of prominent spectral coefficients (Geiger *et al.*, 2001; Neuendorf *et al.*, 2013). However, the MDCT is not designed to be robust for considerable signal adjustments in the time–frequency domain.

The MDCT analyses the signal in half-overlapping frames; that is, for windows of N samples the hop size is $N/2$ samples. The frequency partials of the MDCT represent sinusoids that are odd symmetric in the first half and even symmetric in the second half of the frame, as seen in Figure 15.10. For this reason, the individual MDCT frames generate temporal aliasing, which is countered by the opposite effect of the next frame. The real-valued amplitudes of the MDCT representation determine the amplitudes of the corresponding frequency components within the frame. Figure 15.5 is also an illustration of the functioning of the transform.

Like the STFT, the MDCT formulates a correlation of a windowed signal sequence with a set of tone sequences. With the MDCT, the tone sequences are real-valued. The MDCT can be

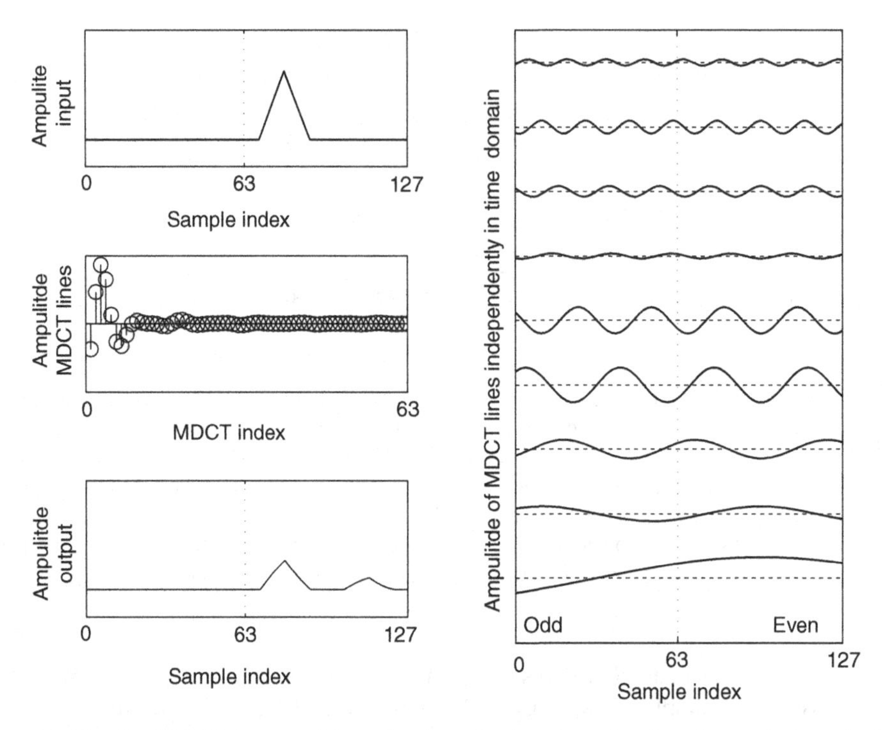

Figure 15.10 The MDCT decomposition of a triangular impulse using the window function in Equation (15.12). Before synthesis windowing, the MDCT frequency partials (right column) correspond to sinusoids that are odd symmetric in the first half of the frame and even symmetric in the second. The property results in the time-aliasing components, which are cancelled out by the opposite effect of the next frame. The output frame (bottom left) is synthesis windowed. Courtesy of Juha Vilkamo.

defined as follows. Let us denote $X(k)$, where $k = 0, \ldots, (N-1)$, as the MDCT components, and $x(n)$, where $n = 0, \ldots, (2N-1)$, as the time-domain sequence. The forward MDCT is

$$X(k) = \sum_{n=0}^{2N-1} w_{\mathrm{a}}(n)x(n)\cos\left[\frac{\pi}{N}\left(n + \frac{1}{2} + \frac{N}{2}\right)\left(k + \frac{1}{2}\right)\right], \tag{15.10}$$

whereas the inverse MDCT used to obtain the time-domain output signal sequence $y(n)$ is

$$y(n) = \frac{2}{N} w_s(n) \sum_{k=0}^{N-1} X(k)\cos\left[\frac{\pi}{N}\left(n + \frac{1}{2} + \frac{N}{2}\right)\left(k + \frac{1}{2}\right)\right]. \tag{15.11}$$

The window sequences must satisfy the amplitude preservation condition in Equation (15.3). Furthermore, they must be designed such that the aliasing cancellation property is preserved. These criteria are fulfilled, for example, by (Geiger *et al.*, 2001):

$$w_{\mathrm{a}}(n) = w_s(n) = \sin\left(\frac{\pi}{4N}(2n+1)\right). \tag{15.12}$$

15.2.4 Pseudo-Quadrature Mirror Filter (PQMF) Bank

The *pseudo-quadrature mirror filter* (PQMF) bank is a real-valued, critically sampled filter bank that has also been applied in audio coding. The working principle of the PQMF is illustrated in the schematic diagram in Figure 15.2. For a PQMF, a low-pass prototype FIR filter response $h_p(n)$ is designed and modulated with tone sequences to obtain the analysis band-pass filter responses $h_k^a(n)$ and the synthesis band-pass filter responses $h_k^s(n)$, where $k = 0, \ldots, K-1$ is the band index. The effect of modulation was explained in Section 15.1.3. The resulting filter bank thus has K filters of equal bandwidth measured in Hz. As illustrated in Figure 15.2, the band-pass signals, which are obtained by applying $h_k^a(n)$ to the signal, are downsampled to obtain the non-redundant form. At the synthesis stage, the signals are upsampled so that the frequency content is duplicated to the original frequencies, followed by the synthesis band-pass filters. Finally, the bands are combined to form the output signal. The processing might be relatively complex computationally, if implemented as shown in the figure. Mathematically equivalent but more efficient implementations also apply for this transform, such as the one outlined by Rothweiler (1983), but are beyond the scope of this book.

The necessary criteria for the prototype low-pass filter response $h_p(n)$ are that energy is preserved between the adjacent bands and that sufficient attenuation in non-adjacent bands is obtained to suppress the aliasing components (Creusere and Mitra, 1995; Cruz-Roldán *et al.*, 2002). For example, Cruz-Roldán *et al.* designed $h_p(n)$ by adjusting the frequency of a windowed sinc-type low-pass filter so that its frequency response approximates the energy-preservation requirement. The analysis and synthesis filters have the same magnitude spectrum, and thus the amplitudes are preserved in total between the adjacent bands. The downsampling–upsampling process is necessary because it reduces redundancy, but it also generates aliasing frequency components that remain after the synthesis filter. These aliasing components between the adjacent bands have opposite phases with respect to each other and cancel each other out if the frequency-band signals are not modified. The aliasing components between the non-adjacent bands are suppressed by the synthesis filter to the point of being negligible. Due to this feature in the design, the reconstruction by a PQMF bank is near-perfect. If frequency decomposition of very narrow bands is desired, the order of the prototype filter must be high.

The PQMF bank band filters are specified as follows. Let N be the prototype filter order, $n = 0, \ldots, N-1$ be the time index, K the number of bands, and $k = 0, \ldots, K-1$ the frequency-band index. The analysis filter responses are

$$h_k^a(n) = h_p(n) \cos\left[\frac{\pi}{2K}(2k+1)\left(n - \frac{N}{2} - \frac{K}{2}\right)\right], \tag{15.13}$$

and the synthesis filter responses are

$$h_k^s(n) = h_p(n) \cos\left[\frac{\pi}{2K}(2k+1)\left(n - \frac{N}{2} + \frac{K}{2}\right)\right]. \tag{15.14}$$

15.2.5 Complex QMF

As shown in Figure 15.6, PQMF is prone to aliasing artefacts if the values of adjacent frequency bands are modified. The ability to process the bands independently without such aliasing can be achieved by applying complex modulators instead of the real modulators in Equations (15.13) and (15.14):

$$h_k^{\mathrm{a}}(n) = h_{\mathrm{p}}(n) \exp\left[\mathrm{j}\frac{\pi}{2K}(2k+1)\left(n - \frac{N}{2} - \frac{K}{2}\right)\right] \tag{15.15}$$

and

$$h_k^{\mathrm{s}}(n) = h_{\mathrm{p}}(n) \exp\left[\mathrm{j}\frac{\pi}{2K}(2k+1)\left(n - \frac{N}{2} + \frac{K}{2}\right)\right], \tag{15.16}$$

where exp $[a]$ means e^a. The complex-modulation process entails a data representation that is oversampled by a factor of two. The difference between the real and complex modulators can be illustrated by their spectral representation, as in Figure 15.11. The real modulators map the low-pass prototype filter spectrum to both positive and negative frequencies, while only

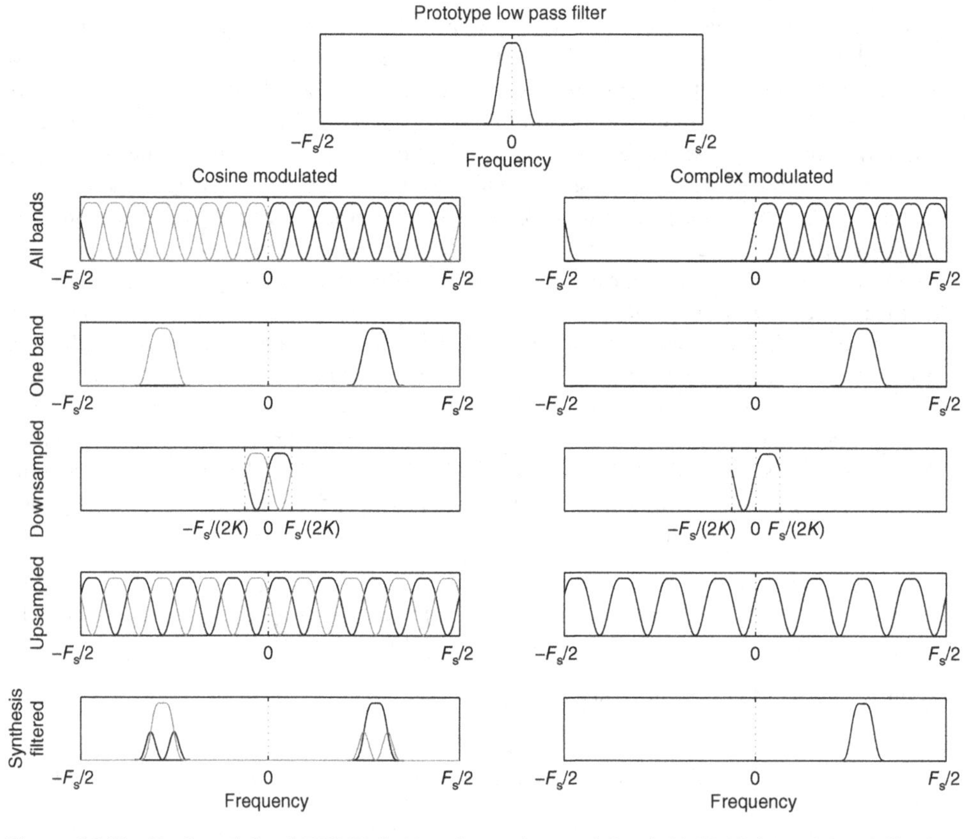

Figure 15.11 Real-modulated PQMF (left) and complex-modulated QMF (right) sub-band filtering. In real-valued processing, the positive and negative frequencies alias on top of each other during the downsampling–upsampling process. The aliased frequencies remain prominent after the synthesis filter. The neighbouring frequency bands have the same aliasing components but with opposite sign, thus cancelling the effect of aliasing. With complex-modulated processing, such aliasing components are absent in the first place, and thus the bands can be processed independently. Courtesy of Juha Vilkamo.

positive frequencies are produced when using complex modulation. In the downsampling process, the positive and negative frequencies of the real modulation wrap on top of each other to become the aliasing components of the neighbouring bands. With both modulators, the aliasing components of the non-neighboring bands are present and suppressed by the synthesis filter. The *complex-modulated QMF* is a typical transform in perceptually motivated time-frequency spatial audio processing techniques, discussed, for example by Breebaart *et al.* (2005, 2007) and Herre *et al.* (2012).

15.2.6 Sub-Sub-Band Filtering of the Complex QMF Bands

A typical resolution in complex QMF processing is $K = 64$ frequency bands, which, with a sampling rate of 44.1 kHz, results in bandwidths of approximately 345 Hz. As discussed earlier, the frequency resolution of hearing functions follows ERB or Bark bands (see Section 9.4.3 on page 167). Thus, this resolution is insufficient for the lowest frequencies. Therefore, the implementations discussed in Breebaart *et al.* (2005, 2007) and Herre *et al.* (2012) have applied a cascaded filter bank at the lowest frequency bands to obtain a higher frequency selectivity, which shown in Figure 15.12. Furthermore, some of the bands are summed together to reduce the complexity.

Different configurations exist for such cascading and summing, and the implementation presented in the figure involves feeding the lowest three bands to filter banks having 8, 4, and 4 bands, respectively, without further downsampling. The resulting bandwidths are then narrower at low frequencies and wider at high frequencies, although the transition is not smooth. In some perceptually motivated applications, the signal analysis is performed with frequency-band signals in which the higher QMF bands are also combined to form a perceptually motivated frequency resolution. The resulting bandwidths of one configuration in such processing are shown in Figure 15.13. The bandwidths follow somewhat the Bark frequency bands. Note that although the perceptual analysis is performed in such combined bands at the higher frequencies, the processing and inverse QMF transforms are applied at the original QMF bands because they contain the full information of the signal content.

15.2.7 Stochastic Measures of Time–Frequency Signals

In several time–frequency processing techniques for spatial audio (for example, Breebaart *et al.*, 2005, 2007; Faller, 2006; and Herre *et al.*, 2012) the spatial sound is synthesized by controlling the energies and interdependencies of the loudspeaker signals in the frequency bands. Let $X_1(k, m)$ and $X_2(k, m)$ be the signals of the two channels of a stereophonic set-up in the frequency domain with frequency band k and frame index m. The following parameters are often used to describe the channel or inter-channel energetic properties within a time–frequency area. Of the computations, the expectation operation $\mathrm{E}[\cdot]$ is typically implemented using a mean or a sum of the samples over a time–frequency area. First, let us define the channel energies and their cross-term:

$$\begin{aligned} E_1 &= \mathrm{E}\left[|X_1|^2\right] \\ E_2 &= \mathrm{E}\left[|X_2|^2\right] \\ \gamma_{12} &= \mathrm{E}\left[X_1 X_2^*\right], \end{aligned} \tag{15.17}$$

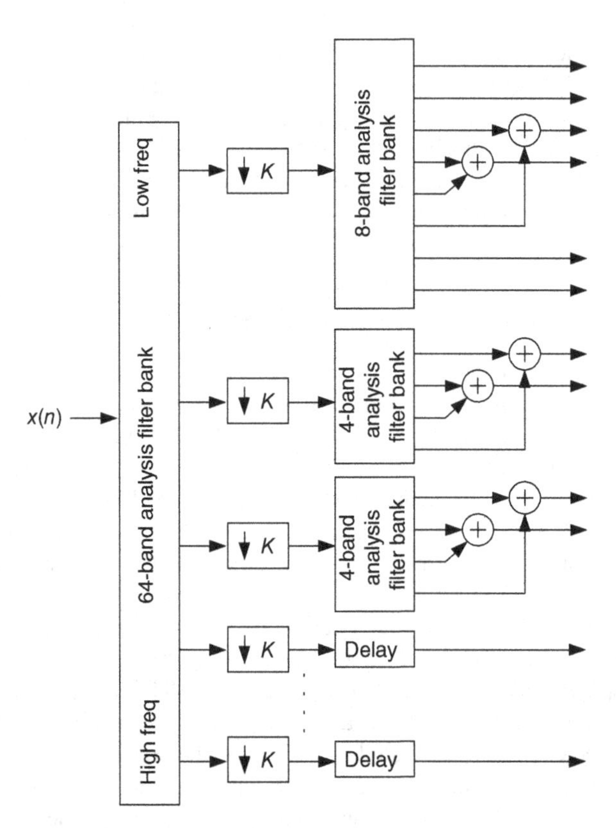

Figure 15.12 Hybrid QMF; that is, the sub-sub-band filtering of the lowest frequencies in a uniform, complex-modulated QMF bank. Some bands are combined to reduce complexity. The order of combination is specific, since the outputs of the secondary filter banks are not in the order of the absolute frequency. Adapted from Herre *et al.* (2005).

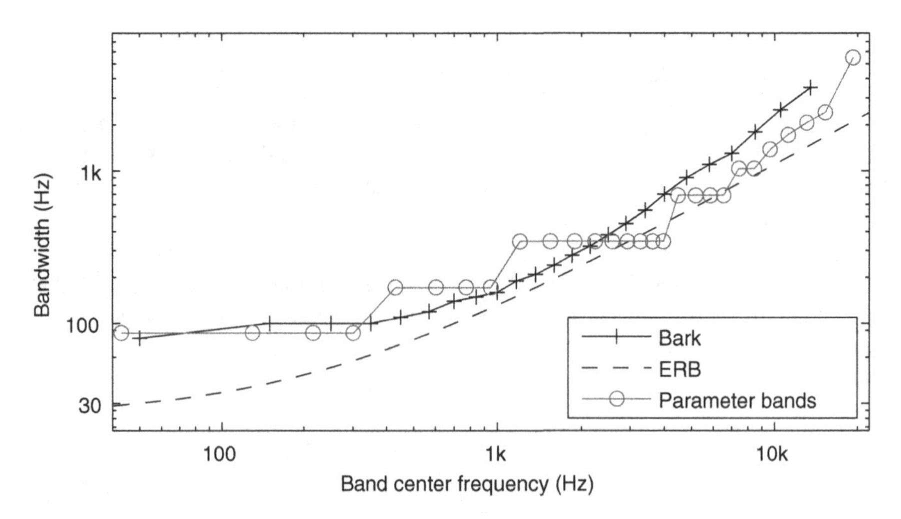

Figure 15.13 The bandwidths of an implementation using a hybrid QMF sub-sub-band filter bank that combines the bands in a perceptual fashion compared with Bark and ERB bandwidths.

where X_2^* is the complex conjugate of X_2. The inter-channel level difference in dB is

$$\text{ICLD} = 10\log_{10}\left(\frac{E_1}{E_2}\right), \tag{15.18}$$

and the inter-channel phase difference is

$$\text{ICPD} = \arg\left(\gamma_{12}\right), \tag{15.19}$$

where the arg() operator determines the angle of a complex value in the complex plane. The inter-channel coherence is

$$\text{ICC} = \frac{|\gamma_{12}|}{\sqrt{E_1 \cdot E_2}}. \tag{15.20}$$

The value ICC is a normalized similarity index between 0 and 1, where ICC = 1 means that the signals are coherent, although potentially with level differences, and ICC = 0 means that the signals are incoherent. A similar measure without the absolute-value operator is

$$\text{ICC}' = \frac{\gamma_{12}}{\sqrt{E_1 \cdot E_2}}, \tag{15.21}$$

which also includes the angle of the complex-valued cross-term. If the imaginary part is ignored, the value ICC ranges from -1 to 1 and determines the inverse or in-phase coherence between the channel pair. Ignoring the imaginary part means that signals that are out of phase by $\pi/2$ are treated as incoherent signals, which may be suitable, for example, for stereo upmixing using direct-ambience decomposition (Avendano and Jot, 2002; Faller, 2006). The task of the processing algorithm dictates how to account for the phase offset between the signals.

15.2.8 Decorrelation

Decorrelation is a method for processing a signal so that its ICC $\approx$ 0 with respect to the original signal as well as with respect to the processed signals from other decorrelators. Ideally, a decorrelator is designed such that the perceptual characteristics of the sound are least affected. Decorrelation is necessary for applications that increase the number of independent channels, such as upmixing or surround sound rendered from a few microphone signals.

There are various implementations of decorrelators. A typical approach is to alter the time or phase structure of the signal over a short time interval. Examples of such processes are different delays in frequency bands, all-pass filters, and convolutions with short noise sequences. It is widely known in the field that decorrelators can cause degradation of the perceived sound quality with certain signal content such as applause (Kuntz *et al.*, 2011; Laitinen *et al.*, 2011), because, as a consequence of the decorrelation, the sharp temporal structure of the signal is altered. Applications employing decorrelators thus often apply specific processes to avoid or reduce such effects, for example, with onset detectors that bypass transients from the decorrelators.

15.3 Time–Frequency-Domain Audio-Processing Techniques

In this section, a set of key applications in the field of perceptually motivated time–frequency audio processing is reviewed.

15.3.1 Masking-Based Audio Coding

With audio coding methods such as MPEG-1 Layer-3 (MP3) (ISO/IEC, 1993) and MPEG-2 Advanced Audio Coding (AAC) (Bosi *et al.*, 1997; ISO/IEC, 1997), the main means of reducing the bit-rate is to transform the signals into the time–frequency domain and to optimize the quantization of the time–frequency samples using a perceptual masking model, as shown in Figure 15.14.

If quantization noise was equally spread over the entire frequency region, it would be easily audible in those frequency regions where the signal level was low. The spectrum of quantization noise can be shaped so that it follows the masking curve created by the signal, but shifted slightly lower in level. See the figures in Section 9.2 for examples of masking curves. In general, if the level is set to about 13 dB lower than that of the signal, quantization noise is no longer audible, although corresponding quantization noise with a flat spectrum is clearly annoying. This effect is known as the '13-dB miracle' from an audio demonstration given by J. D. Johnston and K. Brandenburg at AT&T Bell Labs in 1990. The audio signals used in the demonstration are described by Brandenburg and Sporer (1992).

Critically sampled filter banks are preferred for audio coding, since signal modifications, except those for the quantization, are not intended, and the property of having the least number of data points for the transmission is desired.

15.3.2 Audio Coding with Spectral Band Replication

By reducing the bit rate, a limit is eventually reached when the quantization noise of a traditional audio encoder significantly exceeds the masking threshold, equivalent to exceeding the −13-dB noise level in the previous example. To optimize the quality in scenarios when bit rates of, for example, 24 kilobits per second per channel are applied, the method of spectral band replication (SBR) (Ekstrand, 2002) can be applied to utilize the typical temporal similarities

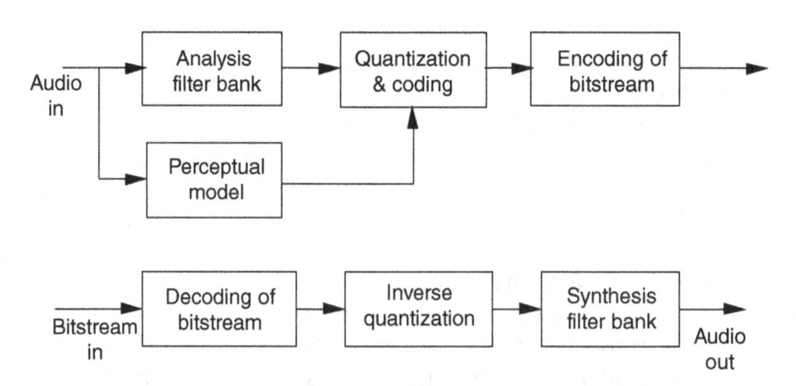

Figure 15.14 Block diagram of an audio encoder and decoder based on perceptual masking. Adapted from Brandenburg (1999).

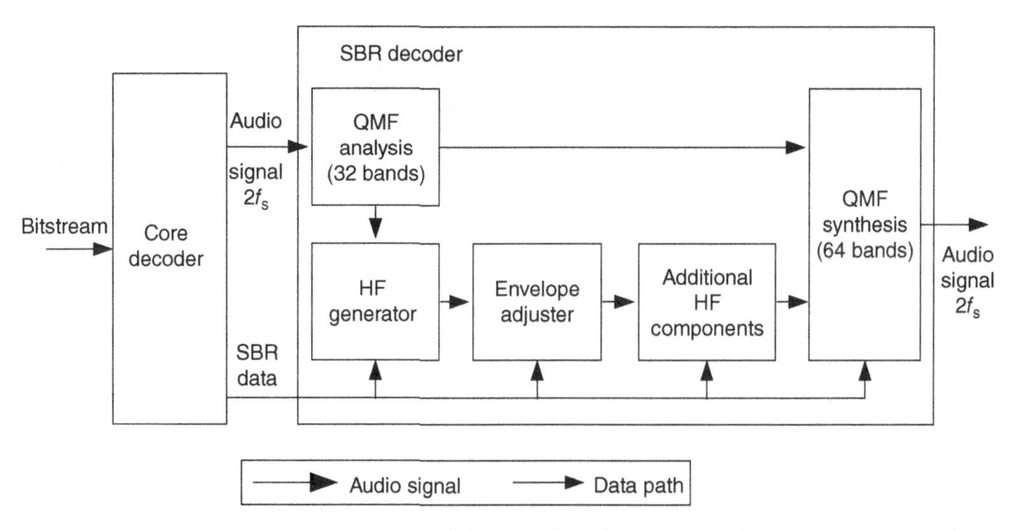

Figure 15.15 Spectral band replication, in which only the lower frequencies are transmitted by the core encoder, for example AAC, and where the higher frequencies are predicted based on the lower frequencies, and adjusted to better match the original higher frequencies based on low-bit-rate side information. In this case, the sampling frequency of the output signal is doubled from the audio signal originating from the core decoder. Other ratios of sampling frequencies are also possible. Adapted from Ekstrand (2002).

between the low and high frequencies. With SBR, the higher frequencies are not transmitted, and the bits are instead allocated to convey the lower frequencies more accurately, from which the higher frequencies are predicted (see Figure 15.15). Side information at a low bit rate is transmitted to adjust the spectral envelope of the higher frequencies to match better that of the original higher frequencies.

15.3.3 Parametric Stereo, MPEG Surround, and Spatial Audio Object Coding

In stereo and multi-channel audio signal transmission with low bit rates, the bit allocation can be optimized by transmitting the spatial aspect as low-bit-rate side information and transmitting only a reduced number of downmixed audio channels (Baumgarte and Faller, 2003; Faller and Baumgarte, 2003; Schuijers *et al.*, 2003). For two-channel stereo signals, Parametric Stereo (PS) (Breebaart *et al.*, 2005; Purnhagen, 2004), the signal-flow diagram of which is shown in Figure 15.16, can be used to convey the channel data using a mono downmix and the parametric side information containing the ICLD, ICPD, and ICC parameters in the frequency bands. MPEG Surround (Breebaart *et al.*, 2007; Herre *et al.*, 2005) and MPEG Spatial Audio Object Coding (SAOC) (Herre *et al.*, 2012) are similar parametric multi-channel techniques. MPEG Surround can be used for efficient transmission of 5.1 surround audio content in stereo or mono downmix channels with spatial metadata. SAOC provides a mixture of audio objects, that is, single-channel audio signals, in the downmix channels, the spatial rendering of which can be manually adjusted at the receiver end based on the parametric side information. An example of an SAOC application is found in the context of virtual reality, where the different talker

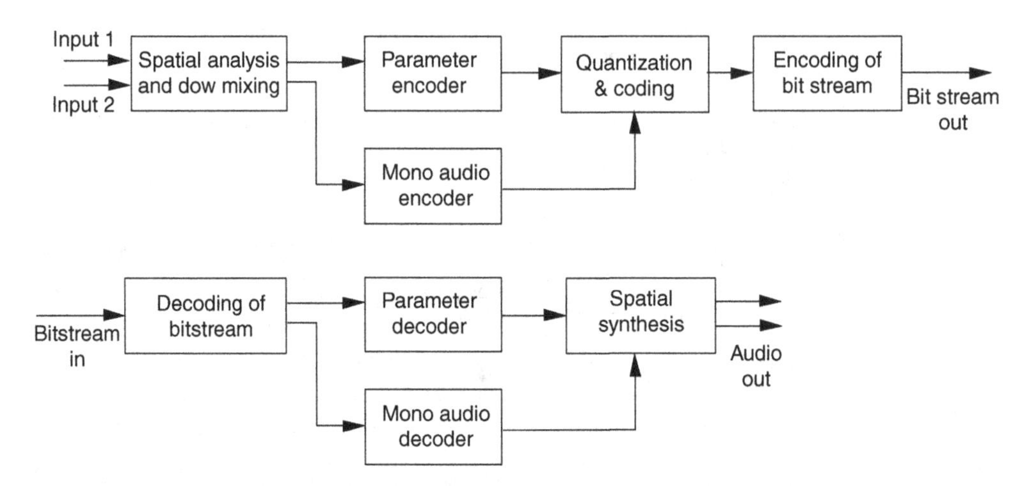

Figure 15.16 Parametric Stereo encoding and decoding. The inter-channel level differences, phase differences, and coherences are measured in the frequency bands for transmission as low-bit-rate side information. The signals are downmixed and encoded with a core encoder, such as AAC. At the receiver end, the AAC bit stream is decoded. The spatial properties of the stereo sound are re-synthesized using amplitude and phase adjustments and decorrelation. Adapted from Breebaart *et al.* (2005).

signals can be combined to save the bit rate, but spatially rendered independently based on the positioning of the talkers in the virtual environment.

The parametric side information can be embedded in the bit stream of the core coder in such a way that a receiver without a parametric decoder can decode the downmix channels. Enhanced decoders can take advantage of the parametric side information using the same bit stream. The downmix channels can be encoded using, for example, AAC or adaptively with a speech codec, which is a technique applied in the recent Unified Speech and Audio Coding (USAC) scheme (Neuendorf *et al.*, 2013).

15.3.4 Stereo Upmixing and Enhancement for Loudspeakers and Headphones

One of the tasks in audio is to present a two-channel stereophonic audio track using more than two loudspeakers. In principle, it is not possible to derive more independent channels than there already are. However, when the low spatial resolution of humans is taken into account, stereophonic signals can be rendered to a higher number of loudspeakers with plausible results.

In adaptive stereo upmixing (Avendano and Jot, 2004; Faller, 2006), the stereo sound is modelled in frequency bands in terms of the direct and ambient signal components that are redistributed to the extended loudspeaker set-up, as shown in Figure 15.17. An example of a direct-ambience model is to assume an amplitude-panned direct source and incoherent equal-energy ambience in the frequency bands, expressed as

$$\begin{bmatrix} X_1(k,m) \\ X_2(k,m) \end{bmatrix} = \begin{bmatrix} \sqrt{g(k,m)} \\ \sqrt{1-g(k,m)} \end{bmatrix} D(k,m) + \begin{bmatrix} A_1(k,m) \\ A_2(k,m) \end{bmatrix}, \tag{15.22}$$

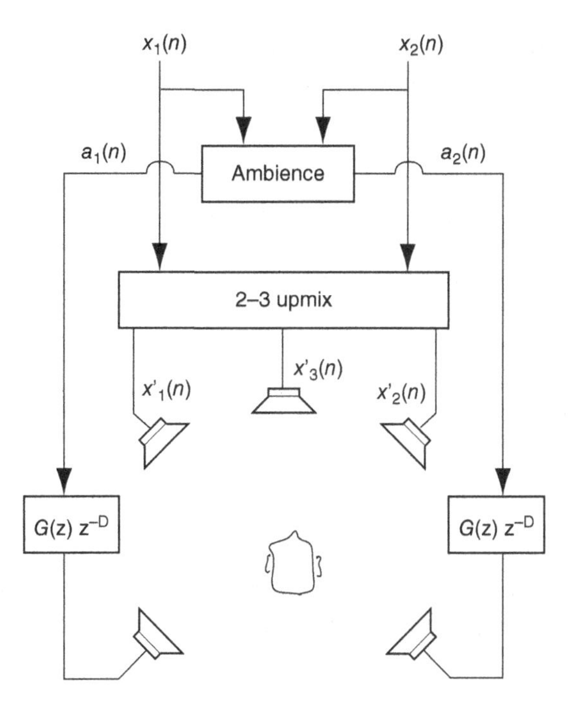

Figure 15.17 In time–frequency-domain upmixing from stereophonic sound to 5.1 surround, a direct-ambience decomposition is applied to obtain the rear ambience channels and a 2–3 upmixer is applied to obtain the centre channel. The operators $G(z)$ are decorrelating all-pass filters. Adapted from Avendano and Jot (2004).

where $0 \leq g(k, m) \leq 1$ is an energy-panning coefficient, $D(k, m)$ is the amplitude-panned signal, and $A_1(k, m)$ and $A_2(k, m)$ are the left and right channel ambience signals, respectively. According to the model, the two ambience components and the amplitude-panned components are incoherent with respect to each other, and the ambience energy is the same in both channels. The model parameters, g, $E_D = \mathrm{E}\left[|D|^2\right]$, and $E_A = \mathrm{E}\left[|A_1|^2\right] = \mathrm{E}\left[|A_2|^2\right]$, can be solved uniquely based on the inter-channel measures described in Section 15.2.7. Different direct-ambience models have been discussed by Merimaa *et al.* (2007). The benefits sought by upmixing stereo to a surround set-up with a bigger number of loudspeakers include improved directional quality of amplitude-panned virtual sources in a larger listening area, and a more evenly surrounding reproduction of the ambience, such as reverberation, also in a larger listening area (Faller, 2006).

A concept similar to upmixing, but for headphone playback, involves adaptively processing the frequency bands of a stereo signal to obtain the natural binaural characteristics. Menzer and Faller (2010) and Faller and Breebaart (2011) applied direct-ambience signal analysis to process the ambience signal to match the frequency-band coherence occurring in a diffuse sound field. The direct signal was processed using head-related transfer functions (HRTFs), which are free-field transfer functions from a source to both ears, or with binaural room impulse responses (BRIRs) (Faller and Breebaart, 2011). The procedure was reported in informal listening to improve the perceived naturalness of the sound over the original stereo (Menzer and

Faller, 2010) and to improve the width and accuracy of the sound stage over the conventional methods of binaural processing of stereo sound (Faller and Breebaart, 2011).

Summary

Various methods exist to map a time-domain audio signal to the frequency domain, and different methods have been developed for different applications. If the signal is to be processed in the frequency domain, the time–frequency transform methods used should be robust against aliasing artefacts. Processing in the time–frequency domain enables the use of efficient coding strategies for monophonic and multi-channel signals.

Further Reading

The reader might find Smith (2011) useful for learning more about time-frequency signal processing techniques. The reader is encouraged to read more on coding of audio signals in Kahrs and Brandenburg (1998). The well-known audio codecs, such as MPEG-1 Layer-3 (mp3), and Advanced Audio Coding (AAC) are explained by Brandenburg (1999). A deeper discussion on multi-channel audio reproduction is given by Breebaart and Faller (2008), and an extensive review of transform-based parametric audio coding can be found in Herre and Disch (2014).

References

Avendano, C. and Jot, J.M. (2002) Ambience extraction and synthesis from stereo signals for multi-channel audio up-mix. *IEEE Int. Conf. Acoustics, Speech, and Signal Proc.*, volume 2, pp. 1957–1960.

Avendano, C. and Jot, J.M. (2004) A frequency-domain approach to multichannel upmix. *J. Audio Eng. Soc.*, **52**(7/8), 740–749.

Baumgarte, F. and Faller, C. (2003) Binaural cue coding – Part I: Psychoacoustic fundamentals and design principles. *IEEE Trans. Speech and Audio Proc.*, **11**(6), 509–519.

Bosi, M,, Brandenburg, K., Quackenbush, S., Fielder, L., Akagiri, K., Fuchs, H., and Dietz, M. (1997) ISO/IEC MPEG-2 advanced audio coding. *J. Audio Eng. Soc.*, **45**(10), 789–814.

Brandenburg, K. (1999) MP3 and AAC explained, *17th Int. Conf. Audio Eng. Soc.* AES.

Brandenburg, K. and Sporer, T. (1992) "NMR" and "Masking Flag": Evaluation of Quality Using Perceptual Criteria. *11th Int. Audio Eng. Soc. Conf.: Test & Measurement* AES.

Breebaart, J. and Faller, C. (2008) *Spatial Audio Processing: MPEG Surround and Other Applications.* John Wiley & Sons.

Breebaart, J., Chong, K.S., Disch, S., Faller, C., Herre, J., Hilpert, J., Kjörling, K., Koppens, J., Linzmeier, K., Oomen, W., Purnhagen, H., and Rödén, J. (2007) MPEG Surround – the ISO/MPEG standard for efficient and compatible multi-channel audio coding. *Audio Eng. Soc. Convention 122.* AES.

Breebaart, J., van de Par, S., Kohlrausch, A., and Schuijers, E. (2005) Parametric coding of stereo audio. *EURASIP J. Appl. Signal Proc.*, **2005**, 1305–1322.

Creusere, C.D. and Mitra, S.K. (1995) A simple method for designing high-quality prototype filters for M-band pseudo QMF banks. *IEEE Trans. Signal Proc.*, **43**(4), 1005–1007.

Cruz-Roldán, F., Amo-López, P., Maldonado-Bascón, S., and Lawson, S.S. (2002) An efficient and simple method for designing prototype filters for cosine-modulated pseudo-QMF banks. *IEEE Signal Proc. Letters*, **9**(1), 29–31.

Ekstrand, P. (2002) Bandwidth extension of audio signals by spectral band replication. *1st IEEE Benelux Workshop on Model based Processing and Coding of Audio.*

Faller, C. (2006) Multiple-loudspeaker playback of stereo signals. *J. Audio Eng. Soc.*, **54**(11), 1051–1064.

Faller, C. and Baumgarte, F. (2003) Binaural cue coding – Part II: Schemes and applications. *IEEE Trans. Speech and Audio Proc.*, **11**(6), 520–531.

Faller, C. and Breebaart, J. (2011) Binaural reproduction of stereo signals using upmixing and diffuse rendering. *Audio Eng. Soc. Convention 131.* AES.

Geiger, R., Sporer, T., Koller, J., and Brandenburg, K. (2001) Audio coding based on integer transforms. *Audio Eng. Soc. Convention 111*. AES.

Herre, J. and Disch, S. (2014) Perceptual audio coding. In Chellappa, R. and Theodoridis, S. (eds) *Image, Video Processing and Analysis, Hardware, Audio, Acoustic and Speech Processing*. Academic Press, pp. 757–800.

Herre, J., Purnhagen, H., Breebaart, J., Faller, C., Disch, S., Kjörling, K., Schuijers, E., Hilpert, J., and Myburg, F. (2005) The reference model architecture for MPEG spatial audio coding. *Audio Eng. Soc. Convention 118*.

Herre, J., Purnhagen, H., Koppens, J., Hellmuth, O., Engdegaard, J., Hilpert, J., Villemoes, L., Terentiv, L., Falch, C., Hölzer, A., Valero, M.L., Resch, B., Mundt, H., and Oh, H.O. (2012) MPEG spatial audio object coding – the ISO/MPEG standard for efficient coding of interactive audio scenes. *J. Audio Eng. Soc.*, **60**(9), 655–673.

ISO/IEC (1993) Coding of moving pictures and associated audio for digital storage media at up to about 1.5 mbit/s – Part 3: Audio. Standard 11172-3.

ISO/IEC (1997) MPEG-2 advanced audio coding, AAC. Standard JTC1/SC29/WG11 (MPEG).

Kahrs, M. and Brandenburg, K. (1998) *Applications of Digital Signal Processing To Audio and Acoustics*, volume 437. Springer.

Kuntz, A., Disch, S., Bäckström, T., and Robilliard, J. (2011) The transient steering decorrelator tool in the upcoming MPEG unified speech and audio coding standard. *Audio Eng. Soc. Convention 131*.

Laitinen, M-V., Küch, F., Disch, S., and Pulkki, V. (2011) Reproducing applause-type signals with directional audio coding. *J. Audio Eng. Soc.*, **59**(1/2), 29–43.

Menzer, F. and Faller, C. (2010) Stereo-to-binaural conversion using interaural coherence matching. *Audio Eng. Soc. Convention 128*.

Merimaa, J., Goodwin, M.M. and Jot, J.M. (2007) Correlation-based ambience extraction from stereo recordings. *Audio Eng. Soc. Convention 123*.

Neuendorf, M., Multrus, M., Rettelbach, N., Fuchs, G., Robilliard, J., Lecomte, J., Wilde, S., Bayer, S., Disch, S., Helmrich, C., Lefebvre, R., Gournay, P., Bessette, B., Lapierre, J., Kjörling, K., Purnhagen, H., Villemoes, L., Oomen, W., Schuijers, E., Kikuiri, K., Chinen, T., Norimatsu, T., Chong, K.S., Oh, E., Kim, M., Quackenbush, S., and Grill, B. (2013) The ISO/MPEG unified speech and audio coding standard – consistent high quality for all content types and at all bit rates. *J. Audio Eng. Soc.*, **61**(12), 956–977.

Purnhagen, H. (2004) Low complexity parametric stereo coding in MPEG-4. *7th Int. Conf. on Digital Audio Effects DAFx04*.

Rothweiler, J. (1983) Polyphase quadrature filters – a new subband coding technique. *IEEE Int. Conf. on Acoustics, Speech, and Signal Proc.*, volume 8, pp. 1280–1283.

Schuijers, E., Oomen, W., and Breebaart, J. (2003) Advances in parametric coding for high-quality audio. *Audio Eng. Soc. Convention 114*.

Smith, J.O. (2011) *Spectral Audio Signal Processing*. W3K.

Vickers, E. (2012) Frequency-domain implementation of time-varying FIR filters. *Audio Eng. Soc. Convention 133*. AES.

16

Speech Technologies

Speech technology covers applications such as the recognition or synthesis of speech, speaker recognition, and optimized coding and enhancement of speech. Major research efforts have been invested in speech technology both in academia and in industry, which have led to several breakthroughs. After several decades of research, coding, synthesis, and recognition of speech are used extensively in several applications. The widespread adoption of speech technologies has been enabled by the advent of sufficiently powerful and relatively inexpensive digital processors.

Many user interfaces based on speech exist. A computer may take commands by recording the speech of the user, thus requiring *speech recognition* abilities, and it may deliver messages to the user by producing intelligible speech, which requires *speech synthesis* technology. This interaction was shown conceptually in Figure I.5 on page 5. In mobile communication, the goal is often to present speech at as low a bit rate as possible, which has led to many methods for *speech coding*, where the special characteristics of speech signals have been taken into account.

This chapter provides a very brief overview of different technologies in speech coding, synthesis, and recognition. The focus is very much on acoustics, signal processing, and audio, and the linguistic and statistical aspects are, in many places, treated superficially. Overall, the aim of this chapter is to give a general description of the main fields in speech technology and to guide interested readers to more comprehensive sources.

Speech enhancement is a wide area, which covers such topics as *error concealment*, *noise reduction*, *bandwidth extension*, and *echo control* in the context of speech communication (Vary and Martin, 2006). The goal of noise reduction and echo control should be clear to the reader. Bandwidth extension refers to techniques used to extend the bandwidth of transmitted narrowband speech to deliver better speech quality to the user. Error concealment refers to applications where the speech decoder is forced to 'fill in' missing data from the received speech signal. For example, when some frames of speech are lost in communication, the content of missing frames has to be guessed through some intelligent solutions to maximize speech quality. These techniques are not discussed in detail in this book.

Communication Acoustics: An Introduction to Speech, Audio, and Psychoacoustics, First Edition.
Ville Pulkki and Matti Karjalainen.

16.1 Speech Coding

The idea in *speech coding* is to transmit or store and replay speech signals using minimal information capacity (number of bits) and with the best possible sound quality. In reality, this implies optimization, which is a trade-off between quality and cost. For an essentially unrestricted channel capacity, the speech signal can be conveyed with such high quality that no degradation is perceived, and there is no need to further improve quality. The transmission in this case is said to be *transparent*. Unfortunately, digital transmission (and storage) of speech requires, in practice, that the information capacity of the channel is limited. The number of bits per second to be transmitted, the bit rate, can be reduced by using proper coding techniques – the proper representation of the speech signal and quantization to a lower rate of bits. In wireless speech communication in particular, the channel capacity is limited, and cost vs. quality optimization is important.

Early analogue telephony transmitted speech as an electrical signal over a wired network. Although this seems a relatively straightforward approach, the system had some technical challenges, and seminal psychoacoustic testing was performed to find the smallest frequency region of speech that would still produce good enough speech quality for the receiver. The first frequency region suggested for telephones ranged from 250 Hz to 2750 Hz (Martin, 1930), but this was later altered to the range between 300 Hz and 3400 Hz, known as the *telephone band*. The telephone band was found to be a good compromise between technical constraints and the intelligibility of speech, and it is still in use today.

A basic technology for digital speech coding is the standard G.711, where the speech signal in the telephone band is sampled at 8 kHz using pulse-code modulation (PCM) with logarithmic quantization using 8 bits (ITU, 1988; Paez and Glisson, 1972). The first version of G.711 was finalized in 1972, after which it has been used widely in communication networks.

The high levels of a signal are thus represented with larger quantization steps than the lower levels. Logarithmic quantization generates unacceptable distortion in a music signal. However, the artefacts are not prominent with speech signals, and logarithmic quantization, in practice, gives a dynamic range of about 12 bits. Thus, when each sample is quantized using 8 bits, and sampled at the rate of 8 kHz, the resulting rate of transmission is 64 kbit/s, which is acceptable in wired communication, but relatively high for wireless telephony.

The demand for further reducing the data rate emerged when the technologies for digital mobile phones were developed in the 1980s. A more effective speech coding principle was developed, which takes advantage of the knowledge of speech production mechanisms. In principle, the method finds the parameters of a simple *source–filter model* of speech production and transmits only the model parameters, possibly with a residual signal not produced by the model (Kleijn and Paliwal, 1995). The filter parameters are often modelled using linear predictive coding, and the source is modelled as noise (unvoiced phonation) or an impulse train (voiced phonation). The linear predictive coding fits an all-pole filter, or an IIR filter, to the signal, as explained in Section 3.3.5 on page 59, and in parallel to this the parameters of the source signal are found, such as pitch, gain, and voicing of the excitation. The parameters of the source and the filter are adapted by measuring how well the system output fits the signal. The parameters are then transmitted and are used at the receiving site to synthesize speech, as shown in Figure 16.1.

There are many kinds of speech codecs designed for different purposes, such as the 13-kbit/s codec with regular pulse excitation with long-term prediction (RPE-LTP) for the first generation GSM mobile phones (ETSI, 1992). RPE-LTP uses an 8th-order linear prediction filter to

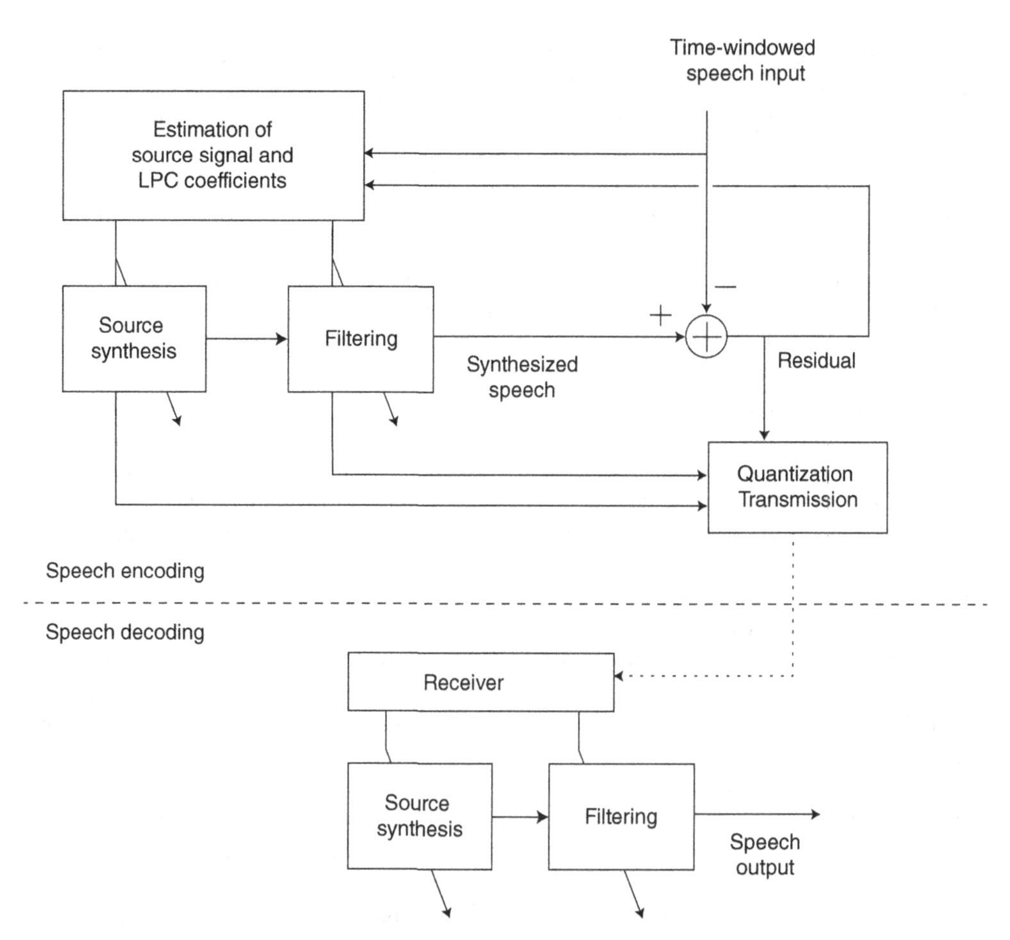

Figure 16.1 A schematic diagram of speech encoding and decoding. The parameters of the source–filter model are estimated iteratively, and the residual-error signal can be weighted perceptually to adapt the system to the most relevant features of speech. The transmitted data consist of filter and source parameters, and possibly the residual signal as well.

model the formant structure of speech, and a single-tap pitch predictor is used to model the periodic structure of voiced speech. Speech compression is achieved by 3:1 downsampling of the residual. Currently, different variants of the code-excited linear prediction (CELP) codec are used widely. In CELP, the residual signal is not sent as a signal, only a few parameters describing the residual are transmitted, which gives a significantly better speech quality at low bit rates. In VoIP applications, the codec commonly used is the conjugate-structure algebraic code-excited linear prediction (CS-ACELP) codec specified in standard G.729 (Salami *et al.*, 1998). The adaptive multi-rate (AMR) codec is commonly used in mobile phones (ETSI, 2011). It was standardized by the European Telecommunications Standards Institute (ETSI) in 1999. It is based on the ACELP technique, and is also able to use in transmission eight different bit rates ranging from 4.75 kbps up to 12.2 kbps. This enables adaptive allocation of bit stream resources to maximize the quality of experience.

The 64-kbit/s data rate obtained with the PCM codec described above can be reduced to rates of 8–12 kbit/s with parametric coding of speech, which is a tolerable data rate in wireless communication. The perceptual quality of speech does not drop noticeably when no strong background sounds are present (Kleijn and Paliwal, 1995).

A trend in speech coding is *wideband coding*, where 'wideband' typically refers to the acoustic bandwidth from 50 Hz to 7000 Hz. This bandwidth gives a substantially better speech quality with 'brighter' and 'fuller' sound compared to narrowband speech. Wideband coding requires, naturally, a higher transmission rate than narrowband codecs, and the use of such codecs is made possible by faster VoIP connections and more powerful mobile networks. In addition to wideband coding, superwideband, with a 14-kHz bandwidth, and fullband, with a 20-kHz bandwidth, have also been introduced. A review of codecs with these wider frequency ranges is given by Cox *et al.* (2009).

Speech transmitted by parametric methods is rated to be intelligible to a high degree, although the listeners may characterize it as being a bit 'synthetic'. Relatively often, the obtained quality depends on the speaker. With some codecs the quality obtained with male speakers is better than with females (Vary and Martin, 2006, p. 285). Since the encoder makes a strong assumption that the encoded sound is a human voice, a natural consequence is that the quality of reproduction degrades for some other signals, such as with music. This also leads to a situation where, in conditions of strong background noise, the encoder may make wrong assumptions with regard to the parameters of the source–filter model, and the speech reproduced may turn out to be unintelligible. However, some codecs are designed to use source–filter models of speech only when the sound evidently is speech; the codec changes the mode of operation to coding of a waveform if the input signal is not single-source speech (Neuendorf *et al.*, 2013).

Speech coding methods thus use the principle behind the *vocoder* (short for voice encoder), which was developed in the 1930s (Dudley, 1940). In the original implementation, the input speech signal was passed through a filter bank, and the signal in each band was passed through an envelope follower. The control signals from the envelope followers was communicated to the decoder. The decoder applied these (amplitude) control signals to corresponding filters in the synthesizer. Since the control signals changed only slowly compared to the original speech waveform, the bandwidth required to transmit speech was reduced, and the parameters could be modelled statistically. Nowadays, the term vocoder is used to mean the principle where speech is dissolved into relatively simple parameters, and synthesized back to a perceptually similar speech signal.

16.2 Text-to-Speech Synthesis

Synthetic speech, in the sense of artificially created speech signals, has a relatively long history, including the acoustic–mechanical speech production by Kratzenstein and Von Kempelen (Schroeder, 1993) in the late 18th century, a mechanically controlled electronic synthesizer called the *Voder* by Dudley in 1939, and more advanced electronic synthesizers since the 1950s (Karjalainen and Laine, 1977; Karjalainen *et al.*, 1980; Klatt, 1987; Schroeder, 1993). These electronic models of human voice generation then evolved to be controlled by computer, enabling intelligible *text-to-speech synthesis*.

The main principle of speech synthesis is to use a signal model of speech production, such as the source–filter method discussed in the previous section, and to control the parameters of the model to create speech, realizing the text input of the system as understandable and providing as natural speech as possible. The success of the source–filter approach, or vocoder approach, in speech synthesis proves that good-quality speech can be synthesized if the parameters are derived from natural speech. Unfortunately, deriving the parameters from text input is far from an easy task, as will be shown below.

This section first reviews the early knowledge-based synthesis methods, where complex rule-based systems were built for speech synthesis. Although such methods have been abandoned, they might still be of interest for historical and educational purposes, and for this reason they are discussed here.

Data-based synthesis methods have largely been adopted in academia and in industry since the 1990s. The increase in power and resources of computer technology has enabled the building of natural-sounding synthetic voices based on the utilization of large, single-speaker databases of natural speech (Zen *et al.*, 2009). In contrast to knowledge-based synthesis, each phonetic unit need not be crafted for each applicable context anymore, but the information in the database is used to train the system on how to produce correct-sounding speech.

Data-based methods can be divided into two subclasses: *unit-selection synthesis* and *statistical parametric synthesis*. Unit selection synthesis does not perform actual 'synthesis' of sound, but the output signal is composed of audio samples taken from the database (Beutnagel *et al.*, 1999). Statistical parametric synthesis, on the other hand, synthesizes the speech signal using a model, the parameters of which are controlled using a system trained with real speech samples.

16.2.1 Early Knowledge-Based Text-to-Speech (TTS) Synthesis

The source–filter signal models of human voice were discussed in Section 5.3 on page 90. In principle, any static voice produced by the human speech organs can be produced with such source–filter models. The goal of text-to-speech synthesis is actually much more complicated, as the system should be able to produce words, sentences, and complete utterances that are not only intelligible, but also indistinguishable from human speech in the ears of a listener. The first efforts to synthesize speech created sets of rules on how to go from text to control parameters of the signal model of voice and finally to intelligible speech.

An overall structure of knowledge-based *text-to-speech synthesis* is characterized in Figure 16.2. The first phase is to read text in and parse it into an internal representation in terms of phonetics and linguistics of the language to be synthesized. Various methods are applied here depending on the complexity and degree of linguistic processing. A simple case might have only *text normalization*, such as expanding abbreviations, numbers, and possibly some application-specific rules. *Letter-to-phoneme mapping* is always needed, which is simple in some languages (Finnish is an example) and complex in others (like English or French).

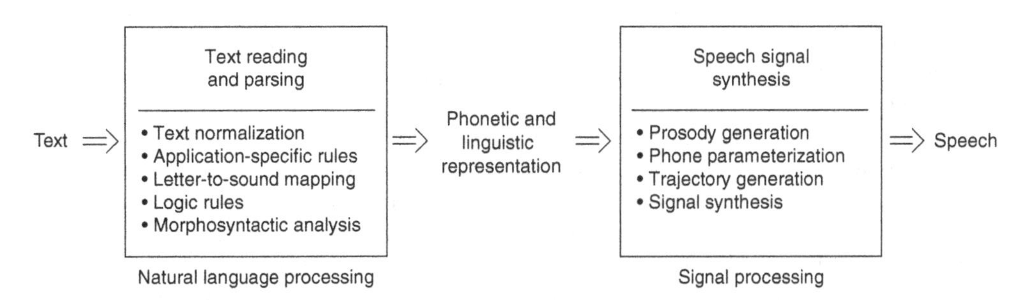

Figure 16.2 A general description of knowledge-based text-to-speech synthesis (TTS).

In a more advanced case, the parsing may include morphological and syntactic analysis of the language being synthesized. These functions are examples of *natural language processing*.

The second phase of knowledge-based text-to-speech synthesis, depicted in Figure 16.2, converts the phonetic and linguistic representation to a speech signal. This part, consisting largely of signal processing, gives a parameterized representation of the phonemes (phones) and attaches prosodic features to them. Continuous-time parameters are produced which control the final speech signal synthesis that may be based on any synthesis model, such as those discussed in Section 5.3.

Figure 16.3 illustrates an example of the organization of the synthesis, and shows the multilevel structure of information and data in speech generation. In this specific case, the linguistic pre-processing is simple, resulting in a phoneme string representation using letter-to-phoneme rules. The linguistic analysis also contains a simple analysis of the syllabic and sentence structure of the message. Based on these representations, the synthesis process proceeds to prosodic feature and segmental parameter computation for the segments of the phonemes. The next step is to convert the segmental parameters into continuous-time trajectories for synthesis model control.

The largest problem with rule-based text-to-speech synthesizers is the quality of speech. When unlimited text is automatically transformed into speech, the synthesized speech almost always sounds 'robotic' and is barely intelligible. These methods have been largely abandoned after the introduction of data-based speech synthesis, which will be discussed below. However, in some special cases, knowledge-based speech synthesizers have their uses. For example, in a situation where data-based speech synthesis would require far more computational power than is available and where a lower quality of speech could be tolerated, the rule-based solutions could still be used.

16.2.2 Unit-Selection Synthesis

A database consisting typically of tens of hours of recorded speech has first to be created for unit-selection synthesis. During the creation of the database, each recorded utterance is segmented into some or all of the following: individual phones, diphones, half-phones, syllables, morphemes, words, phrases, and sentences. The segmentation can be done using automatic systems, although often the segments have to be manually fine-tuned.

The units in the speech database are then indexed, and the segments are associated with both phonetic and prosodic information. The phonetic information labels the phone and also describes its *phonetic context* as the position of the phone in the syllable, word, and sentence. The prosodic parameters are then the acoustic parameters analysed from the phone, such as the fundamental frequency (pitch), voicing, and spectral structure. The *prosodic context* then describes the prosodic parameters in neighbouring segments.

At run time, the text to be synthesized is given, and a set of required units is formed based on the text. The desired target utterance is created by determining the best chain of candidate units from the database, which is the *unit selection* process. An important concept is the *target cost*, which measures how well a candidate unit from the database matches the required unit. It is basically a distance measure. Similarly, the *concatenation cost* defines how well two units combine when presented successively. This evaluation can be performed using a specially weighted decision tree, where the cost is to be minimized. The best-matching units are then concatenated, as shown in Figure 16.4.

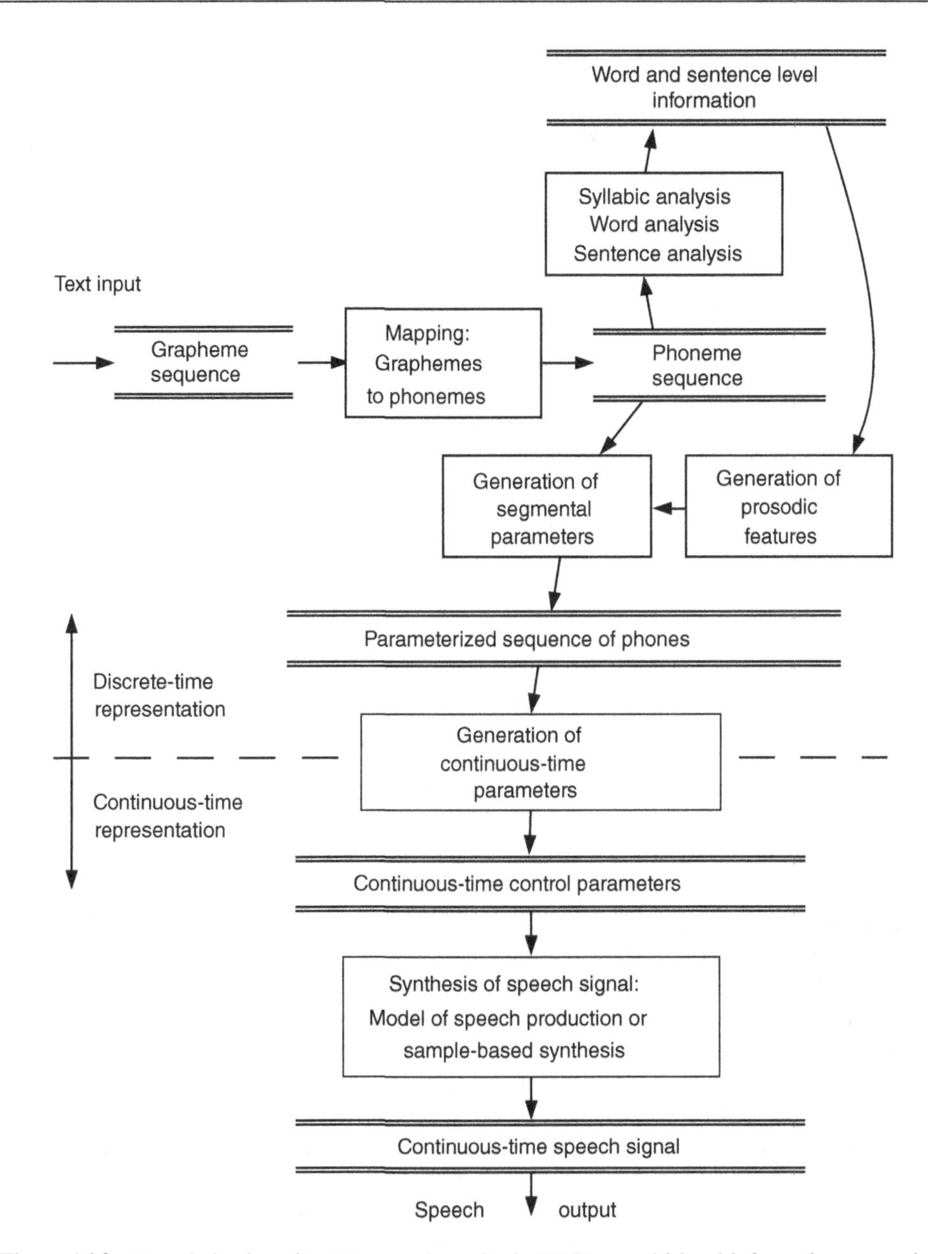

Figure 16.3 Knowledge-based text-to-speech synthesis (TTS) as multi-level information processing.

When designing an implementation of unit-selection synthesis, the size of the units may be set to be shorter or longer. If they are made shorter, more joining points will be available to select the next unit. Since there are more possibilities to select the joining point, there will be a smaller concatenation cost, which leads to better quality. However, too short a length for the units may also cause problems. Kishore and Black (2003) tested the optimal length for the Hindi language, and the best result was obtained with the unit length equalling that of syllables.

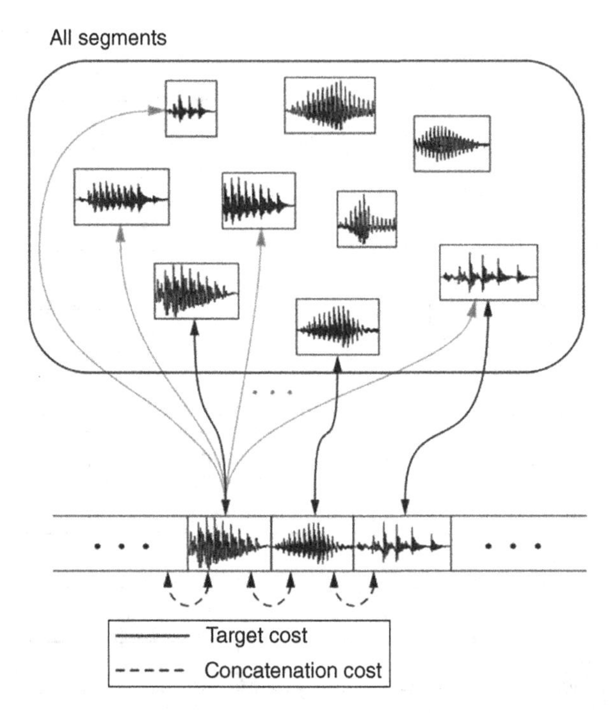

Figure 16.4 An overview of the general unit-selection scheme. The solid lines represent target costs and the dashed lines concatenation costs. Adapted from (Zen *et al.*, 2009), and reprinted with permission from Elsevier.

The units can also be made to have variable lengths, which may at least partly mitigate this issue (Segi *et al.*, 2004).

In principle, unit selection produces a very natural result, because it does not change significantly the recorded speech. The output from the best unit-selection systems can be indistinguishable from real human voices, especially in applications for which the system has been tuned (Beutnagel *et al.*, 1999). Unfortunately, achieving the most natural results typically requires unit-selection speech databases to be very large, representing tens of hours of speech, which can be a limiting factor in some applications. The method does not itself provide strong means to modify the type of speaker nor to change prosodic features dynamically, since in principle it is based on the playback of catenated speech samples. Also, the quality of the system may severely deteriorate if the phonetic and prosodic contexts required in a sentence are under-represented in a database (Zen *et al.*, 2009), and the errors caused by mismatches between concatenated phones reduce the intelligibility of speech.

16.2.3 Statistical Parametric Synthesis

Statistical parametric synthesis methods try to utilize the best properties of signal models and statistical models of natural speech. The source–filter models of speech are often used in statistical parametric synthesis, and they will be used to illustrate the techniques in this section.

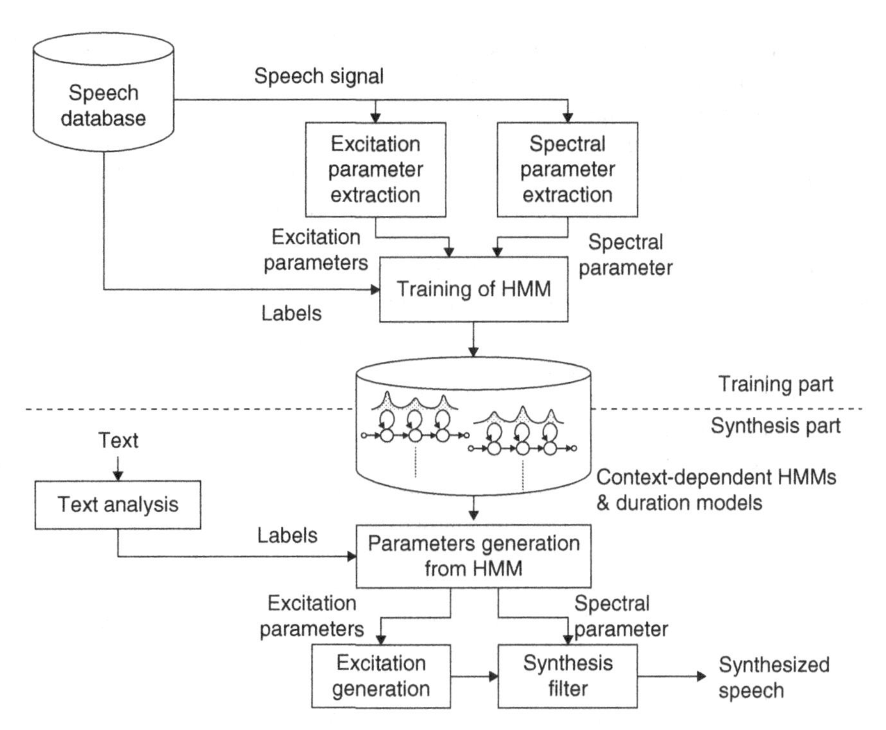

Figure 16.5 A block diagram of statistical parametric speech synthesis implemented with hidden Markov models (HMM). Adapted from Zen *et al.* (2009), and reprinted with permission from Elsevier.

The basic idea behind this type of synthesis does not, in principle, limit the choice of modelling technique; for example, sinusoidal modelling of speech has also been used in the same context (Erro *et al.*, 2014; Masuko *et al.*, 1996). The driving idea is to derive the parameters for source–filter synthesis by training the system using large speech databases. The promise is that the speech synthesized from text should have better quality than with knowledge-based synthesis, but in addition, the flexibility and adaptivity to modify the characteristics of voice obtainable with source–filter synthesis should be gained.

A block diagram of the statistical parametric system based on hidden Markov models (HMM) is shown in Figure 16.5. Before the training part, a large speech database is collected, where, again, a data structure of phonemic and prosodic context is associated with each segment of speech. The size of the speech database is typically much smaller than with unit-selection synthesis, as, with just one hour of speech data, decent synthesis results can be obtained (Yamagishi *et al.*, 2009). The structure contains information such as the current phoneme and its position, adjacent phonemes, the current syllable and its position, adjacent syllables, and information on nearby syllables. The structure may also contain similar data about words, phrases, and utterances in the context of each segment.

In the HMM-based speech synthesis framework, each context-dependent phoneme is modelled as a sequence of HMM states (commonly 3 or 5). Each state models the vocoder parameters using a single Gaussian distribution for each parameter. In the training of the system, the principle of the vocoder is utilized: the most suitable parameters for excitation and spectral filtering are computed that are assumed to produce an output perceptually similar to the original segment when applied to the source–filter model in synthesis. This is done by first decomposing the speech data into vocoder parameters, and then estimating the Gaussian statistics (mean and variance) of the vocoder parameters and the duration for each state for all context-dependent phonemes. Using multi-stream HMM training (Gales and Young, 2008), the system can model the spectrum, excitation, and duration of each segment in a unified framework.

The parameters utilized in the source and filter depend on their actual implementation. For example, the spectral parameters may be represented as mel frequency cepstral coefficients (Fukada *et al.*, 1992) or as line spectral frequencies (Soong and Juang, 1984), and the fundamental frequency of the voiced excitation is often presented as the logarithm of f_0. The computation of mel cepstra was discussed in Section 13.1.1. The logarithm of f_0, in turn, corresponds to the frequency scale used in music (see Section 11.6.2). Commonly, a simple impulse train is used for exciting voiced speech, but the waveform of a voiced glottal excitation, inverse filtered from natural speech, may also be used to improve naturalness (Raitio *et al.*, 2011). The STRAIGHT method is nowadays commonly used as a speech vocoder (Kawahara, 2006), and in other speech processing besides data-based speech synthesis. In STRAIGHT synthesis, the source signal is generated using mixed excitation consisting of impulses and a noise component acting as the aperiodic component of voiced speech. Finally, the pitch-synchronous overlap add (PSOLA) method (Moulines and Charpentier, 1990) is used to reconstruct the excitation signal, which is then applied to excite the filter.

The synthesis part performs an operation that resembles the inverse of speech recognition (Zen *et al.*, 2009). A given word sequence is first converted into a sequence of context-dependent labels. After this, the HMM of an utterance is constructed by concatenating the context-dependent HMMs following the label sequence. Then, smooth sequences of spectral and excitation parameters are generated from the utterance HMMs using the mean and variance values of each state. Finally, a speech signal is synthesized using excitation generation and a speech synthesis filter.

With regard to the quality of speech obtained with statistical parametric synthesis, Zen *et al.* (2009) conclude with the words, 'Although even the proponents of statistical parametric synthesis feel that the best examples of unit-selection synthesis are better than the best examples of statistical parametric synthesis, overall it appears that the quality of statistical parametric synthesis has already reached a level where it can stand in its own right.' The speech quality obtained with statistical parametric synthesis is thus not on the same level with unit-selection synthesis, but better quality is obtained than with knowledge-based speech synthesis.

One benefit of statistical parametric synthesis is that the source–filter-type synthesis of speech signals provides possibilities for changing the characteristics of voice, such as speaking styles and prosodic features, and possibly even provides multilingual support (Zen *et al.*, 2009). With the unit-selection method, such modifications are a bit complicated, as voice conversion techniques have to be utilized (Stylianou *et al.*, 1998), which may degrade the quality of the speech.

However, although understandable speech is obtained with statistical parametric synthesis, its major drawback is still in the quality of speech. The speech is perceived as less authentic than with unit-selection synthesis, and several possible factors affecting this and various suggested refinements are reviewed by Zen *et al.* (2009).

16.3 Speech Recognition

The goal of *speech recognition* (Rabiner and Juang, 1993) is to capture the acoustic human speech signal and process the spoken language into text. It is one of the biggest challenges in speech technology. Evolution and culture have built speech into a fast and robust communication channel over adverse acoustic channels, tolerant to relatively high background noise levels. Speech is not only a rich container of information added to words and non-speech voices, it also carries the identity, age, and gender about the speaker and prosodic features carrying emotions and other meanings. Additionally, the speed of speech may vary substantially, and the vast number of languages, dialects, and speaking styles and disorders make the problem even more challenging.

In principle, speech recognizers should simply mimic brain functions to perform the task as well as humans do. Note that the goal of processing is different in brains and in speech recognizers. Brains turn speech into neural code and speech recognizers to written text. Speech recognition conducted by humans is based strongly on the high-level functions of the brain, such as language, awareness of environment, and assumptions about the content of speech based on previous utterances. Auditory models and models of the brain are clearly not at the level where such functions can be emulated. Thus, the most successful speech recognition techniques are based on generic principles of pattern recognition and data processing.

The complexity of a speech recognition task depends heavily on the definition of the task. The simplest task is to recognize speech from one known speaker uttering temporally separated words from a small, known vocabulary. The current methods perform well in such tasks. The task becomes harder when the vocabulary is made larger or not restricted at all, when words are not separated by silences, when multi-language speech is allowed, and when the content of speech is not limited to any specific topic. Even harder tasks are those involving multiple speakers, and when the level of background noise is high.

The speech recognition process can be divided into different phases. The first phase is to pre-process the speech signal to remove unnecessary redundancy and to describe the signal with *features*. Given the acoustic models of speech sounds, the statistical language models, and the lexicon of words, the next phase is a *pattern recognition* task where the decoder turns the sequence of features directly into the most likely sequence of words in the language.

The most common speech recognition systems are based on HMMs (Rabiner, 1989). The early techniques in speech recognition were dynamic time warping (Vintsyuk, 1968) and neural networks (Kohonen, 1988), but they were largely abandoned in the late 1990s. However, recently, deep neural networks have been proposed for use in speech recognition (Dahl *et al.*, 2012).

The main parts of a typical speech recognizer are shown in Figure 16.6. The speech input is turned into feature vector sequences first. After this, the probability of the vectors or vector sequences belonging to linguistic classes is computed using HMMs. The probability sequences are then decoded into text using a vocabulary and a model of language. Prior to using the recognizer, the HMM has to be trained with natural speech, and the vocabulary and model of language have to be constructed.

Speech recognizers commonly window the input signal at a rate of about 100 Hz and utilize window lengths of about 20 ms. The features computed from the signal segment are typically mel-frequency cepstral coefficients. Differentials between temporally subsequent features are also commonly utilized and have been found to improve the accuracy of recognition.

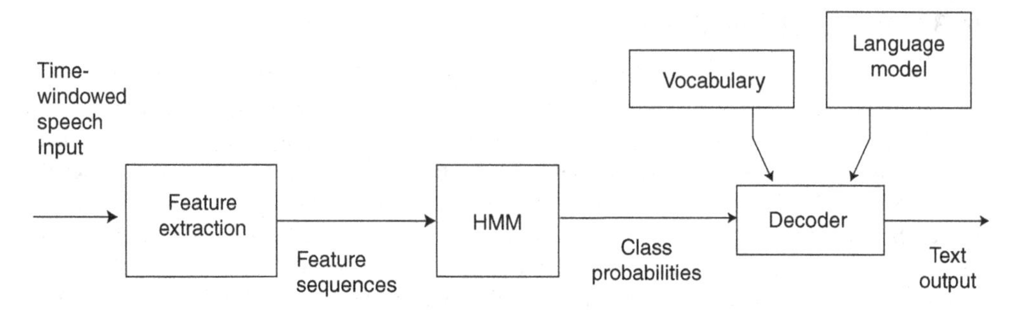

Figure 16.6 A simplified block diagram of speech recognition.

If the recognition task is limited to a small vocabulary and to isolated words, it is possible to construct separate HMMs for all the different words without sharing parameters between the models. The number of states in a word HMM is then determined with some heuristics, such as relating it to the number of phonemes or the average number of observations (Rabiner, 1989). As the size of the vocabulary increases, a word-level approach becomes infeasible due to the large number of states and parameters. Instead, a small set of segments, such as phones or diphones, is modelled with unique HMMs. The word models can then be formed by a simple concatenation of these units. A three-state, left-to-right HMM (Schwartz *et al.*, 1985) is commonly used for the model of a phoneme.

The sequences of probabilities of phonemes are then turned into written text using knowledge about the language they were spoken in. Some information about the language is already embedded in the HMMs of the units of speech, either in the form of word models or as the phone inventory. However, an extensive knowledge of language must be embedded in the recognizer. Most commonly, the grammatical constraints are learned as statistical dependences in the context of words from large text databases (Goodman, 2001). When the vocabulary is small, it may be feasible to make a list of allowed words and define their HMMs. However, when the size of the vocabulary is larger, acoustically similar words are hard to distinguish during recognition. Detecting boundaries of words in continuous speech is also difficult, which adds to the challenge. It becomes necessary to utilize either task-specific or general constraints to overcome these problems.

Speech recognition has already been adopted in many applications. For example, some televisions and handheld devices can already be controlled by voice, and telephone customer services may use speech recognition in automated tasks. Speech recognition is effective in browsing the contents of spoken speech databases (Chelba *et al.*, 2008). One application that is also actively being studied is a telephone service that translates the language of a speaker into another language. Although this service has not yet been achieved, translation in limited cases has been possible. Optimally, the characteristics of the voice of the original speaker should be preserved in the translated speech output. The biggest challenges are, however, in the processing of language, and not in the processing of speech signals.

Summary

Speech coding is perhaps the most mature of the speech technologies. Very good speech quality can be obtained with low bit rates when signal models of speech production mechanisms are utilized. The methods are relatively simple, and no large databases are required.

Speech synthesis and recognition techniques have also been developed fairly extensively. When compared with speech coding, a clear difference is that the systems must have access to large databases of natural language and/or speech in order to obtain good results. Speech synthesis and recognition thus require that the computer has knowledge of the recognized or synthesized language at some level, something that is not needed in speech coding. The performance of the synthesis and recognition techniques is still far from that of human performance, although multiple commercial applications already exist.

Further Reading

The synthesis and recognition of speech is described in more detail by Huang *et al.* (2001), Jurafsky and Martin (2008), and Saon and Chien (2012). A public domain toolkit for research in automatic speech recognition is available, and its documentation also serves as a good explanation of speech recognition (Young *et al.*, 2006). Hidden Markov models are explained in detail by Gales and Young (2008). The enhancement, coding, and error concealment in speech transmission are discussed by Vary and Martin (2006), and more information on audio coding can be found in Chu (2004).

References

Beutnagel, M., Conkie, A., Schroeter, J., Stylianou, Y., and Syrdal, A. (1999) The AT&T next-gen TTS system. *Joint meeting of ASA, EAA, and DAGA*, pp. 18–24.

Chelba, C., Hazen, T.J., and Saraçlar, M. (2008) Retrieval and browsing of spoken content. *Signal Proc. Mag., IEEE*, **25**(3), 39–49.

Chu, W.C. (2004) *Speech Coding Algorithms: Foundation and Evolution of Standardized Coders*. John Wiley & Sons.

Cox, R.V., De Campos Neto, S.F., Lamblin, C., and Sherif, M.H. (2009) ITU-T coders for wideband, superwideband, and fullband speech communication. *Communications Mag., IEEE*, **47**(10), 106–109.

Dahl, G.E., Yu, D., Deng, L., and Acero, A. (2012) Context-dependent pre-trained deep neural networks for large-vocabulary speech recognition. *IEEE Trans. Audio, Speech, and Language Proc.*, **20**(1), 30–42.

Dudley, H. (1940) The carrier nature of speech. *Bell Sys. Tech. J.* **19**(4), 495–515.

Erro, D., Sainz, I., Navas, E., and Hernaez, I. (2014) Harmonics plus noise model based vocoder for statistical parametric speech synthesis. *IEEE J. Selected Topics in Signal Proc.*, **8**(2), 184–194.

ETSI (1992) GSM Full Rate Speech Transcoding. Recommendation ETSI GSM 06.10, European Telecommunication Standards Institute.

ETSI (2011) AMR speech codec, general description. Standard 3GPP TS 26.071, version 10.0.0, 3rd Generation Partnership Project.

Fukada, T., Tokuda, K., Kobayashi, T., and Imai, S. (1992) An adaptive algorithm for mel-cepstral analysis of speech. *IEEE Int. Conf. Acoustics, Speech, and Signal Proc.*, volume 1, pp. 137–140 IEEE.

Gales, M. and Young, S. (2008) The application of hidden Markov models in speech recognition. *Found. Trend. Sig. Proc.*, **1**(3), 195–304.

Goodman, J.T. (2001) A bit of progress in language modeling. *Comp. Speech Lang.*, **15**(4), 403–434.

Huang, X., Acero, A., Hon, H.W., and Foreword By-Reddy, R. (2001) *Spoken Language Processing: A guide to theory, algorithm, and system development*. Prentice Hall.

ITU (1988) *Pulse code modulation (PCM) of voice frequencies*. Recommendation ITU-T G.711, International Telecommunication Union, Geneva, Switzerland.

Jurafsky, D. and Martin, J.H. (2008) *Speech and Language Processing: An Introduction To Natural Language Processing, Computational Linguistics, and Speech*, 2nd edn. Pearson Prentice Hall.

Karjalainen, M. and Laine, U.K. (1977) Speech synthesis project in Tampere: Results and applications. *Proc. of IV Nordic Meeting on Med. and Biol. Eng.*, pp. 30.1–3.

Karjalainen, M., Laine, U., and Toivonen, R. (1980) Aids for the handicapped based on "synte 2" speech synthesizer. *IEEE Int. Conf. Acoustics, Speech, and Signal Proc.*, volume 5, pp. 851–854 IEEE.

Kawahara, H. (2006) Straight, exploitation of the other aspect of vocoder: Perceptually isomorphic decomposition of speech sounds. *Acoust. Sci. Technol.* **27**(6), 349–353.

Kishore, S. and Black, A.W. (2003) Unit size in unit selection speech synthesis. *INTERSPEECH*.

Klatt, D.H. (1987) Review of text-to-speech synthesis of English. *J. Acoust. Soc. Am.*, **82**(3), 737–793.

Kleijn, W.B. and Paliwal, K.K. (1995) *Speech Coding and Synthesis*. Elsevier Science.

Kohonen, T. (1988) The'neural'phonetic typewriter. *Computer*, **21**(3), 11–22.

Martin, W. (1930) Transmitted frequency range for telephone message circuits. *Bell Sys. Tech. J.* **9**(3), 483–486.

Masuko, T., Tokuda, K., Kobayashi, T., and Imai, S. (1996) Speech synthesis using HMMs with dynamic features. *IEEE Int. Conf. Acoustics, Speech, and Signal Proc.*, volume 1, pp. 389–392 IEEE.

Moulines, E. and Charpentier, F. (1990) Pitch-synchronous waveform processing techniques for text-to-speech synthesis using diphones. *Speech Commun.* **9**(5), 453–467.

Neuendorf, M., Multrus, M., Rettelbach, N., Fuchs, G., Robilliard, J., Lecomte, J., Wilde, S., Bayer, S., Disch, S., Helmrich, C., Lefebvre, R., Gournay, P., Bessette, B., Lapierre, J., Kjörling, K., Purnhagen, H., Villemoes, L., Oomen, W., Schuijers, E., Kikuiri, K., Chinen, T., Norimatsu, T., Chong, K.S., Oh, E., Kim, M., Quackenbush, S., and Grill, B. (2013) The ISO/MPEG unified speech and audio coding standard – consistent high quality for all content types and at all bit rates. *J. Audio Eng. Soc.*, **61**(12), 956–977.

Paez, M. and Glisson, T. (1972) Minimum mean-squared-error quantization in speech pcm and dpcm systems. *IEEE Trans. Commun.*, **20**(2), 225–230.

Rabiner, L. (1989) A tutorial on hidden Markov models and selected applications in speech recognition. *Proc. IEEE*, **77**(2), 257–286.

Rabiner, L.R. and Juang, B.H. (1993) *Fundamentals of Speech Recognition*, volume 14. Prentice Hall.

Raitio, T., Suni, A., Yamagishi, J., Pulakka, H., Nurminen, J., Vainio, M., and Alku, P. (2011) HMM-based speech synthesis utilizing glottal inverse filtering. *IEEE Trans. Audio, Speech, and Language Proc.*, **19**(1), 153–165.

Salami, R., Laflamme, C., Adoul, J.P., Kataoka, A., Hayashi, S., Moriya, T., Lamblin, C., Massaloux, D., Proust, S., Kroon, P., and Shoham, Y. (1998) Design and description of CS-ACELP: A toll quality 8 kb/s speech coder. *IEEE Trans. Speech and Audio Proc.*, **6**(2), 116–130.

Saon, G. and Chien, J.T. (2012) Large-vocabulary continuous speech recognition systems: A look at some recent advances. *IEEE Signal Proc. Mag.*, **29**(6), 18–33.

Schroeder, M.R. (1993) A brief history of synthetic speech. *Speech Commun.* **13**(1), 231–237.

Schwartz, R., Chow, Y., Kimball, O., Roucos, S., Krasner, M., and Makhoul, J. (1985) Context-dependent modeling for acoustic-phonetic recognition of continuous speech. *IEEE Int. Conf. Acoustics, Speech, and Signal Proc.*, volume 10, pp. 1205–1208 IEEE.

Segi, H., Takagi, T., and Ito, T. (2004) A concatenative speech synthesis method using context dependent phoneme sequences with variable length as search units. *Fifth ISCA Workshop on Speech Synthesis*.

Soong, F.K. and Juang, B.H. (1984) Line spectrum pair (lsp) and speech data compression. *IEEE Int. Conf. Acoustics, Speech, and Signal Proc.*, volume 9, pp. 37–40 IEEE.

Stylianou, Y., Cappé, O., and Moulines, E. (1998) Continuous probabilistic transform for voice conversion. *IEEE Trans. Speech and Audio Proc.*, **6**(2), 131–142.

Vary, P. and Martin, R. (2006) *Digital Speech Transmission: Enhancement, Coding and Error Concealment*. John Wiley & Sons.

Vintsyuk, T. (1968) Speech discrimination by dynamic programming. *Cybernet. Sys. Anal.*, **4**(1), 52–57.

Yamagishi, J., Kobayashi, T., Nakano, Y., Ogata, K., and Isogai, J. (2009) Analysis of speaker adaptation algorithms for HMM-based speech synthesis and a constrained SMAPLR adaptation algorithm. *IEEE Trans. Audio, Speech, and Language Proc.*, **17**(1), 66–83.

Young, S., Evermann, G., Gales, M., Hain, T., Kershaw, D., Liu, X.A., Moore, G., Odell, J., Ollason, D., Povey, D., Valtchev, V., and Woodland, P. (2006) Htk book (for htk version 3.4). Technical report. `http://htk.eng.cam.ac.uk/docs/docs.shtml`.

Zen, H., Tokuda, K., and Black, A.W. (2009) Statistical parametric speech synthesis. *Speech Commun.*, **51**(11), 1039–1064.

17

Sound Quality

The concept *quality* has two meanings. 'Quality' is used in this book as a synonym for 'excellence', to grade or rank objects on a subjective scale of preferability such as 'good–poor', based on some explicit or implicit criteria. The other common meaning is related to categorization by type or class of objects. When two observations or entities cannot be compared on the same (metric) scale they are said to be qualitatively different. Such category-related sound quality pertains to perceived features, attributes, factors, dimensions, or properties of auditory events, such as loudness or roughness. However, in this book, the term 'sound quality' is limited to the meaning involving preferability or acceptability.

The inherent topic of the discussion on quality after its definition is evaluation. We experience some objects or states of the world as more desirable, valuable, positive, appealing, useful, or what have you than others. Although often weakly formulated and structured, such conceptions and rankings help us to set goals of action and to find better solutions to problems at hand. A widely used term in this context is *quality of experience* (QoE) (Le Callet *et al.*, 2012), which denotes the overall acceptability of an application or service as perceived subjectively by the end user.

The theory of psychoacoustics, discussed in Chapter 8, uses the human as a simple metering device, where sound events evoke auditory events with attributes that can be measured using psychoacoustic techniques. In psychoacoustics, expectations, mood, and other cognitive factors of individual subjects are minimized when the values of attributes are measured. In the context of sound quality, cognitive factors can no longer be disregarded, since 'quality of sound' means the suitability of a sound to a specific situation, and such suitability cannot be judged without cognitive functions. The same sound may produce different sound quality in different contexts, depending on the mode of operation and the expectations of a subject. For example, higher *intelligibility of speech* improves the sound quality in mobile phones, but the ability of a worker to concentrate in an open-plan office is impaired by intelligible speech from neighbouring cubicles. The properties of sound can thus have either negative or positive effects on sound quality.

Communication Acoustics: An Introduction to Speech, Audio, and Psychoacoustics, First Edition.
Ville Pulkki and Matti Karjalainen.

Although the interpretation of 'sound quality' varies widely in different domains of acoustics, audio, and speech, the concept has come into increasingly widespread use (Blauert and Jekosch, 1997), and in one form or another may be considered generally applicable to all sounds that humans encounter.

17.1 Historical Background of Sound Quality

The concept of sound quality has a relatively long history of emergence. Probably the oldest sounds associated with a quality rating have been human speech and singing, then theatre and music-making, including musical instruments. The first quality rating factors were subjective and implicit, based on emerging a esthetic factors and how the sounds had a desired effect in practice. The centuries-long evolution of present-day acoustic musical instruments is an early example of 'product sound quality' development by gradual experimentation. The acoustics of concert halls and other performing spaces is another similar case, where, until the beginning of the 20th century, sound quality evaluation had little, if any, scientific basis.

The development of physics and related mathematics started to enable a relationship between objective factors and subjective quality of sound. The sound spectrum (including sounds from musical instruments) and related hearing processes were studied in the late 1800s and early 1900s by Helmholtz (von Helmholtz, 1954) and the basics of concert hall acoustics by Sabine (Sabine, 1922). Inventions in electronic communications – the telephone, gramophone, and radio – had a strong impact on our understanding of sound quality. Particularly in telephone transmission, there was a practical need to know how the distortion caused by, and the limitations of, the early technology affected the intelligibility of speech and the recognition of individual speakers. Starting from the 1920s, Harvey Fletcher and the Bell Laboratories research group (Fletcher, 1995) made fundamental studies that laid the groundwork for engineering psychophysics (psychoacoustics) as a systematic experimental science by making quantitative formulations for articulation and intelligibility of speech. Subjects in listening tests were used as 'meters' to 'measure' desired factors in speech transmission. This was the basis, for example, for setting the standard of the telephone bandwidth that is still in use today.

The goal of high sound quality was clearly necessary in sound reproduction using microphones, tape recorders, amplifiers, record players, and loudspeakers, nowadays called audio techniques. To maximize the a esthetic experience of reproduced music, the HiFi (high fidelity) movement emerged. It was partly an engineering-oriented attempt to minimize distortion and colouration of sound in a reproduction channel and partly a highly subjective 'golden ear' and 'expensive gadget' hobby. Only the emergence of audio coding in digital audio at the end of the 1980s forced the modelling of auditory perception to become a central engineering challenge. In a similar manner, multi-channel and 3D sound reproduction have elevated studies and modelling in spatial sound perception to a higher scientific and engineering level.

Since the 1980s, the investigation of noise control techniques has been increasingly directed also towards qualitative aspects, that is, *noise quality* (Marquis-Favre *et al.*, 2005b), not only to simple quantitative measures, such as the A-weighted sound level for estimating the risk of noise-induced hearing loss. Earlier studies on the subjective effects of noise also exist, but the signal-analysis-based approach was introduced to understand such effects as annoyance caused by noise. This gradual shift of focus is natural, since, in many cases, hearing loss is not the primary problem anymore and quality-of-life aspects are found to be increasingly important.

The notion of sound quality, applicable to both positive effects (music and speech) and negative effects (noise), finds a generalization in the concept of *product sound quality*

(Blauert and Jekosch, 1997). In its most general sense, this concept also covers traditional sound quality aspects, since a concert hall, a musical instrument, a musical performance (even a music composition), audio equipment, a noisy working machine, or a car making noise are equally 'products' in the wide sense of the term. In all these cases, the goal is that the sound of a product meets the needs and requirements of the customer at hand and optimizes the sound quality factors against the cost of the product.

17.2 The Many Facets of Sound Quality

The discussion above brings up the question of whether a universal approach to assess or evaluate sound quality exists. Furthermore, if this is not possible, what are the different scientific methods and engineering techniques to evaluate sound quality? Different domains of sound quality, as discussed above, turn out to be truly different, so finding a simple general model of sound quality does not seem possible. Perhaps the only common factor is the listener: we must start by using subjects in listening experiments to get data on the factors affecting sound quality and, based on these data, build models and theories of sound quality.

The formalization and quantification of the concept of sound quality may raise conflicting opinions. On the one hand, a subjectivist believes that the experience of quality is highly individual and there are no grounds for generalization and formalization, while on the other hand, an objectivist opines that a coherent general theory for measuring sound quality can be developed. Both views are partly right and partly wrong. A further conceptual discussion can help to understand these issues more thoroughly.

How sound affects us can be categorized as follows:

- *Physical and physiological effects.* Only a very intense sound (above 120 dB) can have a considerable physical effect. Physiologically, the most important factor is the risk of hearing impairment, which typically occurs after long exposure to levels above 85 dB (A-weighted daily equivalent; see Section 19.3 for more details). From this point of view, a criterion for high-quality sound design is to keep the sound level low enough not to harm humans, animals, or nature. Compromises are needed when the cost of noise control becomes excessively high.
- *Information and knowledge.* That sound transmits information and knowledge is a desirable and valuable characteristic, although too much exposure to information can lead to negative effects. Information conveyed by environmental sounds is important for orientation in everyday life. Thus, it is desirable that, for example, appliances and vehicles make sounds that inform us about their existence and functioning, as long as the negative factors of this sound do not exceed the positive ones. The information aspect is most essential in speech communication, where the speech quality provided in transmission techniques is needed for undistorted transfer of information.
- A *esthetic and emotional effects.* These are the most demanding aspects of quantifying sound quality from a scientific and engineering point of view. Using listening experiments and statistical analysis, it is always possible to seek factors affecting the perceived quality of sound. Reactions to a sound are often strongly dependent on the sociocultural background of the subject, the context of presentation, and various other factors. For example, many objective factors from speech transmission as well as from high-quality sound reproduction that affect the a esthetic and emotional aspects of the percept have been identified as also

being properties that make up good musical instruments. In the context of noise, we may study the psychological and emotional factors that have a negative effect on our quality of life.

In the discussion below, sound quality is related primarily to the informational and a esthetic aspects of sound. Physiological (or physical) effects are considered only when needed.

The following sections discuss the concept of sound quality from a methodological point of view and in different problem domains, such as speech transmission, concert hall and auditorium acoustics, audio reproduction of sound, noise quality, and the general concept of product sound quality.

17.3 Systemic Framework for Sound Quality

Sound quality is primarily subjective. The most reliable method to study an informative feature or the a esthetic value of a sound is to conduct psychophysical experiments with a group of human subjects, or *assessors*. Often, this is an implicit activity, like in engineering prototyping or product development, where engineers apply their intuitive or introspective knowledge on what constitutes quality of sound. Sometimes the attributes of sound that make it high quality are obvious and sometimes not. In general, understanding the relations between attributes of physical sound and sound quality requires systematic experimentation and statistical analysis of the gathered data. When optimizing the sound quality of a product for a specific market, an extensive experimental basis is necessary.

The problem with subjective evaluation is that the required experiments are typically very laborious and time consuming. The experiments should be carried out in conditions that correspond to real usage of the sounds or related products. Thus, it would be much easier and less expensive to use objective criteria or models for sound quality. Ideally, a computational model provides an estimate of quality from signal analysis which correlates well with subjective data. The development of such a model (Figure 17.1) is typically based on proper subjective listening experiments from which data are collected and analysed statistically. Using various techniques, the factors and features of sounds that best explain the subjective behaviour are sought. The computational model can be a simple linear regression model or a more complex non-linear model, such as a neural network trained to map the feature parameters to quality indices. Figure 17.2 characterizes the general structure of computational sound quality models.

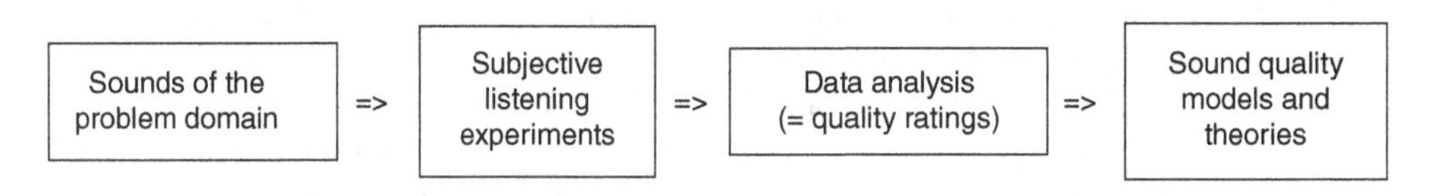

Figure 17.1 The development of sound-quality models and theories.

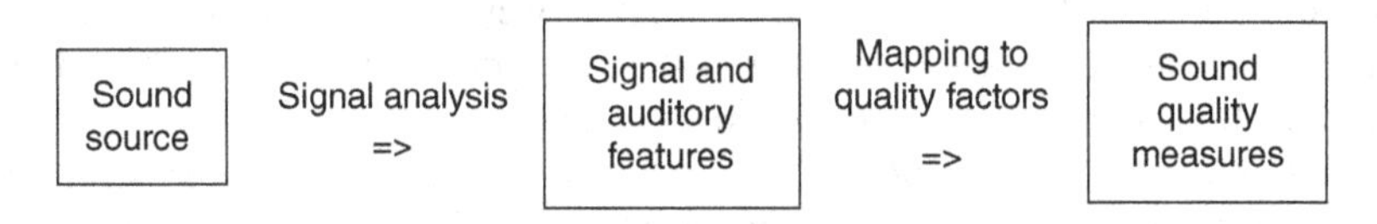

Figure 17.2 A general structure of a computational sound quality model.

The development of such models is also a tedious and demanding task, and objective models can never fully replace subjective evaluation. The advantages of objective models lie in efficiency, rapid evaluation, and repeatability of results. The development of such models may also yield a deeper understanding of the phenomena than subjective results. The disadvantage of objective models is that they never take into account all factors in full detail, and thus their domain of validity (meaning that they correlate well with subjective results) is limited.

17.4 Subjective Sound Quality Measurement

In speech and audio techniques, the final 'truth' about the sound quality achieved lies in the general opinion of the larger public. A number of techniques have been developed to measure the quality of sound associated with a set of sounds. The sounds in the set may be produced, for example, with a set of different audio systems, handheld devices, vacuum cleaners, or concert halls. In the technical development of systems, the choice of parameters for the system often changes the sound output of the system in quite an unpredictable manner. Typically, the only possible method to evaluate the differences between the sounds is to organize subjective tests.

Such tests can be conducted in various ways. A basic approach is to ask the subjects to sort sounds in order of preference, or to ask them directly to rate the quality of sound. The quality can be rated 'in general', or, alternatively, with associated a certain aspect of sound, such as 'rate the quality of speech in terms of intelligibility'. This is perfectly adequate for many applications, and many of the psychoacoustic techniques described in Chapter 8 can be used to measure the value either of the overall quality or of a specific attribute of sound.

In some cases it is not known in advance in which perceptual dimensions the sounds being studied differ from each other. In such cases, *descriptive sensory analysis* techniques can be used to characterize the dimensions, as described in Section 8.8. Furthermore, two concepts often employed in the context of sound quality are the mean opinion score (MOS) scale and the MUSHRA (multiple-stimulus hidden reference with anchors) method for scaling. These methods were not discussed in detail in the chapter on psychoacoustics, and thus they are briefly reviewed here.

17.4.1 Mean Opinion Score

The *mean opinion score* (MOS) value is often used to quantify the sound quality in general or in terms of a specific aspect of sound. Additionally, separate MOS scales have been defined for cases where the degradation of quality or the relative quality is measured. The MOS is a subjective measure obtained using psychoacoustic testing. The selection of the scale and the methodology to measure it depend on the task. There are a number of published and even standardized methods to measure MOS, for example ITU-T P.800 (1996), and this section merely provides a brief introduction to the topic. Detailed methods to measure the MOS in different cases and references to standards for MOS measurements are covered by Bech and Zacharov (2006).

The MOS scale consists of a numerical scale ranging from 1 to 5, either as whole numbers or with fractional intervals, and a descriptor associated with each number. Thus, the scale links a discrete numeric scale to verbal categories. It also enables the measurement of the quality on a

Table 17.1 An example of MOS and DMOS numeric and qualitative scales.

Value	Quality (MOS)	Impairment (DMOS)
5	Excellent	Imperceptible
4	Good	Perceptible, not annoying
3	Fair	Perceptible, slightly annoying
2	Poor	Annoying
1	Bad	Very annoying

Table 17.2 An example of a CMOS numeric and qualitative scale.

Value	Categories (CMOS)
3	Much better
2	Better
1	Slightly better
0	About the same
−1	Slightly worse
−2	Worse
−3	Much worse

very broad scale without a reference sound. The MOS scale is also called an *absolute category rating* (ACR), as defined in ITU-T P.800 (1996) and ITU-T P.910 (2008).

There are many variants of the MOS. The direct measure of quality is simply called the MOS value, and the measure of impairment between reference and test cases is called the *degradation mean opinion score* (DMOS). Example MOS and DMOS scales are presented in Table 17.1.

The *comparative MOS* is the third basic MOS scale, and it is typically defined to have values between −3 and 3. It can thus be used where given test case can be ranked better than reference. This can occur, for example, in speech enhancement techniques. An example of a CMOS scale is give in Table 17.2.

Since MOS contains the word 'mean', it is clear that some kind of average of a large number of subjective ratings is taken. The statistical analysis of the results is an important part of the work. Without proper analysis of the results, the validity of the results cannot be shown. Some general concepts from statistical analysis were introduced in Section 8.9, but a detailed description of relevant methods is beyond the scope of this book. The reader is again referred to Bech and Zacharov (2006) for techniques to analyse MOS values.

17.4.2 MUSHRA

The *multiple-stimulus hidden reference with anchors* (MUSHRA) method is often used in sound quality measurements where multiple stimuli are scaled during the same task onto one MOS scale (ITU-R BS.1534-1, 2003) and was originally developed to test speech and audio codecs. MUSHRA provides reliable results if the sample differences are large, such that the

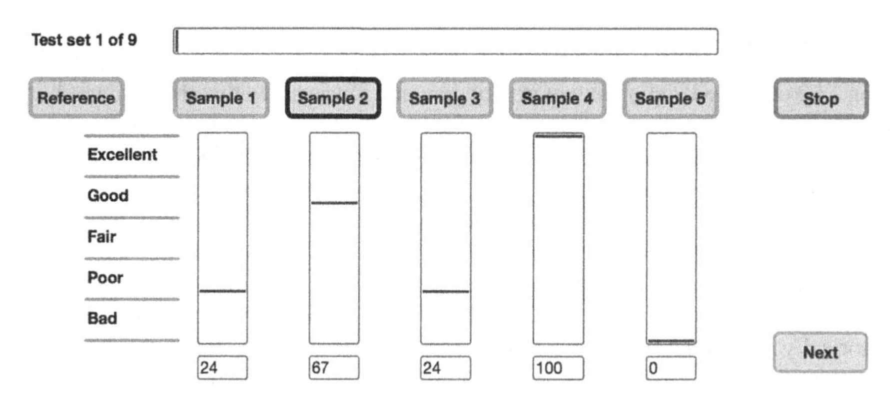

Figure 17.3 A graphical user interface for multiple stimulus hidden reference with anchors (MUSHRA) testing. Courtesy of Tapani Pihlajamäki.

listeners can perceive them without careful concentration. If the differences are small in the test cases, more accurate methods should be used, such as the method described in ITU-R BS.1116-1 (1997).

In a MUSHRA test, subjects listen to different sound samples: the same programme material that has been processed differently. The samples contain a reference known to the subjects, several test samples, a hidden reference, and anchors. The hidden reference is simply an unmodified copy of the original, and the anchors are modified versions of the original, processed by, say, low-pass filtering up to 3.5 kHz, the addition of noise, or the loss of packets during transmission. The subjects can freely listen to the samples, switching from one sample to another by pressing buttons on a user interface, as shown in Figure 17.3. On making a choice, the current sample quickly fades out and the chosen sample fades in. The new sample is played back, continuing from the temporal position where the playback of the current sample ended, so that the programme material continues playing without notable interruptions when the sample is changed. This helps to reveal differences between the samples. The test samples, the hidden reference, and anchors are positioned randomly on the user interface. The subjects rate the samples based on a given task, such as 'rate the quality of reproduction', using the sliders in the user interface.

The number of samples is recommended to be less than 15 (including anchors and references) and the perceptual differences between the samples should not be too small, since otherwise the comparison may be too challenging. The multiple-stimulus methodology has become popular, as relatively reliable results can be obtained faster than with pair-wise comparisons.

The method has also been criticized. For example, if the samples differ from each other in multiple dimensions, the listeners may be confused as to how to rate them. Let us imagine that the sample set consists of audio content with a reference case of 5.1 audio, stereophonic and monophonic down-mixes, and low-pass-filtered versions of the 5.1 audio content. The listeners then have to judge the degree of degradation of sound quality in two dimensions, since both spatial and timbral aspects vary. This may result in data that are difficult to interpret.

Interested readers are referred to Sporer *et al.* (2009) for details on running MUSHRA tests and corresponding data analysis. A revised version of MUSHRA is currently (2014) being standardized, and the final version will include elaborated methods for data analysis and test conduction.

17.5 Audio Quality

Section 14.2 discussed audio content production, and it was noted that the final modifications and final approval of a piece of audio content are conducted typically in the mastering studio. Thus, perfect authenticity in sound reproduction would require an identical listening set-up in a room identical to the mastering studio. Fortunately, this is not necessary, since sufficient quality of the audio experience can also be obtained in other listening conditions. The acoustic differences of listening rooms have an effect on the quality, but the ability of humans to adapt to different acoustic conditions mitigates the differences, as discussed earlier. Impairments in audio devices, on the other hand, often have a significant effect on the quality of the experience.

Historically, the bottleneck in audio quality has been the storing and transmission systems, such as gramophones, vinyl players, and cassette decks. The concept of *high fidelity* was developed to measure and minimize different deviations from perfect responses, such as linear and non-linear distortions. Nowadays, digital transmission has practically removed such problems, and currently the biggest bottlenecks are the microphones and loudspeakers. However, new challenges have emerged with perceptually based lossy audio codecs.

17.5.1 Monaural Quality

Let us first consider deviations that are already audible if only one channel is listened to. The traditional measures affecting sound quality are listed below, and they have already been introduced earlier in this book:

- *Magnitude response* and the perceptual effects of deviations from the ideal (see Sections 4.2.3 and 11.5.1).
- *Phase delay and group delay* and their perceptual consequences (see Sections 4.2.4 and 11.5.2).
- *Non-linear distortion* measures of signal differences between the output and the input with simple signals (see Section 4.2.5).
- *Signal-to-noise ratio* (see Sections 4.2.6 and 17.6.2).

Other traditional concepts concerning analogue audio techniques are, for example, fluttering and rumbling caused by mechanically rotating devices, *dropouts* in magnetic tapes, and pops and crackles in grooves of rotating discs. The advent of digital audio has brought new types of quality degradation that the traditional quality measures fail to detect. These degradation include, for example, quantization noise (see Section 3.3) and aliasing in the time–frequency domain processing of audio (see Chapter 15).

17.5.2 Perceptual Measures and Models for Monaural Audio Quality

The '13-dB miracle' example discussed in Section 15.3.1 shows that human capabilities in listening have to be taken into account when measuring audio quality. This leads directly to the idea of using auditory models (Chapter 13) for quality evaluation. If we can simulate the functioning of the auditory system in all of its relevant stages, theoretically we are able to implement a computational model of hearing that can explain auditory perception when listening to any sounds. Unfortunately, the current status of modelling is far from this, although

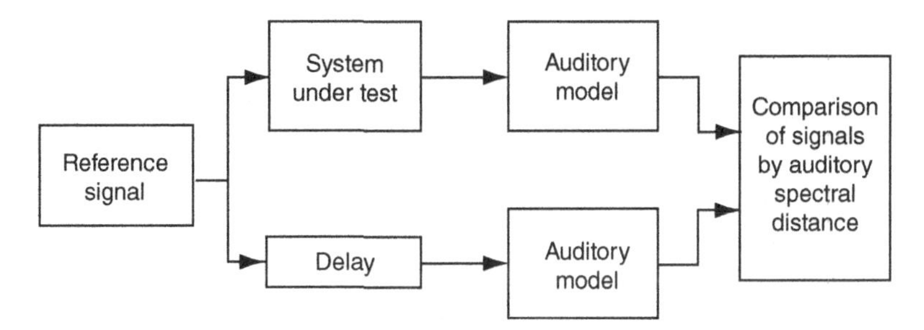

Figure 17.4 The principle of using auditory models to estimate quality degradation in an audio system.

the auditory modelling results explain much better the perceived audio quality of digital codecs than traditional distortion-based measures do.

A method of using auditory models in audio quality evaluation is shown in Figure 17.4; this was suggested by Karjalainen (1985), following pioneering work by Schroeder *et al.* (1979). A reference signal is fed to two identical auditory models, through the system under test to one and suitably delayed to the other so that the two auditory model outputs are aligned in time. The auditory models estimate auditory attributes for both signals, such as *auditory spectrum*, *pitch*, and *localization*. A distance measure is computed between the estimated attributes, from which the degradation of quality caused by the tested system is evaluated. For example, if the auditory spectra differ by more than 1 dB in any critical band, the degradation is audible.

The principle in Figure 17.4 has the advantage that any audio signal can be used as the reference signal. Thus, the audio quality produced by the system can be estimated with real signals, such as music and speech, and not only with simple test signals, such as sinusoids or impulses.

A relatively simple auditory model is presented in Figure 17.5; this implements the principle of audio quality evaluation shown in Figure 17.4 (Beerends and Stemerdink, 1992). The model is based on computing the specific loudness for each critical band using time-windowing, DFT-transfer, and frequency warping, in a similar manner to the Matlab script shown in Section 13.1. The specific loudness spectra are compared, and a time-dependent estimate of the audible difference is obtained, which is then averaged to obtain the final estimate of quality degradation. There are many parameters in the model that are selected to match the estimate with results obtained in MOS listening tests. The measure is called the *perceptual audio quality measure* (PAQM).

The *perceptual evaluation of audio quality* (PEAQ) method (Thiede *et al.*, 2000) is an evolved version of the PAQM computation. PEAQ includes an auditory model implemented either with the computationally lighter DFT processing or with the temporally more accurate filter-bank processing. The computation to estimate the MOS has two stages. First, an auditory model is used to compute auditory features, such as excitation patterns, specific loudness patterns, and modulation patterns, for each auditory frequency band. The features are computed for both the reference signal and the signal reproduced using a device and the simulated or real network. The estimated features and the metrics describing differences in the features between the reference and test cases are provided as an input to a *cognitive model*, which then estimates the MOS value. The cognitive model is, in principle, a pattern recognition algorithm, such as an artificial neural network. The algorithm is trained using a large set of examples from real devices and acoustics conditions which have been analysed by a large panel of listeners.

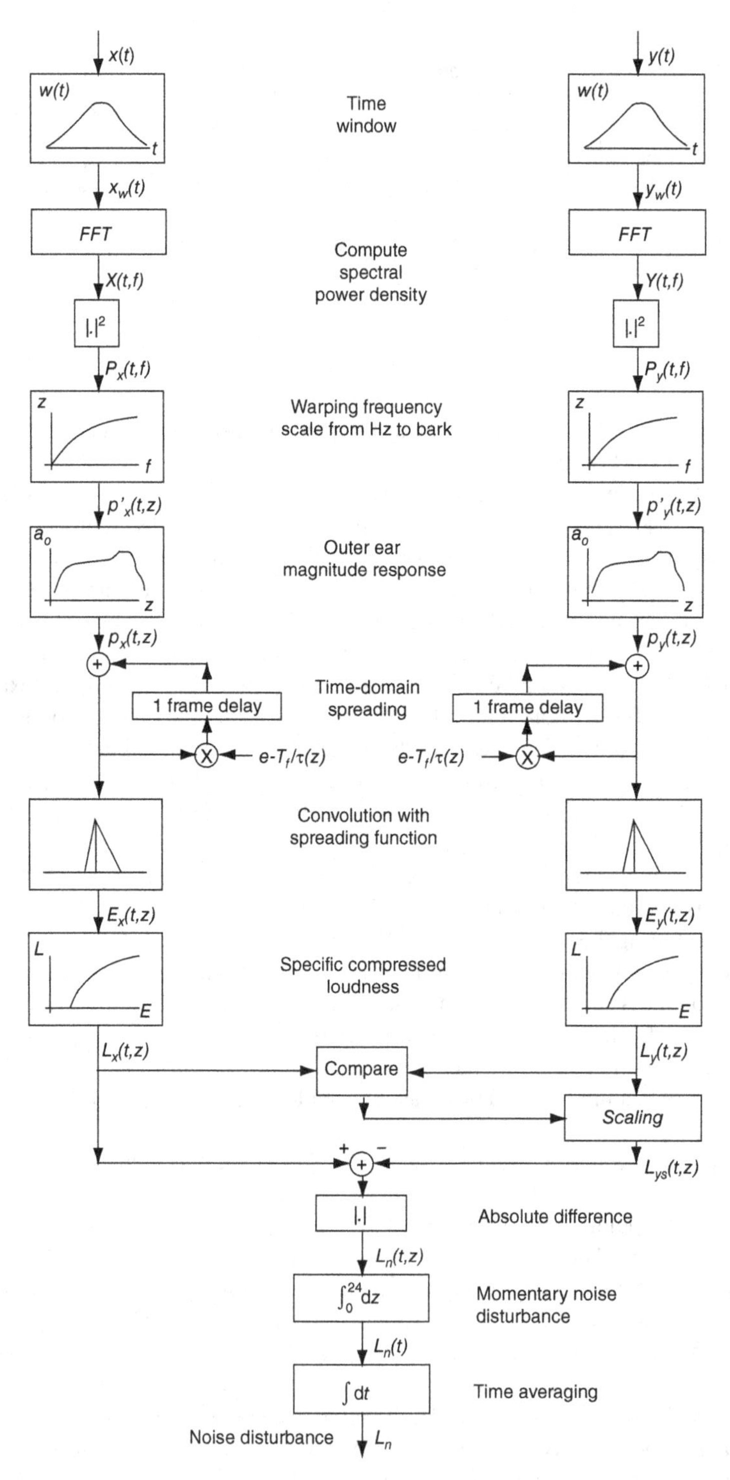

Figure 17.5 The computation of the perceptual audio quality measure (PAQM). Adapted from Beerends and Stemerdink (1992), and reproduced with permission from The Acoustical Society of America.

As the method is trained with a particular set of examples of reference and impaired audio samples, the method has a specific area of applicability. PEAQ has been used successfully to analyse audio quality with masking-based audio codecs without the effect of the listening room. For example, PEAQ estimates low MOS values for signals containing jitter or a slight deviation in the sampling frequency. However, PEAQ cannot operate with acoustically captured signals. It is probable that PEAQ would estimate a degraded MOS if the ear canal signals of a listener in a domestic environment were compared to the ear canal signals of the audio engineer in the mastering studio, although the listener would be satisfied with the audio quality.

Even though current auditory models are not yet able to evaluate all flavours of sound quality, it is clear that similar auditory models will eventually replace traditional audio quality measures.

17.5.3 Spatial Audio Quality

As discussed in Chapter 14, several methods have been developed to reproduce or to synthesize the spatial characteristics of sound as well. The overall advantage in the sound quality resulting when upgrading from monophonic reproduction to spatial audio with multi-channel loudspeaker set-ups has been researched by Rumsey *et al.* (2005). The quality of sound was measured with monophonic, stereophonic, and 5.1 surround reproduction with produced audio content. In addition, the quality was also measured with low-pass timbral degradations of the programme material. It was found that timbral quality corresponded to about 70% of the overall quality, whereas spatial reproduction corresponded to the remaining 30%. An interesting finding was that, when low-pass timbral artefacts were present, the spatial reproduction resulted in no advantage over monophonic reproduction.

This research was conducted with produced audio content, where the reference was 5.1 surround audio reproduction. In the most general case, the reference should be a real acoustic scenario, such as a case where sound sources are located around the listener in 3D, with natural reverberation arriving from all directions to the listener. Should the reproduction of such a case be targeted and the obtained authenticity measured, a major problem emerges. The perception of the original scenario must be compared to reproduction in a listening room, requiring that the subject is moved between the listening room and the original room to make the comparison. Unfortunately, the human auditory memory is too short to make accurate comparisons with delays of more than a few seconds between perceptions. Hence, such an approach is not feasible. In some cases, the reference scenario can be synthesized in an anechoic chamber using a large set of loudspeakers, and then the scenario can be recorded using microphones and reproduced using the same loudspeakers (Vilkamo *et al.*, 2009). This enables direct comparison, although the synthetic nature of the reference casts some doubts on the generalizability of the approach.

A considerably simpler case is the evaluation of spatial sound synthesis methods without aiming for reproduction of the recorded acoustic conditions. In synthesis methods, a specific spatial attribute is to be controlled, and the reference case can also often be created. For example, in virtual source positioning methods, the synthesized position of the virtual source can be measured relatively simply by using appropriate psychoacoustic test methods. The possible degradation of overall quality should also be measured, which is relatively simple to do, since the reference case can be generated by positioning a real source in the panning direction, thus enabling the comparison of the virtual and real sources in listening tests.

Auditory-modelling-based objective quality measurement systems could potentially solve the problem of the impossible comparison of the reproduced spatial sound to the original conditions. Binaural models, which were introduced in Section 13.5, have been suggested for use in measuring the quality of spatial sound reproduction. Indeed, in limited cases, auditory models are applicable for evaluating spatial audio quality (Blauert, 2013b). Although a large number of binaural models have been proposed, the explanation of human perception of spatially complex scenarios still seems challenging. The research in this field is thus incomplete. The need for models for the analysis of spatial audio in industry exists, since an effort to extend the PEAQ standard to the measurement of quality over stereophonic and multi-channel audio formats has been initiated. Unfortunately, the current results in the process are not promising (Liebetrau *et al.*, 2010), and the process is on hold as of now (2014).

17.6 Quality of Speech Communication

This section covers some basic concepts and aspects related to speech communication over different channels which are key factors of sound quality in the context of speech. The discussion touches different layers of speech quality: we start from intelligibility and discuss some slightly higher-level concepts as well. The field is wide, as speech quality is needed in many applications with different needs, such as telephony, voice-over-internet, radio, and public address. The discussion merely scratches the surface of the topic.

Speech intelligibility (Blauert, 2005, Chapter 7; Quackenbush *et al.*, 1988; Steeneken and Houtgast, 1985) is a property referring to how well the meaning of a spoken message is transmitted to a listener. As such, intelligibility depends on three factors:

- the ability of the *speaker* to produce a message with acoustically and linguistically clear contents;
- how well the *transmission channel* is able to transmit the message; and
- how well the *listener* is able to receive and analyse the message.

In speech communication without electrical devices, the transmission channel is the acoustic path from the lips of the speaker to the ears of the listener. When electrical devices are used, a microphone and a headphone and loudspeaker are needed together with an electrical transmission line, possibly with coding and transmission technologies.

The technical interpretation of *speech intelligibility* is related to the attributes of the transmission channel. To measure speech intelligibility subjectively, a set of speech signals is specified and delivered over the channel. The listener reports the message he or she perceived, and the proportion of correct identifications is taken as the subjective measure of speech intelligibility. There also exist objective measures, both instrumental and computational, that correlate well with subjective speech intelligibility.

When the quality of speech is at a level where the intelligibility is relatively good, certain other dimensions in sound quality are of interest. Such attributes are, for example, *speaker recognizability* and *speech naturalness*. For example, in telephony, the minimum requirement on quality is intelligibility of speech, but usually recognizability of the speaker is also required. The quality of synthetic speech also needs to be measured. For example, if speech synthesis is used in announcements in public spaces, the intelligibility of the messages has to be known. The same methods can be used with synthetic speech as with natural speech.

Different subjective and objective methods have been developed to measure the quality of speech, indicating the articulation, intelligibility, and quality of the reproduction of timbre (Quackenbush *et al.*, 1988). We will list some relevant techniques and later present some of them in greater detail.

17.6.1 Subjective Methods and Measures

- *Articulation.* The term articulation here means the overall functioning of the speech transmission channel, not just the functioning of the speech organs, as discussed in Section 5.1.3. A measure for the quantity is obtained from a listening test, where the task of the subjects is to listen to nonsense phoneme sequences composed as a catenation of consonants (C) and vowels (V), such as /CV/ or /CVC/, and to report the sequences perceived. The percentage of correct answers gives the *articulation score*. The *articulation index* is the articulation score modified to obtain additivity, just as the values of loudness are additive but the values of loudness level are not (Fletcher, 1995).
- *Intelligibility* and *intelligibility score.* The articulation test, but this time conducted with real words or sentences measures the intelligibility of the communication channel. The percentage of correct answers is the intelligibility score.
- *Rhyme test.* The test uses rhyming words or one-syllable words where changing the first phoneme changes the meaning of the word, such as pay/may/day/say/way. The percentage of correct answers measured gives this measure of speech quality. Different variations of this test exist, differing in the application and realization.
- *Speech interference test.* Here, noise is added to interfere with a reference speech signal and the speech signal to be tested. The level of noise is first adjusted so that the articulation in the reference speech is 50%, after which that level of noise is sought that produces the same score with the test speech signal. The difference in the levels of noise in the test and reference cases is the quality factor Q.
- *Quality comparison methods.* The quality of multiple speech samples is compared, and the subject is asked to rank them in order of preference. The subject may also be asked to focus on a specific perceptual attribute of the sounds.
- *Isopreference method.* In this method, a set of recordings is first made where a signal at different levels is transmitted through the channel accompanied by additive background noise at different levels. The listener evaluates the different recordings and forms a map of preferences with coordinates defined by the level of speech signal and the level of noise. Numerous variations on this approach exist.
- *Mean opinion score* (MOS). This is a commonly used scale for sound quality in which numerical values from 1 to 5 are associated with verbal category ratings. See Section 17.4.1.
- *Indirect judgement tests* (Quackenbush *et al.*, 1988). These tests aim to evaluate speech quality by measuring factors assumed to affect it. Such methods include, for example, the *paired acceptability rating method* (PARM), the *quality acceptance rating test* (QUART), and the *diagnostic acceptability measure* (DAM). The DAM method utilizes 20 different given scales with values between 0 and 100, and measures such as 'rasping', 'hissing', and 'acceptability' are evaluated. The measures are assumed to be related to the features of the speech signal itself, to features of background sound, or to the general impression, respectively.
- *Communicability tests.* Here, the task of a subject is to communicate with another subject through a channel, and to conduct a defined task together. For example, one of the subjects might instruct the other how to draw a picture. Immediately on completion of the task the

subjects are asked to rate the ease of communication, for example, on a scale from '1 = no meaning understood using reasonable effort' to '5 = completely relaxed communication; no effort required'.

- *Task recall tests.* In these tests, the subject has to remember as many words as possible that he or she hears through a channel. The test measures the ease of communication, since a flawed communication channel makes remembering words difficult.
- *Noise suppression tests.* Many mobile communication devices have algorithms to suppress background noise. Such processing affects 1) overall quality, 2) intrusiveness of background noise, and 3) the quality of the speech signal. A subjective test specified in ITU-T P.835 (2003) is often used in industry, which is discussed in slightly more detail in Section 17.8.2.

The speech material used in the tests is typically a subset of the vocabulary of a language. When the words for the material are chosen, they have to be phonetically balanced for the targeted purpose. For example, for mobile speech communication tests, the material has to contain the phonemes in the same probabilities as in the everyday language of the test subjects.

17.6.2 Objective Methods and Measures

- *Articulation index* (AI). This was developed to measure speech intelligibility over a transmission channel that is assumed to be nearly linear, but, with disturbance caused by additive noise. The method assumes that the loss of articulation can be estimated by summing the AI values over 20 frequency bands, following roughly the Bark scale.
- *Percentage articulation loss of consonants* ($\%\mathrm{AL}_{\mathrm{cons}}$) (Peutz, 1971). This is a simple and relatively often used estimate of speech intelligibility in a room, auditorium, or other large space. The $\%\mathrm{AL}_{\mathrm{cons}}$ value is computed from the basic acoustic parameters of the space. The method is described in Section 17.9.3.
- *Speech transmission index* (STI). The index is based on the *modulation transfer function* (MTF), and it can be used to estimate relatively reliably the effect of reverberation and additive noise of a transmission channel on speech intelligibility. The method is described in Sections 17.7.1 and 17.7.2, and STIPA, a simplified version of STI, is discussed in Section 17.7.4.
- *Signal-to-noise ratio* (SNR). This is a traditional measure of how well a signal differentiates from background noise. It has different variants, such as frequency-weighted SNR and segmental SNR. The classical SNR can be defined as $\mathrm{SNR} = 10\log_{10}\{\sum_n x^2(n)/\sum_n[x(n) - x_{\mathrm{d}}(n)]^2\}$, where $x(n)$ is the original signal and $x_{\mathrm{d}}(n)$ is the distorted signal after going through the communication channel. The SNR is very sensitive to all kinds of differences between $x(n)$ and $x_{\mathrm{d}}(n)$ as well as to differences that are not at all perceivable. Thus, it is a relevant measure only in some simple cases.
- *Spectral distance measures.* These measures are based on the notion that the magnitude spectrum and spectral differences reflect better the perceptual attributes of sound than signals or signal differences in the time domain. The difference between smoothed time–frequency presentations of a signal and the communicated signal often produces usable information about the distortion in the system. Several different versions of these spectral distance measures exist. The cepstrum has also been similarly used to evaluate the difference.

- *Weighted spectral slope distance measure*. This uses the slope of spectra instead of the basic magnitude spectrum to compute the spectral distance, and so it reflects certain phonetic differences between speech sounds more accurately.
- *LPC distance measures*. These measures are based on linear predicton. LPC coefficients are computed for both the original signal and the distorted signal, and a distance measure is computed from the LPC coefficients, reflecting the difference between the signals.
- *Auditory sound quality measures*. These computational methods for signal difference measurement mimic the spectro-temporal resolution of hearing in quality estimation. The sound quality measures for audio are reviewed in Section 17.5.2 and for speech in telecommunication in Section 17.8.1. These methods are widely used in the mobile telecommunication industry.

17.7 Measuring Speech Understandability with the Modulation Transfer Function

The understandability of speech is of crucial importance in public announcement systems, in telephones, and also in auditoria with or without sound reinforcement. The intelligibility can be estimated by computing the *speech transmission index* (STI), which is based on the concept of the *modulation transfer function*. STI methods are well established, and in some countries the STI values for sound systems in public spaces are regulated. Furthermore, the standard that describes sound systems for emergency purposes ISO 7240-19 (2007) defines the minimum STI value to be 0.5. Public announcement systems for public spaces are costly, and if the STI of a new building is measured to be lower than that targeted in the design process, the acoustic and electrical changes that must be made may be expensive. This is one reason why STI simulations and measurement systems are of great importance. Thus, this section makes a relatively detailed overview of the techniques used.

17.7.1 Modulation Transfer Function

The *modulation transfer function* (MTF) is an objective measure of a communication channel required to compute the *speech transmission index* described in the following section. The background of the MTF lies in the basic properties of speech spectra and their variation over time. Figure 3.6 on page 54 shows that the produced spectra change radically between phones. As discussed in Section 11.5.1, human hearing is very sensitive to *temporal changes* in magnitude spectrum, suggesting that the temporal modulations present in each auditory band are very important for successful speech reception.

When speech is communicated over an acoustic space or a technical channel, it seems to better retain understandability the more naturally the modulation in each auditory frequency band is conserved. Based on knowledge from auditory modelling, it would be logical to monitor the temporal changes in specific loudness for each critical band (see Section 10.2.5). However, such an approach has not been adopted in this context, but a slightly simpler and technically more straightforward technique, based on the MTF, is widely utilized instead.

The MTF reflects how well the modulation of the envelope of a narrowband signal is conserved when travelling from the source to the receiver (Houtgast and Steeneken, 1985; Steeneken and Houtgast, 1985). A single MTF value does not estimate the speech intelligibility,

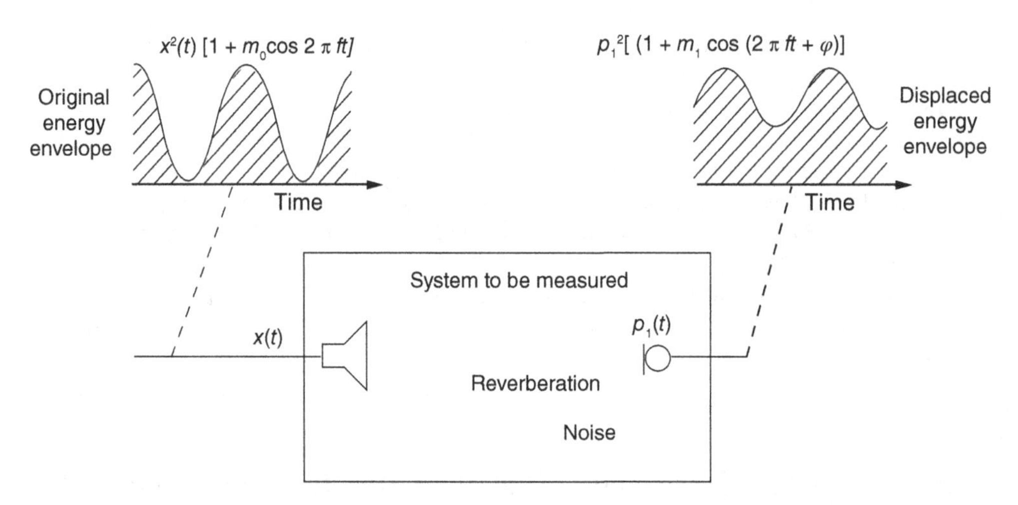

Figure 17.6 A modulated octave-band signal is reproduced in a room or over an arbitrary communication channel and recorded at the position of the listener. The modulation changes depending on reverberation and noise in the system.

but, as will be discussed in the next section, when the MTF is measured for different modulation frequencies for different narrowband signals, the speech transmission index (STI) can be computed. The STI correlates well with different subjective intelligibility measures.

When the modulation is computed from a squared pressure response, representing the energy, the effects of background noise and reverberation are made commensurate. Also, the sinusoidal modulation of a carrier signal, after the effects of noise and reverberation, is preserved as a sinusoid, though with a lower modulation depth. The principle is shown in Figure 17.6, where a modulated signal $x(t)$ is presented to an acoustic system from where the pressure signal $p_1(t)$ is captured. The envelopes related to the squared signals are shown in the figure. Note that, in general in STI literature, pressure squared is associated with 'intensity'. In this book we restrict the term intensity to mean the net flow of energy.

The MTF is most commonly measured by applying the signal

$$x(t) = \sqrt{0.5(1 + m_0 \cos(2\pi f_m t))}\, s(t) \tag{17.1}$$

to the system, where $s(t)$ is an octave-band signal with centre frequency f and f_m is the modulation frequency. When no background noise is present and no reverberations or echoes exist in the room, the degree of modulation in the system is preserved, and the pressure signal

$$p_0(t) = A_0\sqrt{0.5(1 + m_0 \cos(2\pi f_m t + \varphi_0))}\, s(t) \tag{17.2}$$

can be measured, where A_0 is a static gain and φ is a term related to the delay of the phase. The response shares the unmodified modulation depth m_0 in this ideal case. We are interested

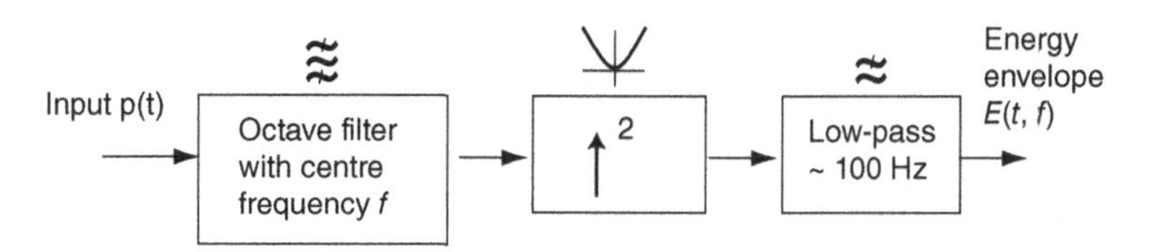

Figure 17.7 The processing of recorded pressure signals $p(t)$ to extract energy envelopes $E(t,f)$, where f is the centre frequency of the octave-band filter.

in how the modulation is transferred when noise and reverberation exist. Let us assume that $p_1(t)$ is recorded in reverberant and noisy conditions:

$$p_1(t) = A_1\sqrt{0.5(1 + m_1 \cos(2\pi f_\mathrm{m} t + \varphi_1))}\; s(t), \tag{17.3}$$

where A_1 is a static gain, and φ_1 is a term related to the delay of the phase.

The depth of modulation m_1 is the variable that is to be measured. To this end, we may process the measured pressure signal as shown in Figure 17.7 to obtain the energy envelope signal $E_1(t,f)$ for the octave band with centre frequency f. Note the similarity of the set-up to the filter-bank-based auditory models in Section 13.2. The frequency content of $s(t)$ is chosen to be located above 100 Hz, and consequently only its envelope remains after the processing. The energy envelope $E_1(t,f)$ obtained from recorded signal $p_1(t)$ has the form

$$E_1(t,f) = E_1\left[1 + m_1 \cos(2\pi f_\mathrm{m} t + \varphi)\right], \tag{17.4}$$

where E_1 is the energy of the signal recorded and m_1 is the depth of modulation. The value of m_1 can be calculated as

$$m_1(f,f_\mathrm{m}) = 2\,\frac{\sqrt{\left|\int_t E_1(t,f)\sin(2\pi f_\mathrm{m} t)\,\mathrm{d}t\right|^2 + \left|\int_t E_1(t,f)\cos(2\pi f_\mathrm{m} t)\,\mathrm{d}t\right|^2}}{\int_t E_1(t,f)\,\mathrm{d}t} \tag{17.5}$$

(IEC, 2011). The equation thus computes the magnitude of modulation at frequency f_m of the signal $E(t,f)$, using the set-up in Figure 17.7 normalized to values between zero and one. Finally, the modulation transfer ratio is given by

$$m(f,f_\mathrm{m}) = \frac{m_1(f,f_\mathrm{m})}{m_0(f,f_\mathrm{m})}, \tag{17.6}$$

which is sometimes also called ‘modulation reduction’. The closer the number is to unity, the better the modulation is transferred, and values near zero imply largely lost modulation.

Figure 17.8 shows the reduction in modulation with reverberation in case A and with added noise in case B computed for a speech signal. The left-most panels show the change in the envelope of the measured signal at the octave-band noise with centre frequency 500 Hz. The reverberation smooths the change of the envelope in time, acting as a low-pass filter for the envelope and thus also for modulation. The effect of additional noise is seen from the increase in the minimum value of the envelope.

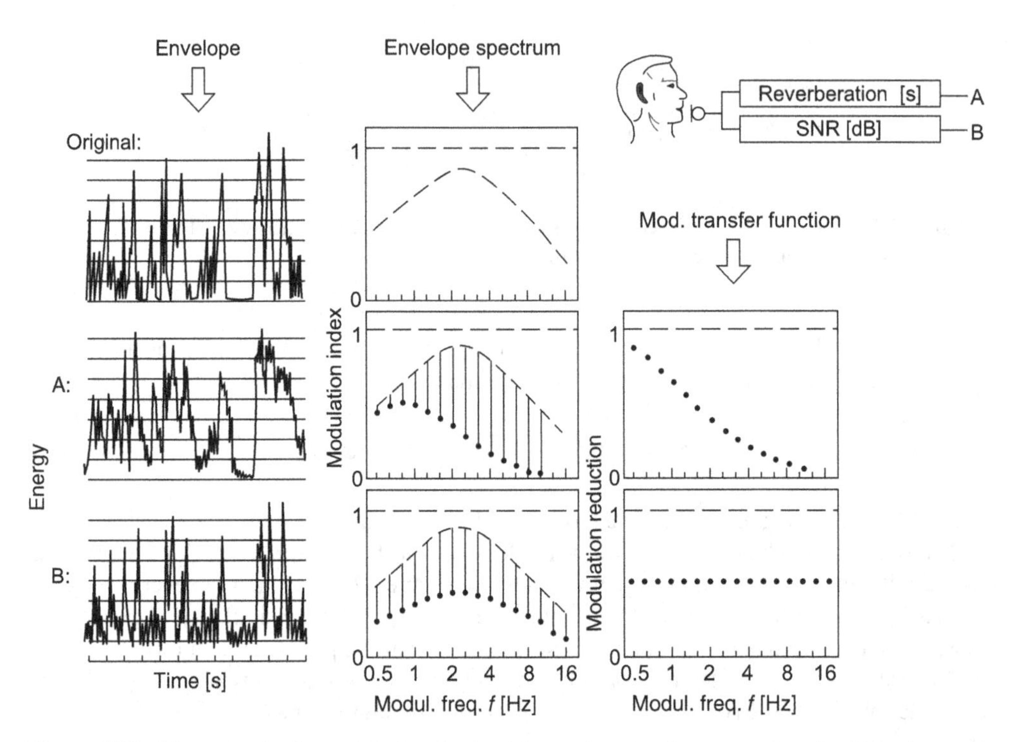

Figure 17.8 The reduction in modulation depth with reverberation in system A and with background noise in system B. The left panels show, from top to bottom, the envelope of the signal in its original form, after reverberation, and after additive noise. The envelope spectrum plots (centre) show the modulation spectra in each case, and the modulation transfer function (right) shows the reduction in modulation for each case depending on the modulation frequency. Adapted from Houtgast and Steeneken (1985). Courtesy of Bruel & Kjær.

The spectra of the modulation in the original and in the transmitted signal are shown in the centre panel at modulation frequencies from 0.5 Hz to 16 Hz. Reverberation reduces the modulation depth more at high modulation frequencies than at low modulation frequencies. Noise, in turn, reduces the modulation depth equally at all modulation frequencies. The right-most plots show the corresponding modulation transfer functions for cases A and B. In case A, the reverberation clearly acts as a low-pass filter in the transfer function, while the noise in case B reduces the modulation evenly at all modulation frequencies.

The MTF can, in principle, be measured with signals that have energy at all audible frequencies and modulations at all modulation frequencies of interest. Speech, music, and other such signals can be used. However, the measurement can be conducted with more accuracy using signals specifically designed for the task. The MTF can also be estimated during the design of the acoustics of halls, or during the design of public address systems, if the impulse response and background noise levels can be estimated.

A straightforward method to measure the MTF is to use a loudspeaker or an artificial mouth in a room or in an auditorium in the position where the speaker would be. The source is used to emit 100% amplitude-modulated octave-band noise (m_0=1.0) at a level corresponding to the average level of speech. The measurement is repeated for each modulation frequency f_m, from 0.63 Hz to 12.5 Hz in steps of 1/3 octave, as shown in Figure 17.9.

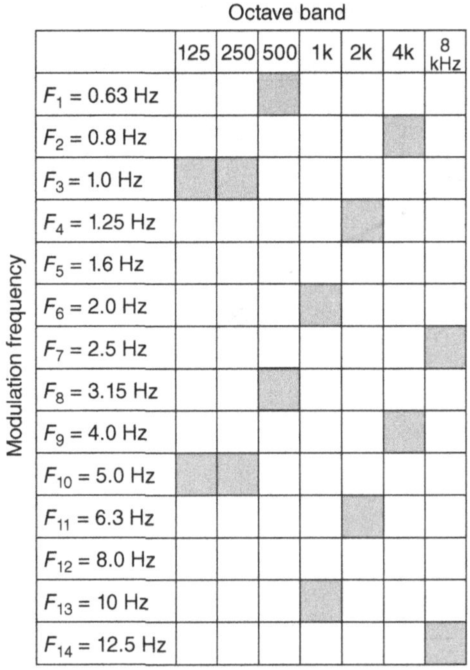

Octave band

Modulation frequency	125	250	500	1k	2k	4k	8 kHz
$F_1 = 0.63$ Hz							
$F_2 = 0.8$ Hz							
$F_3 = 1.0$ Hz							
$F_4 = 1.25$ Hz							
$F_5 = 1.6$ Hz							
$F_6 = 2.0$ Hz							
$F_7 = 2.5$ Hz							
$F_8 = 3.15$ Hz							
$F_9 = 4.0$ Hz							
$F_{10} = 5.0$ Hz							
$F_{11} = 6.3$ Hz							
$F_{12} = 8.0$ Hz							
$F_{13} = 10$ Hz							
$F_{14} = 12.5$ Hz							

Figure 17.9 The octave bands of the carrier signal and the modulation frequencies used in the STI measurement, represented by all the squares in the matrix. The grey squares represent the corresponding values in the STIPA measurement method.

A microphone is placed in the position of the listener to measure the response, and in each case the modulation of the envelope is analysed at frequencies having octave bands of the carrier signal in the range 125 Hz–8 kHz. The reduction in modulation from the original 100% gives the value of the MTF $m(f,f_{\mathrm{m}})$, where f is the centre frequency of the octave band and f_{m} is the frequency of modulation.

The effects of reverberation and noise can, in general, be expressed as

$$m(f,f_{\mathrm{m}}) = \frac{1}{\sqrt{1 + \left(2\pi f_{\mathrm{m}} \frac{T_f}{13.8}\right)^2}} \cdot \frac{1}{1 + 10^{-\mathrm{SNR}_f/10}}. \tag{17.7}$$

Here, the first term corresponds to the effect of reverberation and the second reflects the effect of background noise.

17.7.2 Speech Transmission Index STI

It was proposed that the modulation transfer function (MTF) be measured at seven frequency bands with fourteen modulation frequencies, resulting in $7 \times 14 = 98$ m values (see Figure 17.9). Ideally, the estimate of speech intelligibility should be expressed with a single value.

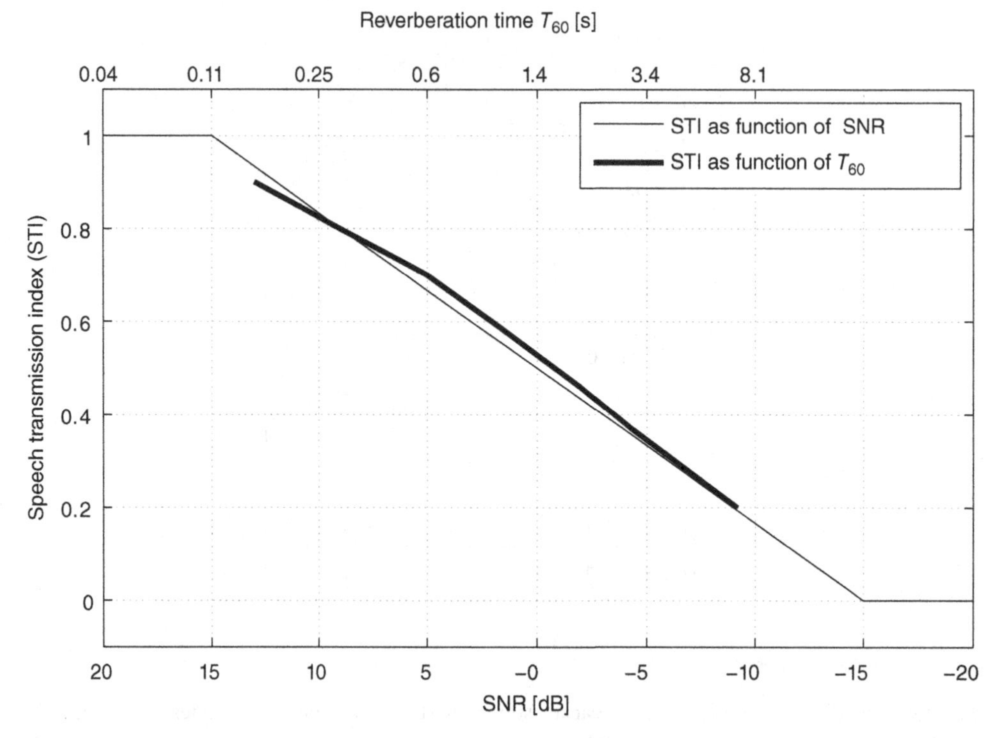

Figure 17.10 The dependency of the STI in the ideal case with noise only as a function of SNR, and with reverberation only as a function of T_{60}. Adapted from Houtgast and Steeneken (1985).

The *speech transmission index* (STI) has been found to serve this purpose. The principle is that the 98 m values are first transformed into apparent SNR values as

$$\text{SNR}_{\text{app}} = \max\left(-15, \min\left(15, 10\log\frac{m}{1-m}\right)\right), \tag{17.8}$$

where SNR_{app} is expressed in dB on a scale from −15 to 15. The values are scaled to lie between 0 and 1, and a weighted average is calculated (IEC, 2011). The weights emphasize the octave bands most relevant for understanding speech. The value of the STI is limited to between 0.0 and 1.0, 0.0 corresponding to estimates with no speech intelligibility and 1.0 to perfect speech intelligibility.

Figure 17.10 shows the effect of the SNR on the STI and also the effect of ideal reverberation with different T_{60} values. Since the curves are on top of each other, the correspondence of the SNR to the reverberation time is easily seen.

17.7.3 STI and Speech Intelligibility

The speech transmission index (STI) cannot directly be associated with the subjective intelligibility of speech, since many other features affect perceptual intelligibility, such as the

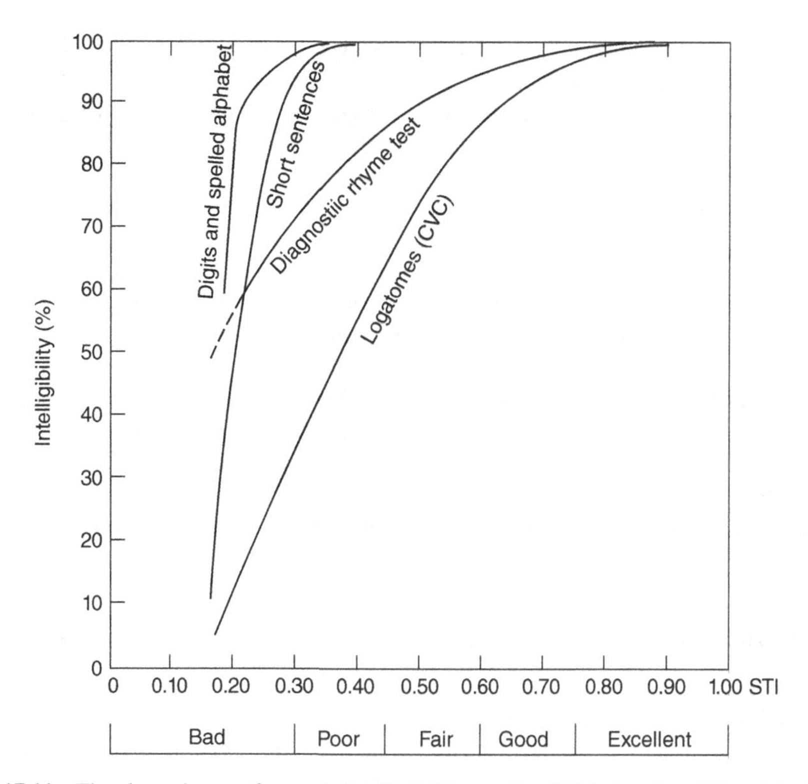

Figure 17.11 The dependence of speech intelligibility on the STI index for different test signals. Adopted from (Houtgast and Steeneken, 1985 and Steeneken and Houtgast, 2002). Courtesy of Bruel & Kjær.

speaker, familiarity with the message, and language. However, the STI has been found to have a monotonic effect on intelligibility: the higher the STI, the higher the intelligibility (Steeneken and Houtgast, 2002). Figure 17.11 shows the dependency between the type of speech and the STI. When the speech content is short words from a known small vocabulary, high intelligibility is obtained even with relatively low STI values in the range 0.2–0.4. With sentences, the linguistic context aids in understanding the message, and thus good intelligibility is obtained with STI values in the range 0.4–0.5. In more challenging cases, such as with logatomes, a considerably higher STI is needed for high intelligibility. In the case of logatomes, it would be more correct to call the measured value articulation instead of intelligibility.

The STI values are also categorized in Figure 17.11. The labels designate quality categories that can be used to interpret the measurement results. The step size for the categories is 0.15. If a measurement of STI is repeated several times, and if the standard deviation is notably smaller than 0.15, the STI measurement and its interpretation can be considered to be reliable.

17.7.4 Practical Measurement of STI

The measurement of the STI is quite a complicated procedure – 98 different $m(f, f_\mathrm{m})$ values must be measured individually, and this takes about 15 minutes to conduct. Often, the reverberation and background noise have such properties that the measurement can be simplified to cover fewer combinations of modulation frequencies and octave bands. STIPA employs two modulation frequencies for each frequency band, as shown in Figure 17.9 (IEC, 2011). Using only two modulation frequencies makes it possible to present all of the octave-band signals simultaneously, in which case each octave-band signal is modulated by the factor $\sqrt{0.5(1 + m_0 \cos(2\pi f_{m1} t) + m_0 \cos(2\pi f_{m2} t))}$, where f_{m1} and f_{m2} are the modulation frequencies at the specific octave band, and the value of m_0 is about 50%. With this method, the measurement time is reduced to between 10 and 15 seconds.

There are a number of dedicated measurement devices available for STIPA. Relatively good results can also be measured using inexpensive applications on handheld devices, available for at least the IOS and Android platforms, and the interested reader is encouraged to test the applications to get hands-on experience of STI measurements.

The STI and STIPA, and the predecessor of STIPA, RASTI, have been standardized in many versions. The different versions and their usage in practical situations are discussed by Steeneken *et al.* (2011). STIPA measurements seem to provide results quite similar to STI measurements in many cases, although the measurement range has been pruned from the STI measurements. However, in some cases, significant differences occur (Mapp, 2005).

17.8 Objective Speech Quality Measurement for Telecommunication

The STI measure considers only the intelligibility of speech. This has not been found to be an adequate analysis system for modern telecommunication systems, where the speech codecs may, in the worst case, for example, make the speaker unidentifiable, although the intelligibility of speech is good. To estimate other factors in transmitted speech, more advanced objective analysis methods have been developed.

There are three main requirements for quality measurement methods:

- *Models for general speech quality* are expected to give a high MOS value only for natural-sounding and intelligible speech.
- *Methods for measuring the perceptual effect of background noise suppression* estimate the performance of the noise-suppression algorithms, and also evaluate whether the methods introduce unwanted side effects affecting the quality of the speech signal.
- *Measures for echo suppression* evaluate the capability of the telecommunication system to reduce unwanted echoes in two-way communication.

The problem definition is quite broad, as three different aspects are to be measured. Also, telecommunication devices are used in various acoustic conditions, and the network connectivity for the devices affects the sound quality. Furthermore, interfering background noise, nearby sources, room acoustics, and acoustics in hands-free use may make the acoustic conditions challenging. The transmission of data over a network may also cause unpredictable delays, jitter, and occasional loss of data. In addition, the acoustic feedback from the receiver to the sender depends on the properties of the device, on the acoustic conditions on the receiver side, and also on the properties of the connection.

A device that enters the market should thus be tested under many different conditions. Unfortunately, conducting such a set of listening tests that would cover all possible combinations of different conditions would be a tedious task. Instead of listening tests, objective methods are used widely in industry, which is reflected in the active standardization of the methods. The methods are signal-processing structures that normally include some parts simulating the properties of human hearing. The methods may also include acoustic measurements of the devices under controlled acoustic conditions.

Quite commonly, when a device is approved by an authority or network operator, it has to achieve certain scores in standardized objective tests. The methods are relatively complicated, and they are only briefly touched on below. Interested readers can find in-depth information on the current criteria (2014) for the properties of devices and technologies for speech communication in the 3GPP measurement techniques (ETSI, 2014a) and in the specifications of the requirements for the devices (ETSI, 2014b).

17.8.1 General Speech Quality Measurement Techniques

The first recommendation of a method to measure objective speech quality (ITU-T, 1998a) was called the perceptual speech quality measure (PSQM) (Beerends and Stemerdink, 1994). The PSQM is almost identical to the PAQM model for audio quality shown in Figure 17.5, with only some minor details being different (Beerends and Stemerdink, 1994). The PSQM was primarily focused on identifying the quality impact of speech codecs. Unfortunately, the method did not turn out to be successful, and the recommendation has since been withdrawn. The PSQM was unable to estimate correctly the quality impairment due to filtering, variable delay, and distortion common in mobile communication. For example, the transmission delay may vary in mobile transmission, and the variations can be fatal in VoIP applications. Subtracting time-varying feature vectors, as shown in Figure 17.5, would estimate a large error in the case where the test case had random variations in the lengths of syllables in speech. Such small delay variations do not cause much deterioration of perceived quality, implying that the model would underestimate the speech quality in this case.

Subsequent to the PSQM, work was initiated to create an algorithm suitable for assessing the additional impact of network impairment, which resulted in a method called the perceptual evaluation of speech quality (PESQ)(Beerends *et al.*, 2002; ITU-T, 1998b). The PESQ method includes processing steps such as automatic compensation of the dynamically varying jitter prior to the auditory model, and other similar steps, which are designed to directly overcome the problems that existed with the PSQM. Many of the features do not stem directly from the neural structure of the auditory system, although they are motivated perceptually. PESQ can be characterized as a signal difference measurement device, which has some features stemming from the human auditory system.

PESQ was validated with results from numerous experiments that specifically tested its performance across combinations of factors such as filtering, coding distortions, variable delay, and channel errors. It is recommended to be used for speech quality assessment of telephone-band handset telephony and narrowband speech codecs. PESQ is measured using an electrical interface (by using a connector between the device and the measurement device), which does not take into account the acoustic degradations of the signal in real listening. The TOSQA (Telecommunications Objective Speech Quality Assessment) system can also take input from an acoustic interface (TOSQA, 2003), which may have a significant effect on the perceived and measured quality. Otherwise, it shares many similar operation principles with PESQ.

The PESQ method has been extended to cover the assessment of wideband speech as well as networks and codecs that introduce time warping. The most recent version (ITU-T, 2011) is known as POLQA (perceptual objective listening quality assessment) (Beerends *et al.*, 2013; ITU-T, 2011), and again, this gives better estimates of subjective evaluations from measured signals.

A specific method has been developed to estimate objectively the speech quality perceived by hearing-impaired users wearing a hearing aid. The Hearing-Aid Speech Quality Index (HASQI) (Kates and Arehart, 2014) is based on a model of the auditory periphery that can incorporate changes due to hearing loss. The model can be used to predict the effect of signal processing in hearing aids to the perceived speech quality of either normally-hearing or hearing-impaired listeners, and it has an interesting application in the development of hearing aids. That is, the engineers often have normal hearing, which means that they cannot test the devices thoroughly themselves. Moreover, the hearing aids should be suited for different types of impairment. Thus, a method to evaluate the speech quality with different severities and types of hearing impairment is potentially helpful in product development.

17.8.2 Measurement of the Perceptual Effect of Background Noise

Telecommunication devices, such as mobile phones, often involve algorithms to suppress background noise using non-linear DSP methods with time-variant processing. An unwanted side effect is that the noise suppression algorithms may also degrade the quality of speech, especially if the background noise is non-stationary. Objective measures have been developed to measure such effects, and the most recent one is 3QUEST, defined by ETSI (2008). The method is targeted for both wide- and narrowband transmission in noisy environments, and it has been calibrated with a large set of subjective tests.

The subjective tests were conducted using a set of noisy recordings with real devices, according to ITU-T P.835 (2003). In the tests, the subjects rated three different factors in the samples:

- Speech MOS (S-MOS): the speech sample was rated 5 – not distorted, 4 – slightly distorted, 3 – somewhat distorted, 2 – fairly distorted, or 1 – very distorted.
- Subjective noise MOS (N-MOS): the background of the sample was 5 – not noticeable, 4 – slightly noticeable, 3 – noticeable but not intrusive, 2 – somewhat intrusive, or 1– very intrusive.
- Overall MOS (G-MOS) on the standard MOS scale.

The method is a relatively complex signal processing structure, which includes parts mimicking human time–frequency resolution, and ultimately uses a trained pattern recognition algorithm to produce final quality estimates. It requires three inputs to the system: a clean speech signal, the unprocessed signal, and the processed signal. The clean signal is the speech signal of a real human subject recorded in a free field. The unprocessed and processed signals are measured in realistic noisy conditions created in the laboratory. A dummy head with a mouth is placed in a noisy environment generated in a laboratory, and the device under test is placed near the dummy head, just as it would be in practice. The noisy environment is generated using an equalized loudspeaker set-up with four loudspeakers and one subwoofer around the head. The sound applied to the loudspeakers originates from recordings from noisy natural conditions.

A separate microphone positioned near the microphone of the device under test is used to capture the unprocessed signal. The processed signal is the signal captured and processed by the device, possibly also aiming to reduce the background noise. The recording of the device is performed separately with all noise signals captured in different natural conditions. The use of different noisy signals simulates the functioning of the device in different acoustic scenarios. The recordings of the processed and unprocessed signals are used to estimate the MOS. The measurement system is described ETSI (2008).

Similarly processed signals were used in listening tests to provide reference data to train the 3QUEST algorithms. Finally, the 3QUEST system estimates the S-MOS, N-MOS, and G-MOS values for any signals recorded in an identical manner from any device under test (3QUEST, 2008).

17.8.3 Measurement of the Perceptual Effect of Echoes

A common unwanted feature of two-way telecommunication is the presence of loud echoes (Appel and Beerends, 2002). In natural conditions, a speaker perceives his or her own voice from the mouth-to-ear path and from the acoustic response of the environment, providing direct feedback that is used subconsciously to control the speech production process. Interference in the feedback can influence the comfort of speaking and also the manner of speaking. A well-known effect is the raising of one's voice in the presence of loud background noise, called the Lombard effect (Lane and Tranel, 1971). In contrast, we lower the volume of our voice when we are played back our voice loudly. Delaying the echo increases the perception of discomfort. For small delays ($<$10 ms) and high levels, the echo interferes with the sound coming directly from the mouth of the speaker, leading to the perception of colouration due to comb filtering. Medium delays (10–30 ms) lead to the perception of hollowness in one's voice, and for larger delays ($>$ 30 ms) we perceive a clear, distinct echo. When the delay is large ($>$ 200 ms) and the level of the echo is high, subjects experience difficulty in producing words (Appel and Beerends, 2002). The difficulties may be simply because, when producing a syllable, one's own voice present in the ear canals due to the echo has the timbre of the previous syllable, which causes confusion in our neural speech production system.

In two-way communication with mobile phone, some cases the sound from the loudspeaker is captured by the microphone in it, causing the phone to send the signal back to the caller. This can happen, for example, when using the phone in the hands-free mode via a loudspeaker. In such a situation, the round trip delay is of the order of 300 ms, a result of the signal processing and other delays in both phones, and also due to transmission delays. In VoIP calls, the round trip delay is typically even longer, often more than 500 ms. Such long delays can thus cause difficulties in speaking, which motivates the implementation of echo cancellation algorithms and subsequently objective measures to measure the positive and negative effects of the algorithms.

A method to measure objectively the degradation of quality due to strong echoes was presented by Appel and Beerends (2002). Another application for the same task is the *echo quality evaluation of speech in telecommunications* test (EQUEST), which is an instrumental method for estimating the annoyance (EQUEST, 2012). EQUEST is, again, a relatively complex signal processing algorithm, which uses psychoacoustic knowledge to perform the evaluation. This time the knowledge of the psychoacoustic temporal masking characteristics of human listeners is injected into the algorithm to estimate the annoyance caused by echoes. An alternative method for this task is described in the 3GPP standard (ETSI, 2014a).

17.9 Sound Quality in Auditoria and Concert Halls

The spaces for presenting arts involving sound, such as theatres, auditoria, and especially concert halls, have a special status with respect to sound quality. The quality of sound created by verbal or music performances for an audience has been of interest for hundreds and even thousands of years (Blauert, 2013a). For this reason, the design of concert hall acoustics has been a showcase of acoustic technologies to the community at large. Concert halls have been designed and built using trial-and-error, and a relatively good consensus exists concerning about 20 halls with great 'acoustics', such as Vienna Musikverain (Vienna, Austria), Metropolitan Opera (New York, USA), Boston Symphony Hall (Boston, USA), and Concertgebow (Amsterdam, The Netherlands). During the late 1900s, a scientific approach was finally adopted for concert hall design, although what is the best method is still open to debate.

Concert halls for music performances, that is, big halls for orchestral music, opera houses, chamber music halls, and halls for electronically reinforced sound, have to be designed keeping in mind the primary use of the hall. The main difference between the halls is the reverberation: too long a reverberation at too high a level compared to the acoustics optimal for the music performed there, and the general impression of the music is degraded. On the other hand, if the hall is too 'dry' in reverberation, the loudness of music may be perceived as too low with acoustic instruments, and the general impression of the music will again be different from the way it should be. The evolution of concert halls can also be assumed to have had an impact on music. Since composers created their music to be performed in specific halls, the composition of the music, and also the composition of orchestras, was adapted to them.

The criteria for speech auditoria and drama theatres, where the target is to maximize speech quality, are different than for halls for music performances. The overall guidelines for speech intelligibility presented in Section 17.6 also hold for auditoria and concert halls, and the STI measure can be used to estimate speech intelligibility. A simpler method than STI to estimate intelligibility in an auditorium is presented in this section, where the proportion of consonants not delivered intelligibly to the listener is approximated using knowledge of reverberation time and hall geometry.

17.9.1 Subjective Measures

A considerable effort has been made to define the vocabulary to describe the main properties of concert halls (Barron, 1993; Beranek, 1996). Beranek (1996) suggests a list of 18 attributes based on his own extensive experience of listening in concert halls, which includes terms describing sound from both the audience and the stage. Lokki (2014) used descriptive sensory analysis with reproduced spatial sound of different concert halls, and ended up with a rather similar, though shorter, list. It can be assumed that Beranek's list contains some attributes with which all listeners do not agree. However, Beranek's list is presented here, as it gives the reader a general impression of the kind of attributes that are at least thought to exist in subjective attribute palettes of concert halls.

- *Intimacy* or *presence*. The hall gives an impression of a small and intimate space.
- *Reverberation* or *liveness*. A long and perceivable reverberant tail makes the hall give the impression of a 'live' hall, and, correspondingly, a short reverberant tail makes the hall 'dry'.
- *Spaciousness: Apparent source width* (ASW). This is the width of the auditory object associated with the sound source itself. The reflections and the reverberation of the hall may make the sound sources seem to be wider than they actually are.

- *Spaciousness: Listener envelopment* (LEV). This is the directional distribution of the auditory object associated with reverberant sound. LEV is judged to be high when the reverberant sound is perceived to arrive from all directions.
- *Clarity*. This attribute is related to how well the sounds generated by instruments in a musical performance stand apart from each other. It depends on the performance and also on the acoustics, according to Beranek.
- *Warmth*. A hall is said to be warm if the reverberation time is longer at low frequencies (below 350 Hz) than at higher frequencies. If the reverberation time is too long, or if the low frequencies are overly strong, the hall may be called 'dark', which is an undesirable feature.
- *Loudness*. This simply means the perceived loudness at the listening position.
- *Acoustic glare*. This is generated if the sound is reflected by flat, smooth side panels to the audience. Rough and irregularly shaped panels reduce glare.
- *Brilliance*. This is the perception when high frequencies are prominent and decay slowly.
- *Balance*. A good balance is obtained when all sound sources are audible to the listener as intended. The balance depends, naturally, on the performers, but also on the acoustics of the hall.
- *Blend*. This is defined by the 'mixing' of sounds at the listening position. With a good blend, the sounds from the instruments are perceived as intervals and chords by the listener, with the intended level of consonance and dissonance.
- *Ensemble*. This refers to the ability of the performers to synchronize their playing as intended in the music. Typically, a better ensemble is obtained when the performers hear each other clearly on the stage.
- *Immediacy of response*. This is related to how performers perceive the hall's response to the played notes. If the response contains significantly delayed and strong reflections, the playing of the performers is affected negatively.
- *Texture*. This is the temporal pattern derived from the early reflections of the hall.
- *Freedom from echo*. An echo is a reflection that is loud enough and delayed sufficiently to be perceived as a separate auditory event, as discussed in Section 12.5.1. Echoes are not desired in concert halls.
- *Dynamic range and background noise level*. The lower end of the dynamic range is, in principle, defined by the level of background noise, or, if extremely low, the hearing threshold. The upper end of the range depends on the loudness of sounds a source may generate depending on the source itself, and also on the room response.
- *Extraneous effects on tonal quality*. No extra sounds should be produced by the hall, such as rattling sounds. Beranek also mentions the shift of localization of the sources in this context, which is referred to as *image shift*.
- *Uniformity of sound*. The sound should have good tonal quality at all listening positions.

17.9.2 Objective Measures

Several studies seeking objective measurements of concert halls that correlate with subjective perception have been conducted (Bradley, 2011). The measurements should thus estimate subjective factors such as those listed in the previous section. Again, in principle, the best objective measurement method would be an auditory model responding to concert hall acoustics in the same way as a real listener would do. Some attempts to use auditory models to assess concert hall acoustics have been made (van Dorp Schuitman, 2011), although no final answers

Table 17.3 The objective measures of concert hall acoustics defined in ISO 3382-1 (2009).

Subjective level of sound	Sound strength G in decibels
Perceived reverberance	Early decay time (EDT)
Perceived clarity of sound	Clarity C_{80} in decibels
Apparent source width (ASW)	Early lateral energy fraction, J_{LF}
Listener envelopment	Late lateral sound level, L_J in decibels

exist. Unfortunately, it seems that the models currently proposed are not able to explain human sensitivity to fine a esthetic details of the responses of concert halls to instrument sounds.

In practice, the perception of concert hall acoustics has traditionally been estimated with the analysis of measured impulse responses. Different metrics have been proposed (Barron, 1993; Beranek, 1996), and a set of measurements has also been standardized (ISO 3382-1, 2009), some of which are discussed below. The measurements discussed here are only a representative subset of the complete set in the standard. The standard proposes that the acoustics of a hall can be described with a few measurements obtained by spatially averaging over several positions. Many aspects of the standard have been criticized: the algorithms to compute the parameters are imprecise the applied frequency range is too narrow compared to human perception; and a single omnidirectional source is an inadequate representation of the sound sources present in a real orchestra (Bradley, 2011; Kirkegaard and Gulsrud, 2011). Furthermore, an impulse response is only a technical measure and does not represent how a human perceives the response of the hall to continuous instrument sounds or voices.

However, because the methods are widely used, they can be assumed to deliver some useful information. Thus, we think the standardized methods based on impulse responses might be of interest to the readers of this book. The measures are outlined in Table 17.3, and the methods to compute the measures are shown below.

- *Strength*: The ratio of the energy at the listening position to the energy measured 10 m in a free field from the source is called the strength G. Mathematically,

$$G = 10 \log_{10} \frac{\int_0^\infty p^2(t)\,\mathrm{d}t}{\int_0^\infty p_A^2(t)\,\mathrm{d}t}, \tag{17.9}$$

  where $p(t)$ is the sound pressure measured at the listener's position and $p_A(t)$ is the sound pressure measured at a distance of 10 m from the source in a free field, when an omnidirectional sound source is used as the excitation.
- *Early decay time* EDT: The time required for the reverberation to decay from 0 dB to −10 dB, scaled to correspond to the decay from 0 dB to −60 dB. This is calculated from the gradient of the energy decay curve (EDC) as introduced by Schroeder (1965) and defined as:

$$\mathrm{EDC}(t) = \int_t^\infty h^2(\tau)\mathrm{d}\tau \tag{17.10}$$

  The EDC function is typically more smooth than the impulse response itself, and so it is more useful than ordinary amplitude envelopes for estimating EDT.

- *Clarity*: This measure expresses the energy ratio between the early and late responses. A strong early response is beneficial to clarity, while a strong late response is harmful. C_{80} is a commonly used measure where the boundary between the early and late responses is set at 80 ms and is defined as

$$C_{80} = 10\log_{10}\frac{\int_0^{80\,\mathrm{ms}} p^2(t)\,\mathrm{d}t}{\int_{80\,\mathrm{ms}}^{\infty} p^2(t)\,\mathrm{d}t}. \tag{17.11}$$

- *Lateral fraction*: This measure, J_{LF}, is obtained from the impulse responses measured using a figure-of-eight microphone signal $p_8(t)$ with the null of the response pointing towards the source and an omnidirectional microphone $p(t)$. It is computed as

$$J_{\mathrm{LF}} = \frac{\int_{5\,\mathrm{ms}}^{80\,\mathrm{ms}} p_8^2(t)\,\mathrm{d}t}{\int_0^{80\,\mathrm{ms}} p^2(t)\,\mathrm{d}t}. \tag{17.12}$$

  J_{LF} reflects the ratio of lateral sound in the overall response.
- *Late lateral sound level*: Defined as

$$L_{\mathrm{J}} = 10\log_{10}\left(\frac{\int_{80\,\mathrm{ms}}^{\infty} p_8^2(t)\,\mathrm{d}t}{\int_0^{\infty} p_{10}^2(t)\,\mathrm{d}t}\right), \tag{17.13}$$

  L_{J} has a higher level in decibels if the reverberation after 80 ms of the arrival of the direct sound has a considerable degree of laterally flowing energy.

17.9.3 Percentage of Consonant Loss

The *percentage articulation loss of consonants* ($\%\mathrm{AL}_{\mathrm{cons}}$) is a simple measure used in the design of auditoria and concert halls to estimate the understandability of speech, based on a relatively simple mathematical formulation (Davis and Patronis, 2006; Peutz, 1971):

$$\%\mathrm{AL}_{\mathrm{cons}} = 200\,r^2(T_{60})^2/(VQ) + \mathrm{k}, \tag{17.14}$$

where r is the distance between the speaker and the listener, T_{60} is the reverberation time, V is the volume of the room, Q is the directivity of the source, and k is a constant describing the individual hearing capabilities of the listener. In the best case k = 1.5 and in the worst k = 12.5. For distances greater than $r = 0.20\sqrt{V/RT}$, the equation becomes

$$\%\mathrm{AL}_{\mathrm{cons}} = 9\,T_{60} + \mathrm{k}. \tag{17.15}$$

The value of $\%\mathrm{AL}_{\mathrm{cons}}$ thus estimates the percentage of consonants that are not perceived correctly in an auditorium. Relatively high values, 25–30%, may be acceptable, since the redundancy in speech makes it possible to ‘guess’ the ‘lost’ phones.

17.10 Noise Quality

Noise can be defined as sound that is disturbing or annoying. In principle, this subjective definition does not exclude any sounds, since basically any sound can be disturbing depending on many listener-related factors. Although this book mostly concerns itself with audio and speech techniques, applications in which the sounds are desired by the listener, noise is also interesting in the context of sound quality and also in the context of psychoacoustics. A basic introduction to the relation of noise to psychoacoustics is given by Marquis-Favre *et al.* (2005b).

Noise will be discussed in Chapter 19 in the context of technical audiology, where issues such as the effects of excessive SPL on the auditory system and limits for noise exposure in work environments are of interest. In the context of sound quality, the most relevant concepts are *annoyance* and *disturbance* (Guski *et al.*, 1999; Öhrström and Rylander, 1982; Ouis, 2001; Pedersen and Waye, 2004). All terms related to noise quality are negative, and as such, annoyance and disturbance should be minimized. We use the term annoyance as a general concept of noise quality, but, also to describe how noise may upset an operation or activity. The term disturbance is connected to negative feelings where the functioning of the subject is not necessarily disrupted, the capacity of the subject to perform any task is merely hampered.

The degree to which noise is annoying is studied primarily with listening tests, similarly to the quality of sound in general. All that has been said earlier in this book about psychoacoustic research methods is valid for noise as well. Often, the tests relating to the quality of noise are conducted with inexperienced listeners, and so the test design has to be simple enough. For example, a two-alternative forced choice test is often used, where the result places the sound samples in the order of annoyance.

The results from listening tests can be compared with psychoacoustic attributes computed using auditory models, such as loudness, fluctuation strength, sharpness, roughness, tonality, and impulsiveness. A common approach is to attempt to create a model from these attributes to estimate the annoyance and disturbance caused by noise, which should be in agreement with results from listening tests. If successful, such a model could be used as an objective measurement system of subjective annoyance (Marquis-Favre *et al.*, 2005a). Unfortunately, these models are usually valid only with a limited set of noise signals, and separate models have to be constructed for different types of noise signals, and in some cases the models do not explain the measured results (Waye and Öhrström, 2002).

Loudness, or loudness level, is usually one of the attributes used to explain annoyance. The sharpness of sound, that is, the high level of high-frequency components, and also the roughness of sound increase annoyance. The narrowband components of sound and certain temporal components, such as buzzing, banging, or screeching, are also perceived as more annoying.

The subjective nature of noise is clearly evident in open-plan offices, where most office desktops are located nowadays. Acoustic noise, and especially speech and laughter, is the most significant source of distraction in the physical work environment in open-plan offices (Helenius *et al.*, 2007; Jensen and Arens, 2005; Pejtersen *et al.*, 2006; Virjonen *et al.*, 2009). In contrast, sounds that are very stable in time and have a nearly constant sound pressure level, like ventilation noise, cause very little distraction. Quite interestingly, in these environments, high speech intelligibility decreases sound quality, which is just the opposite of the requirement in public spaces, drama theatres, and auditoria. Hongisto (2005) suggests that the STI value between desktops in open-plan offices should be below 0.2 to prevent the negative effects of being able to hear each other's discussions.

17.11 Product Sound Quality

Blauert and Jekosch (1997; 2012) define *product sound quality* as 'the adequacy of a sound in the context of a specific technical goal and/or task'. All products that produce a perceivable sound have their product sound quality evaluated every time they are used.

Blauert and Jekosch show that product sound quality is a broader concept than the auditory attributes evoked by a sound event. This characteristic is essential to relate the sound of the product and the subject actively using the product. The evoked auditory attributes are interpreted differently depending on the expectations of the subject. For example, the presence of a buzzing sound is generally not desirable, but when a subject uses an electric shaver, the buzzing sound communicates that the device is on and working. The subject also uses the fine structures of the sound to monitor the inner condition and quality of the device itself. The simple quantitative input–output relationship that psychoacoustics aims to measure has to be extended to cover psychological concepts such as *cognition*, *action*, and *emotion*.

The goal of product sound quality is not only pleasantness of sound, just as in noise control, minimizing the level of sound is not the only goal. A more important factor than the pleasantness of sound is often the informativeness of sound. Communication of the state of functioning of the product is often the factor determining why the perception of the sound is desirable. In particular, if the subject has been exposed many times to the sound of the product, the auditory system serves as a very sensitive indicator of the condition and state of the device producing the sound.

Some examples of product sound quality are discussed below.

- *Vehicles*. The concept of product sound has been strongly affected by the need of the automotive industry to design sounds generated by vehicles to give an impression of high quality in every aspect. The sounds generated by a car indicate certain aspects of how it functions, and these positive sounds are thus enhanced to compete with other vehicle brands.

  In the case of vehicles, besides the sound of the engine, the product sounds also include the sounds generated by the wheels and the turbulence of air. Additionally, they also include the sounds generated when using different parts of the vehicle, such as opening the window, moving the seat, and pressing the buttons. The sounds and audio-tactile interaction when entering a car are important: opening the lock, using the door handle, and closing the door generate both auditory and tactile perceptions, which create an impression of the quality of finishing of the car.

  The sound of the engine when listened to inside the cabin should be designed such that it is not disturbing, although it must be heard over other sounds in the cabin, since it gives information on the functioning of the engine. Some cars even have mode switches where the sound level of the engine can be selected to be higher in the 'sporty' mode and lower in the luxurious' mode. The change in sound level can be implemented either by opening a channel to the engine chamber or, more simply, just by reproducing the engine sound using the car audio system.
- *Household appliances*. Most household appliances produce sound, either continuously or only when used actively. Continuous sounds should typically be almost silent, and they should not cause annoyance. In a device that is used occasionally, louder sounds may be acceptable. For example, a vacuum cleaner may be thought to be less effective if its sound is very soft. However, the sound of the vacuum cleaner should not have disturbing components, such as rattling or high-level, high-frequency sounds. The vacuum cleaner is an example of a product where the sound level of the device itself is decreasing with the evolution of vacuum cleaners. The first vacuum cleaners were really noisy, and very loud sound was an indication of high power. Fortunately, nowadays vacuum cleaners are more silent, and the association between loudness and assumed power is weaker. Consumers have learned that the vacuum cleaner motor can be both powerful and silent, and they pay more attention to the sounds created by the suction of air and by particles entering the suction tube.

- *Personal devices*. Electric shavers and hairdryers are also good examples of devices where the product sound quality has been taken into account. For example, the shaver should have a 'manly' and 'powerful' sound, communicating that the device is designed for a 'real man' to cut a 'strong beard'.

Summary

This chapter has broadly introduced the reader to sound quality. In the course of history, different aspects of sound quality have been of interest at different times. A factor unifying different sound quality trends seems to be the concept of product sound quality, which can be used to investigate sounds from different devices, systems, and from information and entertainment utilities. Although the components of sound quality are different in different cases, they always stem from the properties of the auditory system.

Further Reading

The theory and measurement of speech quality can be studied further by referring to Jekosch (2006), Möller (2000), Quackenbush *et al.* (1988), Raake (2007). Further knowledge of sound quality in audio reproduction can be found in Bech and Zacharov (2006) and Toole (2012). More information on recent trends in sound quality in concert halls can be found in Blesser and Salter (2007) and Pätynen *et al.* (2014).

References

3QUEST (2008) 3-Fold Quality Evaluation of Speech in Telecommunications. Application note, HEAD acoustics.

Appel, R. and Beerends, J.G. (2002) On the quality of hearing one's own voice. *J. Audio Eng. Soc.*, **50**(4), 237–248.

Barron, M. (ed.) (1993) Auditorium Acoustics and Architectural Design, E & FN Span.

Bech, S. and Zacharov, N. (2006) *Perceptual Audio Evaluation – Theory, Method and Application*. John Wiley & Sons.

Beerends, J.G. and Stemerdink, J.A. (1992) A perceptual audio quality measure based on a psychoacoustic sound representation. *J. Audio Eng. Soc.*, **40**, 963–978.

Beerends, J.G. and Stemerdink, J.A. (1994) A perceptual speech-quality measure based on a psychoacoustic sound representation. *J. Audio Eng. Soc.*, **42**(3), 115–123.

Beerends, J.G., Hekstra, A.P., Rix, A.W., and Hollier, M.P. (2002) Perceptual evaluation of speech quality (PESQ) the new ITU standard for end-to-end speech quality assessment part II: Psychoacoustic model. *J. Audio Eng. Soc.*, **50**(10), 765–778.

Beerends, J.G., Schmidmer, C., Berger, J., Obermann, M., Ullmann, R., Pomy, J., and Keyhl, M. (2013) Perceptual objective listening quality assessment (POLQA), the third generation ITU-T standard for end-to-end speech quality measurement part I – Temporal alignment. *J. Audio Eng. Soc.*, **61**(6), 366–384.

Beranek, L. (ed.)(1996) *Concert and Opera Halls – How they Sound*. Acoustical Society of America.

Blauert, J. (2005) *Communication Acoustics*. Springer.

Blauert, J. (2013a) Conceptual aspects regarding the qualification of spaces for aural performances. *Acta Acustica United with Acustica*, **99**(1), 1–13.

Blauert, J. (2013b) *The Technology of Binaural Listening*. Springer.

Blauert, J. and Jekosch, U. (1997) Sound-quality evaluation: a multi-layered problem. *Acta Acustica United with Acustica*, **83**(5), 747–753.

Blauert, J. and Jekosch, U. (2012) A layer model of sound quality. *J. Audio Eng. Soc.*, **60**(1/2), 4–12.

Blesser, B. and Salter, L.R. (2007) *Spaces Speak, Are You Listening?* MIT Press.

Bradley, J. (2011) Review of objective room acoustics measures and future needs. *Appl. Acoust.*, **72**, 713–720.

Davis, D. and Patronis, E. (2006) *Sound System Engineering*. Taylor & Francis.

EQUEST (2012) Echo Quality Evalution of Speech in Telecommunications. Application note, HEAD acoustics.

ETSI (2008) Speech Quality performance in the presence of background noise Part 3: Background noise transmission – Objective test methods. Recommendation EG 202 396-3, European Telecommunication Standards Institute.

ETSI (2014a) Universal mobile telecommunications system (UMTS); LTE; speech and video telephony terminal acoustic test specification. Recommendation 3GPP TS 26.132, European Telecommunication Standards Institute.

ETSI (2014b) Universal mobile telecommunications system (UMTS); LTE; Terminal acoustic characteristics for telephony; Requirements. Recommendation 3GPP TS 26.131, European Telecommunication Standards Institute.

Fletcher, H. (ed.)(1995) *Speech and Hearing in Communication*. Acoustical Society of America.

Guski, R., Felscher-Suhr, U. and Schuemer, R. (1999) The concept of noise annoyance: How international experts see it. *J. Sound Vibr.*, **223**(4), 513–527.

Helenius, R., Keskinen, E., Haapakangas, A., and Hongisto, V. (2007) Acoustic environment in Finnish offices – The summary of questionnaire studies. *19th Int. Congr. Acoust.*

Hongisto, V. (2005) A model predicting the effect of speech of varying intelligibility on work performance. *Indoor Air*, **15**(6), 458–468.

Houtgast, T. and Steeneken, H.J.M. (1985) The modulation transfer function in room acoustics. *B&K Techn. Rev.*, **3**, 3–12.

IEC (2011) IEC 60268-16:2011 sound system equipment – Part 16: Objective rating of speech intelligibility by speech transmission index.

ISO 3382-1 (2009) Acoustics – measurement of room acoustic parameters – Part 1: Performance spaces. Organization for Standardization.

ISO 7240-19 (2007) Design, installation, commissioning and service of sound systems for emergency purposes. Standard 7240-19, International Organization for Standardization.

ITU-R BS.1116-1 (1997) Methods for the subjective assessment of small impairments in audio systems including multichannel sound systems. Recommendation, International Telecommunication Union, Geneva, Switzerland.

ITU-R BS.1534-1 (2003) Method for the subjective assessment of intermediate quality level of coding systems. Recommendation, International Telecommunication Union, Geneva, Switzerland.

ITU-T (1998a) Objective quality measurement of telephone–band (300–3400 Hz) speech codecs. Recommendation P.861, International Telecommunication Union, Geneva, Switzerland.

ITU-T (1998b) Perceptual evaluation of speech quality (PESQ): An objective method for end-to-end speech quality assessment of narrow-band telephone networks and speech codecs. Recommendation P.862, International Telecommunication Union, Geneva, Switzerland.

ITU-T (2011) Perceptual objective listening quality assessment. Recommendation P.863, International Telecommunication Union, Geneva, Switzerland.

ITU-T P.800 (1996) Methods for subjective determination of transmission quality. Recommendation, International Telecommunication Union, Geneva, Switzerland.

ITU-T P.835 (2003) Subjective test methodology for evaluating speech communication systems that include noise suppression algorithm. Recommendation, International Telecommunication Union, Geneva, Switzerland.

ITU-T P.910 (2008) Subjective video quality assessment methods for multimedia applications. Recommendation, International Telecommunication Union, Geneva, Switzerland.

Jekosch, U. (2006) *Voice and Speech Quality Perception: Assessment and Evaluation*. Springer.

Jensen, K. and Arens, E. (2005) Acoustical quality in office workstations, as assessed by occupant surveys. *Indoor Air*, pp. 2401–2405.

Karjalainen, M. (1985) A new auditory model for the evaluation of sound quality of audio systems. *Proc. of IEEE ICASSP'85*, pp. 608–611.

Kates, J.M. and Arehart, K.H. (2014) The hearing-aid speech quality index (HASQI) version 2. *J. Audio Eng. Soc.*, **62**(3), 99–117.

Kirkegaard, L. and Gulsrud, T. (2011) In search of a new paradigm: How do our parameters and measurement techniques constrain approaches to concert hall design? *Acoustics Today*, **7**(1), 7–14.

Lane, H. and Tranel, B. (1971) The lombard sign and the role of hearing in speech. *J. Speech, Lang. Hearing Res.*, **14**(4), 677–709.

Le Callet, P., Moller, S., and Perkis, A. (eds) (2012) *Qualinet White Paper On Definitions of Quality of Experience*. Qualinet.

Liebetrau, J., Sporer, T., Kämpf, S., and Schneider, S. (2010) Standardization of PEAQ-MC: Extension of ITU-R BS.1387-1 to multichannel audio. *40th Int. Audio Eng. Soc. Conf.: Spatial Audio* AES.

Lokki, T. (2014) Tasting music like wine: Sensory evaluation of concert halls. *Physics Today*, **67**(1), 27–32.

Mapp, P. (2005) Is STIPA a robust measure of speech intelligibility performance? *Audio Eng. Soc. Convention 118* AES.

Marquis-Favre, C., Premat, E., and Aubree, D. (2005a) Noise and its effects: a review on qualitative aspects of sound. Part II: Noise and annoyance. *Acta Acustica United with Acustica*, **91**(4), 626–642.

Marquis-Favre, C., Premat, E., Aubree, D., and Vallet, M. (2005b) Noise and its effects a review on qualitative aspects of sound. Part I: Notions and acoustic ratings. *Acta Acustica United with Acustica*, **91**(4), 613–625.
Möller, S. (2000) *Assessment and Prediction of Speech Quality In Telecommunications*. Springer.
Öhrström, E. and Rylander, R. (1982) Sleep disturbance effects of traffic noise: a laboratory study on after effects. *J. Sound Vibr.* **84**(1), 87–103.
Ouis, D. (2001) Annoyance from road traffic noise: A review. *J. Environment. Psych.*, **21**(1), 101–120.
Pätynen, J., Tervo, S., Robinson, P.W., and Lokki, T. (2014) Concert halls with strong lateral reflections enhance musical dynamics. *Proc. Nat. Acad. Sci.*, **111**(12), 4409–4414.
Pedersen, E. and Waye, K.P. (2004) Perception and annoyance due to wind turbine noise: a dose–response relationship. *J. Acoust. Soc. Am.*, **116**(6), 3460–3470.
Pejtersen, J., Allermann, L., Kristensen, T., and Poulsen, O. (2006) Indoor climate, psychosocial work environment and symptoms in open-plan offices. *Indoor Air.*, **16**(5), 392–401.
Peutz, V.M.A. (1971) Articulatory loss of consonants as a criterion for speech transmission in a room. *J. Audio Eng. Soc.*, **19**, 915–919.
Quackenbush, S.R. and Clements, M.A. (eds) (1988) *Objective Measures of Speach Quality*. Prentice–Hall.
Raake, A. (2007) *Speech Quality of VoIP: Assessment and Prediction*. John Wiley & Sons.
Rumsey, F., Zieliński, S., Kassier, R., and Bech, S. (2005) On the relative importance of spatial and timbral fidelities in judgments of degraded multichannel audio quality. *J. Acoust. Soc. Am.*, **118**(2), 968–976.
Sabine, W.C. (1922) *Collected Papers on Acoustics*. Harvard University Press.
Schroeder, M.R. (1965) New method of measuring reverberation time. *J. Acoust. Soc. Am.*, **37**(3), 409–412.
Schroeder, M.R., Atal, B.S., and Hall, J.L. (1979) Optimizing digital speech coders by exploiting masking properties of the human ear. *J. Acoust. Soc. Am.*, **66**, 1647–1652.
Sporer, T., Liebetrau, J., and Schneider, S. (2009) Statistics of MUSHRA revisited. *Audio Eng. Soc. Convention 127* AES.
Steeneken, H.J.M. and Houtgast, T. (1985) A review of the MTF concept in room acoustics and its use for estimating speech intelligibility in auditoria. *J. Acoust. Soc. Am.*, **77**, 1060–1077.
Steeneken, H.J. and Houtgast, T. (2002) Validation of the revised STIr method. *Speech Commun.* **38**(3), 413–425.
Steeneken, H.J., van Wijngaarden, S.J., and Verhave, J.A. (2011) The evolution of the speech transmission index. *Audio Eng. Soc. Convention 130* AES.
Thiede, T., Treurniet, W.C., Bitto, R., Schmidmer, C., Sporer, T., Beerends, J.G., and Colomes, C. (2000) PEAQ - the ITU standard for objective measurement of perceived audio quality. *J. Audio Eng. Soc*, **48**(1/2), 3–29.
Toole, F. (2012) *Sound Reproduction: The Acoustics and Psychoacoustics of Loudspeakers and Rooms*. Focal Press.
TOSQA (2003) Option TOSQA. Telecommunications objective speech quality assessment. Application note, HEAD acoustics.
van Dorp Schuitman, J. (2011) *Auditory modelling for assessing room acoustics*. PhD thesis, Technische Universiteit Delft.
Vilkamo, J., Lokki, T., and Pulkki, V. (2009) Directional audio coding: Virtual microphone-based synthesis and subjective evaluation. *J. Audio Eng. Soc.*, **57**(9), 709–724.
Virjonen, P., Keränen, J., and Hongisto, V. (2009) Determination of acoustical conditions in open-plan offices: Proposal for new measurement method and target values. *Acta Acustica United with Acustica*, **95**(2), 279–290.
von Helmholtz, H. (1954) *On the Sensation of Tone*. Dover Publications.
Waye, K.P. and Öhrström, E. (2002) Psycho-acoustic characters of relevance for annoyance of wind turbine noise. *J. Sound Vibr.*, **250**(1), 65–73.

18

Other Audio Applications

A number of audio and speech techniques have not yet been discussed in the book, and this chapter covers some of them. Four areas of application are briefly discussed in separate sections: *virtual reality*, *sonic interaction design*, *computational auditory scene analysis*, and *music information retrieval*. In addition, the last section of this chapter merely lists some techniques that are not covered at all in this book.

Virtual reality environments reproduce sound to the user using some of the sound reproduction techniques described in Chapter 14. However, the techniques required for virtual reality differ from those for basic sound reproduction because the interaction between the user and the virtual reality has a strong effect on the sound, in contrast to traditional sound reproduction where the listener does not affect the sound content. The audio engine synthesizes and reproduces a meaningful representation of the sound scene in the virtual world on-the-fly, depending on the actions of the user in the world. Another field involving interaction between humans and computers in the context of audio techniques is *sonic interaction design*, where new methods for human–computer interaction by means of sound are sought.

The speech recognition techniques described in Section 16.3 aim to analyse speech signals using complex pattern recognition techniques to obtain performance on a par with, or even better than human listeners are capable of. The last two techniques described in this chapter share some similarities with speech recognition, although the goal here is to recognize or analyse music and/or other auditory scenes in a natural world for different applications. *Computational auditory scene analysis* (CASA) refers to techniques that try to reveal all information from the ear canal signals accessible to a human listener. *Music information retrieval*, which can be seen partly as a subtopic of CASA, covers different techniques on how computers could recognize musical structures in audio tracks in the same manner as humans do and use this information to operate intelligently on music databases.

18.1 Virtual Reality and Game Audio Engines

In *virtual reality* techniques, the perception of physical presence in locations elsewhere in the real world or in imaginary worlds is created for a subject (Sherman and Craig, 2003).

Communication Acoustics: An Introduction to Speech, Audio, and Psychoacoustics, First Edition.
Ville Pulkki and Matti Karjalainen.

Optimally, all information received by the subject via his or her senses would mimic the conditions simulated as if they were real. The different modalities of perception in virtual reality applications have to be covered with different devices. The visuals are displayed either on a computer screen or through special stereoscopic displays, and audio may be delivered over loudspeakers or headphones.

The simulated virtual environment can be similar to the real world in order to create a lifelike experience – for example, for training purposes or for virtual tourism. Alternatively, it can differ significantly from reality, as in virtual reality games. A number of subjects may share the same virtual reality, with the ability to perceive each other's *avatars* (or *characters*), which are virtual representations of the users, and potentially to communicate with each other.

In the scope of this book, consideration of the usage of audio in virtual worlds is of interest (Savioja *et al.*, 1999; Svensson and Kristiansen, 2002; Laitinen *et al.*, 2012; Tsingos *et al.*, 2004; Vorländer, 2007). Virtual worlds include a number of *virtual objects*, which may or may not generate sound, as shown in Figure 18.1. *Virtual reality displays* reproduce the surroundings of the avatar to the user. The visual display is based on what the avatar sees from a particular point, which is restricted to certain directions, called the field of vision. Similarly, the audio engine is used to render all sounds that the avatar would hear in the location. In addition to the virtual objects, the virtual worlds may also contain enclosed spaces and acoustically reflective objects, the effect of which on the sound should be taken into account.

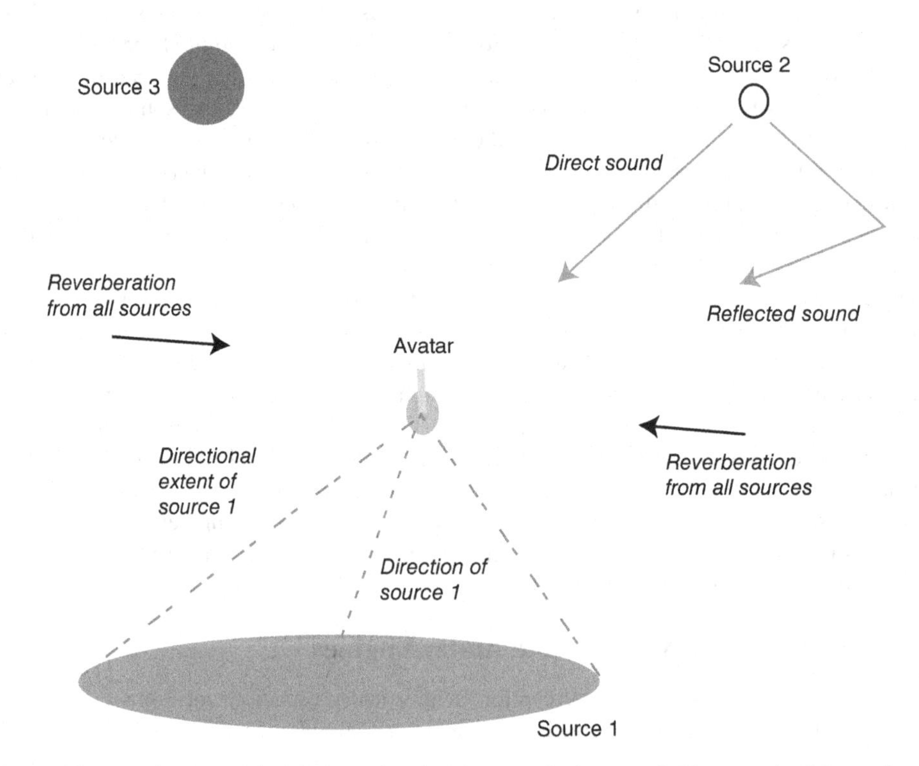

Figure 18.1 A virtual world with three virtual objects producing sound. The user should perceive the sources in the directions relative to the avatar. The spatial extents of the sources should also be perceivable. Furthermore, reflections from nearby surfaces should be rendered, as should the diffuse reverberant field generated by all sources.

Audio content production for virtual realities and especially for computer games differs drastically from production for the music and cinema industries. Both the audio content and the engine rendering it have to be composed in production. Often, background music is also used, and this also changes dynamically. This needs some advanced methods for content production and extensive testing of the system. This is called *dynamic audio* or *adaptive audio*, and the traditional audio can then be referred to as *linear audio*.

The tasks performed by the engine are listed below, not all of which are necessary for plausible rendering of virtual reality audio, since the audio may be implemented with varying levels of detail and accuracy, depending on resources and application.

- *Reproduction or synthesis of source signals*. The sounds emanated by the virtual objects into the virtual space have to be generated somehow. In many cases, sound signals which were produced earlier and stored in the memory of a computer are used. A piece of audio to be rendered in virtual reality is also called an *audio asset*. In some cases, the sounds of the virtual objects may be synthesized using physical models of sound sources, as discussed by Cook (2007) and Farnell (2010).
- *Synthesis of source directivity*. If a realistic acoustic virtual environment is sought, the frequency-dependent directivity of the source should be taken into account. The virtual object then emanates different sound in different directions (Savioja *et al.*, 1999). Quite often this effect is ignored, as the plausibility of the virtual world is often not affected much if it is not implemented. The implementation would also be difficult for computational reasons, as the directional patterns of real sources are complex and frequency-dependent.
- *Simulation of the direct sound path*. The transfer function of the virtual transmission channel between the source and the listener may also be modelled with appropriate delaying, amplification, and filtering, as well as some filtering effects of the atmosphere (Savioja *et al.*, 1999). High-speed movement of the source relative to the listener in the physical world results in the Doppler effect, where the perceived pitch of the sound changes with the change in distance between the listener and the source. This effect can be modelled simply in the reproduction of recorded source signals with dynamic sample rate conversion. Alternatively, the effect can be implemented by updating the sound propagation delay correctly. The delay is computed dynamically from the distance between the avatar and the source.
- *Virtual source positioning*. The virtual source positioning techniques discussed in Section 14.5 can be used to reproduce the direct sound(s) arriving at the listener. The virtual source directions are defined by the positions of the virtual objects relative to the avatar.
- *Spatial extent of virtual sources*. A sound source with a considerable volume, such as a group of talking people, should be perceived as spatially broad when the avatar is near the source, and correspondingly it should be perceived to be spatially narrow when the avatar is far away. If the avatar is inside the group, the subject should perceive the virtual source as surrounding him or her. Some techniques to synthesize the extent of sources were discussed in Section 14.5.8.
- *Room effect simulation*. The image–source-method, discussed in Section 2.4.5, is, in principle, a feasible method to simulate the room effect in virtual reality. The positions of image sources can be computed, and they can be reproduced by applying the appropriate direction in virtual source positioning, as above. The delay and gain for sounds of image sources are specified by their geometric location in the virtual world. However, such accurate modelling often leads to computationally complex solutions. Plausible room effects can also be obtained with digital filter structures that simulate the effect of

room reverberation, as discussed in Section 14.8, with considerably lower computational complexity.

- *Distance rendering*. The sensation of *distance* is largely provided by simulating both the direct sound path and the room effect. When they are reproduced correctly, the distance cues (discussed in Section 12.7) based on loudness, direct-to-reverberant (D/R) ratio, early reflections, and spectral content are correct. In loudspeaker-based audio rendering, the virtual sources can be easily positioned at distances as far as or farther than the loudspeakers are. It is hard to bring the sound source closer than the distance of the loudspeakers without special techniques; only some wave-field-synthesis methods with a massive number of loudspeakers are capable of bringing the virtual source into the listening area (see Section 14.5.6). If accurately head-tracked and well-equalized headphone listening is available (see Section 14.6.2), the virtual sources can be rendered close to and far from the listener, since the distance cues based on range dependence of binaural cues at short distances can also be exploited, and since the listening room has no effect on the sound entering the ear canals, making the task easier. In fact, bringing the virtual sources close with headphones is trivial, since the virtual sources are localized inside the head without special treatment. The correct rendering of distances for virtual sources outside the head, on the other hand, demands special techniques.

18.2 Sonic Interaction Design

A field connected strongly to product sound quality is *sonic interaction design*. In sonic interaction design, methods that use sound to convey information, meaning, and aesthetic or emotional qualities in interactive contexts are exploited (Franinovic and Serafin, 2013), thus extending the field of communication acoustics in the direction of interaction design, where human–computer interfaces are studied. Digital devices communicate with the user using non-speech sounds, which is analogous to product sound quality, where sounds made by devices convey information on the state of the device. In this field, the sounds are specially designed and digitally synthesized to communicate this information.

The main research areas are:

- *Perceptual, cognitive, and emotional study of sonic interactions*. This is an area of the field that addresses human perception of sound in interactive conditions in general (Aglioti and Pazzaglia, 2010; De Lucia *et al.*, 2009).
- *Product sound design*. This area concerns the practices and principles of designing sounds of products to maximize positive effects and minimize negative effects in interaction (Franinovic *et al.*, 2007; Hug, 2008).
- *Sonification*. The goal of sonification is to represent data provided by any process in the form of sound, so that the user can perceive or interpret the conveyed information just by listening (Hermann *et al.*, 2011). In a practical example, the angling of a drilling machine is sonified in order to help the user to drill holes more accurately (Großhauser and Hermann, 2010).

Sonic interaction design is thus approached using knowledge from a number of fields, such as interactive arts, electronic music, cultural studies, psychology, cognitive sciences, and communication acoustics. It shifts the focus from reception-based psychoacoustic studies to studies of perception of sound in active, embodied, and emotionally engaging situations. Multimodality, especially the tight connection between audition, haptics, and gestures, is examined in a unified framework.

18.3 Computational Auditory Scene Analysis, CASA

Auditory scene analysis (ASA) was presented in Section 11.7.2. *Computational auditory scene analysis* (CASA) is defined by Wang and Brown (2006) as 'the field of computational study that aims to achieve human performance in ASA by using one or two microphone recordings of the acoustic scene.' Note that the definition does not say that the human auditory system should be imitated in the processing, only that the performance of the system should match that of a human. Human performance seems difficult to achieve, as we are very good at analysing complex acoustic scenes containing sound from multiple sources through reverberant rooms. The current CASA methods hardly achieve human performance levels in such tasks.

Some of the audio and speech technologies discussed in this book are included within, or at least related to, CASA. For example, speech recognition (Section 16.3), hearing aids and cochlear implants (Sections 19.5 and 19.6), and music information retrieval (Section 18.4) aim to analyse sound scenes as well as humans do. Recent trends in CASA are reviewed by Wang and Brown (2006), where the following topics are discussed:

- *Multiple-f_0 estimation techniques* for revealing the fundamental frequencies of simultaneous harmonic sources.
- *Segregation of monaural signals*, where a mixture of simultaneous speech or other signal sources and interfering noise sources is presented to a system, and different features are analysed from the summed signal in the time-frequency domain, targeting the segregation of individual source signals.
- *Binaural sound localization and segregation*, where the sources are localized and their signals are segregated.
- *Analysis of musical audio signals*, where musical structures are analysed from monophonic signals.
- *Robust speech recognition*, which aims to recognize speech as well as humans do in natural conditions, such as free-form discussions between individuals or speech in noisy conditions with unlimited vocabularies, to name but two.

18.4 Music Information Retrieval

Consider a track of musical content stored as a file containing the signals to be reproduced over each loudspeaker. When humans listen to the track, they perceive musical structures, such as tempo, beat, melodies, harmonies, and genre. The presentation of the music as loudspeaker signals does not *per se* imply anything about the musical content in the track, which is, of course, not a problem when reproducing the sound with audio devices. However, as the databases of music nowadays contain millions of tracks, it would be useful to be able to make automatic queries on the database based on some higher-level specifications. For example, if only audio signals are available, it is not possible to make the request 'I want to hear some sturdy bebop with no trumpet, a female vocalist singing in French, and with electric guitar'. A musically trained listener may be able to pick out such tracks using his or her memory and by listening to available tracks. A computer-based solution would be appealing, since such a task could potentially be automated and computers are much faster at such tasks than humans.

Music information retrieval (MIR) is the area where the strategies for enabling access to music collections, both new and historical, are developed in order to keep up with the expectations of search and browse functionality. These techniques are interesting to end users of music, since they would enable someone to find and use music in a personalized way. Also, different professionals in music, such as music performers, teachers, musicologists, copyright lawyers, and music producers would benefit from the ability to make such structured queries of music databases (Casey *et al.*, 2008).

A widely used application of MIR is content-based music description, where the system identifies what the user is seeking even when he does not know specifically what he or she is looking for. Many commercial applications can identify the original recording from a short and noisy sample taken with a handheld device, for example, in a public space where music is played from recordings. The application identifies the artist, album, and track title by finding the best match to the sample from a large music database.

The driver of MIR systems is metadata. Music tracks are analysed by different systems, and metadata that describes the tracks from different aspects is collected. The simplest type of metadata is *factual metadata*, which describes the objective truths about a track, such as the performer, the name of the piece, and so on. The factual metadata can also be accompanied by *cultural metadata*, which describes properties like mood, emotion, genre, style, and so forth. Most often, these attributes are specified to the database by expert listeners, implying that each track added to a database should first be listened to. The fast pace of music creation motivates the use of automatic systems for these tasks.

MIR systems would thus need automatic methods that analyse musical structures from audio tracks. There are several approaches to this, as summarized by Casey *et al.* (2008). The methods resemble speech recognition techniques, where the input signal is divided into short time frames and complex machine-learning techniques are utilized to decode the content of speech. The task of deciphering the musical structures from a single audio signal is a slightly different task. Music typically has multiple sources that are concurrently active, making the recognition of melody and harmony a complex task. Music also has prominent structures organized in both time and frequency.

Some subtopics that have been researched in MIR are:

- *Beat tracking*. The automatic estimation of the temporal structure of music, such as musical beat, tempo, rhythm, and meter is called beat tracking. Many approaches exist to this end. See, for example, Dannenberg (2005).
- *Melody and bass estimation*. The automatic estimation of melody and bass lines is important because the melody forms the core of Western and many other music genres, and it is also a strong indicator of the identity of a piece of music. Often, the bass line also helps in revealing the harmonic progression. Although the measurement of the lines has, in the past, been problematic, state-of-the-art technologies for automatic melody/bass estimation have reached a state whereby they are able to deal with polyphonic music recordings (Poliner *et al.*, 2007; Ryynänen and Klapuri, 2006).
- *Chord and key recognition*. The information on musical chords and keys is an important part of Western music, and it can be used to understand musical structures. Chord- and key-recognition systems based on the use of HMMs have been found to perform well in the task; these unify recognition and smoothing into a single probabilistic framework (Harte *et al.*, 2006; Lee and Slaney, 2008).

- *Music structure*. Music often includes nested repetitive structures, such as the drum pattern that often repeats itself within each bar and the melodic line that is often repetitive in longer units. The analysis of structures such as verse–verse–chorus–verse is useful for music database applications. Furthermore, such automatic structure extraction can be used in other applications, such as to facilitate the editing of audio in recording workflows (Fazekas and Sandler, 2007).

18.5 Miscellaneous Applications

A number of techniques in audio and speech processing that have not been covered elsewhere in this book are briefly mentioned in the list below, with the intention of communicating to the reader fields related to or within communication acoustics which fall beyond the scope of this book.

- *Beamforming* or *spatial filtering* is a signal processing technique used with arrays of microphones for capturing or enhancing sound sensitive to direction. It processes and combines the acoustic signals of the separate microphones to form a desired beam pattern. Traditional beamforming techniques, such as delay-and-sum or filter-and-sum, combine the signals of the microphones in such a way that signals at particular angles experience constructive interference while others experience destructive interference (Bitzer and Simmer, 2001). More advanced beamforming algorithms offer the capability of focusing in specific directions of a target sound while attenuating interferers and noise originating from other directions, with the most popular algorithm being the *linear constrained minimum variance* (LCMV) or the Frost algorithm (Benesty *et al.*, 2008). Perceptually motivated methods for beamforming have also been proposed to reduce perceivable processing artefacts by humans (Delikaris-Manias and Pulkki, 2013; Faller *et al.*, 2010). Due to the reciprocity between acoustic sources and receivers, loudspeakers can also be used for beamforming.
- *Blind source separation* – the goal here is to separate the components making up a combined signal. The task is relatively simple if the number of sources is at most equal to the number of microphones (Cardoso, 1998). Unfortunately, if the number of sources is higher, and if reflections and reverberation exist in the mixed signal, this task is very difficult. One of the main techniques in this field is *independent component analysis* (ICA) (Hyvärinen *et al.*, 2004).
- *Dereverberation* attempts to remove partially or completely the physical or perceptual effects generated by reverberation from audio or speech signals (Hatziantoniou and Mourjopoulos, 2004).
- *Watermarking* is the process of embedding information into an audio signal so that it is difficult to remove and impossible to perceive by listening. If the signal is copied, this information is also carried in the copy. Watermarking has become increasingly important to enable copyright protection and ownership verification (Bliem *et al.*, 2013; Cvejic and Seppanen, 2008). Watermarking is also used to detect tampering, for copy control, for broadcast monitoring, for transmission of metadata, and in some cases even for echo cancelling (Szwoch *et al.*, 2009).
- *Audio forensics* concerns the acquisition, analysis, and evaluation of sound recordings that may ultimately be presented as admissible evidence in a court of law or some other official venue (Maher, 2009). Its primary characteristics are 1) establishing the authenticity of audio

evidence, 2) enhancing audio recordings to improve speech intelligibility and audibility of low-level sounds, and 3) interpreting and documenting sonic evidence, such as identifying speakers, transcribing dialogue, and reconstructing crime or accident scenes and timelines.
- *Auditory displays* are strongly related to sonic interaction design, though the focus is more on how some information can be delivered from a computer to a human (Kramer, 1994) and less on interaction. A subfield is *sonification*, defined in Section 18.2, though in relation to auditory displays the focus is more on the sonification of complex and multidimensional data sets, as discussed by (Hermann and Ritter, 2004). Another area within auditory displays is *earcons* or *auditory icons*, which are brief, distinctive sounds used to represent a specific event or to convey other information, just like icons in visual displays (McGookin and Brewster, 2004).
- *Semantic audio* – semantic technology involves some kind of understanding of the meaning of the information it deals with. Semantic audio is the field of audio technology covering the development of applications which utilize semantic analysis methods on the audio content and thereby enable new features and functionalities. Examples of such applications are automated mixing of audio, source separation, upmixing, content retrieval, and intelligent audio effects. The definition of semantic audio thus overlaps with music information retrieval, sound reproduction, and audio effects. An overview of the topics in semantic audio can be seen in the programmes of AES conferences containing the title 'Semantic Audio', for example `www.aes.org/publications/conferences/?confNum=53`. The semantic analysis of audio is discussed by Lerch (2012) and Schuller (2013).

Summary

This chapter overviewed techniques for virtual reality audio, sonic interaction design, and music information retrieval. Furthermore, some other techniques, interesting in the scope of this book, were briefly reviewed.

Further Reading

The book by (Vorländer, 2007) might be interesting to readers who want to learn more about the physics behind acoustic virtual reality. More information on music information retrieval and the analysis of music signals is given by (Gold *et al.*, 2011; Muller *et al.*, 2011).

References

Aglioti, S.M. and Pazzaglia, M. (2010) Representing actions through their sound. *Exper. Brain Res.*, **206**(2), 141–151.

Benesty, J., Chen, J., and Huang, Y. (2008) *Microphone Array Signal Processing*. Springer.

Bitzer, J. and Simmer, K.U. (2001) Superdirective microphone arrays. In Brandstein, M. and Ward, D. (eds) *Microphone Arrays*. Springer, pp. 19–38.

Bliem, T., Galdo, G.D., Borsum, J., Craciun, A., and Zitzmann, R. (2013) A robust audio watermarking system for acoustic channels. *J. Audio Eng. Soc.*, **61**(11), 878–888.

Cardoso, J.F. (1998) Blind signal separation: statistical principles. *Proc. IEEE*, **86**(10), 2009–2025.

Casey, M.A., Veltkamp, R., Goto, M., Leman, M., Rhodes, C., and Slaney, M. (2008) Content-based music information retrieval: Current directions and future challenges. *Proc. IEEE*, **96**(4), 668–696.

Cook, P.R. (2007) *Real Sound Synthesis for Interactive Applications*. AK Peters.

Cvejic, N. and Seppanen, T. (eds) (2008) *Difital Audio Watermarking Techniques and Technologies: Applications and Benchmarks*. InormationScience Reference.

Dannenberg, R.B. (2005) Toward automated holistic beat tracking, music analysis and understanding. *Int. Conf. Music Inform. Retrieval*, pp. 366–373.

De Lucia, M., Camen, C., Clarke, S., and Murray, M.M. (2009) The role of actions in auditory object discrimination. *Neuroimage*, **48**(2), 475–485.
Delikaris-Manias, S. and Pulkki, V. (2013) Cross pattern coherence algorithm for spatial filtering applications utilizing microphone arrays. *IEEE Trans. Audio, Speech, and Language Proc.*, **21**(11), 2356 – 2367.
Faller, C., Favrot, A., Langen, C., Tournery, C., and Wittek, H. (2010) Digitally enhanced shotgun microphone with increased directivity. *Audio Eng. Soc. Convention 129* AES.
Farnell, A. (2010) *Designing Sound*. MIT Press.
Fazekas, G. and Sandler, M. (2007) Intelligent editing of studio recordings with the help of automatic music structure extraction. *Audio Eng. Soc. Convention 122* AES.
Franinovic, K. and Serafin, S. (eds) (2013) *Sonic Interaction Design*. MIT Press.
Franinovic, K., Hug, D., and Visell, Y. (2007) Sound embodied: Explorations of sonic interaction design for everyday objects in a workshop setting. *Proc. of the Intl. Conf. on Auditory Display* ICAD.
Gold, B., Morgan, N., and Ellis, D. (2011) *Speech and Audio Signal Processing: Processing and Perception of Speech and Music*. John Wiley & Sons.
Großhauser, T. and Hermann, T. (2010) Multimodal closed-loop human machine interaction. *Human Interaction with Auditory Displays – Proc. Interactive Sonification Workshop*, pp. 59–63.
Harte, C., Sandler, M., and Gasser, M. (2006) Detecting harmonic change in musical audio. *Proc. 1st ACM workshop on Audio and music computing multimedia*, pp. 21–26 ACM.
Hatziantoniou, P.D. and Mourjopoulos, J.N. (2004) Errors in real-time room acoustics dereverberation. *J. Audio Eng. Soc.*, **52**(9), 883–899.
Hermann, T. and Ritter, H. (2004) Sound and meaning in auditory data display. *Proc. IEEE*, **92**(4), 730–741.
Hermann, T., Hunt, A., and Neuhoff, J.G. (2011) *The Sonification Handbook*. Logos Verlag.
Hug, D. (2008) Genie in a bottle: Object-sound reconfigurations for interactive commodities. *Proc. of Audio Mostly Conf.*, Pitea, Sweden, pp. 56–63.
Hyvärinen, A., Karhunen, J., and Oja, E. (2004) *Independent Component Analysis*. John Wiley & Sons.
Kramer, G. (1994) *Auditory Display: Sonification, Audification, and Auditory Interfaces*. Addison-Wesley.
Laitinen, M-V., Pihlajamäki, T., Erkut, C., and Pulkki, V. (2012) Parametric time-frequency representation of spatial sound in virtual worlds. *ACM Trans. Appl. Percep. (TAP)*, **9**(2), 8–20.
Lee, K. and Slaney, M. (2008) Acoustic chord transcription and key extraction from audio using key-dependent HMMs trained on synthesized audio. *IEEE Trans. Audio, Speech, and Language Proc.*, **16**(2), 291–301.
Lerch, A. (2012) *An Introduction to Audio Content Analysis: Applications in signal processing and music informatics*. John Wiley & Sons.
Maher, R. (2009) Audio forensic examination. *Signal Proc. Mag., IEEE*, **26**(2), 84–94.
McGookin, D.K. and Brewster, S.A. (2004) Understanding concurrent earcons: Applying auditory scene analysis principles to concurrent earcon recognition. *ACM Trans. Appl. Percep. (TAP)*, **1**(2), 130–155.
Muller, M., Ellis, D.P., Klapuri, A., and Richard, G. (2011) Signal processing for music analysis. *IEEE J. Select. Topics in Signal Proc.*, **5**(6), 1088–1110.
Poliner, G.E., Ellis, D.P., Ehmann, A.F., Gómez, E., Streich, S., and Ong, B. (2007) Melody transcription from music audio: Approaches and evaluation. *IEEE Trans. Audio, Speech, and Language Proc.*, **15**(4), 1247–1256.
Ryynänen, M. and Klapuri, A. (2006) Transcription of the singing melody in polyphonic music. *Int. Conf. Music Inform. Retrieval*, pp. 222–227 ISMIR.
Savioja, L., Huopaniemi, J., Lokki, T., and Väänänen, R. (1999) Creating interactive virtual acoustic environments. *J. Audio Eng. Soc.*, **47**(9), 675–705.
Schuller, B.W. (2013) *Intelligent Audio Analysis*. Springer.
Sherman, W. and Craig, A. (2003) Understanding virtual reality: interface, application, and design. *The Morgan Kaufmann series in computer graphics and geometric modeling*.
Svensson, P. and Kristiansen, U.R. (2002) Computational modelling and simulation of acoustic spaces. *22nd Int. Audio Eng. Soc. Conf.: Virtual, Synthetic, and Entertainment Audio* AES.
Szwoch, G., Czyzewski, A., and Ciarkowski, A. (2009) A double-talk detector using audio watermarking. *J. Audio Eng. Soc.*, **57**(11), 916–926.
Tsingos, N., Gallo, E., and Drettakis, G. (2004) Perceptual audio rendering of complex virtual environments. *ACM Trans. Graphics*, **23**(3), 249–258.
Vorländer, M. (2007) *Auralization: Fundamentals of Acoustics, Modelling, Simulation, Algorithms and Acoustic Virtual Reality*. Springer.
Wang, D. and Brown, G.J. (2006) *Computational Auditory Scene Analysis: Principles, Algorithms, and Applications*. Wiley-IEEE Press.

19

Technical Audiology

The enhancement of degraded hearing has been around for a long time. The simplest acoustic device used for hearing enhancement is an ear trumpet. An ear trumpet is a horn that collects sound at its large opening and transmits it to the ear, thus amplifying sound, especially from the direction in which the horn is pointing. When the electronic amplification of sound was invented, it was applied to hearing enhancement. However, it was not until the invention of the transistor and miniaturization of devices that *hearing instruments* became practical. Microprocessors and digital signal processing have introduced new possibilities to enhance the performance of hearing instruments. Modern electronics have also been a prerequisite for the development of *cochlear implants*, in which a major part of the peripheral auditory system is bypassed and sound information is fed electronically to the auditory nerve. It would be reasonable to assume that in the future even more complex means will be developed for the enhancement of degraded hearing.

Electronic technology has also enabled better methods for testing hearing performance. Tuning forks or live speech and whisper sounds may still be used to get a rough estimate of whether the auditory system is functioning in general. However, more advanced methods are needed for more accurate information on hearing performance.

Audiology is a branch of medical science and physiology that studies hearing, balance, and disorders related to them (Martin and Clark, 2006). *Technical audiology* is an interdisciplinary field where the expertise and knowledge of technology and audiology are combined in assessing hearing disorders and enhancing hearing abilities.

This chapter discusses the basic concepts of technical audiology. As background, a brief introduction to hearing impairments and disabilities is given. Then, the methods and technology related to measuring hearing abilities are discussed. After this, hearing instruments (hearing aids and cochlear implants) are discussed.

19.1 Hearing Impairments and Disabilities

Based on an estimation by the World Health Organization (WHO, 2013), 360 million people worldwide have disabling hearing loss, which corresponds approximately to 5% of the

Communication Acoustics: An Introduction to Speech, Audio, and Psychoacoustics, First Edition.
Ville Pulkki and Matti Karjalainen.

population. WHO defines disabling hearing loss as a hearing threshold shift of 40 dB in the better hearing ear for adults and 30 dB for children. Naturally, an even larger group of people are affected by milder hearing loss. Furthermore, as the average age and the number of the elderly increases, the number of individuals with hearing loss will increase. Thus, a vast number of individuals with hearing impairments, and consequently many different types of hearing impairments, have to be considered in this field.

Impaired hearing can affect the life of the hearing-impaired individual in many ways, depending on the type and degree of the impairment. Most importantly, a hearing impairment can affect orientation in the environment and communication based on sound. Hearing of speech is a fundamental ability for humans, and therefore speech intelligibility has often been used as the basic criterion of the functionality of hearing. However, even a mild hearing impairment affects the perception of details in sound, for example nuances in music. The consequences of a mild hearing impairment will probably be even more important in the future, along with the increased emphasis on the quality of life.

When hearing is substantially impaired, the ability to react to environmental sounds decreases. The inability to react to sonic warning signals poses a significant risk, for example, in traffic. The use of certain machinery requires the user to monitor the sound generated by the equipment in use. In general, sounds have a considerable effect on the perception of experiences and emotions.

The functioning of hearing is of paramount importance in the early years of life, since a hearing impairment can hinder or restrain early language acquisition. A restraint in language acquisition during a certain period early in life can result in a permanent communication disorder. Before the development of proper hearing diagnostic methods, a hearing-impaired child may have been considered mentally challenged, since it was not possible to discover the hearing impairment that hindered learning.

A hearing disability can be temporary or permanent. A temporary hearing disability can be, for example, due to a disease, short-term noise exposure, or excess ear wax in the ear canal. A disease or excessive noise exposure can damage hearing permanently. Permanent hearing disability can also be inborn. For example, the development of senses in the fetal phase is sensitive to any viral infections carried by the mother during pregnancy.

19.1.1 Key Terminology

Hearing loss is the degradation of the sensitivity of hearing, or, in a broad sense, loss of hearing ability in some dimension. Hearing loss is the primary symptom of hearing disorders. A hearing disorder is a structural or functional impairment of the auditory system. The following terms specify the different aspects related to hearing impairments:

- *Disease* – diagnosed and treated medically.
- *Impairment* – diagnosed with audiometry and treated medically.
- *Disability* – is diagnosed via professional evaluation or self-evaluation and treated with rehabilitation.
- *Handicap* – diagnosed with self-evaluation and treated with rehabilitation.

Hearing losses are typically quantified by comparing the hearing thresholds of a patient to hearing thresholds of normal hearing. Normal hearing is denoted by 0 dB in reference to the *hearing level* (that is, 0 dB HL), and the deviations from 0 dB HL are denoted with a hearing threshold shift. A positive hearing threshold shift denotes degraded hearing. The hearing level is the standardized reference that defines the lowest sound pressure level that causes an auditory event at each frequency. Since the sensitivity of hearing varies with frequency, the sound pressure level needed to produce 0 dB HL varies with frequency. The hearing level has been defined by measuring the hearing thresholds for a large number of young individuals with normal hearing. For further definitions, see ISO 389–1 (1998).

Another reference level is the *sensation level* (SL), which defines the hearing threshold for a given individual. For example, if a person has a hearing threshold shift of +10 dB at a given frequency and is presented a pure tone with a level of 30 dB HL, the level of this tone can be expressed as 20 dB SL.

Although hearing loss is typically quantified in terms of hearing threshold shifts, the term 'hearing loss' in the broad sense accounts for all the phenomena concerned with the degradation of hearing abilities, which will be discussed in more detail later in this chapter.

19.1.2 Classification of Hearing Impairments

Hearing impairments and disabilities can be classified in various ways. Three different classifications are given below based on social, audiological, and medical criteria.

The social classification of hearing disabilities is based on the overall state of the hearing abilities and on the communication method used by the individual:

1. A person who is *hard-of-hearing* has a hearing loss ranging from mild to severe. Some degree of residual hearing is left, and thus speech communication is usually possible, but a hearing aid may be necessary.
2. A *deaf* person has little or no functional hearing. Sign language is often the primary form of communication.
3. A *deafened* person is a deaf individual who has lost the hearing abilities after learning speech. Lip-reading may provide additional cues in communication.

The classification by the European Working Group on Genetics of Hearing Impairment (1996) defines four groups of impairments based on the average of hearing thresholds measured at frequencies 500, 1000, 2000, and 4000 Hz as follows:

- Mild: 20–40 dB HL
- Moderate: 40–70 dB HL
- Severe: 70–95 dB HL
- Profound: equal to or over 95 dB HL

Based on this classification, the boundary between a normally hearing and hard-of-hearing individual is that the normally hearing person has average hearing thresholds below 20 dB HL. However, different definitions for this boundary exist. For example, in the definition by WHO (2013), a normally hearing person has hearing thresholds lower than 25 dB HL at all frequencies.

The medical classification of hearing impairments is based on the location of the impairment in the auditory system. This classification can be presented, for example, as follows (Martin and Clark, 2006):

1. *Conductive impairments* refer to impairments in the conductive path of the auditory system; that is, the outer ear and the middle ear.
2. *Sensorineural impairments* refer to impairments in the inner ear and in the auditory nerve. Sensorineural impairments can be further divided into cochlear and retrocochlear impairments, based on whether they are located in the inner ear or in the auditory nerve.
3. *Central impairments* refer to impairments in the central auditory nervous system. Their treatment and diagnosis is difficult, since the functioning of the higher levels of the auditory system is less known compared to the peripheral auditory system. Decreased speech intelligibility may be an indicator of a central hearing impairment if the peripheral auditory system has been found to function normally.
4. *Psychic impairments* form a group of hearing impairments where no organic cause can be found.

19.1.3 Causes for Hearing Impairments

A hearing impairment can originate from various causes. A typical outer ear problem is occlusion of the ear canal due to ear wax or a foreign object. Middle ear problems can be due to infections or diseases that damage the auditory ossicles or the eardrum. For example, *otosclerosis* is a disease where new growth of bone stiffens the movement of the auditory ossicles and thus attenuates the sound transmitted to the inner ear. Middle ear damage can also be inborn or due to mechanical trauma, such as head injury.

Sensorineural damage is typically caused by excess noise exposure that damages the hair cells in the cochlea. The hair bundles of the hair cells can be damaged, or the cells themselves can suffer from a metabolic disorder, swell up, or be completely destroyed. Cochlear damage can also be inborn or a consequence of a disease, head trauma, or a tumour in the auditory nerve. Some chemical substances (like kanamycin) are ototoxic; that is, harmful to the inner ear.

Hearing degenerates with age, causing *presbyacusis*, also called *age-related hearing loss*. The sensitivity of hearing is also lower in early childhood compared to adulthood.

In addition, several other factors have been noticed to contribute to the onset of hearing loss (Toppila, 2000). It is important to notice that the joint effect of several factors is often larger than the sum of the factors alone. Such factors include:

- *Vibration*. For example, hand vibration can cause Raynaud's syndrome and weaken peripheral blood circulation. This can affect the blood circulation in the cochlea and thus decrease the tolerance of noise.
- *Smoking*. This does not itself cause hearing loss. However, smoking can have a harmful joint effect with noise, evidently due to the weakening of peripheral blood circulation.
- *Genes*. Sometimes the degeneration of hearing is transmitted in the genes, and hearing loss can occur without any clear reason.

19.2 Symptoms and Consequences of Hearing Impairments

A hearing impairment can result in various symptoms. The main symptom is the degraded sensitivity of hearing (hearing loss), which can be in the form of a hearing threshold shift, decreased discrimination of sound events, and/or distortion of sound. Other symptoms are hyperacusis (oversensitivity to sound) and tinnitus (ringing in the ears). Problems with the sense of balance are related to the vestibular system in the inner ear and are thus part of the field of audiology. However, such problems are not discussed in this book.

19.2.1 Hearing Threshold Shift

A hearing threshold shift can occur due to a conductive or sensorineural hearing impairment. In the case of a conductive impairment, the attenuation of sound is typically linear; that is, high and low sound levels are attenuated by the same amount.

For a sensorineural impairment, the type of attenuation depends on whether the inner or outer hair cells are damaged. If the inner hair cells are damaged, the attenuation is not dependent on the sound level. However, if the outer hair cells are damaged, the resulting attenuation of sound is more prominent at low input levels. This is because when the outer hair cells are damaged, the input-level-dependent amplification provided by them does not function properly. Consequently, loud sounds may be heard normally, but quiet sounds are attenuated. This phenomenon is called *recruitment*. Due to recruitment, a change in SPL translates into an abnormally large change in loudness. Moreover, the range of sound pressure levels from the hearing threshold to the loudness discomfort level is decreased. Consequently, dynamic changes in sound level are overemphasized, and sound is often perceived to be either too loud or too quiet. This is shown in Figure 19.1, which indicates the conceptual loudness-matching functions that visualize the SPL needed for a hearing-impaired person to achieve equal loudness as a normally hearing

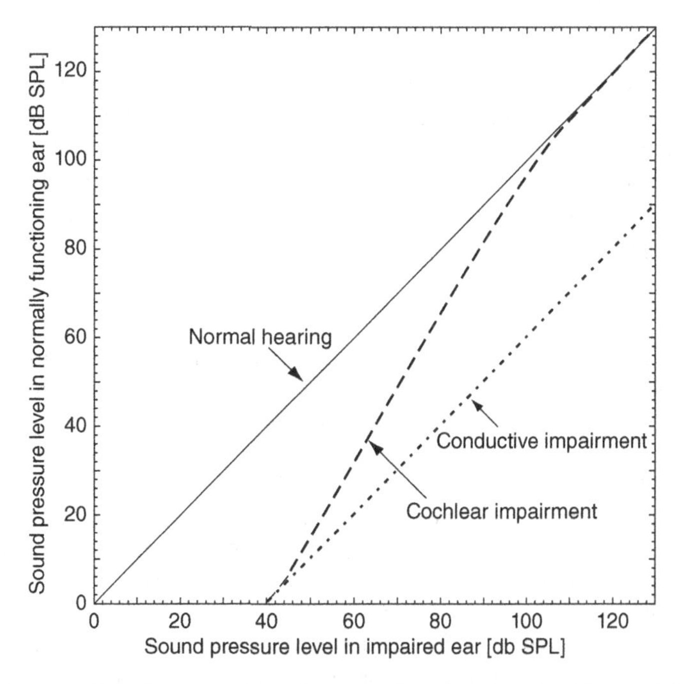

Figure 19.1 Conceptual loudness-matching functions that visualize the SPL needed for a hearing-impaired person to achieve equal loudness as a normally hearing individual: normal hearing, conductive hearing loss, and cochlear hearing loss (recruitment due to outer hair cell damage). Adapted from Moore (2007).

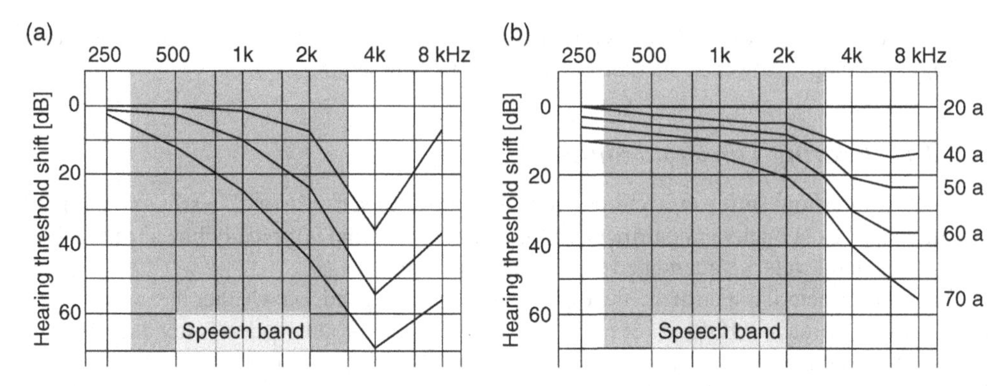

Figure 19.2 Typical hearing threshold shifts induced by (a) excess noise exposure and (b) age. The reference hearing threshold of 0 dB HL denotes normal hearing in young adults.

individual. The curve for patients with cochlear impairment shows a steeper slope, which is a manifestation of recruitment.

A hearing threshold shift, especially when of sensorineural origin, is often frequency dependent. Consequently, hearing can be, for instance, fully functional in one frequency range, have linear attenuation in another frequency range, and recruitment in a third frequency range. This poses certain challenges in the design and individual fitting of hearing instruments, as discussed later in Section 19.5.

Figure 19.2a shows typical hearing thresholds for different degrees of noise-induced hearing impairments. Noise-induced hearing threshold shifts are typically characterized by a notch in the hearing threshold in the 4-kHz range. The effect of age on the degradation of the sensitivity of hearing is individual, but typical hearing threshold shifts due to presbyacusis are shown in Figure 19.2b. The figure shows that age-related hearing loss is typically present at high frequencies. However, the effect of age on hearing loss is difficult to isolate from other factors, such as noise exposure.

19.2.2 Distortion and Decrease in Discrimination

In addition to a hearing threshold shift, a hearing impairment can cause various kinds of distortion. Even at sound pressure levels of normal speech, strong distortion components and echoes may occur, which may decrease speech intelligibility and make listening to music uncomfortable.

Sensorineural hearing impairments often degrade the ability to discriminate sound sources from one another, leading to problems in understanding speech in the presence of background noise or reverberation. The so-called cocktail party effect – the ability to listen to and understand a single speaker in the presence of numerous other speakers – is due primarily to the functioning of the outer hair cells in the cochlea. Thus, if these cells are damaged, the hearing-in-noise ability of the individual is degraded. Problems also arise in reverberant environments even without background noise, since the late reverberation masks the direct sound and early reflections.

Figure 19.3 shows different stages of hair cell damage and the corresponding effect on neural tuning curves. The dashed line in the tuning curve graphs depicts normal hearing. Hearing sensitivity is decreased (that is, the hearing threshold is increased) in a given frequency range

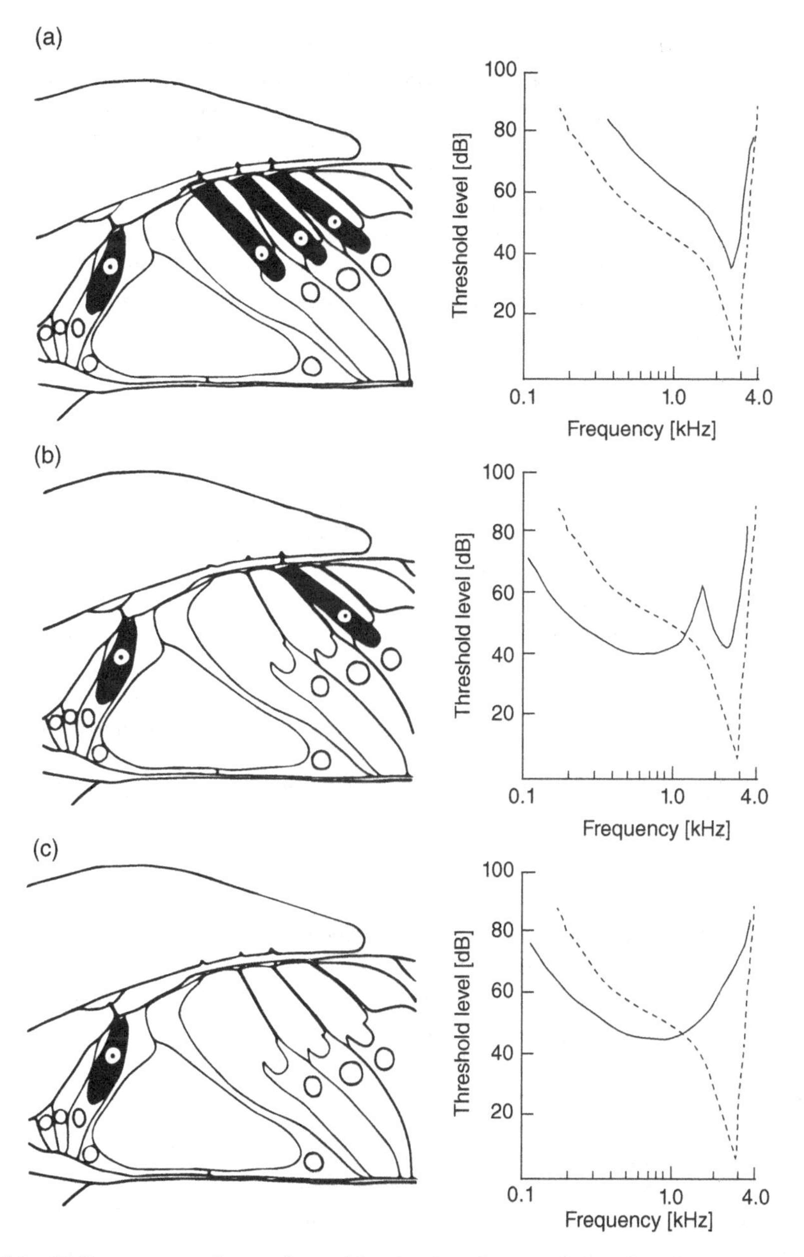

Figure 19.3 Different cases of sensorineural hearing impairments. The hair cell damage in the organ of Corti is depicted on the left. The corresponding neural tuning curve (solid line) compared to normal hearing (dashed line) is depicted on the right. Adapted from Liberman and Dodds (1984).

depending on the degree of hair cell damage in the corresponding place on the basilar membrane. In Figure 19.3a, the inner hair cells are damaged and their sensitivity has decreased, but the outer hair cells function normally. In this case, the threshold level of the neural tuning curves is raised but the frequency selectivity remains normal. A similar tuning curve could result from a conductive hearing impairment. In Figure 19.3b, most of the outer hair cells are damaged, and the corresponding tuning curve has changed shape. In Figure 19.3c, the outer hair cells are completely destroyed, and the corresponding tuning curve is very flat and 'untuned' compared to normal hearing. In this case, the resulting tuning curve depicts the passive amplification characteristics of the basilar membrane without the active effect of the outer hair cells, which, in a normally functioning ear, would increase the frequency selectivity and provide level-dependent amplification, or compression.

The decreased frequency resolution caused by outer hair cell damage effectively makes the critical bands broader with a lower centre frequency and more overlapping. As more energy is added at each critical band, each band receives input from a wider frequency range, the frequency masking effect is stronger. The result is degraded speech intelligibility in noise.

19.2.3 Speech Communication Problems

In addition to the effects of sensorineural impairments discussed above, speech discrimination can also be degraded due to impairments in the processes specifically responsible for speech processing. An individual may hear many sounds without problems but may still be unable to understand speech. For example, there may be problems in analysing the temporal structure of speech. This can depend on the pace of the speech, so that slow speech is intelligible, but as the pace increases, the auditory system cannot structure the speech into linguistic units.

Hearing impairments can also result in problems in speech production due to lack of feedback of the produced sound (see Section 17.8.3 for a review of the effect of feedback on speech production). Moreover, the ability to speak may be entirely restrained due to inborn deafness.

Speech as a form of linguistic communication is a complex phenomenon, which makes measuring the level of speech communication abilities difficult. Since speech is the most important method of communication for humans, speech audiometric methods have been developed to measure these abilities.

19.2.4 Tinnitus

Tinnitus, a ringing in the ears, is a common phenomenon that occurs in most people at some point in their lives, at least for a short while. Tinnitus is an auditory event that is not related to any external sound event. Tinnitus can present itself in various forms of perceived sound. A typical tinnitus sound is a steady pure-tone-like sound, but it can also be, for example, broadband or band-limited noise. The sound may also vary in time, for example, along with heartbeats, in which case the tinnitus is probably due to blood pressure changes in the ear. Another typical cause for tinnitus is excess noise exposure.

Several theories have been put forward explaining the origin of tinnitus in the auditory system (Jastreboff, 1990). Tinnitus may originate from the malfunction of some neural process in the auditory system with no actual vibration. It may also originate in the cochlea, even to such an extent that the basilar membrane vibrates and the tinnitus sound can be measured in the ear canal. In this case, the functioning of the outer hair cells in the cochlea has become unstable and thus a constant oscillation is present even without an acoustic input.

Loud tinnitus is a psychologically difficult impairment and can have a significant effect on the quality of life for the individual. There is no known cure for tinnitus. However, methods exist for treating tinnitus. One approach is to use a tinnitus masker to generate a sound in the ears that makes the tinnitus sound less annoying. Another approach is to use a combination of counselling and sound therapy to habituate the individuals to the tinnitus sound in order to reduce their awareness of it (Jastreboff *et al.*, 1996).

19.3 The Effect of Noise on Hearing

The effect of noise on hearing and the prevention of noise-induced hearing loss are important research subjects in the fields of acoustics and audiology. Noise is the most common cause for work-related impairments (Toppila, 2000). Avoiding exposure to noise is difficult in modern society. Excess noise exposure can cause a hearing impairment, but even lesser exposures can cause mental stress and consequent symptoms.

19.3.1 Noise

Noise is harmful or disturbing sound. This definition embodies two aspects: harmfulness and disturbance. Harmfulness is the more objective of the two: noise can be harmful even if it is perceived as comfortable and non-disturbing. The effects of harmful noise, for example hearing loss, can be measured using audiometric methods. The disturbance caused by sound and especially annoyance are more subjective concepts and can be measured, for example, in terms of a decrease in work efficiency or the ability to concentrate. As already discussed in Section 17.10, *disturbance* can be defined by how much a sound disturbs some action, while *annoyance* is a more subjective concept, which can be estimated based on individual experience. Subjective handicaps may have further indirect consequences, such as mental and physical illnesses.

Good examples of conflicts between subjective and objective points of view of noise are loud music, loud motorsports, shooting, and other noisy hobbies. These can be considered positive experiences even when the risk of hearing loss is obvious.

An essential question in the field of noise control is how to estimate the negative effects of noise. The majority of the related research discusses the risk of hearing loss, while the disturbance and annoyance caused by noise are less studied. There are two reasons for this. First, the risk of hearing loss is primarily a physiological phenomenon, where the induced risk is mainly predicted by the properties of the sound. Second, disturbance and annoyance are more dependent on the individual and on the situation. For example, loud music may be a 'fantastic aesthetic experience' or 'terrible noise' depending on the individual and on the situation.

The primary factor determining the negative effects of noise is *noise exposure*, which depends on the level and on the exposure time. When measuring noise exposure, the frequency-dependent sensitivity of hearing should be taken into account. In practice, this is typically done by measuring the frequency-weighted sound pressure level. Figure 9.3 on page 156 shows the A-, B-, C-, and D-weightings, which approximate the inverted shape of the equal loudness curves at different sound pressure levels (see Figure 9.2). The A-weighting has become the most widely used frequency-weighting, although it is not optimal for all sound pressure levels. However, the A-weighting seems to describe the risk of hearing loss quite well.

Noise exposure can be measured using the *equivalent sound pressure level*, which takes into account the temporal variation of the sound pressure level. The equivalent sound pressure level L_{eq} is the root-mean-square level averaged over a certain time period:

$$L_{eq} = 10 \log_{10} \frac{\sum \Delta t_i \, 10^{L_i/10}}{T}, \tag{19.1}$$

where t_i are the time periods with sound pressure levels of L_i [dB SPL] and T is the total time with which the sound exposure is normalized to the equivalent sound pressure level. T is usually 8 hours, corresponding to a normal working day. An integrated sound pressure level measure L_{eq} can be computed so that the sum becomes an integral when Δt_i tends to zero.

Research has shown that noise with considerable impulsive content has a greater risk of causing hearing loss compared to steady noise. Therefore, it is reasonable that in the case of impulsive noise, this effect is taken into account. A simple rule is that when impulsiveness exceeds a certain threshold, an extra 10 dB is added to the equivalent sound pressure level.

19.3.2 Formation of Noise-Induced Hearing Loss

The threshold of pain in hearing is around 130 dB SPL. Sounds louder than this can cause permanent damage to hearing even over short durations. For example, a starting pistol produces a peak sound pressure level of approximately 140 dB SPL, and more hefty weapons produce even louder sounds. A single strong peak in sound pressure can tear the eardrum, break the auditory ossicles, or destroy the structures of the inner ear immediately. An impulse even at a lower level can cause permanent tinnitus.

Still, typical noise-induced hearing loss develops over the course of time. The loss can remain unnoticed until speech intelligibility begins to suffer. For long-term noise exposures, implying years or decades of exposure, the daily risk threshold has often been quoted to be 85 dB SPL (A-weighted, 8 hours a day), with the criterion of impaired hearing being a hearing threshold shift of 25 dB in middle–high frequencies. This kind of threshold shift already may decrease speech intelligibility. If healthy hearing is to be assured even after years of exposure to noise, the A-weighted risk threshold should be decreased to 80 dB, or even 75 dB. This is because the risk of hearing loss is somewhat individual, and, for example, music listening may suffer even with smaller hearing impairments.

19.3.3 Temporary Threshold Shift

As discussed earlier, a high enough noise exposure can cause the sensitivity of hearing to decrease, and consequently the hearing thresholds to increase. If the exposure is under certain limits, the hearing thresholds can recover after the exposure, fully or to some extent. This phenomenon is called the *temporary threshold shift* (TTS). The level and duration of the TTS are affected by the level and duration of sound. Figure 19.4a shows the increase in hearing threshold as a function of time for different sound levels. Even a sound of 70 dB SPL is capable of generating a minor TTS when the exposure time is several hours. Sound pressure levels of over 100 dB SPL increase the hearing threshold rather quickly.

When the noise exposure has ended, the hearing threshold begins to recover. Figure 19.4b shows the decrease in the hearing thresholds after a TTS caused by a long exposure to sound at different SPLs. The recovery time can be from minutes to several days. As a general rule, one should allow about double the exposure time for recovery, or a time when hearing has recovered to normal. A sufficiently large exposure to noise leads to a permanent hearing threshold shift and the recovery to normal hearing does not happen.

Although hearing recovers to normal after a TTS, it is a sign of a risk of hearing loss, especially in the case of repeated temporary threshold shifts or a single significant TTS. Thus a good rule of thumb is that, for example after a rock concert, hearing should be normal by the next morning at the latest. In addition to TTS, tinnitus or sound distortion effects after an exposure to noise are signs of overload in the auditory system.

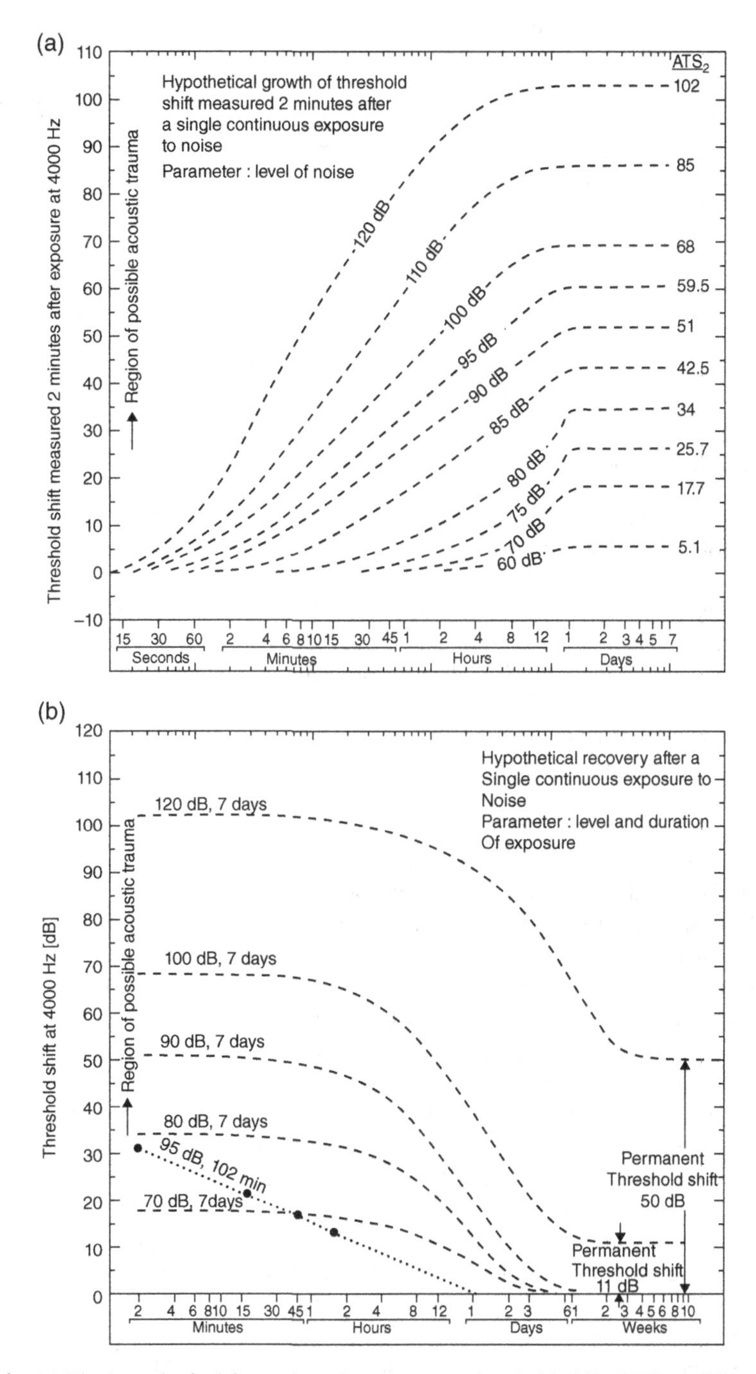

Figure 19.4 (a) The hypothetical formation of a temporary threshold shift (TTS) at different exposure levels as a function of exposure time. (b) The hypothetical recovery from TTS as a function of time, after a long-term noise exposure at different levels. Adapted from Miller (1974), and reprinted with permission from The Acoustical Society of America.

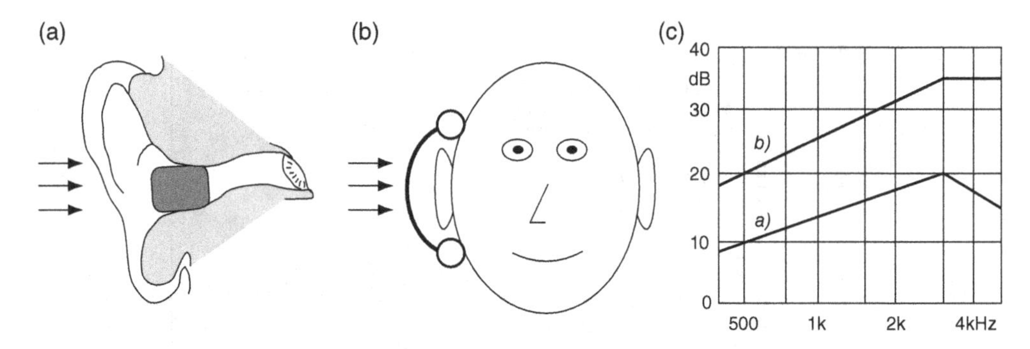

Figure 19.5 (a) Earplug-type of hearing protector, (b) earmuff-type of hearing protector, and (c) typical attenuation characteristics for earplugs and earmuffs. Adapted from Toivanen (1976).

19.3.4 Hearing Protection

When controlling noise that is potentially dangerous to hearing, countermeasures should be taken in the following order of priority:

1. *Decrease the noise emission of the noise source.* For example, design a quieter machine.
2. *Attenuate the transmission path.* For example, encapsulate the machine that produces noise.
3. *Personal hearing protection.* Use when the procedures above are not sufficient.

Hearing protectors attenuate sound that reaches the eardrum. They can be divided into different groups depending on their type:

- *Earplug*: a plug that blocks the ear canal entrance (Figure 19.5a). It is typically made of plastic foam or wax. Individually moulded earplugs exist too. For effective attenuation, the plug should be rigid enough and seal the ear canal properly.
- *Earmuff*: constructed from a stiff cup that is tightly placed against the head with padding so that the whole outer ear is covered (Figure 19.5b). These types of hearing protectors are the most common in situations where significant attenuation is needed. The increase in mass and internal volume of the earmuff increase the attenuation achieved with the protector.
- *Acoustic helmet*: a helmet that covers the whole head or part of the head and can provide extreme attenuation of sound.
- *Active hearing protection*: can be used with any of the above-mentioned passive types of protectors. Active hearing protectors are equipped with an electroacoustic system that can be used in various ways. One method is to reproduce the incoming sound with opposite phase from the earphone, so that this sound cancels out some of the sound coming through the ear protector enclosure. This enables the attenuation of a passive protector to be increased. With well-designed active earplugs, efficient attenuation can also be achieved in the low frequencies. Another approach is to attach active hearing protection to heavy earmuffs and provide amplification only to quiet sounds. In this way, for example, speech communication can be possible while using earmuffs, but loud sounds (e.g., sudden impulsive sounds) are attenuated passively.

Figure 19.5c shows typical attenuation characteristics of earplugs and earmuffs. However, different models vary considerably in their attenuation characteristics. A good, well-fitting earplug may provide better attenuation than a light earmuff. By combining the use of earplugs and earmuffs, even more attenuation can be achieved.

It is hard to achieve good attenuation in the low frequencies unless the protector is very well sealed and of appropriate construction. Thus, the correct use of the hearing protectors is also important. Otherwise, the attenuation performance may be significantly worsened. An earplug should be fitted deep enough and so that it seals the ear canal properly. When using earmuffs, one must ensure that there is no gap between the earmuff padding and the head, for instance, due to hair or glasses. With an improper fit, an earmuff can even form a Helmholtz resonator, which amplifies sound at certain frequencies.

Hearing protection should be used constantly when exposed to loud noise, because even a short time without hearing protection can cause significant noise exposure and thus a considerable risk of hearing loss.

19.4 Audiometry

Audiometry is the science of measuring the functioning of the auditory system. The first measurement methods were based on testing the patient's response to a sound produced, for example, by a tuning fork or by live speech. Nowadays, several dedicated methods and pieces of equipment are used to achieve reliable and repeatable results. The term *audiometer* refers to a pieces of equipment used to conduct audiometry.

The techniques to estimate the functioning of the auditory system can be divided into two categories. Tests where the subject has to listen actively to the sound events and to co-operate in responding to them can be said to be subjective methods. These tests are commonly used, and they can be conducted easily with most people. Some cases require objective methods, where responses do not depend on whether the test subject is listening or not. Such tests can be utilized with infants or with people who are unable or unwilling to co-operate. Furthermore, objective tests can be useful in monitoring the state of the auditory system during, say, a surgical operation.

19.4.1 Pure-Tone Audiometry

Pure-tone audiometry (ISO 8253-1, 2010), as the name suggests, uses pure tones or other narrowband stimuli as test signals to measure the frequency-specific hearing thresholds. A *pure-tone audiometer* is a device that varies, either manually or automatically, the level of the test signal in an adaptive procedure seeking the hearing threshold. Pure-tone audiometry is usually performed by presenting the test stimuli via calibrated headphones, to which the patient responds orally or by pressing a button. This procedure is done separately for both ears, typically at least at the frequencies 250, 500, 1000, 2000, and 4000 Hz, and possibly also at 125 Hz and 6000 Hz and/or 8000 Hz. Békésy audiometry uses a slow frequency sweep, thus measuring a continuous hearing threshold curve as a function of frequency. The steepness of the hearing threshold curve can be obtained from the results of Békésy audiometry.

Pure-tone audiometers are manufactured for different purposes and with different precision requirements. In hearing screening, for example, the aim is only to discover possible significant deviations from normal hearing. Dedicated screening audiometers are available for this purpose. Clinical audiometers (Figure 19.6) are used for more precise measurements. Clinical and diagnostic audiometers can include many special functions. Simple audiometers for home use (for example, computer software) also exist. For reliable results, it should be ensured that the masking effect of background noise does not increase the measured hearing thresholds,

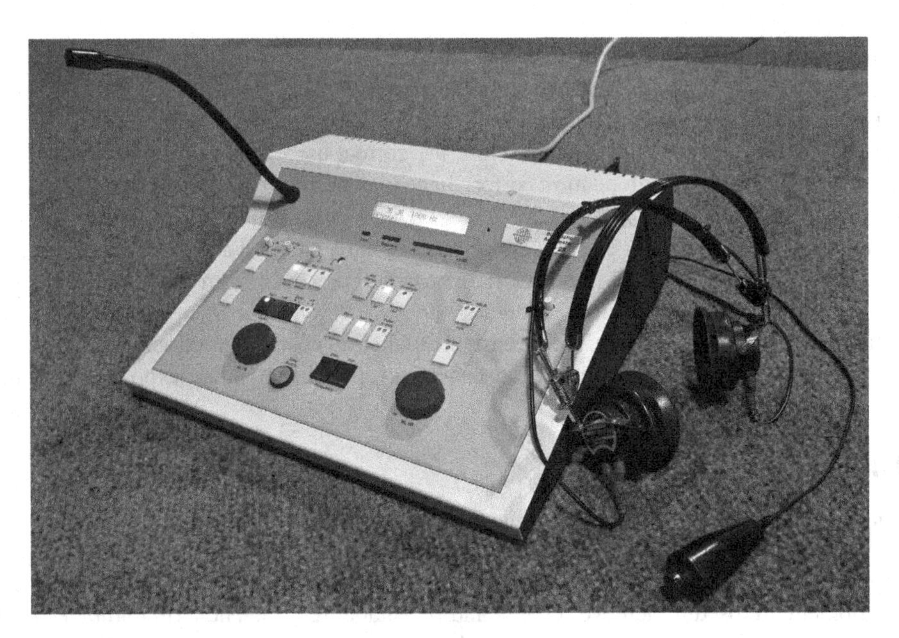

Figure 19.6 An audiometer with audiometric headphones and a patient-response button. Courtesy of Teemu Koski.

and that the transducers are properly calibrated. The hearing threshold curve as a function of frequency measured in pure-tone audiometry is called an *audiogram*.

19.4.2 Bone-Conduction Audiometry

Sound is transmitted to the inner ear not only through the outer and middle ear (that is, air conduction), but also through the bones of the head (that is, via bone conduction). For sound that propagates in air, bone-conducted sound is quieter than air-conducted sound. Thus, air-conducted sound generally dominates in the perception of sound. However, vibration that is coupled mechanically to the bones of the head is effectively transmitted to the inner ear via bone conduction. Consequently, bone conduction has a significant effect, for example, on how one's own voice is perceived.

The bone-conduction phenomenon is utilized in bone-conduction audiometry. In contrast to air-conduction pure-tone audiometry (discussed above), the test stimuli are fed as vibration to the skull at the back of the ear in bone-conduction pure-tone audiometry. The bone-conduction hearing thresholds are then measured in a similar manner to air-conduction pure-tone audiometry. If the bone-conduction threshold is normal but the air-conduction threshold is elevated, the hearing impairment is likely to be conductive. If the bone-conduction threshold has increased by the same amount as the air-conduction threshold, the impairment is likely to be sensorineural or due to a problem at some higher level of the auditory system.

19.4.3 Speech Audiometry

Speech is the most important form of sound-based communication. Degraded speech intelligibility is a common practical consequence of many kinds of hearing impairment. In some cases, these impairments can be diagnosed using pure-tone audiometry. However, speech intelligibility may be decreased even when hearing thresholds are normal and no organic impairment is found. *Speech audiometry* methods (ISO 8253-3, 1996) have been developed to measure speech intelligibility directly.

In speech audiometry, the patient tries to identify words, sentences, or other speech sounds he or she is presented with. Speech can be reproduced to the patient over headphones or loudspeakers. Typically, the target signal level is either constant or varying to measure the percent-correct score or the detection threshold, respectively. Moreover, the speech pace can be varied, because in some cases the individual is able to understand only slow speech. Dedicated speech corpora have been developed for speech audiometry for many languages, for example Kollmeier and Wesselkamp (1997) and Nilsson *et al.* (1994).

Speech audiometry can also be conducted with background noise. These speech-intelligibility-in-noise-measurements aim to assess problems that the hearing impaired commonly have in communication situations involving background noise.

19.4.4 Sound-Field Audiometry

In *sound-field audiometry* (SFA), loudspeakers are used instead of headphones for the reproduction of the test stimuli. Compared to headphone-conducted audiometry, SFA demands more from the equipment and facilities, but in turn enables test conditions that are not possible or practical with headphones. The test stimuli can be narrowband, as in conventional pure-tone audiometry, or, for example, speech in silence or with background noise. The standard ISO 8253-2 (2009) defines the use of narrowband stimuli in SFA.

The primary motivation for using SFA is the limitations in headphone-conducted audiometry. The first limitation, for a certain group of test subjects, is the acoustic coupling between the sound source and the test subject. Depending on the type of hearing instrument and the microphone placement, it is often difficult to achieve a constant and controlled acoustic coupling between the audiometric headphones and the hearing instrument microphone. Small children might not tolerate the use of headphones, again causing uncontrolled acoustic coupling. The second limitation is the realism of the audiometric test. Testing in a sound field, in contrast to the use of headphones, takes into account the spatial attributes of sound and hearing and allows test conditions to more closely resemble real-life situations relevant to the patient.

The simplest SFA set-up consists of one or two loudspeakers. More complex SFA systems aiming to reproduce real or realistic sound scenes have been suggested, for example, by (Favrot and Buchholz, 2010; Koski *et al.*, 2013; Seeber *et al.*, 2010). In addition to audiometry, SFA methods are used to evaluate and compare the performance of hearing instruments (see, for example, (Minnaar *et al.*, 2013)).

19.4.5 Tympanometry

Tympanometry (Campbell and Mullin, 2012; Martin and Clark, 2006) is an objective way to measure the status of the eardrum and the middle ear. It measures the acoustic impedance of the eardrum – the tympanic membrane – to assess the mobility and the pressure difference across the eardrum.

Figure 19.7 shows the measurement set-up for tympanometry. A test sound is presented to the ear canal with an earphone, and the sound pressure in the ear canal is measured with a microphone. The changes in the impedance of the eardrum are reflected in the measured sound pressure. In other words, when much of the sound is reflected back from the eardrum,

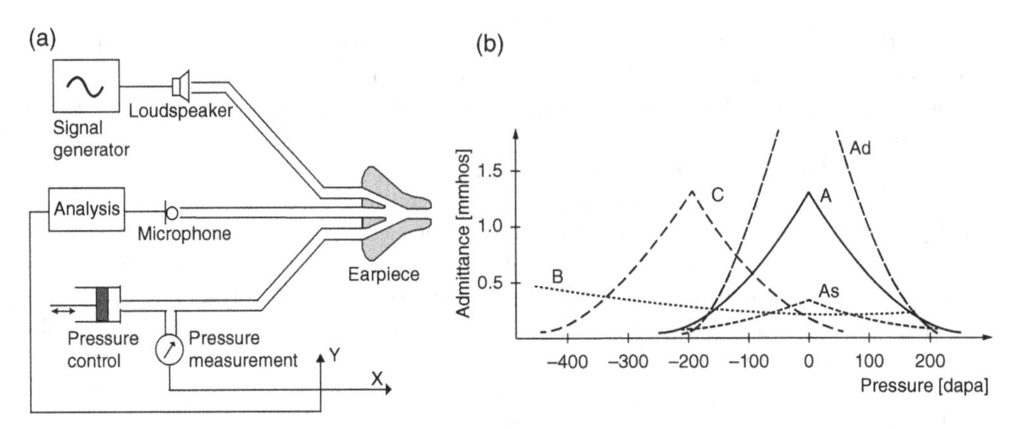

Figure 19.7 (a) The equipment set-up in tympanometry and (b) a conceptual tympanogram, showing examples of different responses to tympanometry measurement. A = normal middle ear function, As (for A-shallow) = stiffened middle ear system, Ad (for A-deep) = flaccid eardrum, B = fluid in the middle ear or perforation of the eardrum, C = negative pressure in the middle ear. Data adapted from Campbell and Mullin (2012).

the acoustic impedance is high, indicating a stiff eardrum (low compliance), and when little sound is reflected back, much of the sound is transmitted to the middle ear, and the acoustic impedance is low, indicating a more mobile eardrum (high compliance). During the playback of the test sound, the external static air pressure is swept over a range of negative and positive pressures and the value of compliance is measured as a function of the static air pressure.

The external pressure where the maximum compliance is found corresponds to the pressure in the middle ear. If this pressure differs considerably from the atmospheric pressure, there is a pressure difference across the eardrum. The absence of a compliance peak may imply fluid in the middle ear. The level of the maximum compliance indicates the condition of the eardrum and the auditory ossicles. For example, a low maximum compliance may indicate stiffness in the auditory ossicles, while a high maximum compliance may indicate a flaccid eardrum.

Even a relatively small pressure difference across the eardrum attenuates the sound transmitted to the inner ear, especially at the low frequencies. A large pressure difference can be very painful. A pressure difference may originate, for example, from an infection or static changes in the air pressure, for example during a flight. The Eustachian tube, which connects the middle ear to the pharynx, normally equalizes the air pressure between the middle ear and the outer ear. However, the tube may be narrow in some individuals, or it may be blocked due to infection, and thus the pressure may not be equalized.

19.4.6 Otoacoustic Emissions

Otoacoustic emissions, discussed in Section 7.5, can be used to objectively evaluate the functioning of the auditory system (Martin and Clark, 2006). An *evoked otoacoustic emission* (EOAE) should be observable if the auditory system is functioning normally; missing EOAEs indicate a conductive or cochlear defect. Such an indication does not specify whether the defect is due to the cochlea not responding to the stimuli or due to attenuation in the conductive path. EOAEs present in cases of sensorineural hearing loss suggest retrocochlear problems.

Transient-evoked otoacoustic emissions (TEOAEs) are measured using brief acoustic stimuli that stimulate a wide area of the cochlea, resulting in a broadband response for a normally functioning auditory system. *Distortion-product otoacoustic emissions* (DPOAEs), in turn, are measured with two tones that differ in frequency, for which a normally functioning cochlea responds with an emission in additional frequencies.

19.4.7 Neural Responses

The functioning of the auditory system can also be assessed by measuring neural responses from different parts of the auditory path. *Electroencephalography* (EEG) and *magnetoencephalography* (MEG) are non-invasive methods to measure such responses. In EEG, the weak electrical potentials produced by brain activity are measured with electrodes that are placed at different locations on the scalp. MEG measures the weak magnetic fields produced by brain activity. Typically, sound stimuli are presented to a subject, and the EEG or MEG responses evoked by the stimuli are measured.

19.5 Hearing Aids

A *hearing aid* (Dillon, 2012) is basically a miniature sound reproduction system with signal processing. Its purpose is to amplify and process sound for the user to compensate for the effects of a hearing impairment. The eventual aim is to enhance the hearing and communication performance of the user. A challenge in hearing-aid design is to achieve a suitable type of amplification that takes into account individual needs in different listening scenarios. Ensuring an adequate SNR for the user is essential. Other design objectives are, for example, comfort and ease of use.

The simplest possible hearing aid is a microphone placed near the ear, a linear amplifier with constant gain over the entire audible frequency range, and a miniature loudspeaker providing the amplified sound to the ear canal. This may suffice to compensate for a conductive hearing loss similar at all frequencies, but not for cases involving frequency- and level-dependent attenuation. Thus, modern hearing aids include frequency-dependent amplification, compression, and other signal processing features.

19.5.1 Types of Hearing Aids

Many types of hearing aids are available, varying in their operating principle and structure. The main types are listed below:

- *Behind-the-ear (BTE) hearing aid*, Figure 19.8a: The instrument enclosure is located behind the ear. BTE aids transmit sound to the ear canal via a tube. The receiver-in-the-canal (RITE) aid is a variation on the BTE, where an earphone is placed in the ear canal and the sound is transmitted there electrically, instead of through a tube as in BTE devices.
- *In-the-ear (ITE) hearing aid*, Figure 19.8b: The instrument is fitted in the concha.
- *In-the-canal (ITC) hearing aid*: The instrument is fitted in the ear canal. A variation on the ITC hearing aid is the *completely-in-the-canal (CIC) hearing aid*, which fits completely in the ear canal (Figure 19.8c).

More hearing aid types exist in addition to the ones listed above. Many of them are variations on the BTE, ITE, and ITC structures, but other approaches are also used. For example, in an *eyeglass instrument*, the aid is embedded into spectacles. This allows, for example, a microphone array to be placed along the frames, thus enabling highly directional sound-capturing

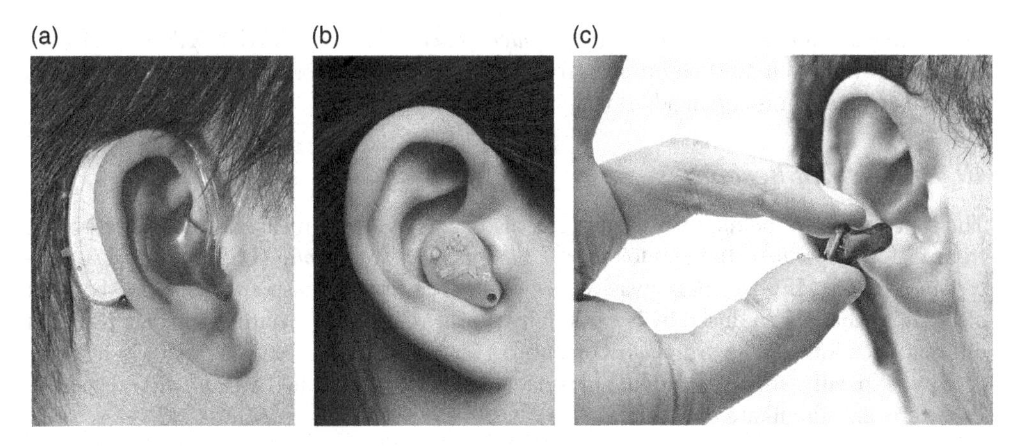

Figure 19.8 Types of hearing aids: (a) behind-the-ear (BTE) hearing aid, (b) in-the-ear (ITE) hearing aid, and (c) completely-in-the-canal (CIC) hearing aid

properties. The main unit of the hearing aid can also be completely detached from the head of the user. This kind of *body-worn instrument* was used especially in the past, when it was not possible to fit all the necessary equipment in a small casing.

19.5.2 Signal Processing in Hearing Aids

Modern hearing aids utilize various signal processing features, aimed at matching the device with the individual needs of the user (Kates, 1998). Digital signal processing enables many features that could not be achieved with analogue technology. Although full coverage of these techniques is beyond the scope of this book, a general overview of the most relevant techniques is given below.

The degree and properties of a hearing impairment can vary considerably for different frequencies. Thus, hearing aids often employ multi-band processing, where the input signal is divided into several frequency bands using a filter bank and the amplification and processing are done separately in the bands. This allows all the parameters related to amplification and other processing to be optimized for each frequency band, based on the measured hearing performance.

Communication situations involving background noise or reverberation often cause problems for the hearing impaired. An individual with impaired hearing might struggle in these kinds of situations even though speech intelligibility in quiet situations is normal. The hearing impaired often need a higher SNR to achieve the same speech intelligibility as normally hearing individuals. Plain amplification amplifies both the desired signal and the unwanted noise, and therefore it does not solve the problem of degraded speech intelligibility in the presence of background noise. Thus, several approaches have been developed to increase the SNR in the hearing aid output and so enhance the intelligibility in acoustically complex environments.

Gain control

Linear amplification increases the output level in the same proportion to the input level. If the input level increases by, say, 1 dB, the output is increased by 1 dB as well. This poses a problem for high input levels: either the level exceeds the dynamic range of the device or the

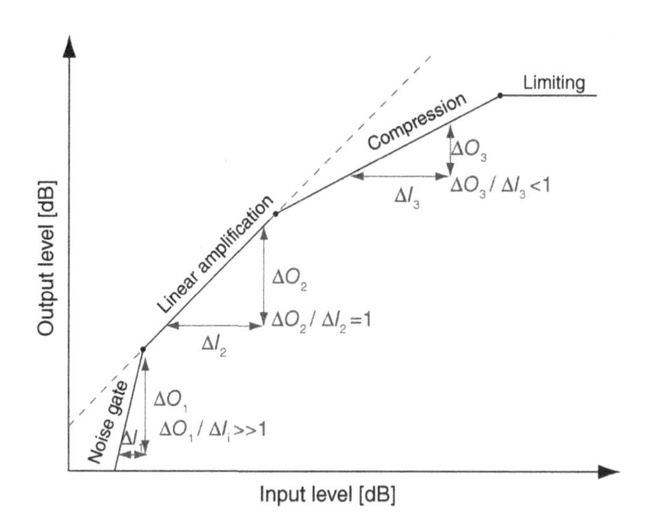

Figure 19.9 An amplification curve with high-level compression, linear amplification, and limiting. The changes in the level of input and output in dB are denoted as ΔI and ΔO, respectively. Courtesy of Teemu Koski.

amplified output can be so loud as to cause a risk of further hearing loss. Thus, the amplifiers in hearing aids are equipped with some kind of *limiter* circuit that limits the output amplitude of the amplifier to a certain value. Unfortunately, a sharp limiter causes the signal to distort as the level exceeds the pre-set limit.

Using a limiter alone with linear amplification is inadequate in hearing aids, since hearing loss is often level-dependent because of recruitment. Thus, hearing aids typically employ *automatic gain control* (AGC) which enables the *level compression* of sound – the amplification of low input levels more than high ones. Consequently, the dynamic range of the input signal is reduced so that it better fits the decreased dynamic range of the impaired hearing.

Figure 19.9 shows an example of an amplification curve that contains a noise gate, a range of linear amplification, compression, and limiting of the highest levels. The noise gate sets the lowest level of sound that is reproduced, thus preventing the amplification of system noise. Compression enables the overall gain to be increased so that low-level sound becomes audible without high-level sound becoming too loud. In addition, applying compression of high levels together with a limiter enables limiting the output level with less distortion than with just a simple limiter. Other strategies for range compression also exist, and these are reviewed by Dillon (2012).

Automatic gain control can be implemented with a feedback loop, as shown in Figure 19.10a. Figure 19.11 shows schematically how the time-domain output signal changes when the input signal is linearly amplified, limited, and compressed. The analysis conducted on the signal always has a finite integration time that the circuit takes to change the gain, as is shown in Figure 19.11.

Automatic gain control can also be implemented with a feedforward loop, as shown in Figure 19.10b. The gain applied to the signal is thus set by analysing either the input or output signals of the amplifier. The difference between the feedback and feedforward loops depends on other processing in the system. For example, if the device has a volume control, the positioning of the loop with respect to the volume control affects the result. A feedforward loop

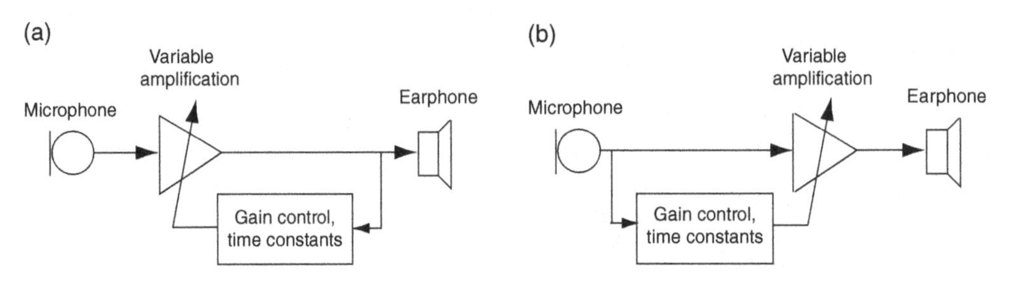

Figure 19.10 Automatic gain control (AGC) implemented with (a) a feedback loop, and (b) a feedforward loop.

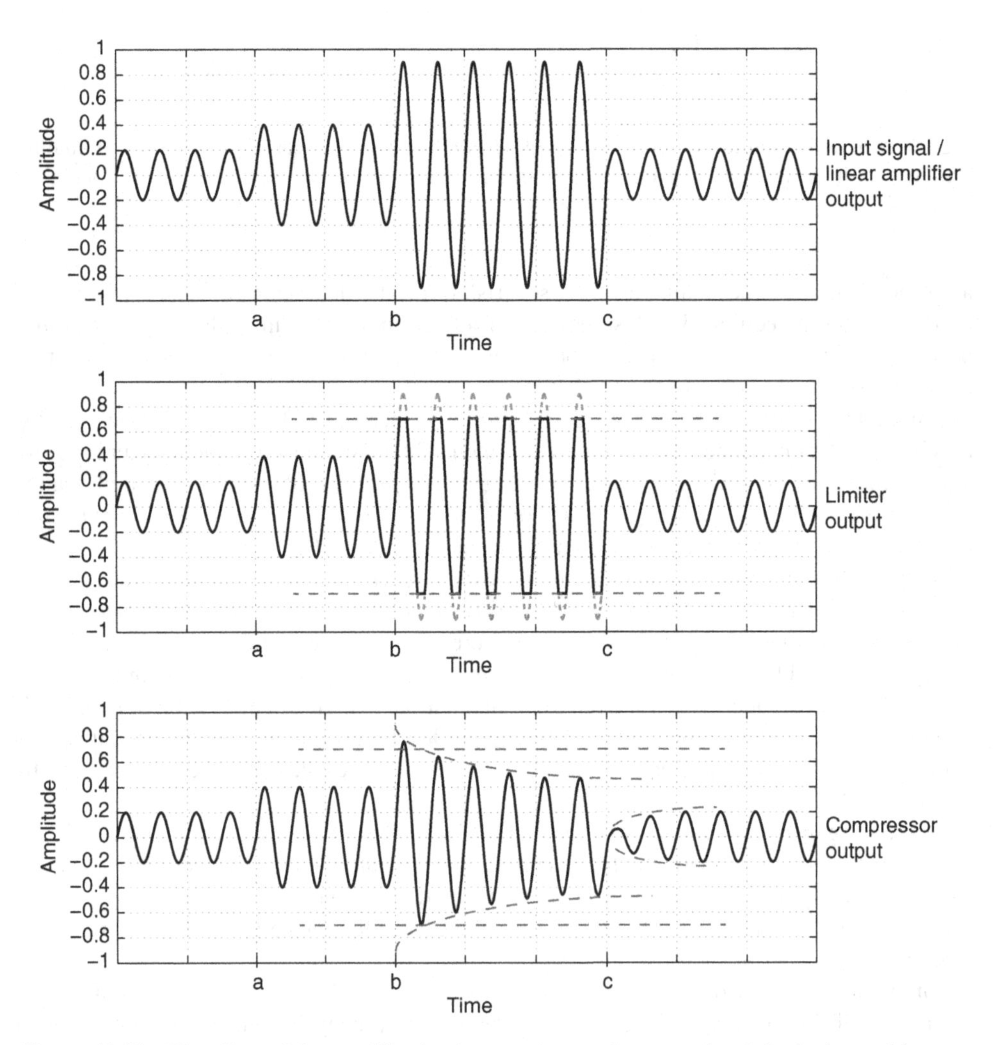

Figure 19.11 The effect of the amplifier implementation on the output level. In the interval between times b and c, the output signal is limited (middle panel) or compressed (the lowest panel). Courtesy of Teemu Koski.

enables global control of the volume without affecting the behaviour of compression, and a feedback loop after the volume control affects the volume only at low input levels and causes a greater proportion of sounds to be compressed heavily and limited.

Directional microphone systems

One approach to increasing the SNR in hearing aids is to use a *directional microphone* to amplify sound in some directions more than others. A simple and widely used design is a first-order subtractive directional microphone, which can be constructed with a single microphone having two ports or with two separate microphones, one port at each. Such a design makes possible first-order directivity patterns including cardioid, hyper-cardioid, super-cardioid, and dipole patterns, as in Figure 14.5 on page 290. In the one-microphone design, sound is fed to both sides of the microphone diaphragm, and the directivity pattern is controlled by the port spacing and an internal acoustic delay. In the two-microphone construction, the signals of the two microphones are subtracted, and the directivity pattern is controlled by the microphone spacing and an electrical delay between the microphone signals. By combining signals from a larger number of microphones, even more directive beams can be achieved. In general, the generation of a directive microphone beam by combining signals from several microphones is called *beamforming*.

It is sensible to generate the directivity pattern so that the direction of highest sensitivity is in the front, since in practical situations people generally look in the direction they want to listen to. Some beamforming algorithms are adaptive, so that the directivity pattern changes depending on the situation in order to minimize the level of unwanted noise. They are typically designed to adaptively set the lowest sensitivity in the direction of the most dominant noise source.

Other signal processing features

In addition to directional microphone systems, the SNR in the hearing aid output can be increased by single-channel noise reduction schemes (Dillon, 2012). These techniques use the temporal, spectral, and statistical information of the incoming sound to suppress the noise. Various *speech enhancement* algorithms assume some properties for speech and use these properties to attenuate unwanted noise and preserve speech-like signals.

Feedback cancellation helps to suppress the loud whistling caused by possible acoustic feedback from the hearing aid earphone to the hearing aid microphone. Feedback cancellation can be implemented, for example, with an adaptive filter that detects the feedback and adjusts the filter response adaptively to attenuate the feedback.

Binaural processing in hearing aids can be advantageous compared to monaural processing. In binaural processing, there is interaction between the left and right hearing aids in order to preserve or enhance the spatial hearing abilities of the user. For example, at high frequencies compression may suppress the sound on the ipsilateral side and leave it untouched on the contralateral side. This biases the ILD cue towards the median plane, which is undesirable. Binaural processing enables, for example, the compression gain to be made equal on both sides and interaural cues to be preserved, which further enhances the spatial hearing performance with hearing aids. Improvements provided with binaural processing have been reported, for example, in terms of speech intelligibility in conditions of noise; see, for example, Moore (2007) for a review. The input from multiple microphones can also be processed non-linearly, combined together with single-microphone processing, to achieve noise reduction or beamforming in hearing aids (Ahonen *et al.*, 2012; Hersbach *et al.*, 2013; Van den Bogaert *et al.*, 2009).

19.5.3 Transmission Systems and Assistive Listening Devices

Hearing aids often have some means to receive sound from an external source. The general aim of such devices is usually to suppress unwanted sounds and make target signals clear and audible, thus enhancing intelligibility even in difficult listening scenarios.

The hearing aid device may also receive sound wirelessly if it is equipped with an induction coil or a wireless receiver. Correspondingly, some public sites are equipped with an induction loop system which can be used to transmit sound wirelessly to the induction coil. Hearing aids may also receive sound via radio-frequency transmission. Various assistive listening devices utilize this channel. For example, an external microphone with a wireless transmitter can be useful when communicating in noisy environments if the microphone can be placed close to the target sound source to maximize the SNR in the hearing aid of the user. Assistive listening devices can also provide the means to connect to other communication devices, such as telephones.

19.6 Implantable Hearing Solutions

19.6.1 Cochlear Implants

A *cochlear implant* (Clark, 2003; Zeng *et al.*, 2003) is a surgically implanted electrical device that transmits sound to the auditory nerve fibres in the cochlea. The conductive part of the auditory path and part of the cochlea are bypassed. A cochlear implant can provide a usable sense of hearing in those cases where the hearing impairment is so severe that a hearing aid cannot provide adequate help. These cases include deafness or severe to profound bilateral hearing loss due to sensorineural impairment.

Figure 19.12 shows a system diagram of a cochlear implant. Microphones and the sound processor are located typically behind the ear, in a similar manner to BTE hearing aids. The sound processor typically employs similar signal processing schemes as in hearing aids, such as beamforming and noise management. After this pre-processing, the sound processor divides the incoming sound into several frequency channels and generates a pulse-train signal representing each channel. Several pulse-coding methods have been developed to implement this (Loizou, 1999). The pulse-train signal is routed to the transmitter coil, which is located on the scalp, behind the ear. The receiver coil of the implant, which receives the signal via induction, is at the same location but under the skin. From the receiver coil, the signal is fed to the electrode array. The electrode array is placed inside the cochlea with a surgical operation that requires high precision. The electrode contacts in the array stimulate the auditory nerve fibres, aiming to trigger patterns of activity in the auditory nerve that imitate the inputs from the hair cells of a functional basilar membrane. Different channels are fed to different electrode contacts, which are placed at different locations on the basilar membrane, thus providing a rough place-code for different frequencies.

The first cochlear implants utilized only one frequency channel. These devices were not able to provide frequency discrimination, but only some sense of the general temporal structure of the signal. Modern cochlear implants employ electrode arrays with multiple frequency channels. However, the cross talk between the channels is a significant factor which limits the frequency resolution that can be achieved with cochlear implants (Schnupp *et al.*, 2011).

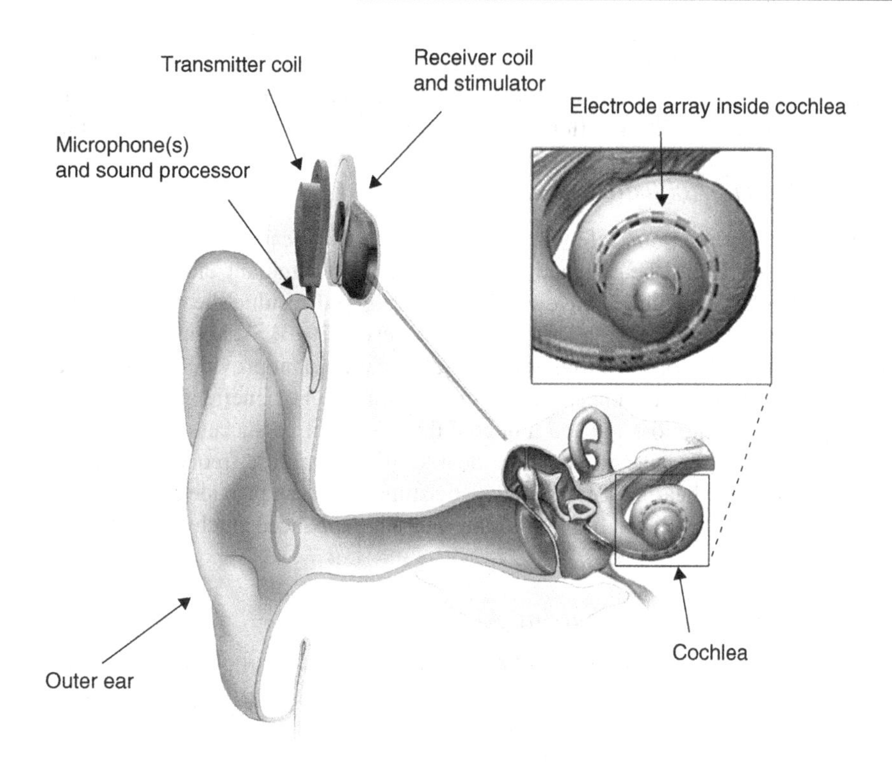

Figure 19.12 System diagram of a cochlear implant. Adapted from National Institutes of Health (2014).

The implication is that the impulses from a single electrode are received by a frequency region of the cochlea larger than one critical band.

By December 2012, approximately 324 200 cochlear implants had been fitted worldwide (National Institutes of Health, 2014). Hence, extensive knowledge has been gained about the outcomes of cochlear implantation. There is individual variation on the hearing performance achieved with cochlear implants. With modern implants and successful rehabilitation, most patients achieve the ability to have a telephone conversation (Zeng, 2004). Visual cues, such as lip-reading, can enhance the communication performance significantly. Even in cases where good speech communication abilities are not achieved, the increased auditory awareness provided by the implant can be advantageous.

The ability to perceive pitch is immensely poorer than that in normal hearing, which naturally makes musical structures such as melody, harmony, and the bass line very difficult if not impossible to perceive. Consequently, the desire to listen to music varies considerably among cochlear implant users. Some users like to listen to music even though it is distorted to the extent that it is difficult to discriminate the instruments from each other. Many instruments producing harmonic tones are generally perceived as unpleasantly distorted, but the basic rhythm of music can be heard almost normally.

In adults, cochlear implants are fitted mainly to those individuals who already spoke before their hearing loss. In adults who were born deaf, the results of cochlear implantation have not been optimal: with no previous sound-evoked input, the adult auditory system may be unable to interpret auditory input and may not have the plasticity to learn it. In contrast, a cochlear implant is well suited for deaf infants – if received at an early enough age – since even unilateral cochlear implantation enables the auditory system to develop to such an extent that spoken language skills are often at age-appropriate levels (Nicholas and Geers, 2007).

19.6.2 Electric-Acoustic Stimulation

Electric-acoustic stimulation refers to electric and acoustic hearing – a cochlear implant with or without an ipsilateral hearing aid. Modern atraumatic implant electrode arrays and so-called soft surgical techniques have resulted in preserved residual hearing in the implanted ear, typically in the low-frequency region (Adunka *et al.*, 2013; Skarzynski *et al.*, 2014). Consequently, in electric-acoustic stimulation, the cochlear implant provides high frequencies and the hearing aid the low frequencies. This combination can be beneficial to patients who have high-frequency hearing loss to such a degree that it cannot be successfully compensated for with a hearing aid, but have an adequate degree of residual low-frequency hearing that can benefit from acoustic hearing. Electric-acoustic stimulation can provide not only better speech intelligibility than a cochlear implant alone, but also enhanced pitch perception and music appreciation (Gantz *et al.*, 2005).

19.6.3 Bone-Anchored Hearing Aids

A *bone-anchored hearing aid* (BAHA) differs in its operating principle from the hearing aid types discussed earlier in this book. In a BAHA system, a small implant is inserted into the skull behind the ear. The sound processor is attached to this implant, and the implant feeds the sound to the skull as mechanical vibrations. Consequently, sound travels to the inner ear via bone conduction. BAHA are suitable for individuals with an impaired outer or middle ear, since these parts are bypassed by the bone-conduction path. BAHA systems can provide better speech intelligibility compared to conventional hearing aids, especially if the air-conduction hearing thresholds differ considerably from the bone-conduction hearing thresholds (de Wolf *et al.*, 2011).

19.6.4 Middle-Ear Implants

In *middle-ear implants*, a transducer is attached to the auditory ossicles or to the round window that vibrates the structures in the same manner that air-conducted sound would. Middle-ear implant systems vary in their way of picking up incoming sound. One approach is to use an external sound processor and an inductive link, similarly to cochlear implants. The microphone can also be implanted under the skin or placed in the ear canal. Reviews by Butler *et al.* (2013) and Kahue *et al.* (2014) conclude that middle-ear implants, on average, provide similar benefits in terms of speech intelligibility to conventional hearing aids. On the other hand, middle-ear implants have several other benefits compared to hearing aids. According to the review by Dillon (2012), the benefits are, to mention just a few, an extended high-frequency response, lower distortion, increased gain before feedback occurs, and benefits due to no obstruction of the ear canal. Furthermore, coupling to the round window can provide usable hearing to patients with a dysfunctional middle-ear system (Colletti *et al.*, 2006).

Summary

This chapter presented a general overview of technical audiology. The chapter discussed the causes and symptoms of hearing impairments, how they are assessed with audiometry, and what solutions have been developed to manage them with hearing aids and implantable hearing devices. Since the topic is broad, the interested reader is encouraged to explore the relevant literature for more in-depth information.

Further Reading

A good amount of literature exists in the field of technical audiology. Sensorineural hearing loss and its perceptual effects are discussed by Moore (2007). A general introduction to audiology is made by Martin and Clark (2006) and to hearing aids by Dillon (2012). An overview of signal processing methods for hearing aids can be found in Kates (1998). For a review of cochlear implant research, see, for example, Wilson and Dorman (2008) and Zeng *et al.* (2003).

References

Adunka, O.F., Dillon, M.T., Adunka, M.C., King, E.R., Pillsbury, H.C., and Buchman, C.A. (2013) Hearing preservation and speech perception outcomes with electric-acoustic stimulation after 12 months of listening experience. *Laryngoscope*, **123**(10), 2509–2515.

Ahonen, J., Sivonen, V., and Pulkki, V. (2012) Parametric spatial sound processing applied to bilateral hearing aids. *45th Int. Conf. Audio Eng. Soc.* AES.

Butler, C.L., Thavaneswaran, P., and Lee, I.H. (2013) Efficacy of the active middle-ear implant in patients with sensorineural hearing loss. *J. Laryng. & Otol.* **127**(Suppl 2), 8–16.

Campbell, K.C.M. and Mullin, G. (2012) Impedance audiometry. *Medscape reference.* `http://emedicine.medscape.com/article/1831254-overview`.

Clark, G. (2003) *Cochlear Implants: Fundamentals and applications*. Springer.

Colletti, V., Soli, S.D., Carner, M., and Colletti, L. (2006) Treatment of mixed hearing losses via implantation of a vibratory transducer on the round window. *Int. J. Audiol.*, **45**(10), 600–608.

de Wolf, M.J., Hendrix, S., Cremers, C.W., and Snik, A.F. (2011) Better performance with bone-anchored hearing aid than acoustic devices in patients with severe air–bone gap. *Laryngoscope*. **121**(3), 613–616.

Dillon, H. (2012) *Hearing Aids*. Boomerang Press.

European working group on genetics of hearing impairment (1996) Info letter II. `http://audiology.unife.it/ www.gendeaf.org/hear/infoletters/Info_02.PDF`.

Favrot, S. and Buchholz, J.M. (2010) LoRA: A loudspeaker-based room auralization system. *Acta Acustica united with Acustica*, **96**, 364–375.

Gantz, B.J., Turner, C., Gfeller, K.E., and Lowder, M.W. (2005) Preservation of hearing in cochlear implant surgery: advantages of combined electrical and acoustical speech processing. *Laryngoscope*, **115**(5), 796–802.

Hersbach, A.A., Grayden, D.B., Fallon, J.B., and McDermott, H.J. (2013) A beamformer post-filter for cochlear implant noise reduction. *J. Acoust. Soc. Am.*, **133**(4), 2412–2420.

ISO 389–1 (1998) Acoustics standard. Reference zero for the calibration of audiometric equipment.

ISO 8253-1 (2010) Acoustics standard. Audiometric testing methods. Part 1: Pure-tone air and bone conduction audiometry.

ISO 8253-2 (2009) Acoustics standard. Audiometric testing methods. Part 2: Sound field audiometry with pure-tone and narrow-band test signals.

ISO 8253-3 (1996) Acoustics standard. Audiometric testing methods. Part 3: Speech audiometry.

Jastreboff, P.J. (1990) Phantom auditory perception (tinnitus): mechanisms of generation and perception. *Neurosci. Res.* **8**(4), 221–254.

Jastreboff, P.J., Gray, W.C., and Gold, S.L. (1996) Neurophysiological approach to tinnitus patients. *Am. J. Oto.* **17**(2), 236–240.

Kahue, C.N., Carlson, M.L., Daugherty, J.A., Haynes, D.S., and Glasscock, M.E. III (2014) Middle ear implants for rehabilitation of sensorineural hearing loss: A systematic review of FDA approved devices. *Otol. Neurotol.*.

Kates, J.M. (1998) *Signal Processing for Hearing Aids*. Kluwer Academic Publishers.

Kollmeier, B. and Wesselkamp, M. (1997) Development and evaluation of a German sentence test for objective and subjective speech intelligibility assessment. *J. Acoust. Soc. Am.*, **102**(4), 2412–2421.

Koski, T., Sivonen, V., and Pulkki, V. (2013) Measuring speech intelligibility in noisy environments reproduced with parametric spatial audio. *Proc. 135th Conv. Audio Eng. Soc.*.

Liberman, M.C. and Dodds, L.W. (1984) Single-neuron labeling and chronic cochlear pathology. III. Stereocilia damage and alterations of threshold tuning curves. *Hearing Res.*, **16**(1), 55–74.

Loizou, P.C. (1999) Signal-processing techniques for cochlear implants. *IEEE Trans. Engineering in Medicine and Biology* (May/June), 34–46.

Martin, F.N. and Clark, J.G. (2006) *Introduction to Audiology*. Prentice-Hall.

Miller, J.D. (1974) Effects of noise on people. *J. Acoust. Soc. Am.*, **56**(3), 729–764.

Minnaar, P., Frølund, A.S., Simonsen, C.S., Søndersted, B., Oakley, S., Dalgas, A., and Bennedbæk, J. (2013) Reproducing real-life listening situations in the laboratory for testing hearing aids. *Proc. 135th Conv. Audio Eng. Soc.* AES.

Moore, B.C. (2007) *Cochlear Hearing Loss: Physiological, psychological and technical issues*. John Wiley & Sons.

National Institutes of Health (2014) Nih publication no. 11-4798. `http://www.nidcd.nih.gov/health/hearing/pages/coch.aspx`.

Nicholas, J.G. and Geers, A.E. (2007) Will they catch up? The role of age at cochlear implantation in the spoken language development of children with severe to profound hearing loss. *J. Speech, Lang. Hearing Res.*, **50**(4), 1048–1062.

Nilsson, M., Soli, S.D., and Sullivan, J.A. (1994) Development of the hearing in noise test for the measurement of speech reception thresholds in quiet and in noise. *J. Acoust. Soc. Am.*, **95**(2), 1085–1099.

Schnupp, J., Nelken, I., and King, A. (2011) *Auditory Neuroscience: Making sense of sound*. MIT Press.

Seeber, B.U., Kerber, S., and Hafter, E.R. (2010) A system to simulate and reproduce audio-visual environments for spatial hearing research. *Hearing Res.* **260**, 1–10.

Skarzynski, H., Lorens, A., Matusiak, M., Porowski, M., Skarzynski, P.H., and James, C.J. (2014) Cochlear implantation with the nucleus slim straight electrode in subjects with residual low-frequency hearing. *Ear and Hearing*, **35**(2), 33–43.

Toivanen, J. (1976) *Teknillinen akustiikka*. Otakustantamo, Espoo.

Toppila, E. (2000) *A systems approach to individual hearing conservation*. PhD thesis, University of Helsinki.

Van den Bogaert, T., Doclo, S., Wouters, J., and Moonen, M. (2009) Speech enhancement with multichannel Wiener filter techniques in multimicrophone binaural hearing aids. *J. Acoust. Soc. Am.*, **125**(1), 360–371.

WHO (2013) Fact sheet n 300: Deafness and hearing loss. `http://www.who.int/mediacentre/factsheets/fs300/en/`.

Wilson, B.S. and Dorman, M.F. (2008) Cochlear implants: A remarkable past and a brilliant future. *Hearing Res.*, **242**(1–2), 3–21.

Zeng, F.G. (2004) Trends in cochlear implants. *Trends in Amplif.*, **8**(1), 1–34.

Zeng, F.G., Popper, A.N., and Fay, R.R. (2003) *Cochlear Implants: Auditory prostheses and electrical hearing*. Springer.

Index

Communication Acoustics: An Introduction to Speech, Audio, and Psychoacoustics, First Edition.
Ville Pulkki and Matti Karjalainen.